STRUCTURE AND MECHANISM
IN PROTEIN SCIENCE

STRUCTURE AND MECHANISM IN PROTEIN SCIENCE

A Guide to Enzyme Catalysis and Protein Folding

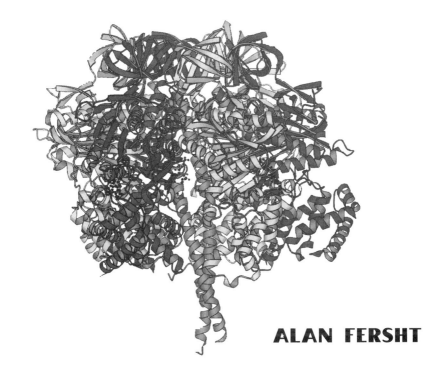

ALAN FERSHT

W. H. Freeman and Company

New York

To My family

W. P. Jencks

M. F. Perutz

Sponsoring Editor: Michelle Russel Julet
Project Editor: Georgia Lee Hadler
Cover and Text Designer: Victoria Tomaselli
Illustration Coordinator: Bill Page
Illustrations: Network Graphics
Production Coordinator: Paul W. Rohloff
Composition: Progressive Information Technologies
Manufacturing: RR Donnelley & Sons Company
Marketing Manager: John Britch

Library of Congress Cataloging in Publication Data
Fersht, Alan, 1943–
     Structure and mechanism in protein science: a guide to enzyme
catalysis and protein folding / Alan Fersht.
          p.   cm.
     ISBN 0-7167-3268-8
     1. Proteins—Structure.   2. Enzymes.   3. Enzyme kinetics.
     4. Protein folding.   I. Title.
     QD431.25.S85F47   1998.
     547′.75—dc21   98-36703
                                                                CIP

Printed in the United States of America

Third printing, 2000

CONTENTS

PREFACE

It is 14 years since I wrote the second edition of *Enzyme Structure and Mechanism,* at a time when we were just entering the age of protein engineering. Since then, the spectacular advances in determining biological structure, manipulating genes, engineering proteins, sequencing whole genomes, and computing have made each successive year more exciting than the last. We are now entering a new age of protein design, a time when the fields of protein folding and enzymology converge. Inspired by these developments, I have expanded *Enzyme Structure and Mechanism* into a more general book about mechanism in protein science that shows the unity of concepts in folding and catalysis.

So much is now known about enzyme mechanisms that the latest compendium on just the chemical aspects takes up four volumes of more than 1500 pages. In the past year alone, more than 20,000 scientific papers on enzymes and more than 2000 on protein folding and design were published. This book is intended to be a guide to the study of structure, activity, and mechanism, and so to provide a path through the vast subject of protein science. I believe that the understanding of mechanism is based on knowing the structures and properties of the reagents and the kinetics and thermodynamics of their reactions, as well as how the kinetics and other properties change when the structures of the reagents are modified and the reaction conditions are altered. Accordingly, the key to understanding proteins and their action is understanding basic chemistry, kinetics, thermodynamics, and structure, and how they are related. I concentrate on the fundamental principles, illustrated by representative examples in depth, rather than providing a comprehensive survey. The emphasis is on simple, direct observation and simple, direct measurements. The method of this book is to discuss basic principles, with an emphasis on the fundamental physical and chemical processes behind them to develop an intuitive feel for the handling of formulas, kinetics, and elementary thermodynamics.

Structure and Mechanism in Protein Science has evolved naturally from *Enzyme Structure and Mechanism,* because the same strategies apply to studying mechanism in both catalysis and folding. The advent of protein engineering allows the structure of proteins to be modified in a manner similar to small molecules so that structure–activity relationships may be studied. There are four new chapters, all based on the impact of protein engineering on protein folding and catalysis. One chapter illustrates the uses of protein engineering in enzymology, with two examples: the tyrosyl-tRNA synthetase as a model case for studying and exemplifying the principles of catalysis; and subtilisin for protein redesign. There are three chapters on protein folding, dealing with protein stability, the kinetics of folding, and the pathways or "energy landscape" of folding. Several chapters—Chapters 1 (The Three-Dimensional Structure of Proteins), 6 (Practical Methods for Kinetics and Equilibria), 10 (Conformational Change, Allosteric Regulation, Motors, and Work), 11 (Forces Between Molecules), 14 (Recombinant DNA Technology), and 16 (Case Studies of Enzyme Structure and Mechanism)—have been extensively rewritten. Other chapters have had varying degrees of updating and pruning, but those on classical enzymology have largely been left alone. I have tried to maintain a balance between historical development and current knowledge.

The ribbon diagrams were drawn with the program MolScript, which Per Kraulis has so generously distributed to the scientific community.

Much of this book was written while I was a Scholar-in Residence at the Fogarty International Center for Advanced Study in the Health Sciences, National Institutes of Health, Bethesda, Maryland, USA. I am indebted to several colleagues for their useful comments, especially those of Attila Szabo on Chapters 17, 18, and 19. I thank in particular Sophie Jackson for her critical reading of the entire manuscript and for her help with the proofs.

Alan Fersht
Cambridge, 1998

The Three-Dimensional Structure of Proteins

Proteins are the major functional molecules of life whose properties are so useful that we employ them as therapeutic agents, catalysts, and materials. Many diseases stem from mutations in proteins that cause them to lose function. In some cases, catalytic activity may be impaired, and so metabolic pathways may be altered (e.g., phenylketonuria). In other cases, structural properties may be impaired, leading to a loss of physical function (e.g., muscular dystrophy). Creutzfeld-Jakob disease and other transmissible encephalopathies result from proteins changing their shape and forming polymers.[1] Similarly, diseases result from amyloidosis, in which proteins gradually convert into long chains of polymerized β sheet and precipitate to form fibrils.[2] Many cancers result from mutations in proteins; some 50% of human cancers are caused by mutations in the tumor suppressor p53 that primarily lower its stability.[3] Enzymes and receptors are the usual targets of drugs, either to restore function or to destroy infectious agents or cancers. The ultimate goal of protein science is to be able to predict the structure and activity of a protein *de novo* and how it will bind to ligands. When this is achieved, we will be able to design and synthesize novel catalysts, materials, and drugs that will eliminate disease and minimize ill health.

There are now significant advances toward this goal. Experimentalists are able to alter the activity and stability of proteins by protein engineering, and the first tentative steps in protein design are under way. Theoreticians are able to simulate many aspects of folding and catalysis with increasing detail and reliability.

The essence of the modern study of proteins is to combine direct observation with structural variation of the reagents: the structures of enzymes and stable reaction intermediates are directly observed by x-ray crystallography and NMR;

transient intermediates are characterized by kinetic spectroscopy; the structures of transition states are inferred from direct kinetic measurements; and the structures of substrates are varied by chemical synthesis and those of enzymes by protein engineering to allow the direct study of the interrelationship between structure and activity. Accordingly, the key to understanding proteins is to understand kinetics and equilibria and their relationship to structure. The starting point is the structure of proteins.

Table 1.1   *The common amino acids*

Amino acid (three- and one- letter codes, M_r)	Side chain R in $RCH(NH_3{}^+)CO_2{}^-$	pK_a's^a
Glycine (Gly, G, 75)	H—	2.35, 9.78
Alanine (Ala, A, 89)	CH_3—	2.35, 9.87
Valine (Val, V, 117)	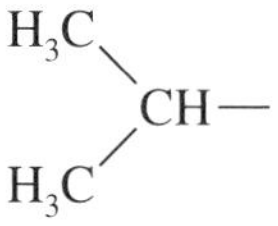	2.29, 9.74
Leucine (Leu, L, 131)	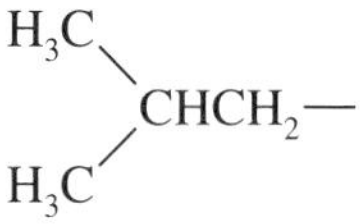	2.33, 9.74
Isoleucine (Ile, I, 131)	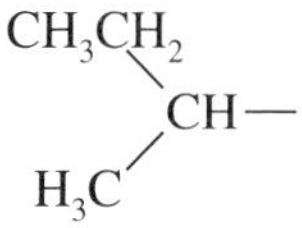	2.32, 9.76
Phenylalanine (Phe, F, 165)	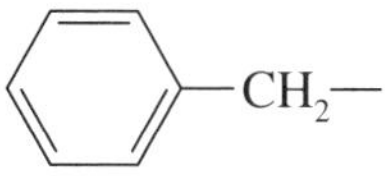	2.16, 9.18
Tyrosine (Tyr, Y, 181)	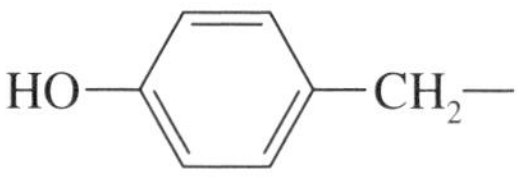	2.20, 9.11, 10.13
Tryptophan (Trp, W, 204)	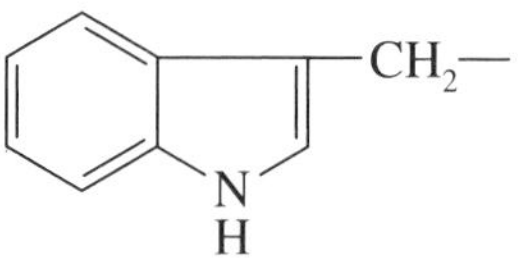	2.43, 9.44
Serine (Ser, S, 105)	$HOCH_2$—	2.19, 9.21

A. The primary structure of proteins

The major constituent of proteins is an unbranched polypeptide chain consisting of L-α-amino acids linked by amide bonds between the α-carboxyl of one residue and the α-amino group of the next. Usually only the 20 amino acids listed in Table 1.1 are involved, although they may be covalently modified after biosynthesis of the polypeptide chain. The primary structure is defined by the

Amino acid (three- and one- letter codes, M_r)	Side chain R in $RCH(NH_3{}^+)CO_2{}^-$	pK_a's[a]
Threonine (Thr, T, 119)	$\begin{array}{c}HO\\\diagdown\\CH-\\\diagup\\H_3C\end{array}$	2.09, 9.11
Cysteine (Cys, C, 121)	$HSCH_2-$	1.92, 8.35, 10.46
Methionine (Met, M, 149)	$CH_3SCH_2CH_2-$	2.13, 9.28
Asparagine (Asn, N, 132	$H_2NC(=O)CH_2-$	2.1, 8.84
Glutamine (Gln, Q, 146)	$H_2NC(=O)CH_2CH_2-$	2.17, 9.13
Aspartic acid (Asp, D, 133)	$^-O_2CCH_2-$	1.99, 3.90, 9.90
Glutamic acid (Glu, E, 147)	$^-O_2CCH_2CH_2-$	2.10, 4.07, 9.47
Lysine (Lys, K, 146)	$H_3N^+(CH_2)_4-$	2.16, 9.18, 10.79
Arginine (Arg, R, 174)	$\begin{array}{c}H_2N^+\\\diagdown\\C-NH(CH_2)_3-\\\diagup\\H_2N\end{array}$	1.82, 8.99, 12.48
Histidine (His, H, 155)	$-CH_2-$ (imidazole ring)	1.80, 6.04, 9.33
Proline (Pro, P, 115)	(pyrrolidine ring with CO_2^-)	1.95, 10.64

[a] From R.M.C. Dawson, D.C. Elliott, W.H. Elliott, and K.M. Jones, *Data for biochemical research,* 2nd ed., Oxford University Press (1969).

sequence in which the amino acids form the polymer. By convention, the sequence is written as follows, beginning with the N-terminus on the left:

$$H_2NCH(R)CO—NHCH(R')CO—NHCH(R'')CO—NHCH(R''')CO—NH\cdots \quad (1.1)$$

The primary structures of almost all intracellular proteins consist of linear polypeptide chains. Many extracellular proteins, however, contain covalent —S—S— cross-bridges in which two cysteine residues are linked by their thiol groups. This either creates intrachain links in the main polypeptide chain or links different chains together (Figure 1.1). The multiply linked chains are generally derived from a single-chain precursor that has been cleaved by proteolysis—examples being insulin from proinsulin and chymotrypsin from chymotrypsinogen. These bridges may be cleaved by reduction with thiols. Proteins consisting of a single polypeptide chain may often be reduced and denatured, then reversibly oxidized and renatured. But reduction and denaturation of proteins derived from the covalent modifications of precursors is generally irreversible. The spontaneous refolding of single polypeptide chains that have not been covalently modified leads to the fundamental principle: *The three-dimensional structure of a protein is determined by its genetically encoded sequence*[4] (see Chapter 19).

B. Methods for determination of three-dimensional structure

1. Structures of crystalline proteins by x-ray diffraction methods

X-ray crystallography has been the single most important factor in the investigation of enzymes because it has provided the experimental basis of our present knowledge of the structure of proteins. The way x-rays are scattered when they strike the electrons of atoms is similar to the way light waves are scattered by the engraved lines of a diffraction grating. The regular lattice of a crystal acts like a three-dimensional diffraction grating in scattering a monochromatic beam of x-rays, giving a pattern in which the diffracted rays reinforce and do not destructively interfere. The structure of the crystal, or more precisely the distribution of its electron density, may be calculated from the diffraction pattern by Fourier transformation. This requires knowledge of the intensities and directions of the diffracted rays and of their phases. Determination of the phase of each diffracted ray (an essential requirement for the Fourier transformation) is the most difficult problem. This is usually done by the method of *isomorphous replacement*.[5] A heavy metal atom is bound at specific sites in the crystal without disturbing its structure or packing. The metal scatters x-rays more than do the atoms of the protein, and its scattering is added to every diffracted ray. Information about the phases of the diffracted rays from the protein can then be obtained from the changes in intensity, depending on whether they are reinforced or diminished by

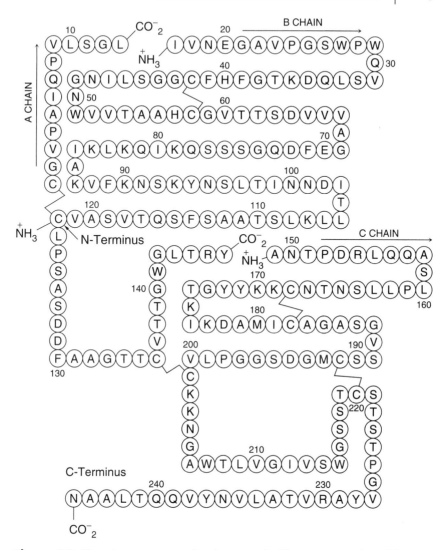

Figure 1.1 The primary structure of α-chymotrypsin. The enzyme consists of three chains linked together by —S—S— bridges between cysteine residues. The chains are derived, however, from the single polypeptide chain of chymotrypsinogen by the excision of residues Ser-14, Arg-15, Thr-147, and Asn-148.

the scattering from the heavy atom. Several different isomorphous substitutions are needed to give an accurate determination of the phases. Once the phase and the amplitude of every diffracted ray are known, the electron density of the protein may be calculated. The structure of the protein is obtained by matching this density to the amino acid residues of the primary structure using a computer display system. The structure may then be refined by computer methods. The structure of a variant of a protein, such as a mutant or a closely related structure, may sometimes be solved by *molecular replacement*. In this method, the known structure is used as a trial structure to compute the electron density of the variant iteratively. The determination of a novel crystal structure may take several months to several years. But the structures of variants of the basic structure are often solved in a matter of days by molecular replacement.

The degree of accuracy that is attained depends on both the quality of the data and the *resolution*. At low resolution, 4 to 6 Å (0.4 to 0.6 nm), the electron density map reveals little more in most cases than the overall shape of the molecule. At 3.5 Å, it is often possible to follow the course of the polypeptide backbone, but there may be ambiguities. At 3.0 Å, it is possible in favorable cases to begin to resolve the amino acid side chains and, with some uncertainty, to fit the sequence to the electron density. At 2.5 Å, the positions of atoms often can be fitted with accuracy of ± 0.4 Å. To locate atoms to 0.2 Å, a resolution of about 1.9 Å and very well ordered crystals are necessary.

The 46-residue protein crambin has been solved at 0.83 Å resolution at 130 K.[6] Some proteins have been crystallized in microgravity in space rockets where the convection-free conditions can produce larger and better crystals. RNase A crystals grown in microgravity diffracted x-rays to 1.06 Å resolution, approximately 0.2 Å higher resolution than previously observed in terrestrially grown RNase A crystals.[7] At these very high resolutions, alternate conformations of some side chains may even be seen.

The isomorphous replacement method becomes ineffective for protein crystals beyond a resolution of 2 to 2.5 Å, because the addition of a heavy atom may cause some alteration in structure and it is difficult to observe small changes in intensities that are already very weak. But, structures may be refined to high resolution, as in the molecular replacement procedure, by using the measured intensities at higher resolution and the model structure that has been determined at resolutions of 2 to 2.5 Å.

2. Neutron diffraction

It is not possible to locate the positions of hydrogen atoms in proteins by x-ray diffraction because the electron density of the H atoms is so weak. They may be observed, however, by neutron diffraction, because the scattering of neutrons by the hydrogen nucleus (the proton) is appreciable even compared with larger nuclei. The hydrogen atoms are seen as regions of negative density due to a change of phase of 180° in the phases of the scattered neutrons. The level of structural detail obtained from neutron diffraction can be much higher than that found from x-ray diffraction because nuclei act as point sources for scattering of neutrons, whereas x-rays are scattered from a diffuse cloud of electrons. The method is time-consuming and requires the use of specialized facilities. The main uses of neutron diffraction are for the location of hydrogen atoms: namely, reaction mechanisms that hinge on the location of a proton; protein dynamics that are monitored by the exchange of NH protons by deuterons from solvent (see section G2b); and protein-water interactions. For example, the position of a proton at the active site of serine proteases was shown conclusively to reside on a histidine and not on an aspartate[8] (see section D2).

3. Structure of proteins in solution from NMR methods

The most important advance in determining the structures of proteins in recent years has been the development of high-field nuclear magnetic resonance methods.[9,10] These methods enable us to determine the structures of moderately sized (< 30 kDa) proteins in solution.[11,12] The principle of the method is quite different from x-ray crystallography in that the overall structure is calculated principally from restraints on the distances between ^{1}H atoms.

The most common nucleus studied is that of ^{1}H, because of its high natural abundance in proteins and because the NMR method is highly sensitive to it. A solution of the protein in a glass tube is placed in a static magnetic field. The higher the field strength, the better the resolution and sensitivity: field strengths of 11.74 tesla for a "500 MHz," 14.09 for a "600 MHz," and 18.78 for an "800 MHz" NMR spectrometer are currently used for structure determination. In a one-dimensional experiment, the sample is subjected to irradiation by a radio frequency field to generate an NMR spectrum. The spectrum consists of a series of lines that vary in their position, their fine structure, and their shape, among other factors. (Some aspects of their shape are dealt with in section G2d.) The *chemical shift* defines the location of the NMR line in the spectrum, generally measured in parts per million (ppm) as a fraction of the static field by which the signal is displaced from that of a reference compound (frequently, 2,2-dimethyl-2-silapentane-5-sulfonate (DSS) for ^{1}H). The chemical shift of a nucleus depends on the atom to which it is bonded and to its environment. Fine structure results from the spins of nuclei that are separated by a small number of covalent bonds (≤ 3) interacting and splitting the resonance line. These through-bond effects are known as *spin-spin* (or *J*) *coupling*. Nuclei can also interact through space; this is called the Nuclear Overhauser Effect (NOE). This results from the coupling of dipoles, the magnitude of which falls off as r^{-6} (see Chapter 11). Hydrogen atoms must be closer than about 5 Å for NOEs to be observed. NOEs are usually classified as "strong," "intermediate," or "weak." A "strong" NOE implies that the ^{1}H nuclei are separated by less than 2.5 Å, an "intermediate" NOE implies less than 3.5 Å, and a "weak" NOE implies less than 5 Å. It is the distance restraints derived from the NOEs that are used to solve structures. The strategy of solving structures in solution by NMR is to assign the resonances in the NMR spectrum to individual protons and then determine which protons are close to each other from the NOEs.

The assignments are made through a battery of techniques to identify the through-bond and through-space interactions. Multidimensional NMR spectra are taken in which the sample is irradiated by two radio frequency fields in

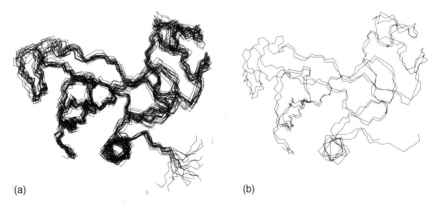

(a)   (b)

Figure 1.2 (a) Superposition of 20 structures calculated from NMR restraints for the polypeptide backbone of the ribonuclease barnase (110 amino acid residues, see Chapter 19). (b) Superposition of the mean polypeptide backbone calculated from NMR and that from x-ray crystallography.

various sequences and pulses. The structure is then calculated from a combination of the restraints on the distances between protons and energy calculations. In practice, there is not a unique solution; a family of possible structures is calculated (Figure 1.2). In regions where there is considerable information and it is consistent, the different solutions give virtually superimposable structures. In other regions, there can be a spread of possible conformations. The more neighbors there are to a particular proton, the more the information about distance restraints, and so its position in space is more closely defined. The structures of proteins up to about 110 amino acid residues have been solved by ^{1}H NMR with very good agreement with crystal structures. For larger proteins, more advanced techniques involving labeling with ^{15}N and ^{13}C and multidimensional NMR are essential.

For very small proteins, NMR has given structural information comparable to the highest resolution studies on crystalline proteins using x-rays. The quality falls off with increasing size. The overwhelming advantage of solving structures by NMR, however, is that it can be used for noncrystalline samples in solution. In crystals, surface residues on one molecule can be distorted by the effects of those on its neighbor because of constraints of crystal packing (see section G1). As will be discussed in more detail later, NMR can be used for measuring the dynamic aspects of proteins in solution (section G2) and characterizing states of proteins not accessible to x-ray crystallography. This is especially important in studying the pathway of protein folding (Chapters 18 and 19).

C. The three-dimensional structure of proteins

A ball and stick model of a protein looks at first glance to be just a mass of atoms. One way of tackling a complex problem is to break it down into smaller parts that can be more readily understood and then reconstruct the whole. A convenient starting point is to remove the side chains. The path of the polypeptide backbone through space can then be easily seen. This is the secondary structure of the protein. It is constructed around planar units of the peptide bond. Closer examination reveals regions where the secondary structure is organized into repetitive, regular elements such as helixes and sheets that are joined by loops. Examination of a large number of proteins reveals further simplifying features, such as the organization of the helixes and sheets into recurrent motifs or supersecondary structure. These are sometimes further organized into structural domains. The side chains can be added back to the backbone, and it is then seen how the tertiary structure of the protein is formed by the packing of the regular elements of secondary structure by way of their side chains. More complex structures are sometimes composed of identical or different polypeptide chains that pack together to give quaternary structure. A logical way of analyzing protein structure is to view it as a hierarchy of substructures.

1.   The structural building blocks

The structures of the basic building blocks of the architecture of proteins were determined by Linus Pauling and R. B. Corey many years before the solution of the structures of globular proteins.[13] They solved the structures of crystalline small peptides to find the dimensions and geometry of the peptide bond. Then, by constructing very precise models, they found structures that could fit the x-ray diffraction patterns of fibrous proteins. The diffraction patterns of fibers do not consist of the lattice of points found from crystals, but a series of lines corresponding to the repeat distances between constantly recurring elements of structure.

a.   The peptide bond

The x-ray diffraction studies of the crystals of small peptides showed that the peptide bond is planar and *trans (anti)* (Figure 1.3). The same structure has been found for all peptide bonds in proteins, with a few rare exceptions. This planarity results from a considerable delocalization of the lone pair of electrons of the nitrogen onto the carbonyl oxygen. The C—N bond is consequently shortened, and it has double-bond character (equation 1.2). Twisting of the bond breaks it and loses the 75 to 90 kJ/mol (18 to 21 kcal/mol) of delocalization energy.

$$\underset{R}{\overset{O}{\big\diagdown}}C-\underset{H}{\overset{\overset{\displaystyle R'}{\diagup}}{N}} \quad \longleftrightarrow \quad \underset{R}{\overset{O^-}{\big\diagdown}}C=\underset{H}{\overset{\overset{\displaystyle R'}{+\diagup}}{N}} \tag{1.2}$$

Proline is an exceptional amino acid residue in that the *cis-trans* equilibrium only slightly favors the *trans* form in peptidyl-proline bonds. Small proline-containing peptides in solution contain some 5 to 30% of the *cis (syn)* isomer, as opposed to less than 0.1% of the *cis* isomers of the other amino acids.[14] The *cis* form is even found in native proteins: two of the four prolines in ribonuclease A

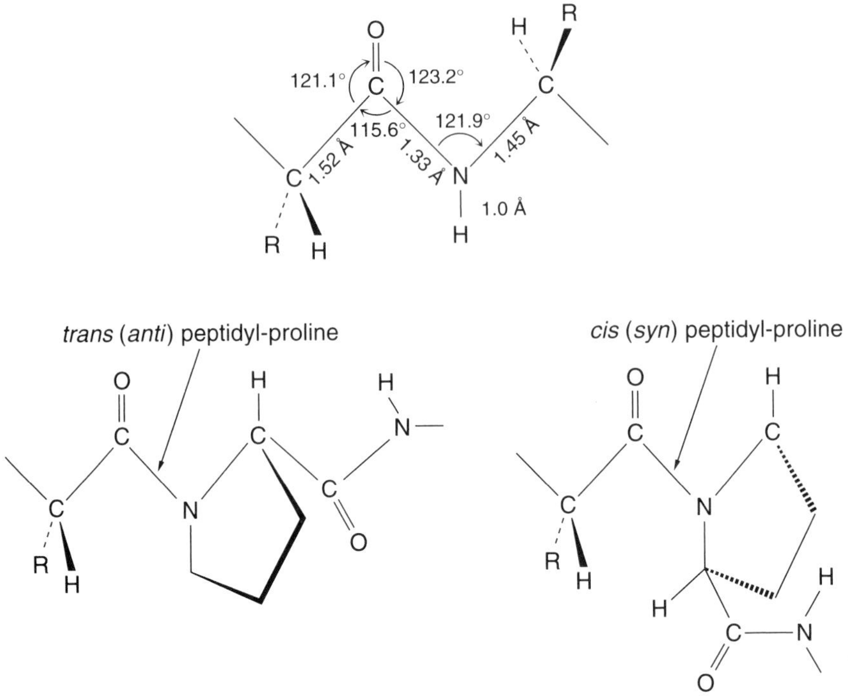

Figure 1.3 The peptide bond. Distances are in Å (1 Å = 0.1 nm). Proline residues are found in both *cis* and *trans* conformations.

are *cis*, as are two of the 17 in carbonic anhydrase. In native proteins, the overall structure of the molecule determines the isomeric form of the amino acid.[15] The slow *cis-trans* isomerization of proline residues causes complications in kinetic experiments on protein folding (see Chapter 18).

b. The α helix and the β sheet

The polypeptide chains of fibrous proteins are found to be organized into hydrogen-bonded structures. In these ordered regions, any buried carbonyl oxygen forms a hydrogen bond with an amido NH group. This is done in two major ways: by forming α helixes (found from the fiber diffraction studies on α-keratin) and by forming β sheets (found in β-keratin). The polypeptide chain of a globular protein also folds upon itself to create local regions of similar α and β structures. The α helix is an important structure; about one-third of amino acid residues in proteins are in α helixes. The α helix (Figure 1.4), is a stable structure, each amide group making a hydrogen bond with the third amide group away from it in either direction — the C=O group of residue i makes a hydrogen bond with the NH of residue $i + 4$. The C=O groups are parallel to the axis of the helix and point almost straight at the NH groups to which they are hydrogen-bonded. The side chains of the amino acids point away from the axis. There are 3.6 amino acids in each turn of the helix. The rise of the helix per turn—the pitch—is about 5.4 Å. The α helixes consist of about four turns on average but can be more than a factor of two longer or shorter.

The first and last four residues differ from the interior ones because they cannot make the intrahelical hydrogen bonds between the backbone C=O groups of one turn and the NH groups of residues in the next. Because of this, the ends of helixes require hydrogen bonds to be made with either solvent or groups in

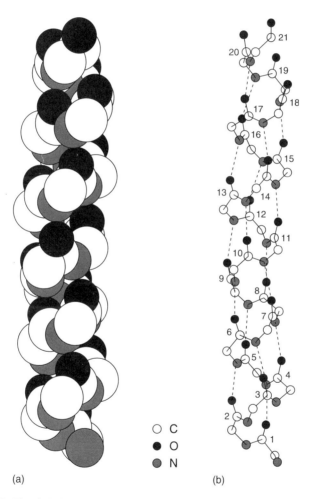

(a)   (b)

Figure 1.4 The right-handed α helix found in proteins. This example is from myoglobin (residues 126–146) showing just the backbone alpha carbon, C=O, and peptide nitrogen atoms. (a) The circles are drawn for the full van der Waals radius of each atom (see Chapter 11). (b) The circles are drawn at 20% of the van der Waals radius of each atom from its center.

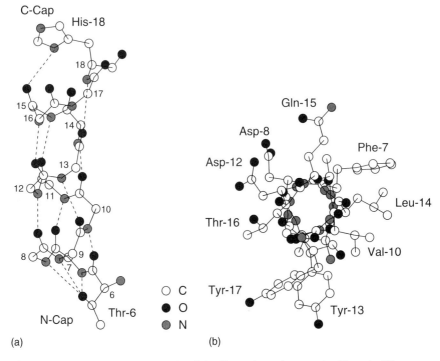

Figure 1.5 A right-handed α helix of the ribonuclease, barnase (residues 6–18). (a) A view perpendicular from the helix axis. The side chains of residues 7–17 are omitted for clarity. This helix has residues at either end, the N- and C-caps, that can make hydrogen bonds with the NH and C=O groups of the first and last turns, respectively. Note that the C-terminus of the helix becomes irregular as the chain leaves the C-cap through a left-handed configuration. (b) The α helix viewed end on. The cylindrical locus of the backbone N, C, and O atoms can be seen clearly. An ideal helix would have a perfect ring of C and N atoms. This real protein helix has bulges as the helix deviates from ideality toward the C-terminus. Note how the hydrophobic side chains of Phe-7, Val-10, and Leu-14, and the hydrophobic part of Tyr-13, are on one side of the helix. This side packs against hydrophobic residues in the nonpolar interior of the protein. The other side of the helix, which is exposed to solvent water, contains hydrophilic side chains. This is a typical amphipathic helix.

the protein. The first and last residue of a helix is frequently a residue that can provide such bonds, and these have been termed the *N*- and *C-caps* (Figure 1.5). It is difficult to define helixes in proteins because the helixes are often distorted at their ends. There is a simple operational definition of the caps. The α helix is a cylindrical structure, and a cylinder may be drawn that contains the alpha carbons of the helix backbone. The caps are the first and last residues whose alpha carbons lie on the cylinder. The factors that stabilize helixes (plus further bibliography) are discussed in more detail in Chapter 17.

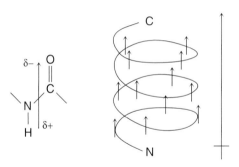

Figure 1.6 The helix dipole.

Helixes are often *amphipathic*: one side has hydrophobic residues that bind to hydrophobic areas in the protein; the other has hydrophilic residues that interact with water (Figure 1.5). The alignment of dipoles of the polypeptide backbone nearly parallel to the axis of an α helix causes a net dipole moment with its positive pole at the N-terminus and negative pole at the C-terminus[16] (Figure 1.6).

An extended polypeptide chain can make complementary hydrogen bonds with a parallel extended chain. This structure in turn can match up with another extended chain to build up a sheet. There are two stable arrangements: the parallel β-pleated sheet in which all the chains are aligned in the same direction and the antiparallel β-pleated sheet in which the chains alternate in direction (Figure 1.7). The repeating unit of a planar peptide bond linked to a tetrahedral carbon produces a pleated structure (Figure 1.8). The parallel β sheet is not found in fibrous proteins but only in globular proteins. Mixed β sheets are also found in globular proteins that contain a mixture of parallel and antiparallel strands. Individual strands generally consist of some 5–10 residues, but can be as short as 2 or 3.

c. Bends in the main chain

Another frequently observed structural unit occurs when the main chain sharply changes direction through a "β bend" composed of four successive residues. In these units, the C=O group of residue i is hydrogen-bonded with the NH of residue $i + 3$ instead of $i + 4$ in the α helix. A 180° change of direction, which can link two antiparallel strands of pleated sheet, can be achieved in two ways (types I and II, Figure 1.9). Type II has the restriction that glycine must be the third residue in the sequence. A distortion of type I gives a bend, type III, which can be repeated indefinitely to form a helix known as the 3_{10} helix (a tighter, less stable helix than the α, with 3.0 residues per turn, forming hydrogen-bonded loops of 10 atoms).

The structures described above are elements of *secondary structure*. The secondary structure of a segment of polypeptide chain is defined as the local spatial arrangement of its main-chain atoms, without regard to the conformation of its side chains or to its relationship with other segments. Sometimes the secondary

Figure 1.7 Parallel *(top)* and antiparallel *(bottom)* β sheet viewed from above. Note the different hydrogen-bonding patterns.

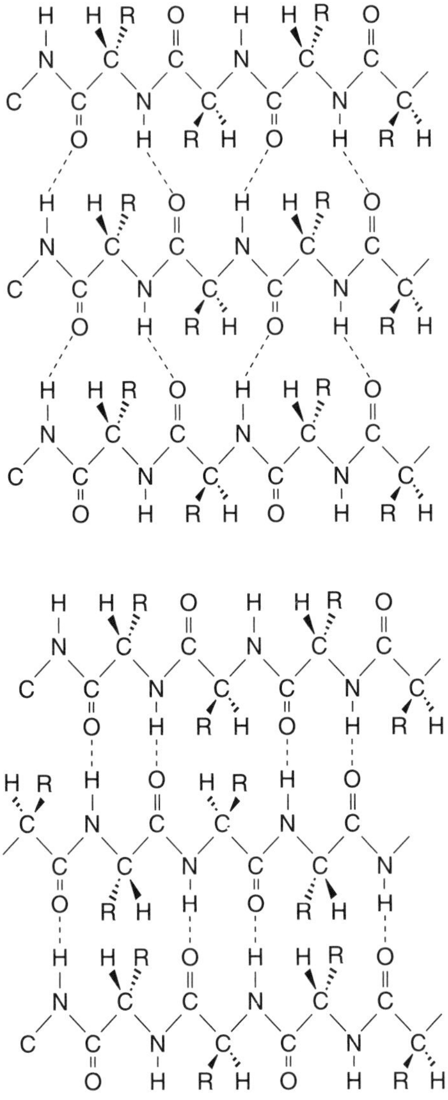

structure is periodic in nature, giving rise to well-defined structure such as α helixes or β sheets.

2. The Ramachandran diagram

a. Secondary structure is defined by angles of rotation

The secondary structure of a region of a protein is defined by the conformation of the polypeptide backbone. The conformation of the polypeptide backbone is defined by the location in space of the three sets of atoms that are linked

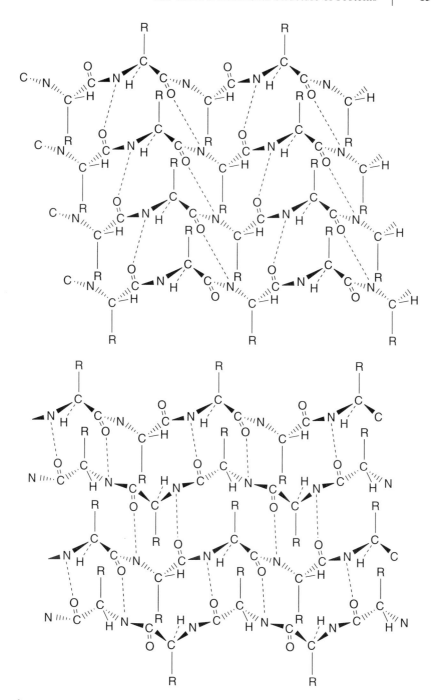

Figure 1.8 Parallel *(top)* and antiparallel *(bottom)* β sheet viewed from an angle to illustrate how the geometry at the alpha carbon causes the sheet to be pleated. The amide H-atoms have been omitted for clarity.

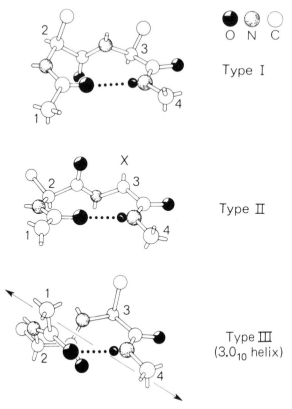

O N C

Type I

Type II

Type III
(3.0$_{10}$ helix)

Figure 1.9 The three types of bends found in proteins. Residue X in type II is always glycine. [From J. J. Birktoft and D. M. Blow, *J. Molec. Biol.* **68**, 187 (1972).]

together; namely, the alpha carbon (C^α), carboxyl carbon (C'), and amide nitrogen atoms. Their positions, and hence the secondary structure, can thus be defined by the angles of rotation about the bonds connecting the three atoms; ϕ, ψ, and ω in Figure 1.10. This figure illustrates a fully extended segment of chain with a residue i, and parts of the following residue $i + 1$ and the preceding residue $i - 1$.

The torsion or dihedral angle ω defines rotation around the peptide bond C'_i—N_{i+1}. This is measured as the rotation of one plane with respect to another. The rotating plane is the one containing N_{i+1}, C^α_{i+1}, and H_{i+1}. ω is measured from the plane in which lie C^α_i, C'_i, and O_i (and N_{i+1}). The angle is defined as 0 when C^α_{i+1} is in the same plane as C^α_i, C'_i, and O_i, and C^α_{i+1} and C^α are *cis*. Values of ω are positive when measured in a clockwise direction for rotation when viewed down the C'_i—N_{i+1} bond from the C'_i. In Figure 1.10, C^α_i and C^α_{i+1} are *trans*, and the value of ω is 180°. This is

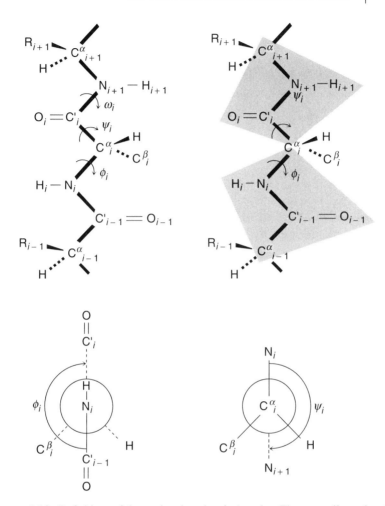

Figure 1.10 Definitions of the torsional angles ϕ, ψ, and ω. These are all equal to 180° for a fully extended polypeptide chain *(top left)*. ω_i defines rotation about the C_i-N_{i+1} bond. The normal *trans* planar peptide bond has $\omega_i = 180°$. ϕ_i describes rotation about the $N_i-C^\alpha_i$ bond, and ψ_i describes rotation about the $C^\alpha_i-C'_i$ bond *(top right)*. The angles may be represented on Newman projection formulas ($\phi = \psi = 180°$, *bottom*): *left*, ϕ_i viewed along $N_i-C^\alpha_i$ bond (N→C); *right*, ψ_i viewed along the $C^\alpha_i-C'_i$ ($C^\alpha_i \rightarrow C'_i$).

of course the standard angle for a peptide bond, which is close to planar and, apart from a few exceptions that are primarily peptidyl-proline bonds, *trans*. In practice, therefore, the secondary structure is defined by the rotation of the planar peptide units around the bonds connecting them to the alpha carbon atoms.

The angle ϕ defines the rotation of the plane containing C^α_i, C'_i, and O_i (and N_{i+1}) around the N_i—C^α_i bond. This is measured from the plane in which lie C^α_i, N_i, and C'_{i-1}. As the peptide bond is generally planar, O_{i-1} and C^α_{i-1} are also in the plane. The angle is defined as 0 when the C'_i is in the same plane as C^α_i, N_i, and C'_{i-1}, and C'_i and C'_{i-1} are *cis* (*syn*). Values of ϕ are positive when measured in a clockwise direction for rotation when viewed down the N_i—C^α_i bond from N to C. In Figure 1.10, C'_i and C'_{i-1} are *trans* (*anti*), and the value of ϕ is 180°.

The angle ψ defines the rotation of the plane containing C'_i, O_i, and N_{i+1} around the C^α_i—C_i bond. This is measured from the plane in which lie C'_i, C^α_i, and N_i (and C'_{i-1}). The angle is defined as 0 when N_{i+1} is in the same plane as C'_i, C^α_i, and N_i, and N_i and N_{i+1} are *cis*. Values of ψ are positive when measured in a clockwise direction for rotation when viewed down the C^α_i—C'_i bond from the C^α. In Figure 1.10, N_i and N_{i+1} are *trans*, and the value of ϕ is 180°. Note that in all cases, an angle of $+180° \equiv -180°$.

b. A restricted number of conformations is accessible

Steric clashes restrict the number of conformations available to the polypeptide chains. The conformational space that is accessible is illustrated in a two-dimensional plot of ϕ against ψ, the Ramachandran diagram (Figure 1.11), named after its inventor, G. N. Ramachandran.[17] Steric effects also distort the geometry of regular elements of secondary structure, and so there is a spread of values. Typical values are listed in Table 1.2.

Table 1.2 *Torsion angles for regular polypeptide conformations*

Structure	Torsion angle (degrees) ϕ	ψ
Hypothetical fully extended	$+180$ ($\equiv -180$)	$+180$
Antiparallel β sheet	-139	$+135$
Parallel β sheet	-119	$+113$
Right-handed α helix	-57	-47
Left-handed α helix	$\sim +60$	$\sim +60$
3_{10} helix	-49	-26
π helix	-57	-70

Note that $+180°$ is the same as $-180°$.

From G. N. Ramachandran and V. Sasisekharan, *Adv. Protein Chem.* **23**, 283 (1968). IUPAC-IUB Commission on Biochemical Nomenclature, *Biochemistry* **9**, 3471 (1970).

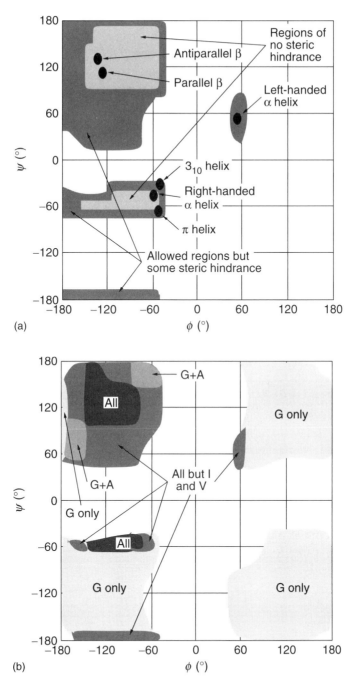

Figure 1.11 Ramachandran diagrams illustrating the preferred conformations of the polypeptide chain. (a) The allowed regions and their relationship to different types of secondary structure. (b) The different conformational restrictions for the different amino acids. Note the high conformational freedom for glycine (G) and the restrictions on isoleucine (I) and valine (V).

3. Motifs or supersecondary structures

Frequently recurring substructures or folds are collectively termed *supersecondary structures* or *motifs*. These are combinations of α and/or β structure. A simple example is a β *hairpin*, consisting of two antiparallel strands joined by a loop of three to five residues (Figure 1.12). This frequently occurs in antiparallel β sheet. Such sheet frequently contains four β strands connected as in Figure 1.13 in a motif called a *Greek key* (or *meander*, which is the Greek word for the pattern) because it is reminiscent of the Greek decorative motif, or six strands described as a *jellyroll*.

The strands in antiparallel sheet can be connected by relatively simple loops. Connections in parallel sheet are, of necessity, more complex because the connectors have to change direction twice. The connections often consist of α helixes in a motif termed a β-α-β (Figure 1.14). This has a strand, followed by a loop, followed by the helix, then another loop followed by the second strand. The β-α-β motif could take up two different stereochemical conformations. The right-handed conformation in Figure 1.14 is the most common. The same is true for three β strands in a β-β-β motif.

The preferred topologies of connection for the β-α-β, β-β-β, Greek key, and jellyroll motifs fit three generalizations found from surveying a large number of structures.

1. Pieces of secondary structure that are adjacent in sequence are often in contact in three dimensions and usually pack in an antiparallel manner.

2. β-X-β units (where the two βs are parallel strands in the same sheet; X is a helix, a strand in a different sheet, or an extended polypeptide) are connected in a right-handed manner (Figure 1.14).

3. The connections between secondary structures neither make knots in the chain nor cross one another.

The structures of most larger proteins ($M_r > 20\,000$, or sequences of greater than ~ 200 residues) appear to be composed of independently folded globular units called *domains*. The definition of a domain is not rigorous. The residues within a domain are seen to interact more with one another than they do with residues in different domains.

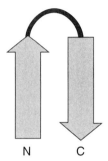

Figure 1.12 β-Hairpin motif.   N   C

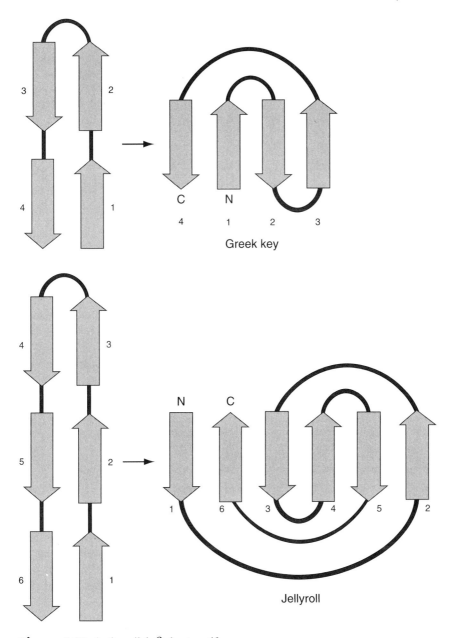

Figure 1.13 Antiparallel β-sheet motifs.

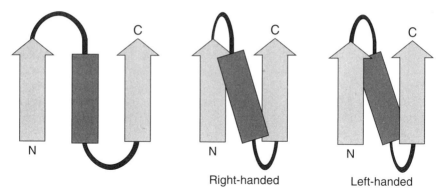

Right-handed   Left-handed

Figure 1.14 β-α-β-Hairpin motif.

4. Assembly of proteins from the building blocks

a. Classification by the packing of secondary structure[15,18,19]

The arrangement in space of all the atoms in a single polypeptide chain or in co-valently linked chains is termed the *tertiary structure*. Most proteins can be considered crudely as layered sandwich structures, with each layer consisting of either α helixes or a β sheet. Four classes of protein molecules have been identified, based upon different combinations of these structures: α/α (all α), β/β (all β), α/β (α and β elements are in a mixed order in the sequence), and $\alpha + \beta$ (the α and β elements are segregated in the sequence, the α occurring in one region and the β in another—see Figure 1.15).

b. Packing of the secondary structure

The major driving force for the folding of proteins appears to be the burying and clustering of hydrophobic side chains to minimize their contact with water (the "hydrophobic effect" discussed in Chapter 11). The basic requirements for folding are, therefore: *i,* that the structures are compact and so minimize the area of hydrophobic surfaces that are exposed to solvent; and *ii,* buried hydrogen-bonding groups are all paired. Formation of α helixes and β sheets maximizes the pairing of the hydrogen bonding groups of the backbone. The helixes and sheets pack by stacking their amino acid side chains. The internal packing is good, with the side chains of one piece of secondary structure fitting between those of another to give a hydrophobic region with a density similar to that of a hydrocarbon liquid or wax. The general shape of secondary structure determines its mode of packing. β sheets form layer structures with helixes or other β sheets packed on their faces. The cylindrical shape of helixes leads them to stack around a central core or form layer structures.

One α helix may pack on another by the mutual intercalation of side chains such that a "ridge" of projecting side chains on one helix fits into a "groove"

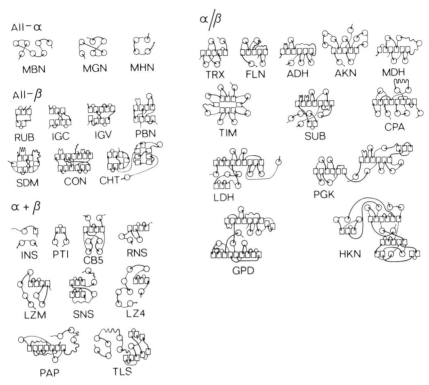

Figure 1.15 Classification of protein structure according to Levitt and Chothia.[15] The protein is viewed from such an angle that most segments of secondary structure are seen end on. Each strand of a β sheet is represented by a square. The front end of each α helix is represented by a circle. The segments that are close in space are close together in the diagram. The segments are connected by bold or thin arrows (from the N- to the C-terminus) to indicate whether the connection is at the near or the far end. The approximate scale is: diameter of α helix = 5 Å; β strand = 5 × 4 Å; separation of helixes = 10 Å; separation of hydrogen-bonded β strand = 5 Å; separation of non-hydrogen-bonded strands = 10 Å.

Abbreviations: MBN, myoglobin; MGN, myogen; MHN, myohemerythrin; RUB, rubredoxin; ICG, immunoglobulin constant region; IGV, immunoglobulin variable region; PBN, prealbumin; SDM, superoxide dismutase; CON, conconavalin A; CHT, chymotrypsin; INS, insulin; PTI, pancreatic trypsin inhibitor; CB5, cytochrome b_5; RNS, ribonuclease; LZM, lysozyme; SNS, staphylococcal nuclease; LZ4, T4 lysozyme; PAP, papain; TLS, thermolysin, TRX, thioredoxin; FLN, flavodoxin; ADH, alcohol dehydrogenase coenzyme domain; AKN, adenyl kinase; MDG, malate dehydrogenase; TIM, triosephosphate isomerase; SUB, subtilisin; CPA, carboxypeptidase; LDH, lactate dehydrogenase; PGK, phosphoglycerate kinase; GPD, glyceraldehyde 3-phosphate dehydrogenase, HKN, hexokinase.

between rows of side chains in the other. Two ideal helixes prefer to stack with their cylindrical axes inclined at either 20° or − 50°. In practice, there is a much wider range of angles because the surfaces of helixes deviate from regularity. There is a common class of proteins, called four-helix bundles, that are composed of four α helixes that stack in antiparallel mode with their axes tilted.

The stacking of an α helix on a β sheet to give an α/β structure nearly always involves a parallel β sheet. The stacking in $\alpha + \beta$ structures often involves antiparallel sheet. β sheet is invariably twisted, with a right-handed twist when viewed parallel to the strands (as in Figure 19.11). Two adjacent rows of residues in an α helix (i, $i + 4$, $i + 8$, etc., and $i + 1$, $i + 5$, $i + 9$, etc.) also form a surface with a right-handed twist. An ideal helix can pack on an ideal sheet with the helix axis parallel to the strands, as the two surfaces are complementary. This is often found in practice. The main region of contact is an even, nonpolar surface made up of side chains of valine, leucine, and isoleucine residues in the β sheet. Antiparallel sheets can pack face to face.

There are few polar interactions across the contacting surfaces. The buried hydrogen-bond donors are invariably paired. Similarly, any buried charged group is paired with an interacting complementary charge to form a salt bridge or with a polar group of complementary partial charge, except in a few rare examples where a charged group is buried for purposes of catalysis.

The irregularly folded segments of polypeptide chain that connect the helixes and sheets—that is, the loops—are usually exposed to solvent and are often short. Globular proteins thus have *hydrophobic cores* with their charged groups on the outside: they are "waxy" on the inside and "soapy" on the outside.

Proteins are not loose and floppy structures; their interior residues are packed as tightly as crystalline amino acids. The packing density (the fraction of the total space occupied by the atoms) is about 0.75 for proteins, compared with values of 0.7 to 0.78 found for crystals. Close-packed spheres have a packing density of 0.74, and infinite cylinders, 0.91. The packing in proteins varies according to the nature of the groups: aliphatic groups—i.e., the protein hydrophobic cores—have smaller volumes in protein interiors than in solution, while peptide and charged groups have larger volumes.[20]

Many proteins are oligomers composed of subunits that are not covalently linked. The overall organization of the subunits is known as the *quaternary structure*. This refers to the arrangement of the subunits in space and the ensemble of subunit contacts and interactions, without regard to the internal geometry of the subunits. The three-dimensional structure of each subunit is still referred to as its tertiary structure. A change in quaternary structure means that the subunits move relative to each other. The subunits may be identical polypeptide chains, or they may be chemically different. As a rule, the contact interfaces of the subunits are as closely packed with each other as are the interiors of proteins. Also, any charged or hydrogen-bonding group on one surface is paired with a complementary partner on the other. Interfaces can have a high number of such hydrogen bonds, since they are important determinants of specificity (Chapter 11).

D. Protein diversity

A wealth of information about protein sequences has accumulated over the past 40 years, initially from direct sequencing of proteins but, for the past 20 years, from the far faster sequencing of DNA and by inferring the protein structure from the genetic code. Since 1995, the sequences of whole genomes have been determined from international efforts (Table 1.3). There are probably more than 4000 proteins encoded in the *E. coli* genome, as judged from the number of open reading frames, and more than 6000 in yeast. This raises some questions: How many different types of structure do we have to consider, and how did such a large number evolve on Earth in a relatively short period of time? Fortunately, there are simplifying features that aid in the classification of structures and point to ways that different proteins could have been generated in nature via shortcuts.

1. Introns, exons, and inteins and exteins

First, there are some breakdowns in the one-to-one correspondence of genome and gene product sequences that later will be seen to have interesting evolutionary consequences. The structural genes of eukaryotic (i.e., higher) organisms are often interspersed with extraneous segments of DNA that do not code for amino acid residues in the protein. The genes of prokaryotes (e.g., bacteria) consist of DNA containing contiguous sequences of nucleotide triplets, each coding for an amino acid (Chapter 14). In eukaryotes, the coding sections, termed *exons*, may

Table 1.3   *Some sequenced genomes*

Cell	Megabases in genome	Number of ORFs	% ORFs of known function[a]
a. Prokarya			
Mycoplasm genitalium	0.6	468	70
Mycoplasm pneumoniae	0.8	677	59
Haemophilus influenzae	1.8	1680	68
Escherichia coli	4.7	4285	72
b. Archaea			
Methanococcus jannaschi	1.7	1735	45
c. Eukarya			
Yeast *(Sacchromyces cerevisiae)*	12.5	6284	60

[a] ORF = open reading frame; i.e. the likelihood that it corresponds to a protein. The percentage of ORFs whose function is known or can be guessed is listed.

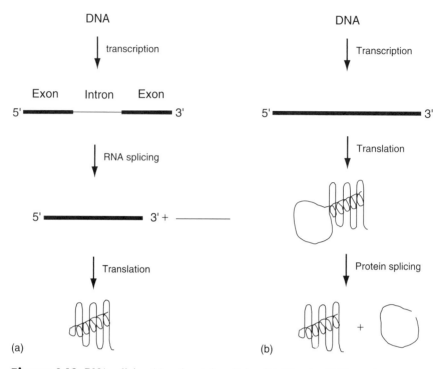

Figure 1.16 RNA splicing (a) and protein splicing (b). [Based on Y. Shao and S. B. H. Kent, *Chemistry & Biology* **4**, 187 (1997).]

be separated by insertion elements, termed *introns*. When the DNA is transcribed, the introns are excised from the mRNA to give a contiguous sequence of exons that is translated into the protein (Figure 1.16). The excision is catalyzed by the RNA itself: sequences of RNA called *ribozymes* act as rudimentary enzymes. The correct protein sequence can be determined by sequencing cDNA, DNA that is produced by copying the mRNA.

There is a further complication in that some proteins rearrange segments of their structure after biosynthesis. An intervening polypeptide domain called an *intein* (*int*ernal prot*ein*) is excised from a precursor polypeptide. The flanking amino-terminal and carboxy-terminal polypeptide domains, *exteins*, are concomitantly ligated together by the formation of a peptide bond[21] (Figure 1.16).

2. Divergent evolution of families of proteins

The mammalian serine proteases have a common tertiary structure as well as a common function. The enzymes are so called because they have a uniquely reactive serine residue that reacts irreversibly with organophosphates such as diisopropyl fluorophosphate. The major pancreatic enzymes—trypsin, chymotrypsin, and elastase—are kinetically very similar, catalyzing the hydrolysis of peptides and synthetic ester substrates. Their activities peak at around pH 7.8 and fall off at low pH with a pK_a of about 6.8. In all three cases, the reaction forms an "acylenzyme" through esterification of the hydroxyl of the reactive serine by the carboxyl portion of the substrate. The major difference among the three enzymes is specificity. Trypsin is specific for the peptides and esters of the amino acids lysine and arginine: chymotrypsin for the large hydrophobic side chains of phenylalanine, tyrosine, and tryptophan; and elastase for the small hydrophobics, such as alanine. The polypeptide backbones of all three are essentially superimposable (Figure 1.17), apart from some small additions and deletions in the chain. The difference in their specificities lies mainly in a pocket that binds the amino

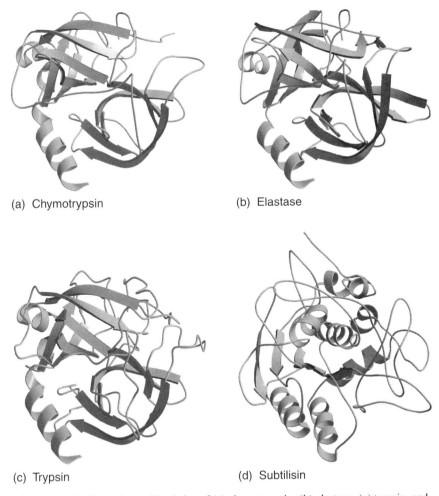

(a) Chymotrypsin

(b) Elastase

(c) Trypsin

(d) Subtilisin

Figure 1.17 The polypeptide chains of (a) chymotrypsin, (b) elastase, (c) trypsin, and (d) subtilisin drawn in MolScript.

acid side chain.[22,23] There is a well-defined binding pocket in chymotrypsin for the large hydrophobic side chains (Figure 1.18). In trypsin, the residue at the bottom of the pocket is an aspartate instead of the Ser-189 of chymotrypsin. The negatively charged carboxylate of Asp-189 forms a salt linkage with the positively charged ammonium or guanidinium at the end of the side chain of lysine or arginine. In elastase, the two glycines at the mouth of the pocket in chymotrypsin are replaced by a bulky valine (Val-216) and threonine (Thr-226). This prevents the entry of large side chains into the pocket and provides a way of binding the small side chain of alanine (Figure 1.18). There is extensive similarity among their primary structures, but only about 50% of the sequences of elastase and chymotrypsin are composed of amino acids that are chemically identical or similar to those in trypsin. This level is highly significant and indicates very similar tertiary structures (see section D8). Closer examination shows that 60% of the amino acids in the interior, but only 10% of the surface residues, are conserved. The major differences occur in exposed areas and external loops. The differences in specificity result from changes in three loops that form the lining of the binding pocket (see Chapter 16, section B1g).[22,23] The active site of chymotrypsin has a catalytic triad of residues. Asp-102, His-57 and Ser-195,

that is present in all serine proteases.

The mammalian serine proteases appear to represent a classic case of *divergent evolution*. All were presumably derived from a common ancestral serine protease.[23] Proteins derived from a common ancestor are said to be *homologous*. Some nonmammalian serine proteases are 20 to 50% identical in sequence with their mammalian counterparts. The crystal structure of the elastase-like protease from *Streptomyces griseus* has two-thirds of the residues in a conformation similar to those in the mammalian enzymes, despite having only 186 amino acids in its sequence, compared with 245 in α-chymotrypsin. The bacterial enzymes and the pancreatic ones have probably evolved from a common precursor.

3. Convergent evolution

The first crystal structure of a bacterial serine protease to be solved — subtilisin, from *Bacillus amyloliquefaciens* — revealed an enzyme of apparently totally different construction from the mammalian serine proteases (Figure 1.17). This was not unexpected, since there is no sequence homology between them. But closer examination shows that they are functionally identical in terms of substrate binding and catalysis. Subtilisin has the same catalytic triad, the same system of hydrogen bonds for binding the carbonyl oxygen and the acetamido NH of the substrate, and the same series of subsites for binding the acyl portion of

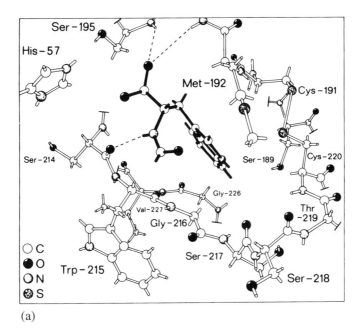

(a)

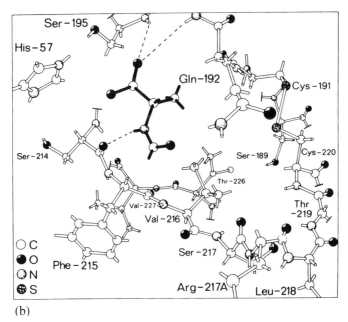

(b)

Figure 1.18 Comparison of the binding pockets in (a) chymotrypsin, with *N*-formyl-L-tryptophan bound, and (b) elastase, with *N*-formyl-L-alanine bound. The binding pocket in trypsin is very similar to that in chymotrypsin, except that residue 189 is an aspartate to bind positively charged side chains. Note the hydrogen bonds between the substrate and the backbone of the enzyme.

the substrate as the mammalian enzymes have. This appears to be a case of *convergent evolution*. Different organisms, starting from different tertiary structures, have evolved a common mechanism. Another example of convergent evolution is the endopeptidase thermolysin from *Bacillus thermoproteolyticus* and the carboxypeptidases. There are no sequence or structural homologies except that the active sites are very similar, containing in each case a catalytically important Zn^{2+} ion. The enzymes consequently appear to have similar catalytic mechanisms.

4. Convergence or divergence?

Our intuition tells us that the mammalian serine proteases have evolved through divergence but their common catalytic mechanism with subtilisin has developed through convergence. Other cases are not so clear-cut. The accepted procedure for distinguishing between convergence and divergence is to count the number of common characteristics. If there are many, then divergence is more likely; if there are few, convergence is the logical interpretation. This has been translated into molecular terms to provide the following six criteria for testing whether two proteins have evolved from a common precursor:[24]

1. The DNA sequences of their genes are similar.
2. Their amino acid sequences are similar.
3. Their three-dimensional structures are similar.
4. Their enzyme-substrate interactions are similar.
5. Their catalytic mechanisms are similar.
6. The segments of polypeptide chain essential for catalysis are in the same sequence (i.e., not transposed).

These criteria are in descending order of strength. If criteria 1 and 2 hold, the rest will follow, in most—but not all—cases. Note that in the serine proteases, tertiary structure is more conserved than primary structure. Sometimes structure has been conserved through evolution, but function has changed; that is, criteria 3 and 4 do not hold. For example, the binding protein haptoglobin appears to have diverged from the serine proteases.

5. α/β barrel (or TIM barrel) proteins

The enzyme triosephosphate isomerase, abbreviated to TIM, was found to have an important type of structure, now called an α/β or TIM barrel, consisting of at least 200 residues. In its idealized form, the barrel consists of eight parallel β strands connected by eight helixes (Figure 1.19). The strands form the staves of the barrel while the helixes are on the outside and are also parallel (Figure 1.20). β strands 1 and 8 are adjacent and form hydrogen bonds with each other. The center of the barrel is a hydrophobic core composed of the side chains of alternate residues of the strands, primarily those of the branched

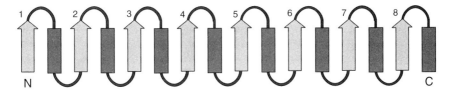

Figure 1.19 β-Barrel motif.

amino acids Val, Ile, and Leu. The enzymes having such barrels catalyze a variety of reactions, have a variety of subunit compositions, and show no homology apart from those that catalyze the same reaction in different organisms. The active site is always in the same region of the enzyme, being formed from the eight loops that connect the carboxyl end of each strand with the amino end of each helix.

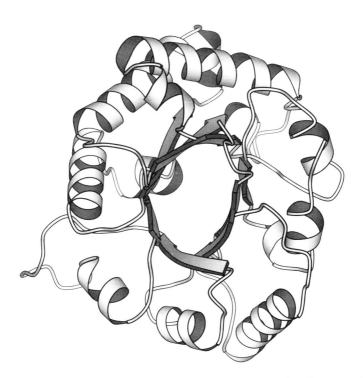

Figure 1.20 The archetypal β barrel, a monomer of triosephosphate isomerase (TIM) drawn in MolScript.

As there appears to be little sequence identity at the gene level and the active sites frequently use different regions of the loops, it would seem that the TIM barrels have arisen by convergent evolution. But in half the TIM barrel proteins, the phosphate binding site is conserved, whereas the percentage of identical residues was barely significant for the remaining parts of the enzymes. This suggests that they could be evolutionarily related after all.[25]

6.   Dehydrogenases and domains[26]

One would expect that the NAD^+-dependent dehydrogenases, a class of enzymes that have the same chemical function and that bind the same cofactor, would form a structurally related family. They appear to do this, but not in the same clear-cut way as do the serine proteases. The tertiary structures of the first two crystal structures to be solved, dogfish lactate dehydrogenase and soluble porcine malate dehydrogenase, were found to be almost superimposable, if the first 20 residues of lactate dehydrogenase are discounted. It seems likely that the two dehydrogenases have evolved from a common precursor dehydrogenase. But the subsequent solution of the structures of horse liver alcohol dehydrogenase and lobster glyceraldehyde 3-phosphate dehydrogenase complicated the picture, because they are extensively different. It was noticed, however, that the structure of each of the four enzymes consists of two domains, one of which is similar in all four. This domain binds the nucleotide NAD^+ (Chapter 8, section B).

The nucleotide-binding fold is a complicated structure that differs in detail from one dehydrogenase to another. In its idealized form, it consists of six strands of parallel β sheet with four parallel helixes running antiparallel to the sheet. This structure occurs in other nucleotide-binding proteins, such as phosphoglycerate kinase. But it is also found in proteins, such as flavodoxin, that are not involved in nucleotide binding. It is not known whether these similarities are evidence for a common evolutionary precursor protein or they are caused by there being only a limited number of ways of folding a polypeptide chain.

7.   Evolution of proteins by fusion of gene fragments

The evolution of proteins by a random choice of individual amino acids seems a most unlikely process. For one thing, the number of different sequences possible for a protein containing 250 amino acid residues (e.g., trypsin or chymotrypsin) is 20^{250} or 2×10^{325}. However, the domain structure of the dehydrogenases points to a very simple means of generating a family of enzymes with diverse specificity. Suppose that there is a gene coding for the dinucleotide-binding domain. Then the fusion of this gene with a series of genes each coding for a separate "catalytic domain" could have generated the family of dehydrogenases.

It has been suggested that the exons could correspond to functional units of proteins and that new proteins could have evolved by the recombination of different exons.[27] Early circumstantial evidence was consistent with this idea. For example, hen egg white lysozyme DNA contains four exons. Exon 2 has the catalytic center of the enzyme and could perhaps have been a primitive glycosidase. Exon 3 codes for substrate-orienting residues. The lysozyme from bacteriophage T4 is prokaryotic and is thus not derived from DNA containing introns. The catalytic center of the molecule appears, however, to be equivalent to the fusion of exons 2 and 3 of the eukaryotic enzyme. A diagonal plot is used to search for compact structural units, "modules," in proteins.[28] There is a good correspondence between these units in hen egg white lysozyme and hemoglobin and the exons in their genes. The modules are linked in lysozyme by disulfide bridges, but the folding of each module may be independent so that the bridges may not be essential for the individual folding.[28]

Subsequent analysis suggests that the original fundamental unit for evolution by fusion of fragments is the domain.[29] The exons are smaller and do not generally correspond to independently stable folding units, as do the domains. It is proposed that insertion of introns and their dispersal through exon shuffling was a later event in evolution, which became important in higher eukaryotes.[29] This is a matter of dispute, and it is possible that there are both early and late introns in evolution.[30]

A related phenomenon is the repeat of a domain along the same polypeptide chain, as found in the immunoglobulins. Such repeats could have evolved by the fusion of two copies of the same gene. The proteases of coagulation, fibrinolysis, and complement activation appear to have arisen in a modular manner.[31]

The evolution of proteins in a modular fashion by fusion of segments of genes, each coding for a module of a compact structural unit of polypeptide, is thus a credible and attractive hypothesis for explaining the rapid generation of enzyme diversity.

It is not easy to mimic the shuffling of domains *in vitro* by manipulation of genes. For example, each catalytic polypeptide chain of the multimeric *E. coli* aspartate transcarbamoylase (ATCase) is composed of two globular domains connected by two interdomain helixes. The *E. coli* enzyme ornithine transcarbamoylase (OTCase) is 32% identical in sequence and thus of presumably similar structure (see section D8). None of the chimeras in which a domain from one enzyme was attached to the corresponding partner in the other is active. The specific intrachain and interchain side-chain interactions also have to evolve for the correcting packing.[32]

Some proteins can oligomerize by *domain swapping*.[33] The monomeric protein has one domain paired with another (intramolecular pairing). In the oligomer, the intramolecular contacts separate so that one of the domains pairs with the equivalent partner in another subunit (intermolecular pairing). This gives rise to dimers and higher order polymers. Examples are bovine seminal ribonuclease and plasminogen activator inhibitor.

8. Homology, sequence identity, and structural similarity

The term *homology* is frequently misused. Two proteins are homologous when they are derived from a common precursor and so are similar in structure but not

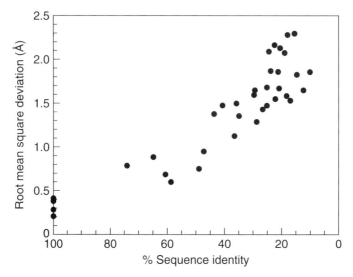

Figure 1.21 Relationship between three-dimensional structure and similarity of sequence. [After C. Chothia and A. Lesk, *EMBO Journal* **5**, 823 (1986).]

necessarily in function. They are either homologous or not. If 50% of the residues in two proteins are the same, they have 50% sequence identity and not 50% homology: they are homologous.

An important question is: What is the relationship between percent sequence identity and similarity of tertiary structure? This depends on the length of the protein: the longer the protein, the lower the percent identity that implies identical structure. For a protein of 85 residues, a 25 to 30% sequence identity implies an identical three-dimensional structure. The more the percent identity of structure, the smaller the root mean square deviation of the coordinates of the two structures[34] (Figure 1.21).

For those interested in statistics, in September 1998, there were 7657 structures deposited in the Brookhaven Protein Database (http://www.pdb.bnl.gov). These are analyzed on the scop database (http://scop.mrc-lmb.cam.ac.uk) into 435 folds, 640 superfamilies, 948 families, and 14 903 domains. There are probably about 1000 different superfamilies.[35]

E.   Higher levels of organization: Multienzyme complexes

Certain enzymes, notably those involved in sequential steps in a biosynthetic pathway or those involved in a complex biochemical process, are organized into physical aggregates. These vary in size and complexity from the simple association of two enzymes, through multienzyme complexes, to large units such as the ribosome. Well-documented examples are tryptophan synthase and the 2-oxoacid dehydrogenase complexes, discussed below, and the fatty acid synthetases. There is evidence that the glycolytic enzymes in *Escherichia coli* are physically associated into a large complex. The group of diverse enzymes responsible for the priming of DNA replication in *E. coli* appears to be organized into a "primosome," with each enzyme having its defined role in the team.

1. Multiheaded enzymes and the noncovalent association of different activities

The enzyme tryptophan synthase is an $\alpha_2\beta_2$ tetramer (i.e., it contains two pairs of identical chains).[36] In the genome, the α chain ($M_r = 29\,000$) and the β chain ($M_r = 44\,000$) are the two most distal genes of the tryptophan operon. (An operon is a set of consecutive genes that may be induced or repressed as a group. The entire set of enzymes of a biosynthetic pathway is frequently encoded in this way so that the pathway can be turned on or off as a whole—coordinate expression.) The tetrameric enzyme catalyzes the following reaction:

$$\text{Indole 3-glycerol phosphate } + \text{ serine } \xrightarrow{\alpha_2\beta_2}$$
$$\text{tryptophan } + \text{ glyceraldehyde 3-phosphate} \qquad (1.3)$$

The enzyme can be resolved into active α and β_2 subunits that separately catalyze the partial reactions 1.4 and 1.5, respectively. The overall reaction (1.3) is the sum of the two:

$$\text{Indole 3-glycerol phosphate } \xrightarrow{\alpha} \text{ indole } + \text{ glyceraldehye 3-phosphate} \qquad (1.4)$$

$$\text{Indole } + \text{ serine } \xrightarrow{\beta_2} \text{ tryptophan} \qquad (1.5)$$

The intermediate (indole) is not released into solution[37] but is shuttled directly between the subunits in the complex through a 25–30 Å long tunnel between the α and β subunits.[38]

Besides having a noncovalent association of subunits as in tryptophan synthase, some enzymes are *double-headed*, in that they contain two distinct activities in a single polypeptide chain. A good example of this is the indole 3-glycerol phosphate–synthase-phosphoribosyl anthranilate isomerase bifunctional enzyme from the tryptophan operon of *E. coli*. The crystal structure of the complex has been solved at 2.0 Å resolution.[39] The two enzymes have been separated by genetic manipulation.[40] The activity of the two separate monomeric monofunctional constituents is the same as in the covalent complex so there is no catalytic advantage of having the proteins fused.

2.   The *arom* complex

The shikimate pathway (Figure 1.22) is one of the two major biosynthetic routes for aromatic products. It synthesizes the precursors for the aromatic amino acids.[41] The five enzymes in the center of the pathway in *E. coli* are separate monofunctional enzymes, and their genes are dispersed about the genome. The corresponding genes are also dispersed in the genome of other prokaryotes, such as *Salmonella typhimurium* and *Bacillus subtilis*. In *Neurospora crassa*, *Aspergillus nidulans*, and other fungal and yeast species, there is an *arom*

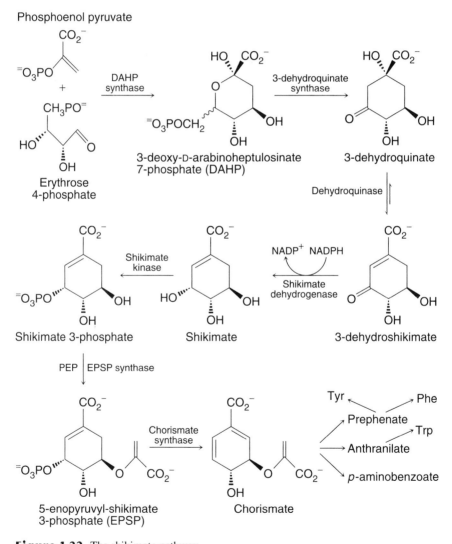

Figure 1.22 The shikimate pathway.

gene that codes for all five. In the common yeast *Sacchromyces cerevisiae*, the *arom* gene codes for a single polypeptide of M_r 174 555 and 1588 amino acids in length. There are five functional regions, each one of which is homologous with one of the enzymes in *E. coli*. Limited proteolysis of the enzyme from *N. crassa* gives a fragment (M_r = 68 000) that carries two of the activities. It is thought that the *arom* gene of the yeast and fungi has evolved by fusion of separate activities. By having all activities on a single polypeptide chain, it is possibly easier for the organism to coordinate their expression during protein biosynthesis.

3. The pyruvate dehydrogenase complex[42]

There are two 2-oxoacid dehydrogenase multienzyme complexes in *E. coli*. One is specific for pyruvate, the other for 2-oxoglutarate. Each complex is about the size of a ribosome, about 300 Å across. The pyruvate dehydrogenase is composed of three types of polypeptide chains: E1, the pyruvate decarboxylase (an α_2 dimer of M_r = 2 × 100 000); E2, lipoate acetyltransferase (M_r = 80 000); and E3, lipoamide dehydrogenase (an α_2 dimer of M_r = 2 × 56 000). These catalyze the oxidative decarboxylation of pyruvate via reactions 1.6, 1.7, and 1.8. (The relevant chemistry of the reactions of thiamine pyrophosphate [TPP], hydroxyethylthiamine pyrophosphate [HETPP], and lipoic acid [lip-S_2] is discussed in detail in Chapter 2, section C3.)

$$\text{Pyruvate} + \text{TPP}—\text{El} \longrightarrow \text{HETPP}—\text{El} + CO_2 \tag{1.6}$$

$$\text{HETPP}—\text{El} + (\text{lip-}S_2)—\text{E2} \longrightarrow \text{TPP}—\text{El} + (\text{acetyl-S-lip-SH})—\text{E2} \tag{1.7}$$

$$(\text{Acetyl-S-lip-SH})—\text{E2} + \text{CoA} \longrightarrow [\text{lip-}(SH)_2]—\text{E2} + \text{acetyl-CoA} \tag{1.8}$$

The complex is constructed around a core of 24 E2 molecules arranged in octahedral symmetry. The lipoic acid residues that transfer the important acetyl group are attached to E2 on the ε-NH groups of two lysine residues per chain. This enables the lipoates to be on a swinging arm of radius 14 Å. The correlation time of an arm that has been spin-labeled is 0.2 ns, compared with 0.01 ns for the residue in solution and 10 μs for the rotational correlation time of the total complex, indicating the free rotation of an acetylated chain. ^{1}H NMR measurements show that the polypeptide chain of E2 is itself very mobile. This increases the spatial freedom of the swinging arm and so facilitates the rapid transacetylation from one E2 chain to another. By this mechanism, a single El dimer can acetylate the lipoic acid residues of possibly 12 E2 chains—the process of "servicing."

4. DNA polymerases

The previous enzymes catalyze sequential steps in biosynthetic pathways. There are other multifunctional enzymes that have multiple activities for other purposes. DNA polymerase I from *E. coli* (see Chapter 14) contains three activities. The enzyme consists of 928 amino acids. Residues 324–518 contain the po-lymerizing activity. The domain consisting of residues 1 to ~320 has a $5' \rightarrow 3'$ exonuclease activity that removes nucleotides in advance of replication. The C-terminal domain of residues ~520 to 928 has a $3' \rightarrow 5'$ exonuclease activity that proofreads errors of misincorporation (see Chapter 13). The N-terminal domain may be removed by proteolysis or genetic manipulation to leave a functional polymerase. The reverse transcriptase (DNA polymerase) from HIV contains a ribonuclease domain that removes RNA from the DNA-RNA hybrid that is formed when the enzyme replicates the RNA genome of the virus.

5.   Reasons for multiple activities and multienzyme complexes

In some cases, there are fairly obvious and specific reasons for the association of different activities of the complex structures. For example, as discussed in section E4 and Chapters 13 and 16, certain DNA polymerases have a $3' \rightarrow 5'$ hydrolytic activity in addition to the polymerization activity. This is an error-correcting mechanism in which hydrolysis must be coupled with synthesis. For the complex series of reactions in the "primosome," there are so many different activities that require coordination that there has to be an organized system enzymes at the point of priming of DNA replication rather than random pathways of association and dissociation of the constituents. There is a dearth of evidence for catalytic rate enhancement per se in having multiple activities on a single chain or in a complex. One piece of evidence is that concentrations of coenzyme A and its esters in chloroplasts are insufficient to account for rates of chloroplast fatty acid synthesis, implying that the substrates are channelled within the chloroplast fatty acid synthase.[43] To summarize, the following possible advantages of multienzyme complexes over individual activities in solution have been proposed.

1. *Catalytic enhancement:* The reduction of diffusion time of an intermediate from one enzyme to the next.
2. *Substrate channeling:* The control over which biosynthetic route an intermediate should follow, by directing it to a specified enzyme rather than allowing competition from other enzymes in solution.
3. *Sequestration of reactive intermediates:* The protection of chemically unstable intermediates from aqueous solution.
4. *Servicing:* As described for pyruvate dehydrogenase, where one subunit can pass a reagent to many different other subunits.
5. *Coordinate regulation:* All may be regulated in a coordinated manner.
6. *Coordinate expression:* All may be biosynthesized in a coordinated manner.

F.   The structure of enzyme–substrate complexes

The outstanding characteristic of enzyme catalysis is that the enzyme specifically binds its substrates, with the reactions taking place in the confines of the enzyme–substrate complex. Thus, to understand how an enzyme works, we need to know not only the structure of the native enzyme, but also the structures of the complexes of the enzyme with its substrates, intermediates, and products. Once these have been determined, we can see how the substrate is bound, what catalytic groups are close to the substrate, and what structural changes occur in the substrate and the enzyme on binding. There is one obvious difficulty: enzyme–substrate complexes react to give products in fractions of a second, whereas the acquisition of x-ray data usually takes minutes to several hours. For this reason, it is usual to determine the structures of the complexes of enzymes with the reaction products, inhibitors, or substrate analogues. There are now, however, methods available for studying reactions in crystals.

1. Methods for examining stable enzyme–substrate complexes

a. Binding of small molecules to protein crystals

Protein crystals generally contain about 50% solvent, and never less than 30%. Often there are channels of solvent leading from the exterior of the crystal to the active site, so that substrates may be diffused into the crystals and bound to the enzyme. Alternatively, it may be possible to co-crystallize the enzyme and substrates. Provided that there are no drastic changes in the structure or packing of the enzyme when it binds the ligand, it has been traditional to solve the structure of the complex by the difference Fourier technique. This involves measuring the *differences* between the diffraction patterns of the native crystals and those soaked in a solution of the ligand. Better still, the structures can be determined using the molecular replacement procedure. The electron density of the bound ligand and any minor changes in the structure of the enzyme may be obtained without solving the whole structure from scratch.

b. Production of stable complexes

The first attempts to determine the structure of a productively bound enzyme–substrate complex were based on extrapolation from the structures of stable enzyme–inhibitor complexes. (The classic example of this, lysozyme, is discussed in section F3.) Such extrapolation may be done in several ways. For example, a portion of the substrate may be bound to the enzyme and the structure of the remainder determined by model building. An alternative method is to use a substrate analogue that is unreactive because its reactive bond is modified. Typical examples are the binding of phosphoglycolohydroxamate, a substrate analogue, to triosephosphate isomerase,[44] or a piece of DNA, which lacks the reactive 2'-OH groups, to a ribonuclease.[45]

Methods are also available in some cases for the direct study of productively bound enzyme–substrate complexes under unreactive conditions. These include the use of a substrate that is weakly reactive because it is at a pH at which the enzyme is largely inactive due to a residue being in the wrong ionic state (e.g., indolylacryloyl-chymotrypsin, section F2), or because it is at a very low temperature (e.g., an elastase acylenzyme, section F2; and ribonucleases, Chapter 16), or because the wrong metal ion is present (e.g., Al^{3+} with xylose isomerase). On occasion, a reactive complex may be stable in the absence of one of the reagents. For example, the productive complex of the tyrosyl-tRNA synthetase with tyrosyl adenylate has been solved because the complex is stable in the absence of tRNA (Chapter 15). An enzyme that has been inactivated by site-directed mutagenesis may also be used.

c. Time-resolved crystallography

A major recent innovation has been to use substrates that are unreactive but may be activated by, for example, photolysis (see Chapter 6). This has necessitated the introduction of crystallographic procedures that can gather data in fractions of a second, rather than the minutes or hours conventionally used. Conventional protein crystallography uses a beam of monochromatic x-rays. An older technique that has been reintroduced is that of *von Laue*, which uses a spectrum of polychromatic radiation. An intense beam from a synchrotron, spanning wavelengths from 0.25 to 2.5 Å, enables data for the Laue method to be taken over a fraction of a second.[46,47]

Results from these methods are discussed in later chapters. The serine proteases and lysozyme are discussed below because of their historical importance in the development of the ideas and techniques involved, and because they are used throughout this text to exemplify basic principles.

2. Example 1: The serine proteases

Peptide and synthetic ester substrates are hydrolyzed by the serine proteases by the acylenzyme mechanism (Figure 1.23).[48] The enzyme and substrate first associate to form a noncovalent *enzyme–substrate* complex held together by physical forces of attraction. This is followed by the attack of the hydroxyl of Ser-195 on the substrate to give the first *tetrahedral intermediate*. The intermediate then collapses to give the *acylenzyme* releasing the amine or alcohol. The acylenzyme then hydrolyzes to form the *enzyme–product* complex via a second tetrahedral intermediate. Crystallographic studies have given the structures of most of the complexes. The rest can be obtained by model building. Working backwards, the structure of the enzyme–product complex, *N*-formyl-L-tryptophan chymotrypsin (see Figure 1.18) was solved by diffusing the product into the crystal and then using the difference Fourier technique.[49] The structure of the nonspecific acylenzyme indolylacryloyl-chymotrypsin was solved at pH 4, where it is deacylated only very slowly. The substrate, indolylacryloylimidazole, an activated derivative of the acid, was diffused into the crystal, where it acylated Ser-195. The structure of carbobenzoxyalanylelastase has been solved at $-55°C$.[50]

Nature has provided a rare opportunity for determining the structures of the enzyme–substrate complexes of trypsin and chymotrypsin with polypeptides.

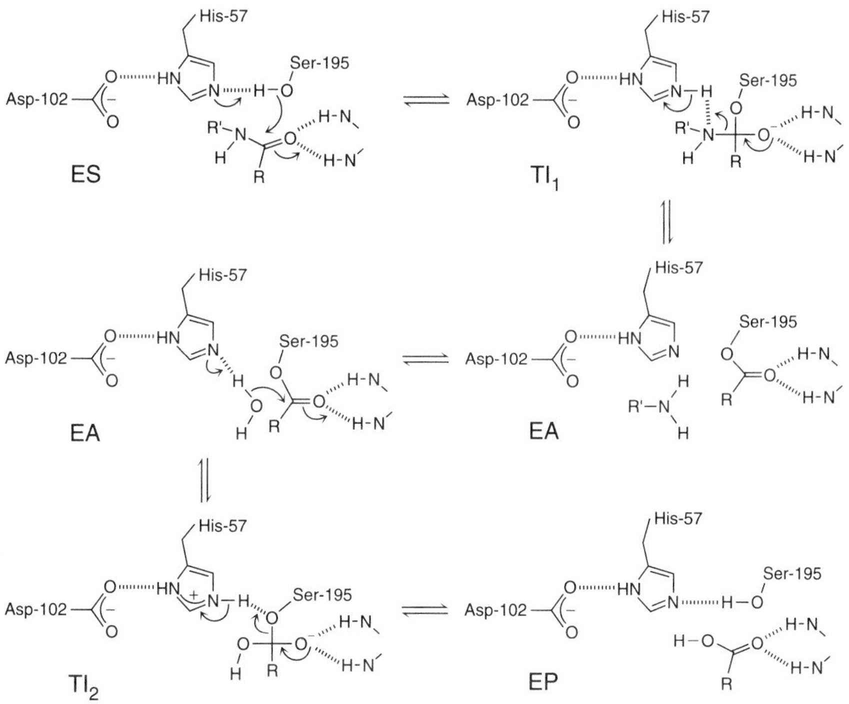

Figure 1.23 Reaction mechanism of chymotrypsin.

There are many naturally occurring polypeptide inhibitors that bind to trypsin and chymotrypsin very tightly because they are locked into the conformation that a normal flexible substrate takes upon binding. They do not hydrolyze under normal physiological conditions because the amino group that is released on the cleavage of the peptide is constrained and cannot diffuse away from the active site of the enzyme. On removing the constraints in the pancreatic trypsin inhibitor by reducing an —S—S—bridge in its polypeptide chain, the peptide bond between Lys-15 and Ala-16 is readily cleaved by trypsin. The structures of trypsin, its complex with the basic pancreatic trypsin inhibitor complex, and the free inhibitor have given the following information about the binding of substrates.

a. The binding site

The binding site for a polypeptide substrate consists of a series of *subsites* across the surface of the enzyme. By convention, they are labeled as in Figure 1.24. The substrate residues are called P (for peptide); the subsites, S. Except at the primary binding site S_1 for the side chains of the aromatic substrates of chymotrypsin or the basic amino acid substrates of trypsin, there is no obvious, well-defined cleft or groove for substrate binding. The subsites run along the surface of the protein.

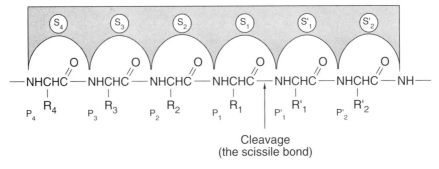

Figure 1.24 Schechter-Berger notation for binding sites in serine proteases.

b.   The primary binding site (S_1)

The binding pocket for the aromatic side chains of the specific substrates of chymotrypsin is a well-defined slit in the enzyme 10 to 12 Å deep and 3.5 to 4 by 5.5 to 6.5 Å in cross section. This gives a very snug fit, since an aromatic ring is about 6 Å wide and 3.5 Å thick. A methylene group is about 4 Å in diameter (Chapter 11), so the side chain of lysine or arginine is bound nicely by the same-shaped slit in trypsin. Of course, as mentioned in section D2, there is a carboxylate at the bottom of the pocket in trypsin to bind the positive charge on the end of the side chain, and also the mouth of the pocket is blocked somewhat in elastase.

The binding pocket in chymotrypsin may be described as a "hydrophobic pocket," since it is lined with the nonpolar side chains of amino acids. It provides a suitable environment for the binding of the nonpolar or hydrophobic side chains of the substrates. The physical causes of hydrophobic bonding and its strength are discussed in Chapter 11.

The hydrophobic binding site in subtilisin is not a well-defined slit as in the pancreatic enzymes, but more like a shallow groove. However, certain hydrogen bonds are found for all the enzymes (see Figure 1.18): the carbonyl oxygen of the reactive bond has a binding site between the backbone NH groups of Ser-195 and Gly-193. The hydrogen bonds made there are very important because the oxygen becomes negatively charged during the reaction. There is also a hydrogen bond between the NH part of the N-acylamino group of the substrate and the C=O of Ser-214.

c.   Sites S_1—S_2—S_3

The hydrogen bond between the N-acylamino NH and the carbonyl of Ser-214 initiates a short region of antiparallel β sheet between the residues Ser-214, Trp-215, and Gly-216 of the enzyme and the amino acids P_1, P_2, and P_3 of the substrate (structure 1.9).

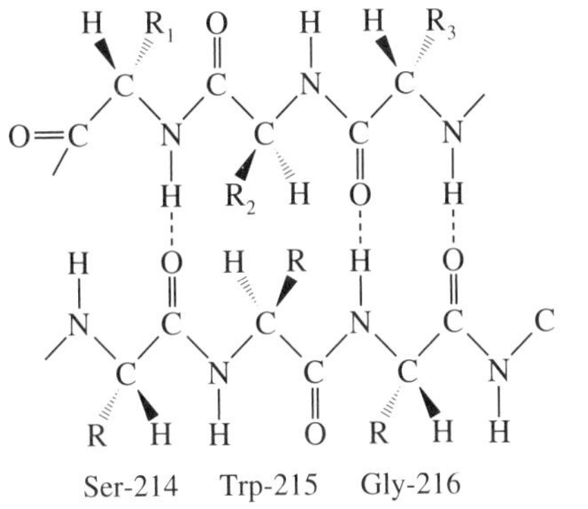

(1.9)

Ser-214   Trp-215   Gly-216

d. Site S_1': The leaving group site

There is a leaving group site that is constructed to fit L-amino acids. The contacts with the enzyme are predominantly hydrophobic, which accounts for the lack of exopeptidase activity with the enzyme, since this would require binding a $-CO_2^-$ in a nonpolar region. ("Leaving group" is chemists' jargon for a group displaced from a molecule—in this case from an acyl group.) The binding in subsites can have profound effects on catalysis, even though they are not directly involved in the catalytic process. This is discussed in Chapter 12.

3. Example 2: Lysozyme

Lysozyme catalyzes the hydrolysis of a polysaccharide that is the major constituent of the cell wall of certain bacteria. The polymer is formed from $\beta(1\rightarrow 4)$-linked alternating units of N-acetylglucosamine (NAG) and N-acetylmuramic acid (NAM) (Figure 1.25). The solution of the structure of lysozyme

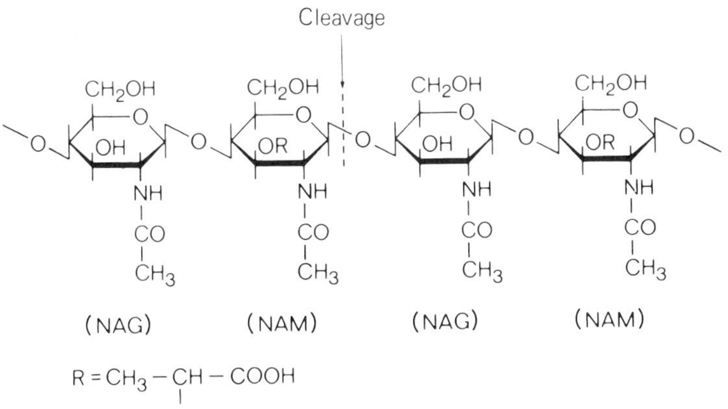

Figure 1.25 The polysaccharide substrate of lysozyme found in bacterial cell walls.

is one of the triumphs of x-ray crystallography.[51] Whereas solving the crystal structures of the serine proteases represented the culmination of a long series of studies stretching back through the history of enzymology, the structure of the little-known enzyme lysozyme stimulated the solution studies. Also, the mechanism of the enzyme, which was previously unknown, was guessed from examining the crystal structure of the native enzyme and the complexes with inhibitors. Unlike chymotrypsin, lysozyme has a well-defined deep cleft, running down one side of the ellipsoidal molecule, for binding the substrate. This cleft is partly lined with nonpolar side chains of amino acids for binding the nonpolar regions of the substrate, and it also has hydrogen-bonding sites for the acylamino and hydroxyl groups. The cleft is divided into six sites, A, B, C, D, E, and F. NAM residues can bind only in B, D, and F, whereas NAG residues of synthetic substrates may bind in all sites. The bond that is cleaved lies between sites D and E. There is no means at present of solving the structure of a productively bound enzyme–substrate complex in the same way as for chymotrypsin and trypsin, since there are no similar natural inhibitors. The method of determining the structure of lysozyme complexes was based on model building, which has been a typical approach. The structure of the enzyme and the inhibitor (NAG)$_3$ was solved first. This is a very poor substrate of the enzyme, since the structural studies show that it is bound nonproductively in the A, B, and C sites, avoiding the D and E sites where cleavage takes place. The structure of a productively bound complex was then obtained by building a wire model of the complex of (NAG)$_3$ with the enzyme and extending the polysaccharide chain by adding further NAG units, using chemical intuition about the contacts made with the enzyme. Nowadays, this procedure is done by using a computer program to optimize the fit between the enzyme and the substrate. It was found that the bond that is cleaved is located between the carboxyl groups of Glu-35 and Asp-52 (which were later proven to be in the un-ionized and ionized forms, respectively). How these contribute to the mechanism will be discussed in Chapter 16. Small distortions occur in the enzyme structure at the active-site binding cleft when the inhibitor is bound. The enzyme is said to undergo a small *conformational change*.

To summarize, the binding sites of lysozyme and the serine proteases are approximately complementary in structure to the structures of the substrates: the nonpolar parts of the substrate match up with nonpolar side chains of the amino acids; the hydrogen-bonding sites on the substrate bind to the backbone NH and CO groups of the protein and, for lysozyme, to the polar side chains of amino acids. The reactive part of the substrate is firmly held by this binding next to acidic, basic, or nucleophilic groups on the enzyme.

G.   Flexibility and conformational mobility of proteins

Although globular proteins are generally close-packed, they do have certain degrees of flexibility. Some enzymes are known to undergo conformational changes on binding ligands or substrates. These changes may be small, as with lysozyme, or they may involve large movements. It will be seen in Chapter 10 that such changes are important in a certain class of enzymes (allosteric) for modulating activity: certain ligands (allosteric effectors) can alter the shape of the protein. This clearly affects the status of crystal structures, since, although protein crystals contain about 50% water, the lattice forces constrain the structure of the protein. The following question is thus often raised.

1. Are the crystal and solution structures of an enzyme essentially identical?

The answer in general, is yes. The identity of the crystal and solution structures is supported by the following evidence:

1. The structures of several proteins have been solved by both NMR in solution and x-ray diffraction from the crystal, and the correspondence is excellent. Sometimes there are discrepancies at the contacts between neighboring molecules in the crystal lattice. Crystal packing can lead to such problems.

2. Some enzymes have been crystallized from different solvents and in different forms, but their structures remain essentially identical.

3. Enzymes in families (e.g., the serine proteases) have structures that are similar.

4. In many cases, the crystal retains enzymatic activity. In some cases, the activity of the enzyme in the crystal is the same as that in solution. The methods used for initiating reactions for study by the Laue method are used to measure activity. For example, pH-jump: the acylenzyme indolylacryloyl-chymotrypsin was crystallized at a pH at which it is stable. On changing the pH to increase the reactivity, the intermediate was found to hydrolyze with the same first-order rate constant as occurs in solution; the reactions of crystalline *ras* p21 protein, glycogen phosphorylase, and chymotrypsin have been initiated by photolysis.[52] Glyceraldehyde 3-phosphate dehydrogenase has also identical reaction rates in the crystal and solution under some conditions.[53]

But where there is an equilibrium among two or more conformations of the enzyme in solution, crystallization may select out only one of the conformations. α-Chymotrypsin has a substantial fraction of an inactive conformation present under the conditions of crystallization, but only the active form of the enzyme crystallizes. An allosteric effector molecule that changes the conformation of the protein in solution may have no effect on the crystalline protein, as, for example, with phosphorylase b.[54] The enzyme is frozen in one conformation, with the crystal lattice forces preventing any conformational change. On the other hand, the addition of an effector to phosphorylase α causes the crystals first to crack and then to anneal, giving crystals of the enzyme in a second conformation.

2. Modes of motion and flexibility observed in proteins

a. Molecular tumbling

Globular proteins are found to rotate in solution at frequencies close to those calculated for rigid spheres. The frequencies are usually expressed in terms of a *rotational correlation time*, ϕ, which is the reciprocal of the rate constant for the randomization of the orientation of the molecule by Brownian motion. For a rigid sphere, ϕ is given by

$$\phi = V\eta/kT \tag{1.10}$$

where V is the molecular volume, η is the viscosity of the medium, k is Boltzmann's constant, and T is the absolute temperature. ϕ is the time taken to rotate through a defined angle. Substituting the values of η, k, and T, and using an approximate relationship between the relative molecular mass M_r of a globular protein and V, gives an approximation that holds at ambient temperatures:

$$\phi = M_r/2000 \text{ ns} \tag{1.11}$$

The rate of molecular tumbling affects the shapes of lines observed in EPR (electron paramagnetic resonance) and NMR spectra. Slow tumbling causes the line widths to broaden. In conventional EPR spectroscopy, the lines undergo a transition from narrow to broad as the correlation time increases through the nanosecond time region. This enables correlation times of proteins of appropriate size to be calculated from the shape of the lines. For example, chymotrypsin ($M_r = 25\,000$) that has been spin-labeled has a measured correlation time of 12 ns, which is close to that expected from equations 1.10 and 1.11. In conventional NMR spectroscopy, on the other hand, the lines broaden indefinitely as the correlation time increases. The consequence is that for correlation times greater than 20 ns or so (i.e., $M_r > \sim 40\,000$), the lines begin to broaden to such an extent that they cannot be readily analyzed. A useful corollary of this is that the presence of sharp NMR (or EPR) lines in the spectra of large molecules with long correlation times indicates that the groups being observed have their own independent mobility. Molecular tumbling also affects fluorescence polarization. Fluorescence (Chapter 6) occurs when a photon is absorbed by a molecule and re-emitted at a longer wavelength. The lifetime of the excited state varies from 1 to 20 ns or so, depending on the nature of the fluorophore. If a fluorescent molecule is excited by a pulse of plane-polarized light, then the degree of polarization of the emitted light (the *anisotropy*) will decay exponentially according to the rotational correlation time. Through use of time-resolved fluorescence polarization spectroscopy and excitation of the tryptophan residue in staphylococcal nuclease B ($M_r = 20\,000$) and in serum albumin ($M_r = 69\,000$), rotational correlation times of 9.9 ns and 31.4 ns, respectively, were measured; these are close to those calculated for the rigid spheres. Just as the presence of sharp lines in NMR spectra implies the existence of independent mobility, so the finding of a correlation time for a fluorophore shorter than that expected for the overall molecular tumbling implies that the fluorophore has additional mobility.

b. "Breathing"

The compact globular regions of proteins have structural fluctuations that have been observed by a variety of techniques. The accessibility of backbone NH groups to solvent has been traditionally measured by rates of isotopic exchange (Chapter 18).[55] The amide hydrogen atoms exchange with tritium from ^3H_2O or deuterium from ^2H_2O in an acid- or base-catalyzed reaction. The rates were originally measured by radioactive incorporation or by analysis of infrared spectra. These methods have been superseded by NMR, which can measure the rate of exchange of *individual* protons of assigned groups in the structure (Chapter 18, section F). The exposed backbone NH hydrogen atoms exchange rapidly. The majority of the buried NH groups also exchange, albeit at lower rates, despite their apparently not being exposed to solvent. There is strong evidence that exchange occurs in native structures by a "breathing" of the protein molecule that allows solvent to penetrate without a mandatory unfolding of the protein (although unfolding does provide a further pathway). Perhaps the nicest demonstration of this comes from measurements of the exchange of ^1H by ^2H in crystalline myoglobin in neutron diffraction studies. Soaking the crystals in ^2H_2O leads to the exchange of 95% of the amide hydrogen atoms under conditions in which the protein cannot possibly unfold because of the constraints of the crystal lattice.[56] Another useful probe of exposure is the quenching of fluorescence by the direct collision of the excited fluorophore with a solute molecule. All the tryptophan residues, both exposed and buried, in a series of proteins are quenched by dissolved oxygen. The protein matrix is penetrated by O_2 at 25 to 50% of the rate of diffusion of O_2 in water.[57] This implies fluctuations in the protein structure on the nanosecond time scale. Computer simulations by molecular dynamics (Chapter 19) of proteins predict such modes.

c. Rotation of side chains

As well as providing a means of measuring $^1H/^2H$-exchange in proteins, NMR is a most powerful technique for studying the mobility of individual amino acids. For example, the rotational freedom of the aromatic side chains of tyrosine and phenylalanine about the C^β—C^γ bond is readily studied by various NMR methods. ^1H NMR can detect whether or not the aromatic ring is constrained in an anisotropic environment. In an isotropic environment or where there is rapid rotation on the NMR time scale, the 3 and 5 protons of phenylalanine and tyrosine are symmetrically related, as are the 2 and 6 (structures 1.12). The resultant spectrum is of the AA'BB' type, containing two pairs of closely separated doublets. But if there is slow rotation in an anisotropic environment, the symmetry breaks down to give four separate resonances (an ABCD spectrum), since the 5 and 6 protons are in different states from the 2 and 3. At an intermediate time range for rotation, the two spectra coalesce and the lines are said to undergo *exchange broadening*. This occurs when the rate constant k for the exchange

$$(1.12)$$

between the two states is approximately equal to $\nu_A - \nu_B$, where ν_A is the NMR resonance frequency in one state and ν_B is its counterpart in the other state. Exchange broadening may be observed and analyzed to give k when $\nu_A - \nu_B$ is in the region 10 to 10^4 Hz; i.e., for $k = 10$ to 10^4 s^{-1}. Thus, "slow" rotation means less than 1 to 10 s^{-1}, and "fast" rotation means greater than 10^4 to 10^5 s^{-1}. Three of the four phenylalanine and three of the four tyrosine residues in the basic pancreatic trypsin inhibitor are rotating at greater than 100 s^{-1}. Yet all four phenylalanine and three of the tyrosine residues are buried in the interior of the protein. The remaining buried one tyrosine and one phenylalanine ring in the trypsin inhibitor are immobilized. This distribution of immobilized and rotating aromatic rings is quite typical. The motion of the aromatic side chains is best described as a series of *jumps* or 180° flips at greater than 10^6 Hz, rather than as a continuous rotation. In contrast to the frequent flipping of the aromatic side chains of buried tyrosine residues, the side chains of buried tryptophan residues are usually immobile. The difference between them may be rationalized when it is realized that the benzene ring is close to being cylindrical because of the thickness of the π–electron cloud, whereas the indole ring of tryptophan is quite asymmetric along the axis of the C^β—C^γ bond. The tryptophan side chain can act as a rigid platform for the construction of hydrophobic regions. Surface amino acids are more mobile than interior ones, and many surface side chains have no unique conformation.

d. Domain movements: Shear and hinge motions, and segmental flexibility

There are larger scale movements in proteins that have low energy barriers. Some segments of proteins are flexibly attached to the bulk of the molecule. Their rapid movement may be detected by time-resolved fluorescence polarization spectroscopy (e.g., the Fab fragment of an immunoglobulin molecule) or by NMR and x-ray crystallography, as described in the following sections. The types of structural changes leading to the movement of domains have been classified into hinge motions, in which two elements of structure open and close as if connected by a hinge, and shear motions, in which one element of structure slides relative to the other.[58] All domain movements may be constructed from a combination of hinge and shear motions.

e. Protein mobility in solution from ^{15}N-NMR-relaxation and other NMR methods

More recently, NMR has proven to be a powerful tool for determining the mobility of individual atoms in proteins. The relaxation of ^{15}N-NMR signals in isotopically labeled protein can be analyzed using the model-free procedure of Lipari and Szabo,[59] which gives an order parameter, S, that varies from 0 for fully disordered to 1 for completely constrained.

The backbone dynamics of 4-oxalocrotonate tautomerase, a 41-kDa homohexamer with 62 residues per subunit, and its complex with a substrate analogue have been analyzed by the model-free formalism.[60] Binding of the analogue freezes the motion of some of the backbone NH vectors in the active site, leading to a loss of entropy (Chapter 2).

Chymotrypsin inhibitor 2 (Chapter 17) is one of the polypeptide inhibitors that was described in section F2 as not being cleaved by serine proteases. ^{15}N-NMR studies have provided an explanation for the resistance to hydrolysis. The model-free formalism has been applied to the inhibitor, which had been cleaved artificially in the binding loop that occupies the binding site of chymotrypsin.[61] The residues in the half of the loop region that contains the NH_2 group that was released on cleavage have significantly lower order parameters than those in the second half. This implies that the NH_2 group that is released on cleavage of the scissile bond in the complex with chymotrypsin remains anchored in its original position, inhibiting the attack of water on the acylenzyme that is formed between the protease and the cleaved inhibitor. Further, the NH_2 group is poised for reversing the formation of the acylenzyme so that the equilibrium between the cleaved and uncleaved inhibitor, bound to the protease, greatly favors the uncleaved complex.

The motion of the flexible loop of precipitated triosephosphate isomerase has been measured by solid-state deuterium NMR in the presence and absence of substrate and transition-state analogs.[62] The loop jumps between two conformations at a rate of 3×10^4 s^{-1} (from the predominant to the less populated form) irrespective of whether a substrate is bound, showing that it is a natural motion of the protein.

The advent of very high fields has allowed the observation of the magnetic alignment of proteins with the magnetic field, which has the potential of giving further information on mobility.[63]

f. Static disorder and real motion in crystals from x-ray diffraction studies

Although protein crystallography has traditionally been used to give only a static picture of protein structure, it is also a powerful technique for studying the mobility of every residue in a protein chain. This is because, in addition to locating the positions of atoms from their electron density, it is also possible to measure their *mean square displacement* from the smearing out of the electron density.

Some regions of the electron density are very weak in certain structures. In other words, the structure in that region is not uniquely defined in space during the time of the experiment. This can be due to either *static disorder* (i.e., the atoms are occupying a number of different conformations in space) or *real motion* (i.e., the atoms are rapidly vibrating about their mean positions). It is possible in theory to distinguish between these modes by studying the dependence of the mean displacement on temperature. The apparent mean amplitudes of true vibrations increase with increasing temperature in accordance with normal thermal motions, whereas those of static disorder do not change. The pioneering study was on sperm whale myoglobin at four temperatures between 200 and 300 K.[54] The greatest displacement of the backbone occurs at the C-terminus. The greatest average root mean square displacements of side chains are about 0.4 to 0.5 Å, for the charged chains on the surface. Buried nonpolar side chains have average values of 0.2 to 0.25 Å. The core of the protein is, in general, more rigid than the outer regions. Examination of the periodicities of the amplitudes suggests that not only do the helixes move as rigid units, but they also experience breathing or rippling modes of vibration. The experimental data are in broad agreement with predictions from molecular dynamics calculations. The residues of highest apparent motion in lysozyme crystals are in the active-site region where the conformational change is observed on substrate binding: specifically, in the lips of the binding crevice.[54] Residues around the active site of α-lytic protease are able to move significantly to accommodate different substrates. The crystal structure at 120 K reveals static disorder from different conformational substates that have been trapped by freezing.[64] An example of movement detected by crystallography is illustrated in Chapter 15, Figure 15.6. In summary, then, small movements and distortions of proteins are possible, especially at the surface, because of their inherent flexibility.

3. Protein mobility and enzyme mechanism

A major question is: Are the modes of mobility observed in enzymes just incidental? That is, are they simply an inherent property of proteins that must always be borne in mind, or are they essential for catalysis? A crude analogy of the problems involved may be made by comparing the properties of a grandfather clock and a digital quartz watch. In the grandfather clock, the large swing of the pendulum is all-important for regulating the time, whereas the vibrations of the atoms in the pendulum rod are incidental. However, in that the vibrations of the atoms are responsible for the thermal expansion of the rod, they do exert a second-order effect on the accuracy of the clock. On the other hand, in the quartz watch, the vibrations are the central mechanism. Enzymes often resemble the grandfather clock. The larger movements, such as hinge bending, are important for specific biological processes such as muscle contraction, for specific purposes such as the "swinging-arm" mechanism for the pyruvate dehydrogenase complex (section E3), and for the structural changes in allosteric enzymes

(Chapter 10), Flexible loops also make large movements to envelope the substrates in triosephosphate isomerase, lactate dehydrogenase (Chapter 16), and the tyrosyl-tRNA synthetase (Chapter 15). This flexibility is probably necessary to allow access of the substrate to be buried in active sites that are deep in the enzyme, and so the loop is a molecular trapdoor. Similarly, in order for hexokinase to bind the whole molecule, the protein structure has to open and close. The importance of the vibrational and breathing modes and the internal rotations is not clear. It has not been demonstrated to date that the vibrational modes can be coupled with the chemical steps of catalysis to enhance reaction rates. However, these modes do contribute to the flexibility of the protein as regards distortion of its structure. This is relevant to theories of enzyme catalysis such as "strain" (Chapter 12), which depends on whether an enzyme can distort a substrate or vice versa. Flexibility could be useful in aiding the access of ligands to active sites. The binding of O_2 to myoglobin and hemoglobin appears to require some movement of the protein to allow access to the heme. One set of examples in which some flexibility of protein structure does appear necessary is in some cases of electron transfer, where the catalyzing protein—for example, a cytochrome—is reversibly oxidized and reduced. It is essential for the structure to be able to relax in order to be able to accommodate the changes of geometry around the iron atom as it changes its oxidation state. In this case, the protein is one of the reactants. The protein could not be rigid, since the intermediates have different geometries. This latter point must apply to some extent to any enzymatic reaction in which covalent bonds are formed with an intermediate.

References

1. A. L. Horwich and J. S. Weissman, *Cell* **89**, 499 (1997).
2. D. R. Booth, M. Sunde, V. Bellotti, C. V. Robinson, W. L. Hutchinson, P. E. Fraser, P. N. Hawkins, C. M. Dobson, S. E. Radford, C. Blake, and M. B. Pepys, *Nature* **385**, 787 (1997).
3. D. Sidransky and M. Hollstein, *Annual Review of Medicine* **47**, 285 (1996).
4. C. B. Anfinsen, *Science* **181**, 223 (1973).
5. D. W. Green, V. Ingram, and M. F. Perutz, *Proc. R. Soc.* **287**, (1954).
6. B. Stec, R. S. Zhou, and M. M. Teeter, *Acta Cryst. Section D Biological Crystallography* **51**, 663 (1995).
7. L. Sjolin, A. Wlodawer, G. Bergqvist, P. Holm, K. Loth, H. Malmstrom, J. Zaar, L. A. Svensson, and G. L. Gilliland, *J. Crystal Growth* **110**, 1 (1991).
8. A. A. Kossiakoff and S. A. Spencer, *Biochemistry* **20**, 642 (1981).
9. K. Wüthrich, *Acta Cryst. Section D Biological Crystallography* **51**, 249 (1995).
10. K. Wüthrich, *NMR of proteins and nucleic acids*, Wiley (1986).
11. D. S. Garrett, Y. J. Seok, D. I. Liao, A. Peterkofsky, A. M. Gronenborn, and G. M. Clore, *Biochemistry* **36**, 2517 (1997).
12. G. M. Clore and A. M. Gronenborn, *Nature Structural Biology* **4**, 849 (1997).
13. L. Pauling, *The nature of the chemical bond*, Cornell University Press (1960).
14. J. F. Brandts, H. R. Halvorson, and M. Brennan, *Biochemistry* **14**, 4953 (1975).

15. M. Levitt and C. Chothia, *Nature, Lond.* **261**, 552 (1976).

16. W. G. J. Hol, *Prog. Biophys. Mol. Biol.* **45**, 149 (1987).

17. G. N. Ramachandran and V. Sasisekharan, *Adv. Prot. Chem*, 283 (1968).

18. C. Chothia and A. V. Finkelstein, *Annual Review of Biochemistry* **59**, 1007 (1990).

19. C. Chothia, T. Hubbard, S. Brenner, H. Barns, and A. Murzin, *Annual Review of Biophysics and Biomolecular Structure* **26**, 597 (1997).

20. Y. Harpaz, M. Gerstein, and C. Chothia, *Structure* **2**, 641 (1994).

21. Y. Shao and S. B. H. Kent, *Chemistry & Biology* **4**, 187 (1997).

22. J. J. Perona and C. S. Craik, *Prot. Sci.* **4**, 337 (1995).

23. J. J. Perona and C. S. Craik, *J. Biol. Chem.* **272**, 29987 (1997).

24. B. W. Matthews, S. J. Remington, M. G. Grutter, and W. F. Anderson, *J. Molec. Biol.* **147**, 545 (1981).

25. M. Wilmanns, C. C. Hyde, D. R. Davies, K. Kirschner, and J. N. Jansonius, *Biochemistry* **30**, 9161 (1991).

26. M. G. Rossman, D. Moras, and K. W. Olsen, *Nature, Lond.* **250**, 194 (1974).

27. W. Gilbert, *Nature, Lond.* **271**, 501 (1978).

28. M. Go, *Proc. Natl. Acad. Sci. USA* **80**, 1964 (1983).

29. L. Patthy, *Current Opinion In Structural Biology* **4**, 383 (1994).

30. M. G. Tyshenko and V. K. Walker, *Biochim. Biophys. Acta Gene Structure and Expression* **1353**, 131 (1997).

31. L. Patthy, *Methods in Enzymology* **222**, 10 (1993).

32. L. B. Murata and H. K. Schachman, *Prot. Sci.* **5**, 719 (1996).

33. M. P. Schlunegger, M. J. Bennett, and D. Eisenberg, *Adv. Prot. Chem.* **50**, 61 (1997).

34. C. Chothia and A. M. Lesk, *EMBO Journal* **5**, 823 (1986).

35. C. Chothia, *Nature* **357**, 543 (1992).

36. S. Rhee, K. D. Parris, C. C. Hyde, S. A. Ahmed, E. W. Miles, and D. R. Davies, *Biochemistry* **36**, 7664 (1997).

37. W. H. Matchett, *J. Biol. Chem.* **249**, 4041 (1974).

38. M. F. Dunn, V. Aguilar, P. Brzovic, W. F. Drewe, K. F. Houben, C. A. Leja, and M. Roy, *Biochemistry* **29**, 8598 (1990).

39. M. Wilmanns, J. P. Priestle, T. Niermann, and J. N. Jansonius, *J. Molec. Biol.* **223**, 477 (1992).

40. M. Eberhard, M. Tsai-Pflugfelder, K. Bolewska, U. Hommel, and K. Kirschner, *Biochemistry* **34**, 5419 (1995).

41. K. M. Herrmann, *Plant Cell* **7**, 907 (1995).

42. L. J. Reed and M. L. Hackert, *J. Biol. Chem.* **265**, 8971 (1990).

43. P. G. Roughan, *Biochem. J.* **327**, 267 (1997).

44. Z. D. Zhang, S. Sugio, E. A. Komives, K. D. Liu, J. R. Knowles, G. A. Petsko, and D. Ringe, *Biochemistry* **33**, 2830 (1994).

45. A. M. Buckle and A. R. Fersht, *Biochemistry* **33**, 1644 (1994).

46. G. K. Farber, *Curr. Biol.* **7**, R352 (1997).

47. K. Moffat, *Laue diffraction*, in *Macromolecular Crystallography*, C. W. Carter and R. M. Sweet, eds., Academic Press, 433 (1997).

48. J. Fastrez and A. R. Fersht, *Biochemistry* **12**, 2025 (1973).

49. T. A. Steitz, R. Henderson, and D. M. Blow, *J. Molec. Biol.* **46**, 337 (1969).

50. X. C. Ding, B. F. Rasmussen, G. A. Petsko, and D. Ringe, *Biochemistry* **33**, 9285 (1994).

51. C. C. F. Blake, D. F. Koenig, G. A. Mair, A. C. T. North, D. C. Phillips, and V. R. Sarma, *Nature, Lond.* **206**, 757 (1965).

52. I. Schlichting and R. S. Goody, *Triggering methods in crystallographic enzyme kinetics*, in *Macromolecular crystallograph*, C. W. Carter and R. M. Sweet, eds., Academic Press, 467 (1997).

53. M. Vas, R. Berni, A. Mozzarelli, M. Tegoni, and G. L. Rossi, *J. Biol. Chem.* **254**, 8480 (1979).

54. L. N. Johnson, E. A. Stura, K. S. Wilson, M. S. P. Sansom, and I. T. Weber, *J. Molec. Biol.* **134**, 639 (1979).

55. S. W. Englander, T. R. Sosnick, J. J. Englander, and L. Mayne, *Current Opinion in Structural Biology* **6**, 18 (1996).

56. J. C. Norvell, A. C. Nunes, and B. P. Schoenborn, *Science* **190**, 568 (1975).

57. J. R. Laskowicz and G. Weber, *Biochemistry* **12**, 4161 (1971).

58. M. Gerstein, A. M. Lesk, and C. Chothia, *Biochemistry* **33**, 6739 (1994).

59. G. Lipari and A. Szabo, *J. Am. Chem. Soc.* **82**, 4546 (1982).

60. J. T. Stivers, C. Abeygunawardana, A. S. Mildvan, and C. P. Whitman, *Biochemistry* **35**, 16036 (1996).

61. G. L. Shaw, B. Davis, J. Keeler, and A. R. Fersht, *Biochemistry* **34**, 2225 (1995).

62. J. C. Williams and A. E. Mcdermott, *Biochemistry* **34**, 8309 (1995).

63. A. Bax and N. Tjandra, *Nature Structural Biology* **4**, 254 (1997).

64. S. D. Rader and D. A. Agard, *Prot. Sci.* **6**, 1375 (1997).

Recommended Reading

C. Branden and J. Tooze, *Introduction to protein structure,* 2nd ed. Garland Publishing (New York and London) (1999).

Chemical Catalysis **2**

The clues to understanding enzyme catalysis are hidden in the last sentence of section F3: "To summarize, the binding sites of lysozyme and the serine proteases are approximately complementary in structure to the structures of the substrates: the nonpolar parts of the substrate match up with nonpolar side chains of the amino acids; the hydrogen-bonding sites on the substrate bind to the backbone NH and CO groups of the protein and, for lysozyme, to the polar side chains of amino acids. The reactive part of the substrate is firmly held by this binding next to acidic, basic, or nucleophilic groups on the enzyme." Chemists use acids, bases, and nucleophiles as catalysts. Enzymes use the same chemistry but greatly enhance and fine-tune it by binding the reagents, transition states, and products in subtle ways. Chapters 12 and 15 are devoted to how enzymes use enzyme-substrate binding energy to enhance and fine-tune their reaction rates.

In this chapter, we discuss the fundamental principles of chemical reactivity and catalysis to understand the organic chemistry of catalysis and how to analyze it. We begin with *transition state theory* because it provides a simple framework for understanding much about reactivity and kinetics. We progress to structure-activity relationships and also discuss some fundamental concepts in analyzing mechanisms.

A. Transition state theory[1-4]

There are several theories to account for chemical kinetics. The simplest is the collision theory, which will be used in Chapter 4 to calculate the rate constants for the collision of molecules in solution. A more sophisticated theory, one that

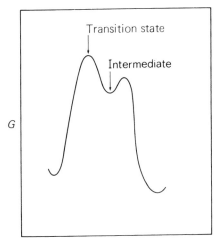

Transition state

Intermediate

G

Reaction coordinate

Figure 2.1 Transition states occur at the peaks of the energy profile of a reaction, and intermediates occupy the troughs.

is particularly useful for analyzing structure–reactivity relationships, is the transition state theory. The processes by which the reagents collide are ignored: the only physical entities considered are the reagents, or ground state, and the most unstable species on the reaction pathway, the *transition state*. The transition state occurs at the peak in the reaction coordinate diagram (Figure 2.1), in which the energy of the reagents is plotted as the reaction proceeds. In the transition state, chemical bonds are in the process of being made and broken. In contrast, *intermediates*, whose bonds are fully formed, occupy the troughs in the diagram. A simple way of deriving the rate of the reaction is to consider that the transition state and the ground state are in thermodynamic equilibrium, so that the concentration of the transition state may be calculated from the difference in their energies. The overall reaction rate is then obtained by multiplying the concentration of the transition state by the rate constant for its decomposition.

The analysis for a unimolecular reaction is as follows. Suppose that the difference in Gibbs free energy between the transition state, $X^{\ddagger}$, and the ground state, X, is $\Delta G^{\ddagger}$. Then, from a well-known relationship from equilibrium thermodynamics:*

---

* The two thermodynamic equations that are most useful for simple kinetic and binding experiments are: (1) the relationship between the Gibbs free energy change and the equilibrium constant of a reaction.

$$\Delta G = -RT \ln K$$

where R is the gas constant and T is the absolute temperature, and (2) the relationship between the Gibbs free energy change and the changes in the enthalpy and entropy,

$$\Delta G = \Delta H - T \Delta S$$

(In this text, the terms *Gibbs free energy* and *free energy* are synonymous.)

$$[X^{\ddagger}] = [X] \exp\left(-\frac{\Delta G^{\ddagger}}{RT}\right) \tag{2.1}$$

The frequency at which the transition state decomposes is the same as the vibrational frequency ν of the bond that is breaking. This frequency is obtained from the equivalence of the energies of an excited oscillator calculated from quantum theory ($E = h\nu$) and classical physics ($E = kT$); that is,

$$\nu = \frac{kT}{h} \tag{2.2}$$

where k is the Boltzmann constant and h is the Planck constant. At 25°C, $\nu = 6.212 \times 10^{12}\ s^{-1}$.

The rate of decomposition of X is thus given by

$$\frac{-d[X]}{dt} = \nu[X^{\ddagger}] \tag{2.3}$$

$$= [X]\left(\frac{kT}{h}\right)\exp\left(\frac{-\Delta G^{\ddagger}}{RT}\right) \tag{2.4}$$

The first-order rate constant for the decomposition of X is given by

$$k_1 = \left(\frac{kT}{h}\right)\exp\left(\frac{-\Delta G^{\ddagger}}{RT}\right) \tag{2.5}$$

The Gibbs free energy of activation, $\Delta G^{\ddagger}$, may be separated into enthalpic and entropic terms, if required, by using another relationship from equilibrium thermodynamics,

$$\Delta G^{\ddagger} = \Delta H^{\ddagger} - T\,\Delta S^{\ddagger} \tag{2.6}$$

(where $\Delta H^{\ddagger}$ is the enthalpy, and $\Delta S^{\ddagger}$ the entropy, of activation). The rate constant becomes

$$k_1 = \left(\frac{kT}{h}\right)\exp\left(\frac{\Delta S^{\ddagger}}{R}\right)\exp\left(\frac{-\Delta H^{\ddagger}}{RT}\right) \tag{2.7}$$

A more rigorous approach includes a transmission coefficient, κ, which is a multiplication factor equal to the fraction of the transition state that proceeds to products. The transmission coefficient is generally close to 1.0 for simple reactions. There can be a large attenuation factor for complex reactions because of viscous drag of the medium. We use transition state theory only qualitatively or to look at changes in rates where the transmission coefficient and other effects cancel out, and so we can ignore them.

1. The significance and the application of transition state theory

The importance of transition state theory is that it relates the rate of a reaction to the difference in Gibbs energy between the transition state and the ground state. This is especially important for comparing the relative reactivities of pairs of substrates, or the rates of a given reaction under different sets of conditions. Under some circumstances changes in rates may be calculated quantitatively; or, more generally, the trends in reactivity may be estimated qualitatively. For example, the alkaline hydrolysis of an ester, such as phenyl acetate, involves the attack of the negatively charged hydroxide ion on the neutral ground state. This means that in the transition state of the reaction some negative charge must be transferred to the ester. We can predict the p-nitrophenyl acetate will be more reactive than phenyl acetate, since the nitro group is electron-withdrawing and will stabilize the negatively charged transition state with respect to the neutral ground state. Consider also the spontaneous decomposition of *tert*-butyl bromide into the *tert*-butyl carbonium ion and the bromide ion. The transition state of this reaction must be dipolar. Therefore, a polar solvent, such as water, will stabilize the transition state, and a nonpolar solvent, such as diethyl ether, will destabilize the transition state with respect to the ground state.

In Chapters 12 and 13, it will be seen how the transition state theory may be used quantitatively in enzymatic reactions to analyze structure reactivity and specificity relationships involving discrete changes in the structure of the substrate. In Chapters 18 and 19, transition state theory is similarly applied to protein folding.

2. The Hammond postulate[5]

A useful guide in the application of transition state theory or in the analysis of structure–reactivity data is the Hammond postulate, which states that if there is an unstable intermediate on the reaction pathway, the transition state for the reaction will resemble the structure of this intermediate. The reasoning is that the unstable intermediate will be in a small dip at the top of the reaction coordinate diagram. This is a useful way of guessing the structure of the transition state for predicting the types of stabilization it requires. For example, an oxocarbenium ion may be an intermediate in the reaction catalyzed by lysozyme (Figure 2.2) (see Chapter 16). Since these are known to be unstable high-energy compounds, the transition state is assumed to resemble the structure of the oxocarbenium ion. One cannot really apply the Hammond postulate to bimolecular reactions. Because these involve two molecules condensing to form one transition state, a large part of the Gibbs free energy change is caused by the loss in entropy (section B4). The Hammond postulate applies mainly to energy differences so it works best with unimolecular reactions. The postulate is sometimes extended to

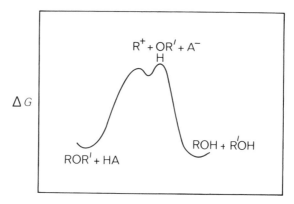

Figure 2.2 The transition state for the general-acid-catalyzed cleavage of an acetal resembles the carbonium ion intermediate that occupies the small "dip" at the top of the energy diagram.

where there are large energy differences between the reagents and products. If the products are very unstable, the transition state is presumed to resemble them; the same applies if the reagents are very unstable.

3.   Chemical basis of the Hammond postulate

A very simple analysis illustrates the chemical principles behind the Hammond postulate. Substrates, intermediates and products sit at the bottom of U-shaped energy wells in Figure 2.1 and 2.2. The shapes at the very bottom are parabolic for small changes in the reaction coordinate. We can see the basis of the Hammond postulate by drawing a reaction coordinate diagram as the intersection of two parabolic curves (Figure 2.3).[6] The energy curves of the substrate and product intersect at the position of the transition state on the reaction coordinate. Suppose we make a small change in the structure of the substrate so that it raises its energy by $\Delta\Delta G$. This raises its curve so that it intersects with the product curve at a higher energy. This also causes the point of intersection to move closer to the bottom of the energy well of the substrate so that the transition state comes earlier in the reaction. Clearly, as the substrate becomes more unstable, the transition state approaches it in structure. The movement along the reaction coordinate is known as the *Hammond effect.*

Figure 2.3 also explains the basis of many of the linear free energy relationships (LFERs) that we use later. The change in equilibrium stability leads to a change in the Gibbs free energy of activation, $\Delta\Delta G^{\ddagger}$. The value of $\Delta\Delta G^{\ddagger}$ depends on where the curves intersect and on their shape. The ratio $\Delta\Delta G^{\ddagger}/\Delta\Delta G$ is called the Brønsted β value. It is approximately constant for small changes in $\Delta\Delta G$. That is, there is a linear relationship between changes in free energy of activation and equilibrium free energies. It is always between 0 and 1.0 for the simple reactions in Figure 2.3. Later on, we use β as an indication of the extent of bond making and breaking in the transition state.

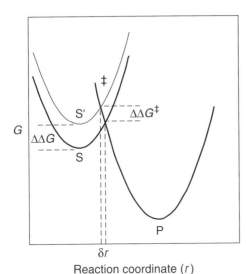

Figure 2.3 The basis of the Hammond postulate and Hammond effect. The Gibbs free energy curves for a substrate and product (heavy lines) intersect on the reaction coordinate at the position of the transition state. A change of structure in S, away from the seat of reaction, destabilizes it, displacing its energy curve (lighter line) higher. The point of intersection moves closer to the substrate.

B. Principles of catalysis

The catalytic power of enzymes is awesome (Table 2.1). A most spectacular example is that of the decarboxylation of orotic acid. It spontaneously decarboxylates with $t_{1/2}$ of 78 million years at room temperature in neutral aqueous solution. Orotidine 5′-phosphate decarboxylase enhances the rate of decarboxylation enzyme-bound substrate by 10^{17} fold. The classical challenge is to explain the magnitude of the rate enhancements in Table 2.1. We will not ask why enzymatic reactions are so fast but instead examine why the uncatalyzed reactions are so slow, and how they can be speeded up.

1. Where, why, and how catalysis is required

In order to understand why enzymes are such efficient catalysts, it is necessary to understand first why uncatalyzed reactions in solution are so slow. As illustrations, we consider the reactions that may be catalyzed by chymotrypsin or lysozyme.

The uncatalyzed attack of water on an ester leads to a transition state in which a positive charge develops on the attacking water molecule, and a negative charge on the carbonyl oxygen (equation 2.8).

$$R'-C{\overset{\displaystyle O}{\underset{\displaystyle OR}{}}} \quad \longrightarrow \quad \left[R'-C{\overset{\displaystyle \overset{\delta-}{O}}{\underset{\displaystyle OR}{}}} \right]_{TS} \tag{2.8}$$

Table 2.1 *Rate enhancements by enzymes*

Enzyme[a]	k_{cat}/k_{uncat}[b]
Sweet potato β-amylase	7.2×10^{17}
Orotidine 5′-phosphate decarboxylase	1.4×10^{17}
Fumarase	3.5×10^{15}
Mandelate racemase	1.7×10^{15}
Staphylococcal nuclease	$5.6 \times 10^{14} \, (>6 \times 10^{16})$[c]
Carboxypeptidase B	1.3×10^{13}
AMP nucleosidase	6.0×10^{12}
Adenosine deaminase	2.1×10^{12}
Ascites tumor dipeptidase	1.2×10^{12}
Cytidine deaminase	1.2×10^{12}
Ketosteroid isomerase	3.9×10^{11}
Phosphotriesterase	2.8×10^{11}
Triosephosphate isomerase	1.0×10^{9}
Carbonic anhydrase	7.7×10^{6}
Chorismate mutase	1.9×10^{6}
Cyclophilin (rotamase)	4.6×10^{5}
Catalytic antibodies	$10^{2} - 10^{5}$[d]

[a] Data from A. Radzicka and R. Wolfenden, *Science* **267**, 90 (1995); S. L. Bearne and R. Wolfenden, *J. Amer. Chem. Soc.* **117**, 9588 (1995); A. Radzicka and R. Wolfenden, *J. Am. Chem. Soc.* **118**, 6105 (1996); S. L. Bearne and R. Wolfenden, *Biochemistry* **36**, 1646 (1997); and R. Wolfenden, X. Lu, and G. Young, *J. Am. Chem. Soc.* **120**, 6814 (1998).
[b] The ratio of the value of k_{cat} for the enzymatic reaction (see Chapter 3) and the corresponding un-catalyzed reaction in aqueous solution under the same conditions.
[c] The nonenzymatic reaction goes less than 1% by P/O cleavage, the route followed by the enzyme.
[d] Catalytic antibodies are antibodies raised against transition state analogues (see Chapter 12) or se-lected for catalysis. One corresponding to chorismate mutase is 10^{4} times less active than the en-zyme (M. R. Haynes, E. A. Stura, D. Hilvert, and I. A. Wilson, *Science* **263**, 646 (1994). Antibodies catalyzing the hydrolysis of a nitrobenzyl ester have rate enhancements of up to 10^{5} (J.-B. Charbon-nier, B. Golinelli-Pimpaneau, B. Gigant, D. S. Tawfik, R. Chap, D. G. Schindler, S.-H. Kim, B. S. Green, Z. Eshhar, and M. Knossow, *Science* **275**, 1140 (1997). This is about 2.5×10^{7} lower than the k_{cat} for the acylation of chymotrypsin by N-acetyl-L-phenylalanine *p*-nitrophenyl ester, and k_{cat}/K_{M} for the overall reaction is 2×10^{5} times lower (k_{cat} for acylation estimated to $1.5 \times 10^{6} \, s^{-1}$ from the data of M. Renard and A. R. Fersht, *Biochemistry* **12**, 4713 (1974) and A. R. Fersht and M. Renard, *Biochemistry* **13**, 1416 (1974)).

The uncatalyzed hydrolysis of an acetal involves a transition state that is close in structure to an oxocarbenium ion and an alkoxide ion (equation 2.9).

$$\tag{2.9}$$

In both reactions the transition state is very unfavorable because of the unsta-ble positive and negative charges that are developed. Stabilization of these charges catalyzes the reaction by lowering the energy of the transition state. Such stabilization can be achieved in the case of the positive charge developing on the attacking water molecule by transferring one of the protons to a base during the reaction. This is known as *general-base catalysis* (equation 2.10).

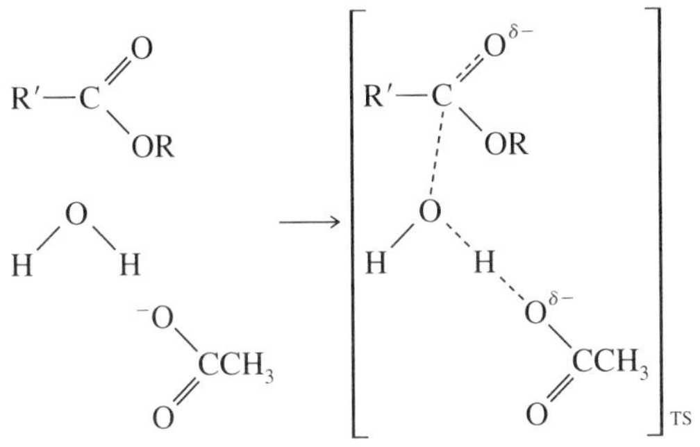

$$(2.10)$$

General-base catalysis by acetate ion

Similarly, the negative charge developing on the alcohol expelled from the acetal can be stabilized by proton transfer from an acid. This is known as *general-acid catalysis* (equation 2.11).

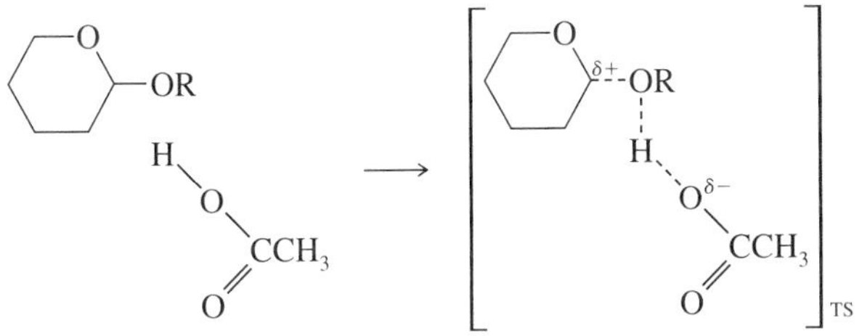

$$(2.11)$$

General-acid catalysis by acetic acid

The acid-base catalysis illustrated in these equations is termed *general* to distinguish it from *specific* acid or base catalysis in which the catalyst is the proton or hydroxide ion.

Positive and negative charges may also be stabilized by *electrostatic catalysis*. The positively charged oxocarbenium ion cannot be stabilized by general-base catalysis because it does not ionize. But it can be stabilized by the electric field from a negatively charged carboxylate ion. The negative charge on an oxyanion may also be stabilized by a positively charged metal ion, such as Zn^{2+} or Mg^{2+}. The stabilization of a negative charge, i.e., an electron, is known as *electrophilic catalysis*.

The above types of catalysis function by stabilizing the transition state of the reaction without changing the mechanism. Catalysts may also involve a different reaction pathway. A typical example is *nucleophilic catalysis* in an acyl transfer or hydrolytic reaction. The hydrolysis of acetic anhydride is greatly enhanced by pyridine because of the rapid formation of the highly reactive acetylpyridinium ion (equation 2.12). For nucleophilic catalysis to be efficient, the nucleophile must be more nucleophilic than the one it replaces, and the intermediate must be more reactive than the parent compound.

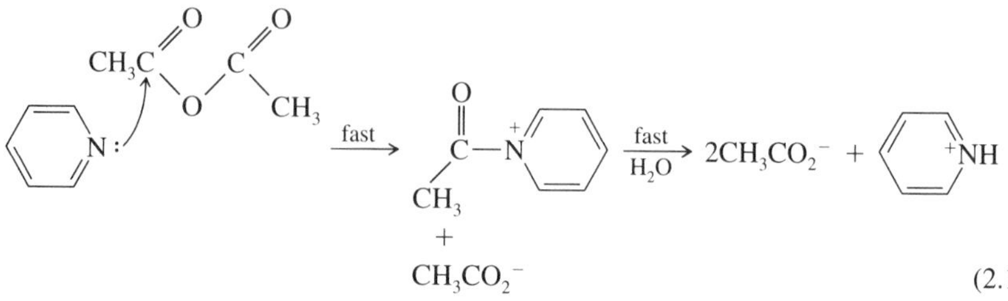

$$(2.12)$$

Nucleophilic catalysis is a specific example of *covalent catalysis*: the substrate is transiently modified by formation of a covalent bond with the catalyst to give a reactive intermediate. There are also many examples of electrophilic catalysis by covalent modification. It will be seen later that in the reactions of pyridoxal phosphate, Schiff base formation, and thiamine pyrophosphate, electrons are stabilized by delocalization.

There is a further important factor that is responsible for slowing down *multimolecular* reactions in solution. These require the bringing together of many molecules in the transition state. This is in itself an unfavorable event because it requires the right number of molecules simultaneously colliding in the correct orientation. The problem is exacerbated by acid-base or covalent catalysis, because even more molecules have to collide in the transition state. The magnitude of this factor is considered later, in the discussions on intramolecular catalysis and entropy (sections B3 and B4).

2. General-acid-base catalysis

a. Detection and measurement

The general-acid-base catalysis of the hydrolysis of an ester is measured from the increase in the hydrolytic rate constant with increasing concentration of the acid or base. This is usually done at constant pH by maintaining a constant ratio of the acidic and basic forms of the catalyst. It is important to keep the ionic strength of the reaction medium constant because many reactions are sensitive to changes in salt concentration. In order to tell whether the catalysis is due to the acidic or basic form of the catalyst, it is necessary to repeat the measurements at a different buffer ratio. For ester hydrolysis, it is found that the increase in rate is generally proportional to the concentration of the basic form, so the catalysis is general-base. The slope of the plot of the rate constant against the concentration of base gives the second-order rate constant k_2 for the general-base catalysis (Figure 2.4).

b. The efficiency of acid-base catalysis: The Brønsted equation[7]

It is found experimentally that the general-base catalysis of the hydrolysis of an ester is proportional to the basic strength of the catalyst (Figure 2.5).[8] The

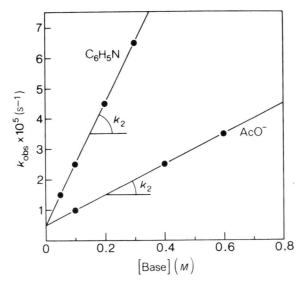

Figure 2.4 Determination of the rate constants for the general-base catalysis of the hydrolysis of ethyl dichloroacetate. The first-order rate constants for the hydrolysis are plotted against various concentrations of the base. The slope of the linear plot is the second-order rate constant (k_2). The intercept at zero buffer concentration is the "spontaneous" hydrolysis rate constant for the particular pH. A plot of the spontaneous rate constants against pH gives the rate constants for the H^+ and OH^- catalysis. It is seen that pyridine is a more effective catalyst than the weaker base acetate ion. [From W. P. Jencks and J. Carriuolo, *J. Am. Chem.* Soc. **83**, 1743 (1961).]

second-order rate constant k_2 for the dependence of the rate of hydrolysis on the concentration of the base is given by the equation

$$\log k_2 = A + \beta \, pK_a \tag{2.13}$$

Equation 2.13 is an example of the Brønsted equation (see section A3). Here, the Brønsted β value measures the sensitivity of the reaction to the pK_a of the conjugate acid of the base. The A is a constant for the particular reaction.

Brønsted equations are also common in general-acid catalysis, as, for example, in the hydrolysis of certain acetals.[8-10] In acid catalysis, α rather than β is used:

$$\log k_2 = A - \alpha \, pK_a \tag{2.14}$$

The values of α and β are always between 0 and 1 for acid-base catalysis (except with some peculiar carbon acids),[11] because complete transfer of a proton gives a value of 1, and no transfer a value of 0. The usual values for ester hydrolysis are 0.3 to 0.5, and for acetal hydrolysis about 0.6.

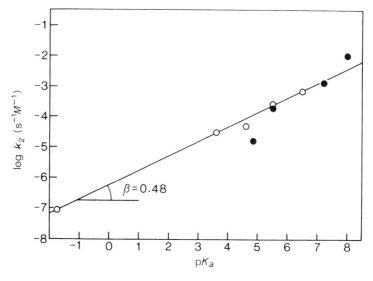

Figure 2.5 The Brønsted plot for the general-base catalysis of the hydrolysis of ethyl dichloroacetate. The logarithms of the second-order constants obtained from the plot of Figure 2.4 are plotted against the pK_a's of the conjugate acid of the catalytic base. The slope is the β value. Note that the points for amine bases (●) fall on the same line as those for oxyanion bases (○), showing that the catalysis depends primarily on the basic strength of the base and not on its chemical nature.

Some values of the rate enhancement by general-base and general-acid catalysis are listed in Tables 2.2 and 2.3. These numbers are obtained from the Brønsted equation by using pK_a's of 15.74 and -1.74 for the ionization of H_2O and H_3O^+, respectively (i.e., from $[H^+][OH^-] = 10^{-14}\ M^2$, and the concentration of water $= 55\ M$). The magnitude of the catalysis depends strongly on α and β and the pK_a of the catalyst. In simple terms: *The stronger the base, the better the general-base catalysis; the stronger the acid, the better the general-acid catalysis.*

c.   The efficiency of acid-base catalysis: The ionization state
      of the catalyst

A crucial factor in whether an acid-base catalyst is effective is whether it is in the correct ionization state under the reaction conditions: an acid has to be in acidic form to be an acid catalyst, and a base in its basic form. For example, an acid of pK_a 5 is a much better general-acid catalyst than one of pK_a 7, but at pH 7 only 1% of an acid of pK_a 5 is in the active acid form while the remaining 99% is ionized. An acid of pK_a 7 is only 50% ionized at pH 7 and still 50% active. Table 2.2 shows that for $\alpha = 0.85$ or less, an acid of pK_a 7 is a better catalyst at pH 7 than is an acid of pK_a 5. Similarly, a base of pK_a 7 is a more effective catalyst than one of pK_a 9 at pH 7 (for $\beta = 0.85$ or less), due to the inherently more reactive base being mainly protonated at the pH below its pK_a. The most effec-

Table 2.2 *Influence of β on general-base catalysis*

β	(Rate in 1-*M* solution of base) ÷ (rate in water)	
	$pK_a = 5$	$pK_a = 7$
0	1	1
0.3	2.9	8.6
0.5	44	427
0.7	951	2.4×10^4
0.85	9.7×10^3	4.9×10^5
1	1×10^5	1×10^7

tive acid-base catalysts at pH 7 are those whose pK_a's are about 7. This accounts for the widespread involvement of histidine, with an imidazole pK_a of 6 to 7, in enzyme catalysis.

3. Intramolecular catalysis: The "effective concentration" of a group on an enzyme

Acid-base catalysis is seen to be an effective way of catalyzing reactions. We should now like to know the contribution of this to enzyme catalysis, but there is a fundamental problem in directly applying the results of the last section to an enzyme. The crux of the matter is that the rate constants for the solution catalysis are second-order, the rate increasing with increasing concentration of

Table 2.3 *Influence of α and state of ionization on general-acid catalysis*

α	(Rate in 1-*M* solution of acid) ÷ (rate in water)			(Rate in 1-*M* solution at pH 7) ÷ (rate in water)[a]		
	$pK_a = 5$	$pK_a = 7$	$pK_a = 9$	$pK_a = 5$	$pK_a = 7$	$pK_a = 9$
0	1	1	1	1	1	1
0.3	31	8.6	2.9	1.3	4.8	2.9
0.5	4.3×10^3	427	44	42	214	43.2
0.7	6×10^5	2.4×10^4	951	6×10^3	1.2×10^4	940
0.85	2.5×10^7	4.9×10^5	9.7×10^3	2.4×10^5	2.4×10^5	9.7×10^3
1	1×10^9	1×10^7	1×10^5	1×10^7	5×10^6	9.9×10^4

[a] The rate of a 1-*M* solution of both the acid and the base forms compared with the uncatalyzed water reaction. It should be noted that the proton becomes an efficient catalyst at higher values of α. For α = 0.3, 0.5, 0.7, 0.85, and 1.0, the reaction rate increases at pH 7 by factors of 1.0003, 2, 3×10^3, 1.3×10^6, and 5.5×10^8, respectively, thus swamping out the catalysis by other acids at the higher values.

catalyst; whereas reactions in an enzyme–substrate complex are first-order, the acids and bases being an integral part of the molecule. So what is the concentration of the acid or base that is to be used in the calculations? The experimental approach is to synthesize model compounds with the catalytic group as part of the substrate molecule, and to compare the reaction rates with the corresponding intermolecular reactions.

A typical example of an *intramolecularly catalyzed* reaction is the hydrolysis of aspirin (equation 2.15).[12] The hydrolysis of the ester bond is achieved by intramolecular general-base catalysis. Comparison with the uncatalyzed hydrolysis rate of similar compounds gives a rate enhancement of some 100-fold from the catalysis.[13] This may be extrapolated to a figure of 5000-fold if the pK_a of the base is 7 rather than the value of 3.7 in aspirin.

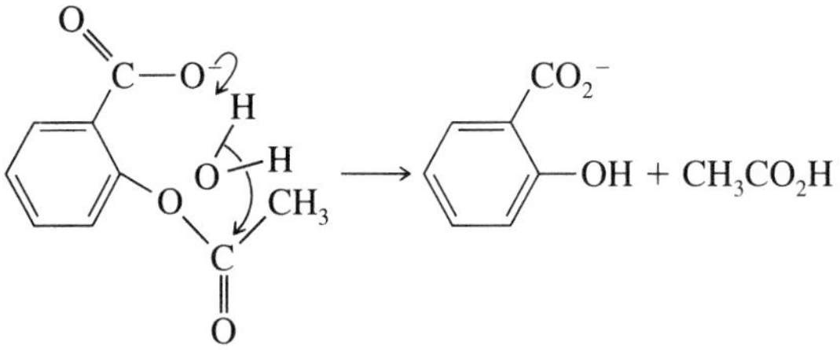

$$ (2.15) $$

The intermolecular general-base catalysis of the hydrolysis may also be measured. Comparing the rate constants for this with those of the intramolecular reaction shows that a 13-M solution of an external base is required to give the same first-order rate as the intramolecular reaction has.[12] The "effective concentration" of the carboxylate ion in aspirin is therefore 13 M. This is a typical value for intramolecular general-acid-base catalysis.

The effective concentrations of nucleophiles in intramolecular reactions are often far higher than this. The examples that follow are for "unstrained" systems. The chemist can synthesize compounds that are strained; the relief of strain in the reaction then gives a large rate enhancement. In the succinate and aspirin derivatives that follow, the attacking nucleophile can rotate away from the ester bond to relieve any strain. The observed rate enhancements are due entirely to the high effective concentration of the neighboring group:

1. *Rates of acyl transfer in succinates:*[14]

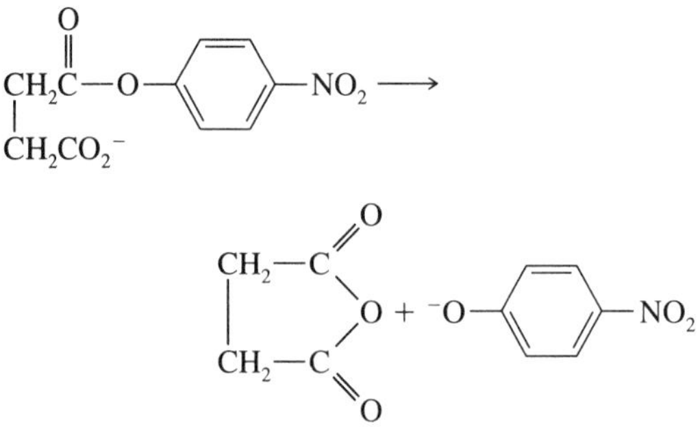

$$ k_1 = 0.8 \text{ s}^{-1} \qquad (2.16) $$

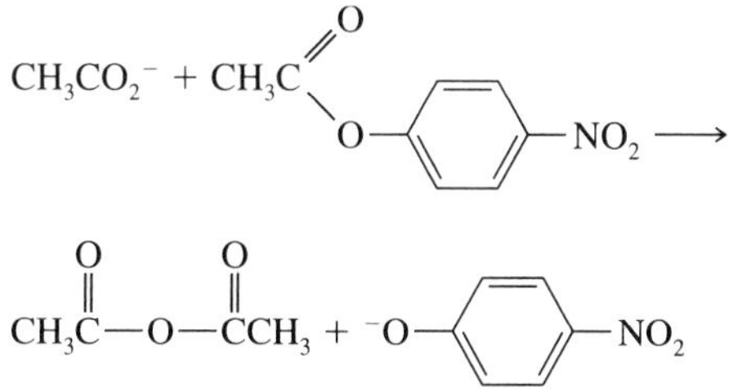

$$CH_3CO_2^- + CH_3C(=O)O\text{-}C_6H_4\text{-}NO_2 \longrightarrow$$

$$CH_3C(=O)\text{-}O\text{-}C(=O)CH_3 + {}^-O\text{-}C_6H_4\text{-}NO_2 \qquad k_2 = 4 \times 10^{-6}\ s^{-1}\ M^{-1} \qquad (2.17)$$

The effective concentration of $-CO_2^-$ is $k_1/k_2 = 2 \times 10^5\ M$.

2. *Rates of acyl transfer in aspirin derivatives:*[15]

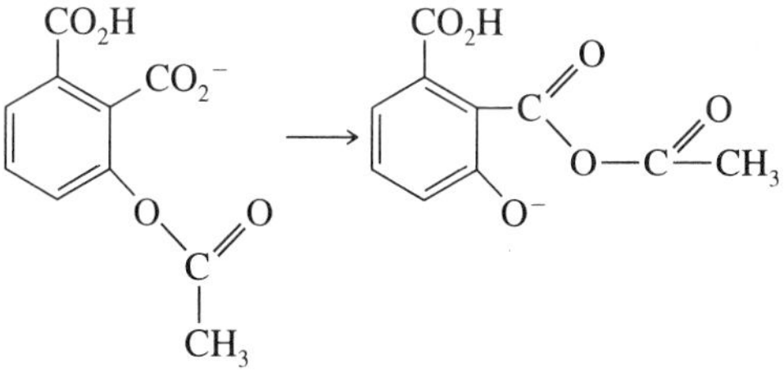

$$k_1 \geq 0.02\ s^{-1} \qquad (2.18)$$

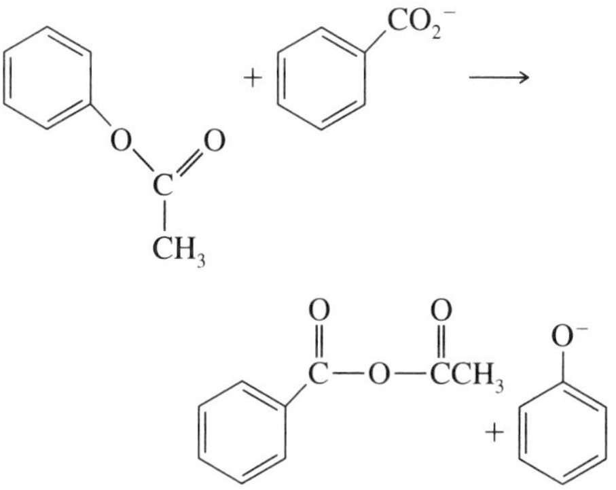

$$k_2 \approx 10^{-10}\ s^{-1}\ M^{-1} \qquad (2.19)$$

The effective concentration of $-CO_2^-$ is $k_1/k_2 > 2 \times 10^7\ M$.

3. *Equilibria for acyl transfer in succinates:*[16,17]

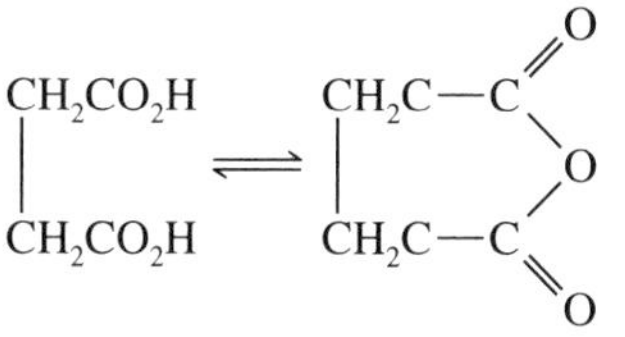

$$K_{eq} = 8 \times 10^{-7} \qquad (2.20)$$

$$2CH_3CO_2H \rightleftharpoons CH_3C(=O)\text{-}O\text{-}C(=O)CH_3 \qquad K_{eq} = 3 \times 10^{-12}\ M \qquad (2.21)$$

The effective concentration of $-CO_2H$ is $3 \times 10^5\ M$.

These examples show that enormous rate enhancements come from intramolecular nucleophilic catalysis.

4. Entropy: The theoretical basis of intramolecular catalysis and effective concentration[18,19]

The high effective concentration of intramolecular groups is one of the most important reasons for the efficiency of enzyme catalysis. This can be explained theoretically by using transition state theory and examining the entropy term in the rate equation (2.7). It will be seen that effective concentrations may be calculated by substituting certain entropy contributions into the $\exp(\Delta S^{\ddagger}/R)$ term of equation 2.7.

a. The meaning of entropy

The most naive explanation of entropy, much frowned upon by purists but adequate for this discussion, is that entropy is a measure of the degree of randomness or disorder of a system. The more disordered the system, the more it is favored and the higher its entropy. Entropy is similarly associated with the spatial freedom of atoms and molecules.

The catalytic advantage of an intramolecular reaction over its intermolecular counterpart is due to entropy. The intermolecular reaction involves two or more molecules associating to form one. This leads to an increase in "order" and a consequent loss of entropy. An effective concentration may be calculated from the entropy loss.

b. The magnitude of entropy

The entropy of a molecule is composed of the sum of its translational, rotational, and internal entropies. The translational and rotational entropies may be precisely calculated for the molecule in the gas phase from its mass and geometry. The entropy of the vibrations may be calculated from their frequencies, and the entropy of the internal rotations from the energy barriers to rotation.

The translational entropy is high: about 120 J/deg/mol (30 cal/deg/mol) for a 1-M solution of a small molecule. This is equivalent to about 40 kJ/mol (9 kcal/mol) at 25°C (298 K). The translational entropy is proportional to the volume occupied by the molecule; the smaller the volume, the more the molecule is restricted and the lower the entropy. Similarly, the entropy decreases with increasing concentration, since the average volume occupied by a molecule is inversely proportional to its concentration. It is important to note that the dependence on mass is low (Table 2.3). A 10-fold increase in mass on going from, say, a relative molecular mass of 20 to one of 200 leads to only a small increase in translational entropy.

The rotational entropy is also high: up to 120 J/deg/mol (30 cal/deg/mol) for a large organic molecule. It, too, increases only slowly with increasing mass, but it is independent of concentration.

Stiff vibrations, as found in most covalent bonds, make very low individual contributions to the entropy. Low-frequency vibrations, where the atoms are less constrained, can contribute a few entropy units. Internal rotations have entropies in the range of 13 to 21 J/deg/mol (3 to 5 cal/deg/mol) (Table 2.4).

J. D. Dunitz[19] has estimated the cost in entropy of tying up solvent water. The entropy of a water molecule of hydration in a crystal or mineral is 42 J/mol/K (10 cal/mol/K), which represents the lower limit for a tightly bound molecule. The entropy of water in liquid water is 67–71 J/mol/K (16–17 cal/mol/K), which represents the upper limit for the least constrained water molecule in solution. Thus, the energetic cost of immobilizing a water molecule is between 0 and 8.3 kJ/mol (0 and 2 kcal/mol) at 25°C (298 K).

c. The loss of entropy when two molecules condense to form one

The combining of two molecules to form one leads to the loss of one set of rotational and translational entropies. The rotational and translational entropies of the adduct of the two molecules are only slightly larger than those of one of the original molecules, since these entropies increase only slightly with size (Table 2.4). The entropy loss is up to 190 J/deg/mol (45 cal/deg/mol) or 55 to 59 kJ/mol (13 to 14 kcal/mol) at 25°C for the small molecules. This may be offset somewhat by an increase in internal entropy due to new modes of internal rotation and vibration (Figure 2.6).

Table 2.4 *Entropy of translation, rotation, and vibration (298 K)[a]*

	Entropy	
Motion	(J/deg/mol)	(cal/deg/mol)
3 degrees of *translational freedom* for M_r 20–200 (standard state = 1 M)	120–150	29–36
3 degrees of *rotational freedom:*		
Water	44	10.5
n-Propane	90	21.5
endo-Dicyclopentadiene	114	27.2
Internal rotation	13–21	3–5
Vibrations (cm^{-1}):		
1000	0.4	0.1
800	0.8	0.2
400	4.2	1.0
200	9.2	2.2
100	14.2	3.4

[a] From M. I. Page and W. P. Jencks, *Proc. Natl. Acad. Sci. USA* **68,** 1678 (1971).

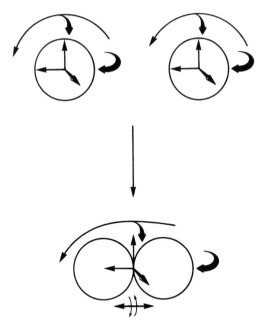

Figure 2.6 A free molecule has three degrees of translational entropy and three degrees of overall rotational entropy. When two molecules condense to form one, the resulting adduct has only three degrees of translational and three degrees of rotational entropy overall, a loss of three degrees of each. A compensating gain of internal vibrational and rotational entropy partly offsets this loss.

The association of two proteins has a different energy balance. The molecules are significantly larger than the small molecules in Table 2.4, and so the complex loses more overall rotational and translational entropy. It does not gain significant internal rotation because the two components are not free to rotate, but it gains more vibrational entropy. A normal mode analysis by B. Tidor and M. Karplus[20] suggests that formation of an insulin dimer from two monomers at 300 K loses 180 J/deg/mol (43 cal/deg/mol) of translational entropy and 200 J/deg/mol (47 cal/deg/mol) of rotational entropy, which are offset by a gain of 100 J/deg/mol (24 cal/deg/mol) of vibrational entropy. The net loss in entropy is 280 J/deg/mol (66 cal/deg/mol). A. Finkelstein and J. Janin[21] estimate from the average amplitude of translational motion of crystalline lysozyme that 210 J/deg/mol (50 cal/deg/mol) of entropy is lost in the immobilization of a small protein molecule.

These losses are for a standard state of 1 *M*. If the solutions are more dilute, the loss will be correspondingly greater since the translational entropy is concentration-dependent.

d. *The entropic advantage of a unimolecular over a bimolecular reaction*

Let us compare the reaction of two molecules A and B combining together to form a third, with its intramolecular counterpart:

$$A + B \longrightarrow AB^{\ddagger} \longrightarrow AB \tag{2.22}$$

$$\underset{\frown}{A \ B} \longrightarrow \underset{\frown}{AB^{\ddagger}} \longrightarrow \underset{\frown}{AB} \tag{2.23}$$

The formation of the transition state $AB^{\ddagger}$ leads to the loss of translational and rotational entropy as described above, although there are some compensating gains in internal rotation and vibration. The intramolecular cyclization in equation 2.23 involves the loss of only some entropy of internal rotation.

Depending on the relative gains and losses in internal rotation, the intramolecular reaction is favored entropically by up to 190 J/deg/mol (45 cal/deg/mol) or 55 to 59 kJ/mol (13 to 14 kcal/mol) at 25°C. Substituting 190 J/deg/mol (45 cal/deg/mol) into the exp $(\Delta S^{\ddagger}/R)$ term of equation 2.7 gives a factor of 6×10^9. Taking into account the difference in molecularity between the second-order and first-order reactions, this may be considered as the maximum effective concentration of a neighboring group, i.e., 6×10^9 *M*. In other words, for B in equation 2.22 to react with the same first-order rate constant as $\underset{\frown}{A \ B}$ in equation 2.23, the concentration of A would have to be 6×10^9 M.

The loss of internal rotation lowers the effective concentration quite considerably. Three internal rotations are lost for succinate:

M. I. Page and W. P. Jencks suggest that the entropy is lowered by about 13 J/deg/mol (3 cal/deg/mol) by the loss of the rotation about the methylene group, and by 25 J/deg/mol (6 cal/deg/mol) by the loss of rotation about each methylene-carboxyl bond. This decrease is equivalent to a factor of 2×10^3 in effective concentration. If the free rotations in the succinate compounds were frozen out, the effective concentration of the neighboring carboxyl group would be increased to about 5×10^8 *M*. This could be increased by a further factor of 10 by allowing for the unfavorable energy associated with the eclipse of the methylene hydrogens on formation of the five-membered ring. Considering these factors, the theoretical value for the maximum effective concentration appears to be quite realistic.

*e. The dependence of effective concentration on "tightness"
of the transition state*

The lower effective concentrations found in intramolecular base catalysis are due
to the loose transition states of these reactions. In nucleophilic reactions, the
nucleophile and the electrophile are fairly rigidly aligned so that there is a large
entropy loss. In general-base or -acid catalysis, there is considerable spatial
freedom in the transition state. The position of the catalyst is not as closely
defined as in nucleophilic catalysis. There is consequently a smaller loss in
entropy in general-base catalysis, so that the intramolecular reactions are not
favored as much as their nucleophilic counterparts.

An early treatment of the problem of calculating effective concentrations was
to consider the concentration of an intramolecular group to be approximately the
same as that of water in aqueous solution, since a molecule in solution is completely surrounded by water.[22] This gives an upper limit of 55 M for effective
concentration, equivalent to 34 J/deg/mol (8 cal/deg/mol) of entropy. That figure
does represent the probability of two molecules being next to each other in solution. But, as soon as the two molecules are tightly linked, there is a large loss of
entropy. A loose transition state may, perhaps, be interpreted as two molecules
that are in close juxtaposition but that retain considerable entropic freedom.

In summary, one of the most important factors in enzyme catalysis is entropy.
Catalyzed reactions in solution are slow because the bringing together of the catalysts and the substrate involves a considerable loss of entropy. Enzymatic reactions take place in the confines of the enzyme–substrate complex. The catalytic
groups are part of the same molecule as the substrate, so there is no loss of translational or rotational entropy in the transition state. One way of looking at this is
that the catalytic groups have very high effective concentrations compared with
bimolecular reactions in solution. This advantage in entropy is "paid for" by the
enzyme–substrate binding energy; the rotational and translational entropies of
the substrate are lost on formation of the enzyme–substrate complex, and not
during the chemical steps. (The loss of entropy on formation of the
enzyme–substrate complex increases its dissociation constant.)

5. "Orbital steering"[22]

Attempts have been made to account for the rate enhancements in intramolecular
catalysis on the basis of an effective concentration of 55 M combined with the
requirement of very precise alignment of the electronic orbitals of the reacting
atoms: "orbital steering." Although this treatment does have the merit of emphasizing the importance of correct orientation in the enzyme–substrate complex, it
overestimates this importance, because, as we now know, the value of 55 M is an
extreme underestimate of the contribution of translational entropy to effective
concentration. The consensus is that although there are requirements for the satisfactory overlap of orbitals in the transition state, these amount to an accuracy
of only 10° or so.[23,24] The distortion of even a fully formed carbon-carbon bond

by 10° causes a strain of only 11 kJ/mol (2.7 kcal/mol). A distortion of 5° costs only 2.8 kJ/mol (0.68 kcal/mol).

6. Electrostatic catalysis[25]

a. Solvent water obscures electrostatic catalysis in aqueous solution

Chemical studies of model compounds do not show large effects from electrostatic catalysis of reactions in water because of its high dielectric constant (Chapter 11). The electrostatic interaction energy between two point charges e_1 and e_2 separated by a distance r in a medium of dielectric constant D is given by

$$E = \frac{e_1 e_2}{Dr} \tag{2.24}$$

For a proton and an electron separated by 3.3 Å (0.33 nm) *in vacuo* (D = 1), this gives -418 kJ/mol (-100 kcal/mol). But in water of dielectric constant 79, the value drops to -5.4 kJ/mol (-1.3 kcal/mol).

It should be noted that the water does not have to be inserted between the two charges to lower the interaction energy. This may be illustrated by the pK_a's of the two following amine bases in water:

$$H_2NNH_2 \underset{H^+}{\overset{pK_a = 8}{\rightleftharpoons}} H_2NNH_3{}^+ \underset{H^+}{\overset{pK_a = -1}{\rightleftharpoons}} {}^+H_3NNH_3{}^+ \tag{2.25}$$

$$\tag{2.26}$$

In the protonation of hydrazine (equation 2.25), the juxtaposition of the two positive charges in the dication should destabilize it by 920 kJ/mol (220 kcal/mol), according to two positive charges separated by 1.5 Å *in vacuo*. Instead, the two pK_a's differ by only 9 units, reflecting a difference of only 53.3 kJ/mol (12 kcal/mol). In equation 2.26, the two nitrogen atoms of triethylenediamine are separated by 2.6 Å. The two positive charges in the dication would be expected to have an unfavorable interaction energy of 546 kJ/mol (130 kcal/mol) *in vacuo*, but the second pK_a is perturbed by only 33.4 kJ (8 kcal). In both cases there appears to be an effective dielectric constant of about 17 between the two nitrogen atoms. This results from the positively charged ions polarizing the solvent and inducing dipoles. The electrostatic field from these dipoles and from any counter ions partially neutralizes the positive field from the cations. For this reason, the surrounding of ions by a dielectric without interposing it between them lowers the interaction energy between the ions (Chapter 11). Model studies in water greatly underestimate the importance of electrostatic catalysis in proteins.

b.   Enzymes may stabilize polar transition states better than water does

Electrostatic interactions are much stronger in organic solvents than in water because of lower dielectric constants. It has been felt for some time now that this property could be used by enzymes as a means of stabilizing polar transition states: positive or negative charges built into the low dielectric medium of the protein structure would have strong electrostatic interactions with charges on polar transition states.[26,27] The crux of the matter is that a protein is a very heterogeneous medium as far as its dielectric constant is concerned; nonpolar alkyl side chains are juxtaposed with polar backbone amide groups. Although this clearly opens up exciting possibilities of electrostatic catalysis, it renders calculation most difficult. Calculations[28,29] suggest that

1. Two or three fixed dipoles in an enzyme (e.g., backbone $\supset$NH groups) can stabilize a charge as effectively as bulk water does.

2. An ion pair may be stabilized more effectively by fixed dipoles in an enzyme than by bulk water.

It is instructive to consider the reasons for these statements. Water solvates an ion by forming a tight solvation shell in contact with the ion; this shell is surrounded by further shells that interact through their electrostatic dipoles. But these outer shells also have to interact with the dipoles and hydrogen bonds of the bulk solvent. This has the effect of randomizing the orientation of the dipoles in the outer shells and partly neutralizing them. In consequence, the calculations suggest that an ion has a higher solvation energy when it is surrounded by just 10 water molecules than when it is in bulk water. (Perhaps primitive enzymes stabilized charged transition states by simply enveloping a hydrated substrate in a hydrophobic pocket?) In a protein, on the other hand, the dipoles are rigidly held in a fixed orientation, pointing toward the substrate.

Although the calculations need experimental testing, it is very pertinent, as illustrated in Figure 2.7, that enzymes always use parts of their own structure or bound ions to solvate transition states, and do not use bulk water. This means that enzymes contain parallel dipoles—that is, dipoles in a high-energy geometry—and so have an "electrostatic strain" built in. A further source of electrostatic energy may come from the oriented dipoles of the hydrogen-bonded backbone groups of an α helix.[30]

To summarize, it seems likely that enzymes can stabilize ion pairs and other charge distributions more effectively than water can because the enzyme has dipoles that are kept oriented toward the charge, whereas water dipoles are randomized by outer solvation shells interacting with bulk solvent.[28]

7.   Metal ion catalysis[31]

a.   Electrophilic catalysis

One obvious role for metals in metalloenzymes is to function as electrophilic catalysts, stabilizing the negative charges that are formed. In carboxypeptidase

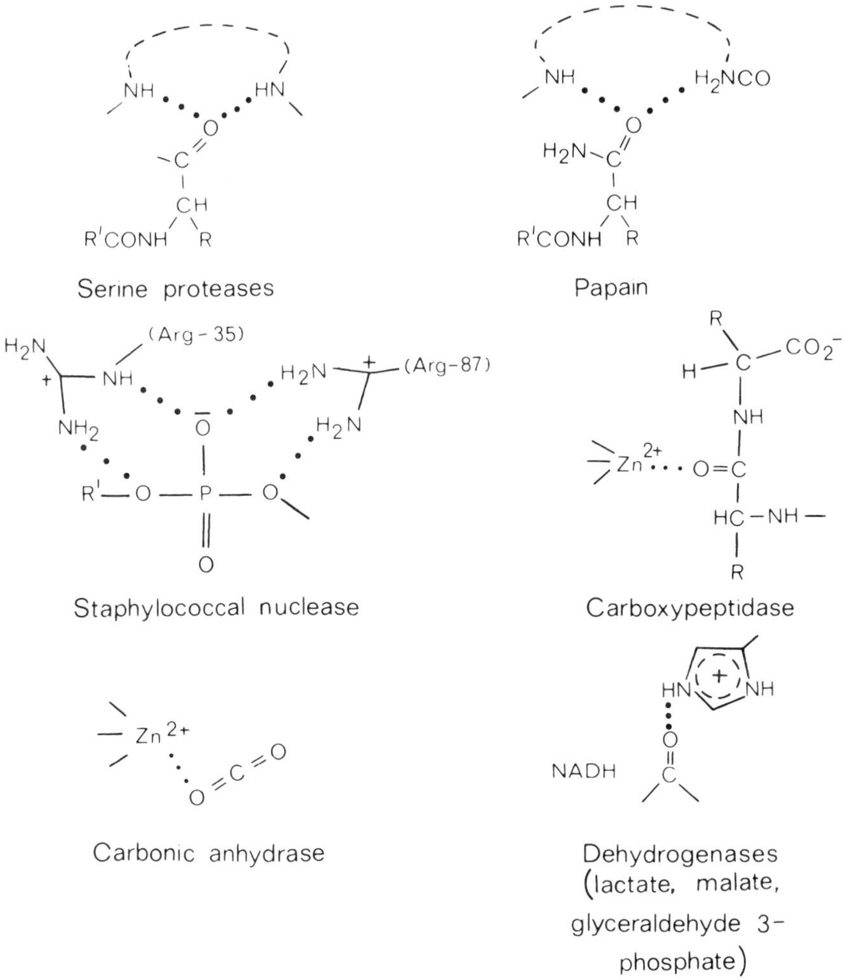

Serine proteases

Papain

Staphylococcal nuclease

Carboxypeptidase

Carbonic anhydrase

NADH

Dehydrogenases
(lactate, malate,
glyceraldehyde 3-
phosphate)

Figure 2.7 Solvation of substrates by enzymes.

(Chapter 16), the carbonyl oxygen of the amide substrate is coordinated to the Zn^{2+} of the enzyme (Figure 2.7). The coordination polarizes the amide to nucleophilic attack and strongly stabilizes the tetrahedral intermediate (equation 2.27). This type of complex formation has been mimicked in model compounds

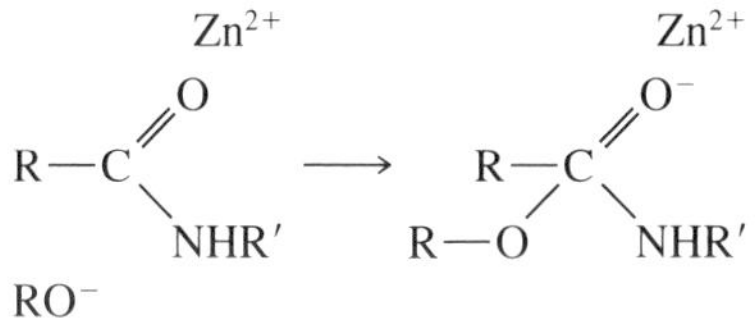

$$R-C\underset{NHR'}{\overset{O}{\diagup}}\overset{Zn^{2+}}{} \longrightarrow R-C\underset{R-O}{\overset{O^-}{\diagup}}\overset{Zn^{2+}}{\underset{NHR'}{}}$$
$$RO^-$$

(2.27)

to give rate enhancements of 10^4 to 10^6.[32,33] For example, the base-catalyzed hydrolysis of glycine ethyl ester is increased 2×10^6-fold when the compound is coordinated to (ethylenediamine)$_2$Co^{3+} (equations 2.28 and 2.29).

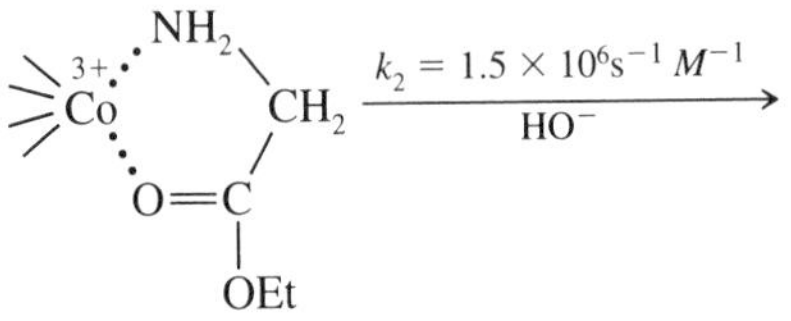

$$k_2 = 1.5 \times 10^6 \text{s}^{-1} M^{-1} \qquad \xrightarrow{\text{HO}^-} \qquad (2.28)$$

$$\text{H}_2\text{NCH}_2\text{CO}_2\text{Et} \quad \xrightarrow[\text{HO}^-]{k_2 = 0.6 \text{s}^{-1} M^{-1}} \qquad (2.29)$$

b. A source of hydroxyl ions at neutral pH

Metal-bound hydroxyl ions are potent nucleophiles.[34–36] The cobalt-bound water molecule in equation 2.30 ionizes with a pK_a of 6.6,

$$(\text{NH}_3)_5\text{Co}^{3+} \text{OH}_2 \rightleftharpoons (\text{NH}_3)_5\text{Co}^{2+} \text{OH} + \text{H}^+ \qquad (2.30)$$

a value 9 units below the pK_a of free H_2O; yet the cobalt-bound hydroxyl group is only 40 times less reactive than the free hydroxide ion in catalyzing the hydration of carbon dioxide.[34] This insensitivity of the high reactivity to pK_a is quite general and independent of the metal involved.[36] Thus, metal-bound water molecules provide a source of nucleophilic hydroxyl groups at neutral pH. Just as a base of pK_a 7 is most effective in general-base catalysis (section B2b), so a metal-bound water molecule of pK_a 7 is most effective for nucleophilic attack because it combines a high reactivity with a high fraction in the correct ionization state.

This is of relevance to the mechanism of carbonic anhydrase. This enzyme, which catalyzes the hydration of CO_2, has at its active site a Zn^{2+} ion ligated to the imidazole rings of three of its histidines. The classic mechanism for the reaction is that the fourth ligand is a water molecule which ionizes with a pK_a of 7.[37] The reactive species is considered to be the zinc-bound hydroxyl. Chemical studies show that zinc-bound hydroxyls are no exception to the rule of high reactivity. The H_2O in structure 2.31 ionizes with a pK_a of 8.7 and catalyzes the hydration of carbon dioxide and acetaldehyde.[38]

$$\text{RZn}^{2+}\text{OH}_2 \quad R = \qquad\qquad (2.31)$$

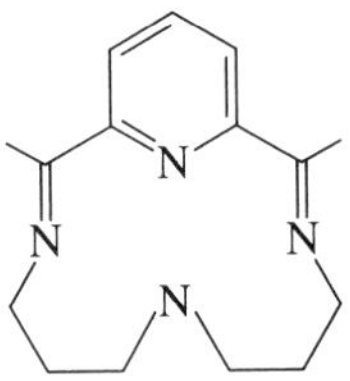

The carbonic anhydrase mechanism probably involves the step in equation 2.32. (The Zn^{2+} ion possibly also polarizes the carbonyl oxygen atom.)[39] See also the mechanism of DNA polymerase (Chapter 13, section B2b, and Figures 13.7 and 13.8).

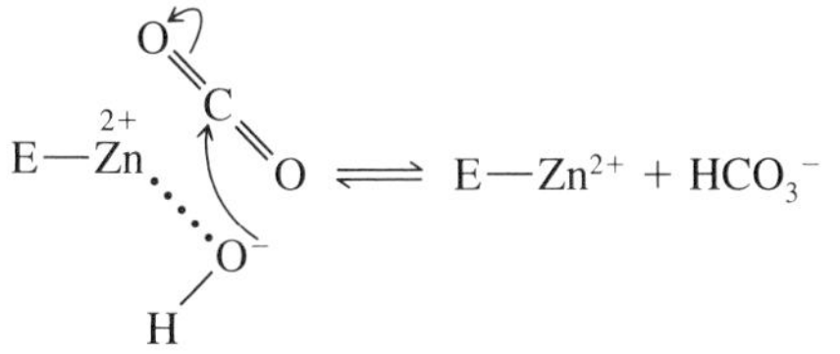

$$E{-}\overset{2+}{Zn}\cdots\underset{\underset{H}{O}}{\overset{O}{\underset{\Vert}{C}}}{-}O \rightleftharpoons E{-}Zn^{2+} + HCO_3^{-} \qquad (2.32)$$

The combination of a metal-bound hydroxyl group and an intramolecular re-action provides some of the largest rate enhancements that can be found in a strain-free system. The complex of glycylglycine with (ethylenediamine)$_2$Co^{3+} and a *cis*-hydroxyl (equation 2.33) hydrolyzes at pH 7 nearly 10^{10} times faster than the free glycylglycine.[40]

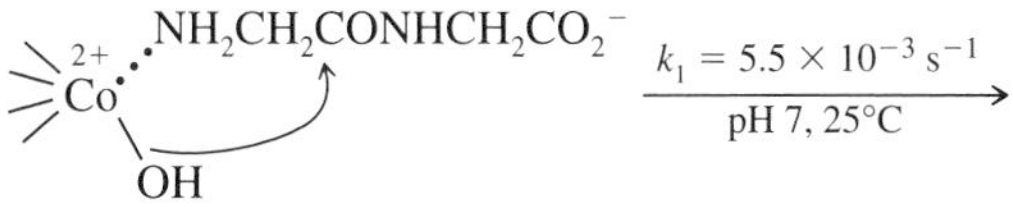

$$\overset{2+}{Co}\cdots\underset{OH}{\overset{NH_2CH_2CONHCH_2CO_2^{-}}{}} \xrightarrow[\text{pH 7, 25°C}]{k_1 = 5.5 \times 10^{-3}\,s^{-1}} \qquad (2.33)$$

There are more complex examples of metal ion catalysis. Cobalt in vitamin B$_{12}$ reactions forms covalent bonds with carbons of substrates.[41,42] Metals can also act as electron conduits in redox reactions. For example, in cytochrome *c* the iron in the heme is reversibly oxidized and reduced.

C. Covalent catalysis

1. Electrophilic catalysis by Schiff base formation

A good example of how transient chemical modification can activate a substrate is Schiff base formation from the condensation of an amine with a carbonyl compound (equation 2.34). The Schiff base may be protonated at neutral pH.

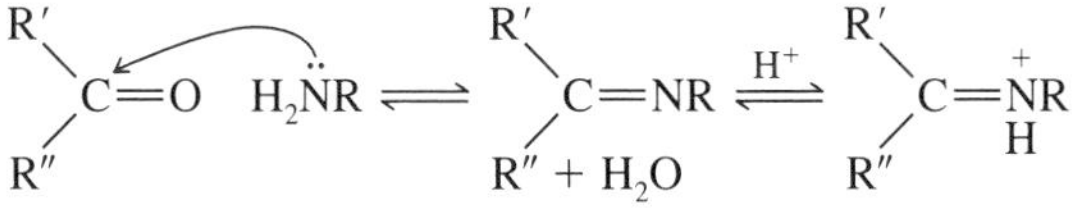

$$\underset{R''}{\overset{R'}{>}}C{=}O \quad H_2\ddot{N}R \rightleftharpoons \underset{R'' + H_2O}{\overset{R'}{>}}C{=}NR \underset{}{\overset{H^+}{\rightleftharpoons}} \underset{R''}{\overset{R'}{>}}C{=}\underset{H}{\overset{+}{N}}R \qquad (2.34)$$

This acts as an *electron sink* to stabilize the formation of a negative charge on one of the α carbons (equation 2.35). After tautomerization to form the enamine, the methylene carbon is activated as a nucleophile.

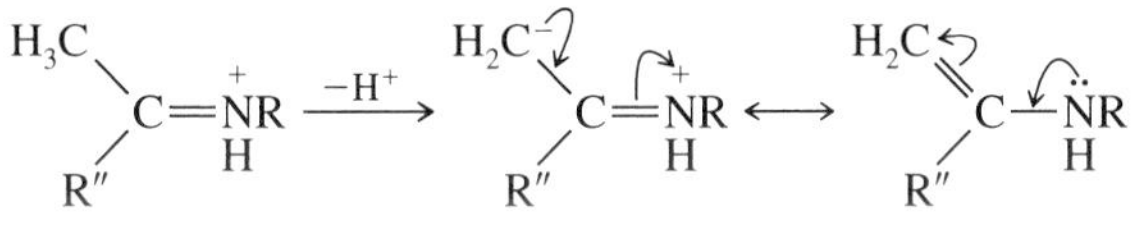

$$\underset{R''}{\overset{H_3C}{>}}C{=}\underset{H}{\overset{+}{N}}R \xrightarrow{-H^+} \underset{R''}{\overset{H_2C}{>}}C{=}\underset{H}{\overset{+}{N}}R \longleftrightarrow \underset{R''}{\overset{H_2C}{>}}C{-}\underset{H}{\overset{..}{N}}R \qquad (2.35)$$

(Enamine)

Another consequence of Schiff base formation is that the carbonyl carbon is activated toward nucleophilic attack because of the strong electron withdrawal of the protonated nitrogen.

a. Acetoacetate decarboxylase[43,44]

This enzyme catalyzes the decarboxylation of acetoacetate to acetone and carbon dioxide. The nonenzymatic reaction involves the expulsion of a highly basic enolate ion at neutral pH (equation 2.36), but the enzymatic reaction circumvents this by the prior formation of a Schiff base with a lysine residue. The protonated imine is then readily expelled.

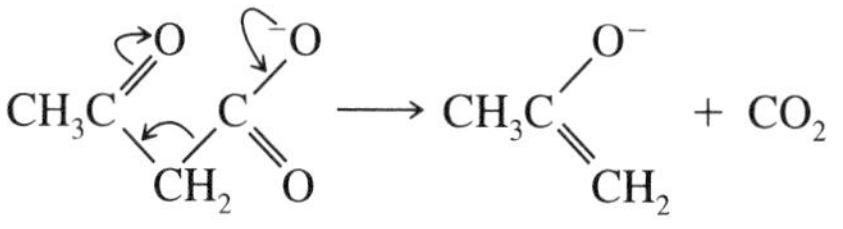

$$CH_3C \overset{O}{\underset{CH_2}{\diagup}} \overset{O}{\underset{O}{C}} \longrightarrow CH_3C \overset{O^-}{\underset{CH_2}{\diagup}} + CO_2 \qquad (2.36)$$

This process may be mimicked in solution by using aniline as a catalyst (equation 2.37).

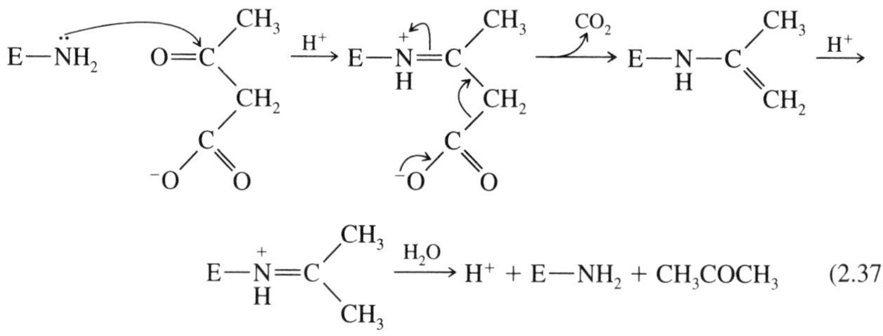

$$E-NH_2 \quad O=C \overset{CH_3}{\underset{CH_2}{\diagdown}} \overset{H^+}{\longrightarrow} E-\overset{+}{N}=C \overset{CH_3}{\underset{CH_2}{\diagdown}} \overset{CO_2}{\longrightarrow} E-N-C \overset{CH_3}{\underset{CH_2}{\diagdown}} \overset{H^+}{\longrightarrow}$$

$$E-\overset{+}{N}=C \overset{CH_3}{\underset{CH_3}{\diagdown}} \overset{H_2O}{\longrightarrow} H^+ + E-NH_2 + CH_3COCH_3 \qquad (2.37)$$

The evidence for the intermediate is that the enzyme is irreversibly inhibited when sodium borohydride is added to the complex with the substrate. Borohydride is known to reduce Schiff bases, and the hydrolysate of the inhibited protein is found to contain isopropyl-lysine (equation 2.38). The carbon in the Schiff base is activated to the attack of

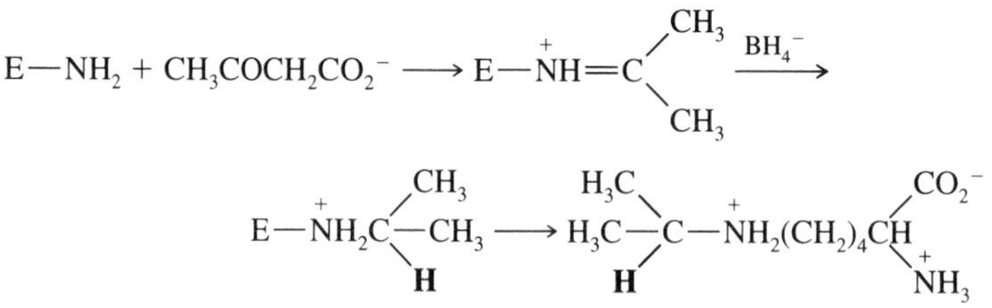

$$E-NH_2 + CH_3COCH_2CO_2^- \longrightarrow E-\overset{+}{N}H=C \overset{CH_3}{\underset{CH_3}{\diagdown}} \overset{BH_4^-}{\longrightarrow}$$

$$E-\overset{+}{N}H_2\overset{CH_3}{\underset{H}{C}}-CH_3 \longrightarrow H_3C-\overset{H_3C}{\underset{H}{C}}-\overset{+}{N}H_2(CH_2)_4\overset{CO_2^-}{\underset{NH_3}{CH}} \qquad (2.38)$$

an H^- ion from the borohydride:

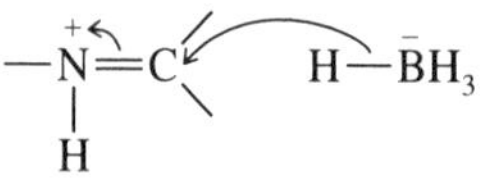

$$-\overset{+}{\underset{H}{N}}=C \overset{}{\diagdown} \quad H-\bar{B}H_3 \qquad (2.39)$$

b. Aldolase and transaldolase

The aldol condensation and the reverse cleavage reaction catalyzed by these enzymes both involve a Schiff base. The cleavage reaction is similar to the acetoacetate decarboxylase mechanism, with the protonated imine being expelled. The condensation reaction illustrates the other function of a Schiff base, the activation of carbon via an enamine (equation 2.40).

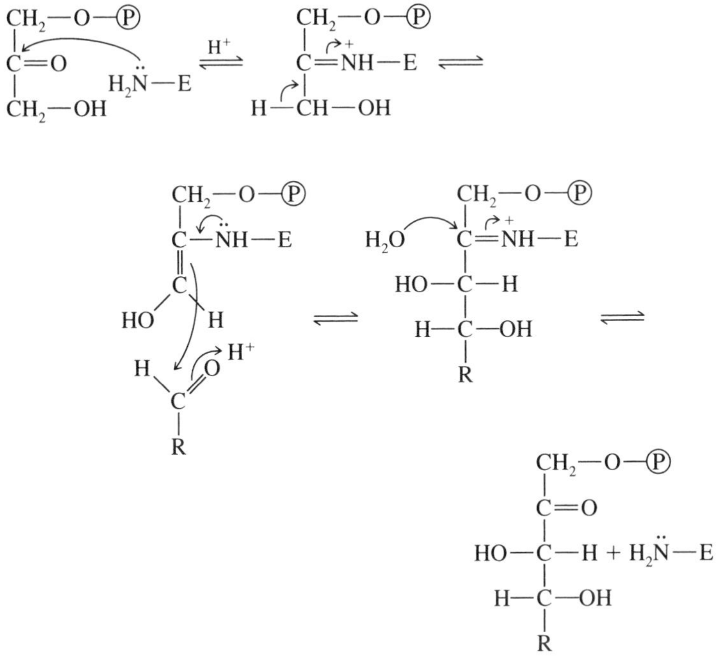

$$(2.40)$$

The intermediate may be trapped as before.

2. Pyridoxal phosphate — Electrophilic catalysis

The principles of the above reactions form the basis of a series of important metabolic interconversions involving the coenzyme pyridoxal phosphate (structure 2.41). This condenses with amino acids to form a Schiff base (structure 2.42). The pyridine ring in the Schiff base acts as an "electron sink" which very effectively stabilizes a negative charge.

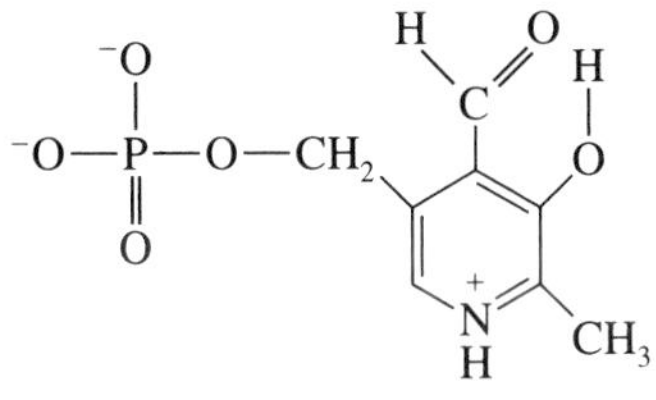

$$(2.41)$$

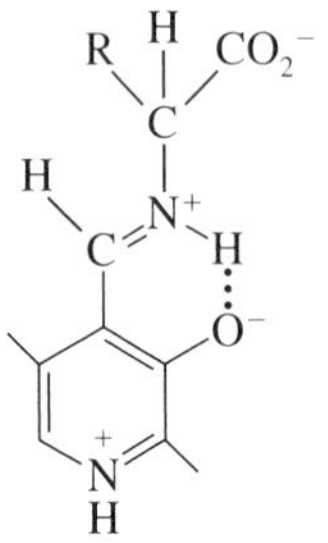

(2.42)

Each one of the groups around the chiral carbon of the amino acid may be cleaved, forming an anion that is stabilized by the Schiff base with the pyridine ring. (See Chapter 8, section F1, for a stereochemical explanation of why a particular bond is cleaved.)

a. Removal of the α hydrogen

The removal of the α hydrogen (equation 2.43) gives a key intermediate

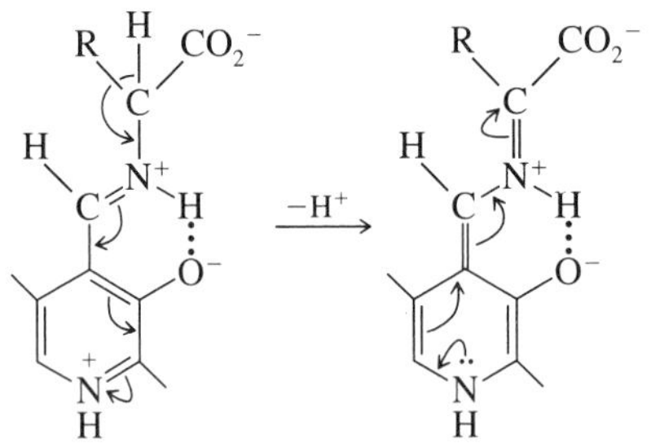

(2.43)

that may react in several different ways:

1. *Racemization.* Addition of the proton back to the amino acid will lead to racemization unless it is done stereospecifically.

2. *Transamination.* Addition of a proton to the carbonyl carbon of the pyridoxal leads to a compound that is the Schiff base of an α-keto acid and pyridoxamine. Hydrolysis of the Schiff base gives the α-keto acid and pyridoxamine, which may react with a different α-keto acid to reverse the sequence:

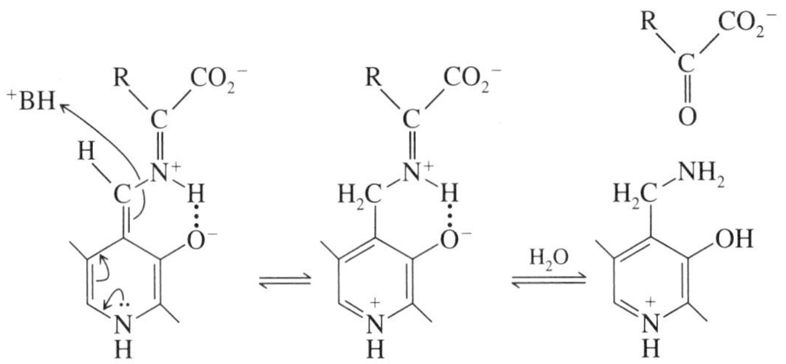

(2.44)

The overall reaction is

$$R'CH(NH_3^+)CO_2^- + R''COCO_2^- \rightleftharpoons$$
$$R'COCO_2^- + R''CH(NH_3^+)CO_2^- \quad (2.45)$$

3. β-Decarboxylation. When the amino acid is aspartate, the second compound in equation 2.44 is analogous to the Schiff base in the acetoacetate decarboxylase reaction and may readily decarboxylate:

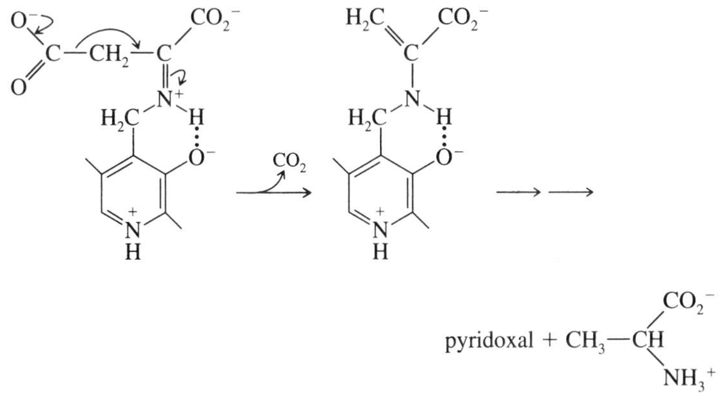

$$\text{pyridoxal} + CH_3-\underset{\underset{NH_3^+}{|}}{CH}\overset{CO_2^-}{\diagup} \quad (2.46)$$

4. Interconversion of side chains. When RX— is a good leaving group, it may be expelled as in equation 2.47.

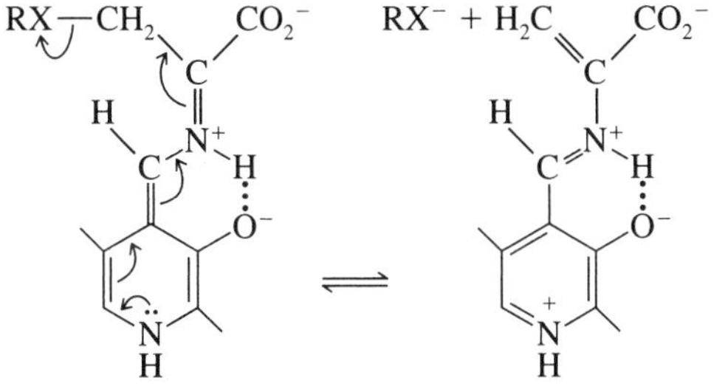

$$(2.47)$$

RX— may be a thiol, a hydroxyl, or an indole group. In this way, serine, threonine, cysteine, tryptophan, cystathionine, and serine and threonine phosphates may be interconverted or degraded.

b. α-Decarboxylation

The "electron sink" allows facile decarboxylation (equation 2.48). The decarboxylated adduct will add a proton to the amino acid carbonyl carbon and

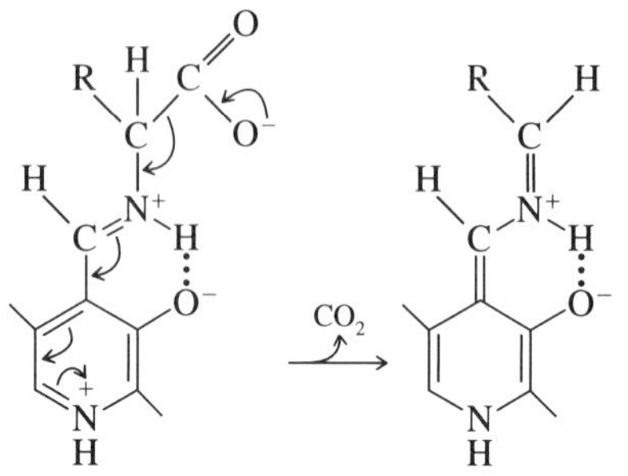

(2.48)

then hydrolyze to give the amine and pyridoxal (equation 2.49),

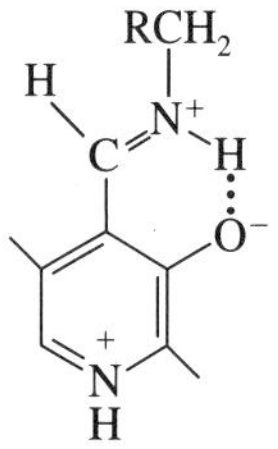

$\xrightarrow{H_2O}$ RCH_2NH_2 + pyridoxal (2.49)

or else it will add the proton to the pyridoxal carbonyl carbon and then hydrolyze to give the aldehyde and pyridoxamine (equation 2.50).

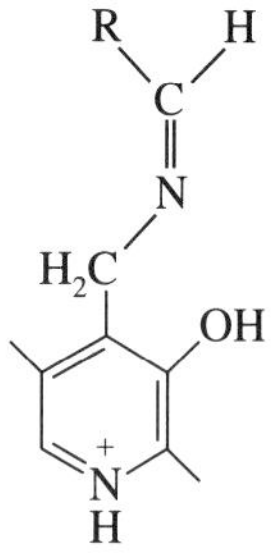

$\xrightarrow{H_2O}$ RCHO + pyridoxamine (2.50)

3. Thiamine pyrophosphate—Electrophilic catalysis

Thiamine pyrophosphate (structure 2.51) is another coenzyme that covalently bonds to a substrate and stabilizes a negative charge.

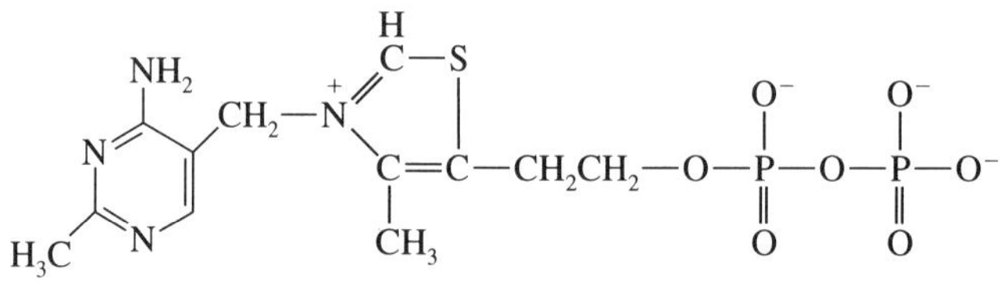

(2.51)

The positive charge on the nitrogen promotes the ionization of the C-2 carbon by electrostatic stabilization. The ionized carbon is a potent nucleophile (equation

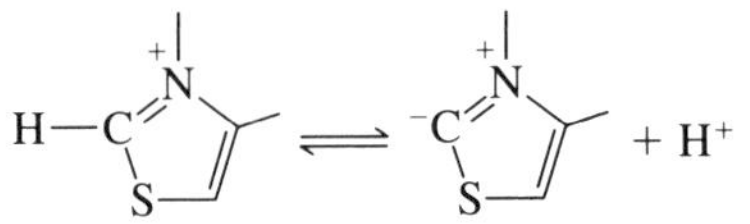

$$H-C \overset{\overset{+|}{N}}{\underset{S}{\rule{0pt}{1em}}} \rightleftharpoons \ ^{-}C \overset{\overset{+|}{N}}{\underset{S}{\rule{0pt}{1em}}} + H^{+} \tag{2.52}$$

2.52). The nitrogen atom can also stabilize by delocalizing a negative charge on the adduct of thiamine with many compounds, as, for example, in *hydroxyethylthiamine pyrophosphate*, a form in which much of the coenzyme is found *in vivo* (equation 2.53).

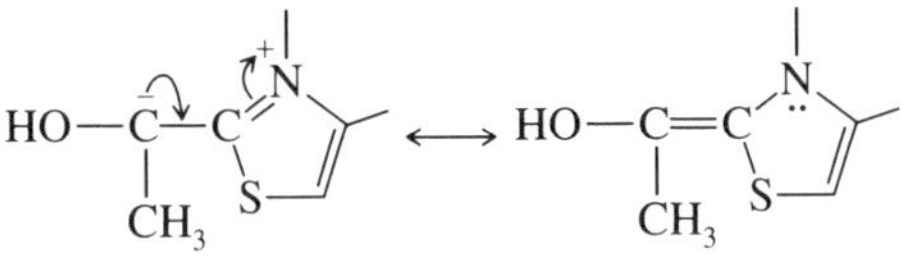

$$HO-\overset{\overset{}{C}}{\underset{CH_3}{\rule{0pt}{1em}}}-C \overset{\overset{+|}{N}}{\underset{S}{\rule{0pt}{1em}}} \longleftrightarrow HO-C=C \overset{\overset{|}{N}}{\underset{S}{\rule{0pt}{1em}}} \tag{2.53}$$

The combination of these reactions allows the decarboxylation of pyruvate by the route shown in equation 2.54. Other carbon-carbon bonds adjacent to a carbonyl group may be cleaved in the same manner.

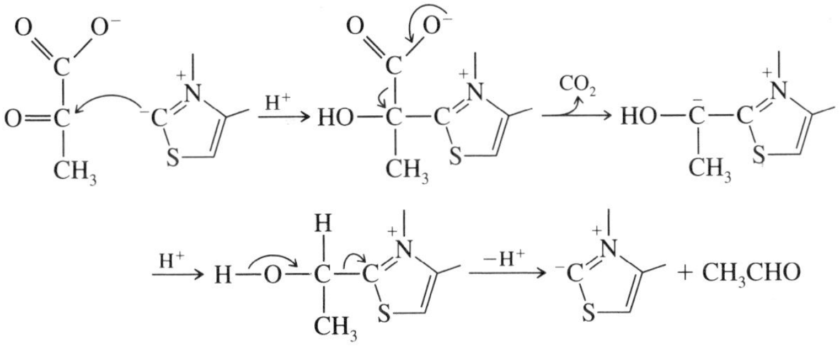

$$\xrightarrow{H^+} H-O-\overset{\overset{H}{|}}{\underset{CH_3}{C}}-C \overset{\overset{+|}{N}}{\underset{S}{\rule{0pt}{1em}}} \xrightarrow{-H^+} \ ^{-}C \overset{\overset{+|}{N}}{\underset{S}{\rule{0pt}{1em}}} + CH_3CHO \tag{2.54}$$

The hydroxyethylthiamine pyrophosphates are potent nucleophiles and may add to carbonyl compounds to form carbon-carbon bonds. A good illustration of carbon-carbon bond making and breaking occurs in the reactions of transketolase. The enzyme contains tightly bound thiamine pyrophosphate and shuttles a dihydroxyethyl group between D-xylulose 5-phosphate and D-ribose 5-phosphate to form D-sedoheptulose 7-phosphate and D-glyceraldehyde 3-phosphate (equations 2.55 and 2.56).

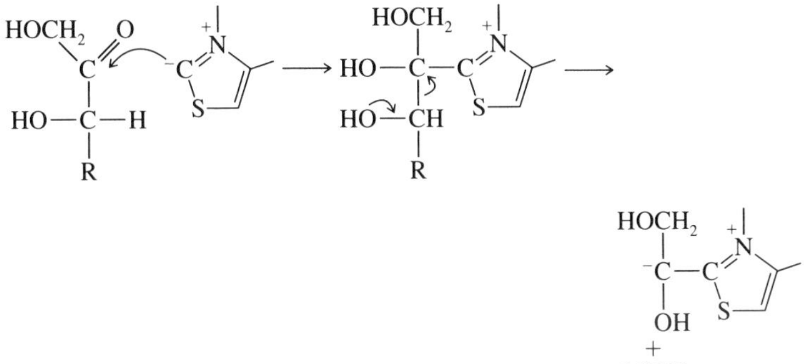

$$\tag{2.55}$$

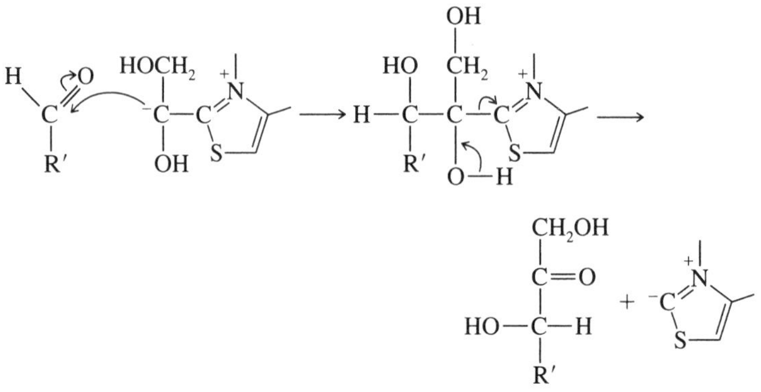

$$\text{(reaction scheme)} \tag{2.56}$$

Hydroxyethylthiamine pyrophosphate is also nucleophilic toward a thiol of oxidized lipoic acid. A hemithioacetal is formed, and this decomposes to give a thioester:

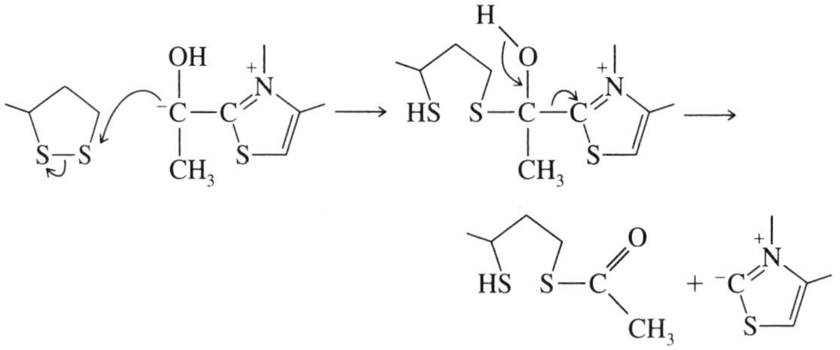

$$\text{(reaction scheme)} \tag{2.57}$$

4. Nucleophilic catalysis

In enzymes, the most common nucleophilic groups that are functional in catalysis are the serine hydroxyl—which occurs in the serine proteases, cholinesterases, esterases, lipases, and alkaline phosphatases—and the cysteine thiol—which occurs in the thiol proteases (papain, ficin, and bromelain), in glyceraldehyde 3-phosphate dehydrogenase, etc. The imidazole of histidine usually functions as an acid-base catalyst and enhances the nucleophilicity of hydroxyl and thiol groups, but it sometimes acts as a nucleophile with the phosphoryl group in phosphate transfer (Table 2.5).

The hydrolysis of peptides by these proteases represents classic nucleophilic catalysis. The relatively inert peptide is converted to the far more reactive ester or thioester acylenzyme, which is rapidly hydrolyzed. The use of the serine hydroxyl rather than the direct attack of a water molecule on the substrate is favored in several ways: alcohols are often better nucleophiles than the water molecule in both general-base-catalyzed and direct nucleophilic attack; the serine

Table 2.5 *Nucleophilic groups in enzymes*

Nucleophile	Enzyme	Intermediate
—OH (serine)	Serine proteases	Acylenzyme
	Alkaline phosphatases, phosphoglucomutase	Phosphorylenzyme
—OH (threonine)[a]	Proteasome, amidases	Acylenzyme
OH⁻ (zinc-bound)	Carbonic anhydrase, liver alcohol dehydrogenase	—
—SH (cysteine)	Thiol proteases, glyceraldehyde 3-phosphate dehydrogenase	Acylenzyme
—CO₂⁻ (aspartate)	ATPase (K⁺/Na⁺, Ca²⁺)	Phosphorylenzyme
—NH₂ (lysine)	Acetoacetate decarboxylase, aldolase, transaldolase, pyridoxal enzymes	Schiff base
	DNA ligase	Adenylenzyme (phosphoamide)
Imidazole (histidine)	Phosphoglycerate mutase, succinyl-CoA synthetase, nucleoside diphosphokinase, histone phosphokinase, acid phosphokinase	Phosphorylenzyme
—OH (tyrosine)	Glutamine synthetase	Adenylenzyme
	Topoisomerases	Nucleotidylenzyme (phosphotyrosine)

[a] The nucleophile resides in an N-terminal threonine. The general-base is the free N-terminal —NH₂ group.

reaction is intramolecular and hence favored entropically; and the arrangement of groups is more "rigid" and defined for the serine hydroxyl compared with a bound water molecule.

D. Structure – activity relationships

One of the most fruitful approaches in the study of organic reaction mechanisms has been to measure changes in reactivity with changes in the structures of the reagents. These studies have given considerable information on the electronic structures of transition states and on the features that determine reactivity, nucleophilicity, and leaving group ability. Structure – reactivity studies with enzymes tend to measure the effects of changes in the structure of the substrate on its interaction with the enzyme, rather than the effects of these changes on the distribution of electrons in the transition state. In general, it is difficult to obtain useful data on the electronic requirements of enzymatic reactions from structure–reactivity studies, both because the range of changes that can be made in the substrate is restricted and because the inductive effects of substituents are often obscured by the effects on binding. However, the lessons learned from the chemical studies have been invaluable to our understanding of the mechanisms of enzymatic reactions.

1. Nucleophilic attack at the carbonyl group

Structure–activity studies have been used to give information concerning the charge distribution in the transition state by noting the effects of electron-withdrawing and electron-donating substituents on the reaction rate. For example, it has been found that the rate of nucleophilic attack on esters increases with (1) electron withdrawal in the acyl portion ($CHCl_2CO_2Et$ is far more reactive than CH_3CO_2Et), (2) electron withdrawal in the leaving group (*p*-nitrophenyl acetate is more reactive than phenyl acetate), and (3) increasing basic strength of the nucleophile—that is, electron donation in the nucleophile (the hydroxide ion is far more reactive than the acetate ion). Using the idea from transition state theory that the reaction rate depends on the energy difference between the transition state and the ground state, we can deduce that the reaction involves an increase of negative charge on the substrate (since the reaction rate is increased by electron withdrawal), and a decrease in charge on the nucleophile (since the rate is increased by electron donation, i.e., electron repulsion). This is consistent either with mechanism 2.58,

$$
\begin{array}{c}
R-C\diagup\!\!\!\!\!\diagdown\substack{O\\OR'}\\Y^-
\end{array}
\longrightarrow
\left[
\begin{array}{c}
R-C\substack{\delta-\\O\\\delta-\,Y\quad OR'}
\end{array}
\right]_{TS}
\tag{2.58}
$$

where the rate-determining step is the formation of the tetrahedral intermediate, or with mechanism 2.59, where the rate-determining step is the breakdown.

$$
\begin{array}{c}
R-C\diagup\!\!\!\!\!\diagdown\substack{O\\OR'}\\Y^-
\end{array}
\rightleftharpoons
\begin{array}{c}
R-C\substack{O^-\\Y\quad OR'}
\end{array}
\longrightarrow
\left[
\begin{array}{c}
R-C\substack{\delta-\\O\\\delta-\\Y\quad OR'}
\end{array}
\right]_{TS}
\tag{2.59}
$$

a. Linear free energy relationships and the Brønsted equation for nucleophilic reactions

A quantitative assessment of the sensitivity of the reaction to electron withdrawal and donation in the attacking nucleophile may be made by measuring the second-order rate constants for the attack of a series of nucleophiles on a particular ester. A plot of the logarithms of the rate constants against the pK_a's of the nucleophiles may be made in the same way as for general-base catalysis (Figure 2.8). Generally, when the measurements are restricted to nucleophiles with similar chemical natures and to a not-too-wide range of pK_a's, a straight-line relationship is found. The slope of the line is termed β, as for general-base catalysis.

These linear relationships between the logarithms of rate constants and the pK_a's are further examples of linear free energy relationships, since the logarithm of a rate constant is proportional to its Gibbs free energy of activation, and the logarithm of an equilibrium constant (such as a pK_a) is proportional to the Gibbs free energy change of a reaction. The relationship between the nucleophilicity of a nucleophile and its basic strength shows that the Gibbs free energy of activation of bond formation with the carbonyl carbon is proportional to the Gibbs free energy of transfer of a proton to the nucleophile.

b. Interpretation of Brønsted β values

Those linear free energy relationships that are derived from plotting the rate constants from one reaction against the equilibrium constants for another are clearly different from those arising from the chemistry of section A3. As such, the values of β are not restricted to $0 < \beta < 1.0$. The sign and magnitude of β are an indication of the charge developed in the transition state. Consider, for example,

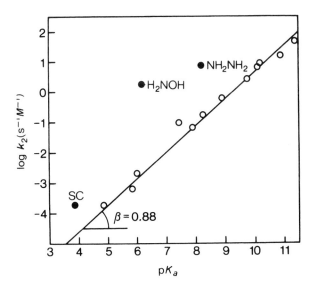

Figure 2.8 A Brønsted plot for the nucleophilic attack of primary and secondary amines of *p*-nitrophenyl acetate. Note that the "*α*"-effect nucleophiles—semicarbazide (SC), hydroxylamine, and hydrazine—are more reactive than would be expected from their pK_a's. [From W. P. Jencks and M. Gilchrist, *J. Am. Chem. Soc.* **90**, 2622 (1968).]

the attack of nucleophiles on esters. The β value for the equilibrium constants for the transfer of acetyl groups between oxyanions and also between tertiary amines is 1.6 to 1.7. The value is greater than 1.0 because the acetyl group is more electron-withdrawing than the proton (for which $\beta = 1.0$ by definition), and is more sensitive to the pK_a of the alcohol or amine. Now it is found that the value of β for the attack of tertiary amines on esters of very basic alcohols is $+1.5$ for the variation of the nucleophile's pK_a, and -1.5 for the variation of the alcohol's.[45] This shows that the transition state for the reaction is very close to the structure of the products; that is, close to complete acyl transfer from the alcohol to the amine:

$$R_3N + CH_3C\overset{O}{\underset{OEt}{\diagdown}} \longrightarrow \left[CH_3 - C\overset{\overset{\delta-}{O}}{\underset{\underset{R_3N}{+}\quad OEt}{}} \right]_{TS} \qquad (2.60)$$

At the other extreme, the reaction of basic nucleophiles with esters containing activated leaving groups exhibits β values of only $+0.1$ to 0.2 for the variation of the nucleophile's pK_a, and -0.1 to 0.2 for the variation of the leaving group's. This shows that the transition state involves little bond making and breaking and is close to the starting materials:

$$RO^- + CH_3C\overset{O}{\underset{OAr}{\diagdown}} \longrightarrow \left[CH_3C\overset{\overset{\delta-}{O}}{\underset{\underset{RO}{\delta-}\quad OAr}{}} \right]_{TS} \qquad (2.61)$$

(Ar = aromatic ring)

The β value is a measure of the charge formed in the transition state rather than of the extent of bond formation. However, in equations 2.60 and 2.61 and in other examples where there is no acid-base catalysis, charge and bond formation are linked, so β does also give a measure of the extent of bond formation. But when there is also acid-base catalysis partly neutralizing the charges formed in the transition state, there is no relation between β and the extent of bond formation.[46]

2. Factors determining nucleophilicity and leaving group ability

The magnitude of general-acid-base catalysis by oxygen and nitrogen bases depends only on their pK_a's, and is independent of their chemical natures (apart from an enhanced activity of oximes in general-acid catalysis). Nucleophilic reactivity depends markedly on the nature of the reagents. These reactions may be divided into two broad classes: nucleophilic attack on *soft* and on *hard* electrophilic centers.[47]

a. Nucleophilic reactions with the carbonyl, phosphoryl, sulfuryl, and other hard groups

The attack of a nucleophile on an amide, ester, or carbonyl carbon involves the formation of a "real" chemical intermediate, and the valency of carbon is not extended beyond its normal value of 4. The attack on a phosphate ester is similar; a pentacovalent phosphate transition state is formed. The transition state of the reaction involves the formation of a normal bond. This is a characteristic of hard centers. The dominant feature controlling nucleophilicity in these reactions is the basic strength of the nucleophile: the stronger the basic strength, the greater the nucleophilicity. There are, however, differences in reactivity among different classes of nucleophiles: amines and thiolate anions tend to be more nucleophilic than oxyanions.[48,49] Also, certain nucleophiles that have two electronegative atoms next to each other, such as NH_2OH, NH_2NH_2, $NH_2CONHNH_2$, HOO^-, and CH_3OO^-, are more reactive than would be expected from their pK_a's (Figure 2.9). This is known as the α effect.

The *ease of expulsion* of a group depends both on its pK_a and on its state of protonation. Basically, a "good" leaving group is one that is stable in solution. For example, the p-nitrophenolate ion is a good leaving group because it is weakly basic; the pK_a of p-nitrophenol is 7.0. The chemical cause of the stability of the ion is that the negative charge is delocalized around the aromatic ring and onto the nitro group (equation 2.62). The ion is readily and directly expelled

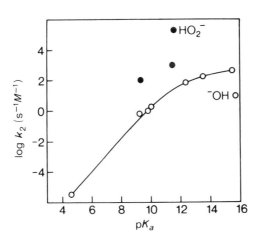

Figure 2.9 A Brønsted plot for the attack of oxyanion nucleophiles on p-nitrophenyl acetate. As in Figure 2.8, the α-effect nucleophiles (●) are unusually reactive. Note how the linear plot breaks down with increasing pK_a for the more reactive nucleophiles. In general, the Brønsted relationships hold only over a limited range of pK_a's in these reactions. The curvature is not often seen in practice because of the limited range of bases used.

from a tetrahedral intermediate (equation 2.63). In this class of leaving group,

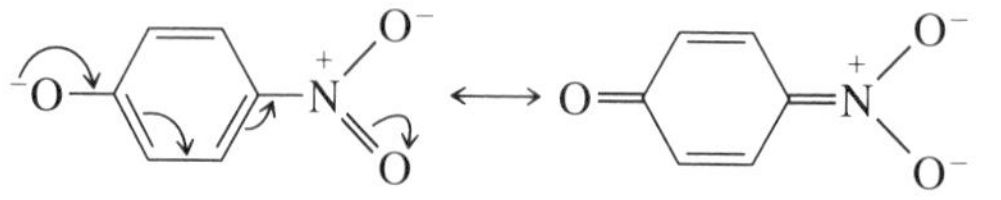

$$ \text{(2.62)} $$

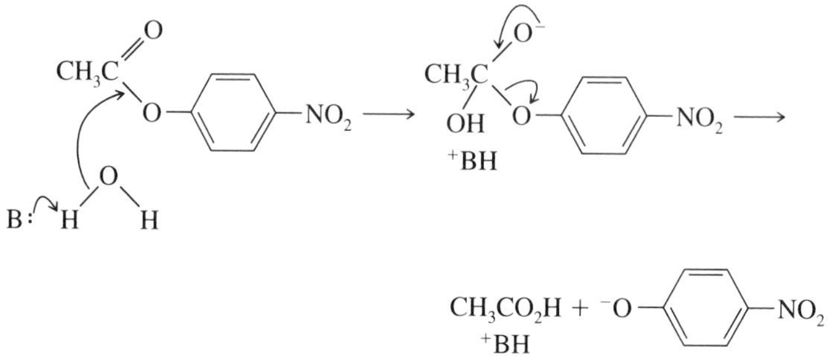

$$ \text{(2.63)} $$

$$ CH_3CO_2H + {}^-O-\langle\text{ring}\rangle-NO_2 \qquad \text{(2.63)} $$
$$ {}^+BH $$

the lower the basic strength, the greater the ease of expulsion. Acetate (the pK_a of acetic acid = 4.76) is a better leaving group than p-nitrophenol and phosphate (with pK_a's ~7), which are better leaving groups than OH^- (the pK_a of water = 15.8).

Alcoholate ions are difficult to expel because they are strongly basic; the pK_a's of simple alcohols are about 16. The expulsion of alcohols is aided by general-acid catalysis:

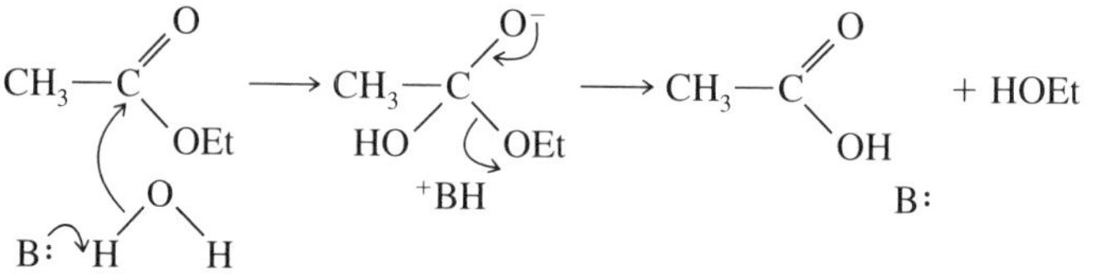

$$ \text{(2.64)} $$

Nitrophenyl esters are often used as synthetic substrates for two reasons: (1) the nitrophenolate ion is a very good leaving group so it forms a reactive substrate; and (2) it has a characteristic absorption at 400 nm and is thus easily assayed spectrophotometrically. Both these factors are caused by the delocalization shown in equation 2.62. But, the use of the nitrophenyl group can give rise to misleading results on the requirements for catalysis since, as discussed later (equation 2.70), it is expelled without acid catalysis.

Amines have to be protonated to be expelled from a molecule, since the amide ions, RNH^-, are far too unstable to be released directly into solution.

b. Nucleophilic reactions with saturated carbon

The attack of a nucleophile on saturated carbon—for example, the bimolecular attack of a thiol on the methyl carbon of S-adenosylmethionine—involves a transition state in which five groups surround the normally tetravalent carbon (equation 2.65).

$$(2.65)$$

This is not a "normal" bond with carbon, and it is peculiar to the transition state. The reaction is typical of a "soft" center. Large, polarizable atoms such as sulfur and iodine (i.e., "soft" ligands) react more rapidly in these reactions, whereas the small atoms of low polarizability, oxygen and nitrogen (i.e., "hard" ligands), are less reactive. The dominant factor in nucleophilicity toward alkyl groups and other soft centers is polarizability. Within any particular class of compounds, increasing basic strength increases the nucleophilicity, but between classes, polarizability is all important (Table 2.6).

As with reactions at the carbonyl group, weakly basic leaving groups are more readily displaced than strongly basic ones.

c.   Leaving group activation

It was seen in the last section that highly basic groups are not readily displaced from carbonyl compounds and from saturated carbon. An extreme example of this is the esterification of an alcohol by a carboxylate ion. This would require

Table 2.6   *Nucleophilic reactivity toward saturated carbon*[a]

Nucleophile	pK_a	Relative reactivity toward CH_3Br
H_2O	-1.74	1.00
NO_3^-	-1.3	11
F^-	3.17	100
$CH_3CO_2^-$	4.76	525
Cl^-	-7.0	1.1×10^3
C_5H_5N	5.17	4.0×10^3
HPO_4^{2-}	7.21	6.3×10^3
Br^-	-9.0	7.8×10^3
OH^-	15.74	1.6×10^4
$C_6H_5NH_2$	4.62	3.1×10^4
I^-	-10.0	1.1×10^5
CN^-	9.40	1.3×10^5
SH^-	7.00	1.3×10^5

[a] From C. G. Swain and C. B. Scott, *J. Am. Chem. Soc.* **75**, 141 (1953).

the formation of a tetrahedral intermediate with two negatively charged oxygens, and the subsequent expulsion of O^{2-} (equation 2.66).

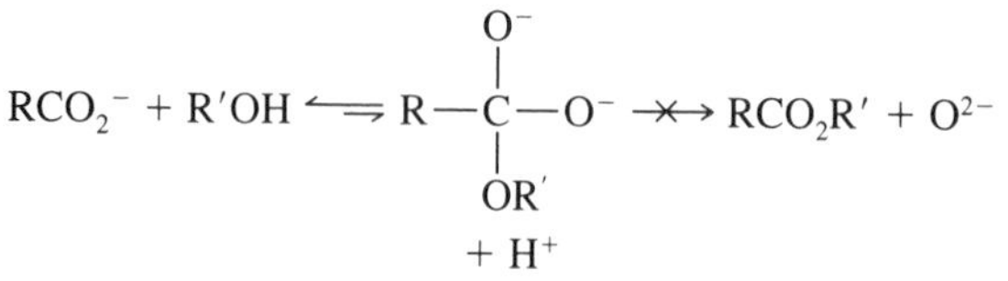

$$RCO_2^- + R'OH \rightleftharpoons R-\overset{\displaystyle O^-}{\underset{\displaystyle \underset{+ H^+}{OR'}}{\underset{|}{\overset{|}{C}}}}-O^- \xrightarrow{\times} RCO_2R' + O^{2-} \tag{2.66}$$

When an aminoacyl-tRNA synthetase catalyzes the esterification of tRNA with an amino acid, the amino acid is activated by formation of an enzyme-bound mixed anhydride with AMP (equation 2.67) in the same way that an organic chemist activates a carboxylic acid by forming an acyl chloride or a mixed anhydride. (Note that in equation 2.67 the substrate is the magnesium complex of ATP, with the metal ion acting as an electrophilic catalyst. The Mg^{2+} binds primarily to the β,γ-phosphates.[50,51])

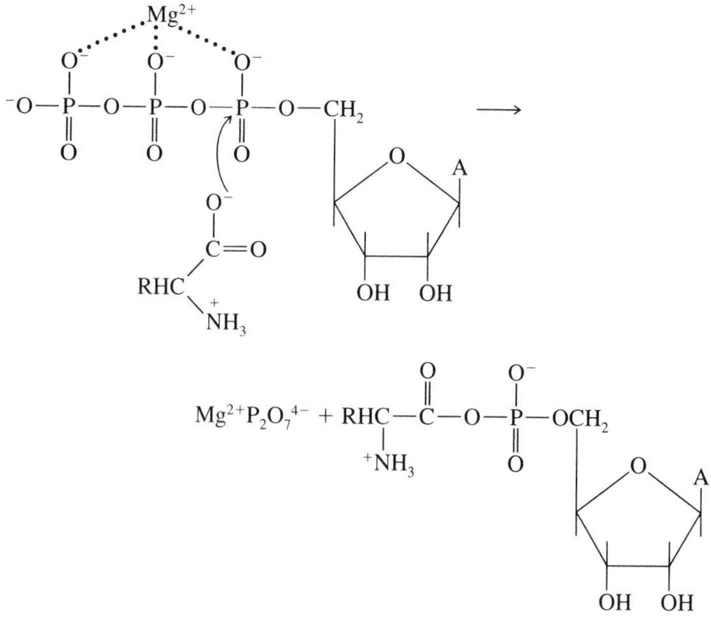

$$ \tag{2.67}$$

The carbonyl group of the amino acid is activated in equation 2.67 because it is bound to a good leaving group (the phosphate of AMP has a pK_a of about 6 or 7) and it may be readily attacked by one of the hydroxyl groups of the ribosyl ring of the terminal adenosine of the tRNA.

Another example of leaving group activation is the utilization of S-adenosyl-methionine rather than methionine in methylation reactions. A relatively basic thiolate anion has to be expelled from methionine, while the nonbasic neutral sulfur is displaced from the activated derivative (equation 2.68):

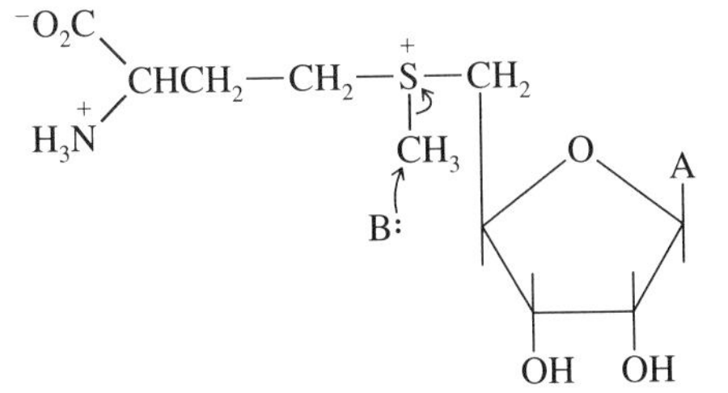

(2.68)

E. The principle of microscopic reversibility or detailed balance

The principle of microscopic reversibility or detailed balance is used in thermo-dynamics to place limitations on the nature of transitions between different quantum or other states. It applies also to chemical and enzymatic reactions: each chemical intermediate or conformation is considered as a "state." The principle requires that the transitions between any two states take place with equal frequency in either direction at equilibrium.[52] That is, the process A → B is exactly balanced by B → A, so equilibrium cannot be maintained by a cyclic process, with the reaction being A → B in one direction and B → C → A in the opposite. A useful way of restating the principle for reaction kinetics is that the reaction pathway for the reverse of a reaction at equilibrium is the exact opposite of the pathway for the forward direction. In other words, the transition states for the forward and reverse reactions are identical. This also holds for (nonchain) reactions in the steady state, under a given set of reaction conditions.[53]

The principle of microscopic reversibility is very useful for predicting the nature of a transition state from a knowledge of that for the reverse reaction. For example, as the attack of ethanol on acetic acid is general-base-catalyzed at low pH, the reverse reaction must involve the general-acid-catalyzed expulsion of ethoxide ion from the tetrahedral intermediate (equation 2.69).

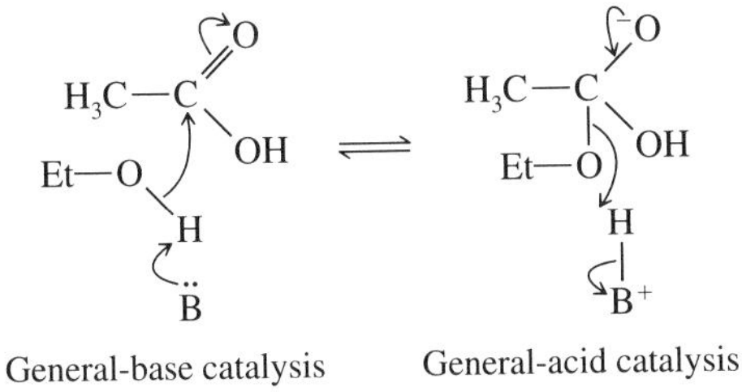

(2.69)

General-base catalysis   General-acid catalysis

Similarly, since p-nitrophenol is ionized above pH 7, its attack on a carbonyl compound cannot be general-base-catalyzed: not only is the p-nitrophenolate ion carrying a full negative charge a more powerful nucleophile than p-nitrophenol, but the ion is present at a higher concentration. Therefore, the ion directly attacks a carbonyl compound, and, by the principle of microscopic reversibility, the expulsion of p-nitrophenolate ion from a tetrahedral intermediate is uncatalyzed (equation 2.70).

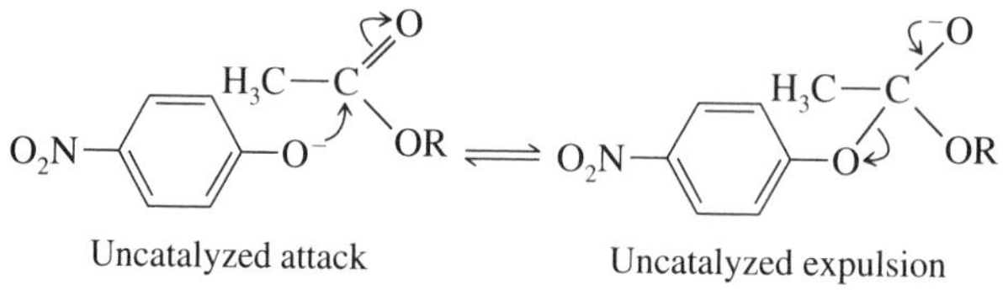

$$O_2N-\underset{}{\bigcirc}-O-\overset{\overset{O}{\parallel}}{\underset{}{C}}-CH_3 \cdots OR \rightleftharpoons O_2N-\underset{}{\bigcirc}-O \cdots \overset{\overset{O^-}{\mid}}{\underset{}{C}}-CH_3 \quad OR \qquad (2.70)$$

Uncatalyzed attack   Uncatalyzed expulsion

Similar arguments have been used to show that the attack of thiols upon esters is not general-base-catalyzed but involves the direct attack of the thiolate ion even at 10 pH units below the pK_a of the thiol, where only 1 part in 10^{10} is ionized.[46]

Care must always be taken to verify that a proposed reaction mechanism satisfies the principle of microscopic reversibility. Periodically, someone publishes a mechanism that contravenes the principle because the reverse reaction uses a different pathway from the forward reaction, under the same set of reaction conditions at equilibrium or in the steady state.

F. The principle of kinetic equivalence

There is an inherent ambiguity, the principle of kinetic equivalence, in interpreting the pH dependence of chemical reactions. When a rate law shows, for example, that the reaction rate is proportional to the concentration of an acid HA, it means that the net ionic charge of the acid appears in the transition state of the reaction; that is, either as the undissociated HA *or* as the two ions H^+ and A^-. Similarly, if the reaction rate varies as the concentration of A^-, the transition state contains either A^- *or* HA and OH^-. This may be shown algebraically as follows.

Rearranging equation 5.2 for the ionization of an acid gives

$$[HA] = \frac{[A^-][H^+]}{K_a} \qquad (2.71)$$

The concentration of the acid is related to the product of the concentrations of its conjugate base and the proton. Because of this, it is not possible to tell whether a reaction that depends on the concentration of HA really involves the undissociated acid or whether it involves the combination of an H^+ with the conjugate base in the transition state. Similarly, since the concentration of the basic form is related to that of the hydroxide ion and the acid by

$$[A^-] = \frac{[HA][OH^-]K_a}{K_w} \tag{2.72}$$

(where K_w is the ionic product of water), it is not possible to distinguish a reaction involving A^- from one involving a combination of HA and OH^- by examining the concentration dependence. This is the principle of kinetic equivalence. For example, mechanisms 2.73 and 2.74 for general-base catalysis follow the same rate law:

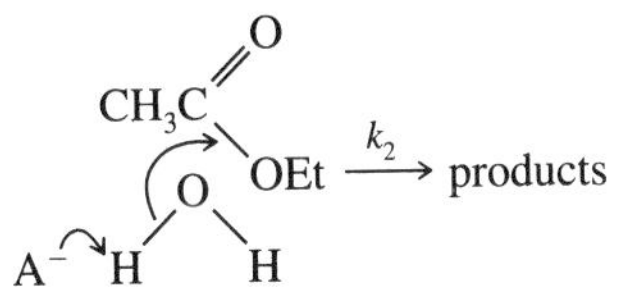

$$\tag{2.73}$$

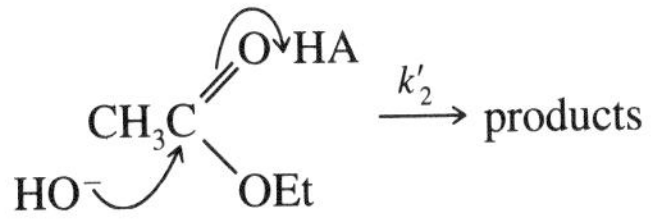

$$\tag{2.74}$$

For mechanism 2.76,

$$v = k_2[A^-][\text{ester}] \tag{2.75}$$

For mechanism 2.77,

$$v = k_2'[HA][OH^-][\text{ester}] \tag{2.76}$$

$$= \frac{k_2'[A^-][\text{ester}]K_w}{K_a} \tag{2.77}$$

$$= k_2''[A^-][\text{ester}] \tag{2.78}$$

Because equations 2.75 and 2.78 are equivalent, the two mechanisms cannot be distinguished by the concentration or pH dependence. Nor can the two mechanisms be distinguished by the sign of the Brønsted β for the variation of the pK_a of the catalyst. One can see this intuitively from transition state theory, since both reactions involve a dispersion of negative charge in the transition state. Alternatively, this result can be derived mathematically. Mechanism 2.73 involves a positive value of β since the catalysis will be stronger for stronger bases. Mechanism 2.74 also involves a positive β value, because although the general-acid component involves a negative value of β, the term $1/K_a$ in equation 2.77 means that a component of $1 \times pK_a$ must be added to log v, which more than compensates for the fractionally negative β value for the chemical step.

The only time that kinetically equivalent mechanisms can be distinguished is when one of the mechanisms involves an "impossible" step and generates, say, a second-order rate constant that is faster than is feasible for a diffusion-controlled reaction (Chapter 4), or, as in the case of mechanism 2.74 for aspirin hydrolysis, a negative energy of activation.[12]

G. Kinetic isotope effects

Information about the extent and nature of the bond making and breaking steps in the transition state may sometimes be obtained by studying the effects of isotopic substitution on the reaction rates. The effects may be divided into two classes, depending on the position of the substitution.

1. Primary isotope effects

A primary isotope effect results from the cleavage of a bond to the substituted atom. For example, it is often found that the cleavage of a C—D bond is several times slower than that of a C—H bond. Smaller decreases in rate, up to a few percent, are sometimes found on the substitution of ^{15}N for ^{14}N, or of ^{18}O for ^{16}O. The magnitude of the change in rate gives some idea of the extent of the breaking of the bond in the transition state.

A simple way of analyzing isotope effects is to compare an enzymatic reaction with a simple chemical model whose chemistry has been established by other procedures. In this section, we are interested primarily in the empirical results of the model experiments, but the following oversimplified account of the theoretical origins of the effects is helpful in understanding their nature. The plot of the energy of a carbon-hydrogen bond against interatomic distance gives the characteristic curve shown in Figure 2.10. The carbon-deuterium bond gives an identical plot, since the shape is determined by the electrons in the orbitals. According to quantum theory, the lowest energy level is at a value of $\frac{1}{2}h\nu$ above the bottom of the well, i.e., above the *zero-point energy*, where ν is the frequency of vibration. The value of ν may be found from the infrared stretching frequencies to give values of 17.4 and 12.5 kJ/mol (4.15 and 3 kcal/mol) for the zero-point energies of the C—H and C—D bonds, respectively. If during the transition state for the reaction the hydrogen or deuterium atom is at a potential energy maximum rather than in a well, there will be no zero-point energy. It is therefore easier to break the C—H bond by 17.4 to 12.5 kJ/mol (4.15 to 3.0 kcal/mol), a factor of 7 at 25°C. In practice, the kinetic isotope effect may be higher than this because of quantum mechanical tunneling, or lower because there are compensating bending motions in the transition state. As a general rule, values of 2 to 15 are good evidence that a C—H bond is being broken in the transition state. It will be seen in Chapter 16 that values of 3 to 5 are found for k_H/k_D in hydride transfer between the substrate and NAD^+ in the reactions of alcohol dehydrogenase.

Carbon-tritium bonds are broken even more slowly than C—D bonds because of the greater mass of tritium. A simple relationship between the deuterium and

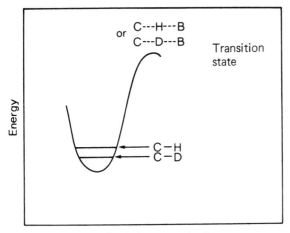

Figure 2.10 The energy changes during the transfer of hydrogen or deuterium from carbon. The energy of the transition state is the same for both (subject to the provisos in the text), but the hydrogen is at a higher energy in the starting materials because of its higher zero-point energy. The activation energy for the transfer of hydrogen is therefore less than that for deuterium.

tritium isotope effects holds at the temperatures at which enzymatic reactions occur.[54]

$$\frac{k_H}{k_T} = \left(\frac{k_H}{k_D}\right)^{1.442} \tag{2.79}$$

Equation 2.79 stems from classical dynamics, based on the relative reduced masses of H, D, and T. Similarly, a further relationship may be calculated,[55] based on classical properties

$$\frac{k_H}{k_T} = \left(\frac{k_D}{k_T}\right)^{3.26} \tag{2.80}$$

This relationship was found to break down for some reactions of yeast alcohol dehydrogenase[56] such that $(k_H/k_T)_{obs} > (k_D/k_T)^{3.26}$. This is unambiguous evidence for quantum mechanical tunneling of the hydrogen. The small mass of the H isotope makes it proportionately more susceptible to tunneling than D or T, and so the "classical" equation 2.80 underestimates the substitution. The behavior of hydrogen is poised between classical and quantum mechanics.[57,58]

2. Multiple isotope effects

One of the important mechanistic uses of isotopic substitution is that it can selectively affect different steps in a stepwise mechanism and so help resolve mechanisms and pathways. For example, in reactions where there is hydride transfer coupled to C—C bond formation, the measurement of both $^{13}C/^{12}C$ and D/H kinetic isotope effects can distinguish whether the steps are concerted or stepwise. Consider, for example, the oxidative decarboxylation of malate catalyzed by the malic enzyme, which could occur either via an intermediate mechanism (2.81) or in a concerted mechanism (2.82).[59]

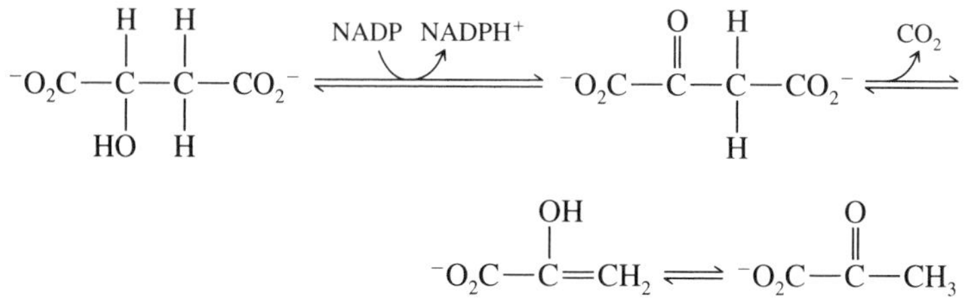

$$\text{(2.81)}$$

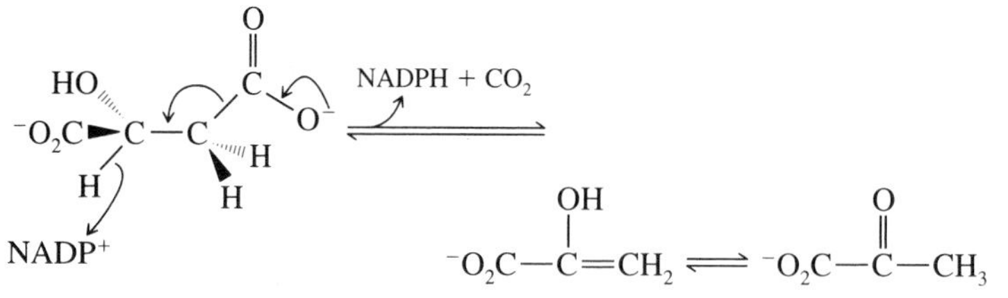

$$\text{(2.82)}$$

The deuterium isotope effect is measured for the hydrogen that is transferred and the ^{13}C-isotope effect is measured for the C of the CO_2 that is expelled. If neither the C—C bond breaking nor the hydride transfer is completely rate determining, then a stepwise mechanism is affected differently from a concerted mechanism. Deuteration slows the reaction down, so if the mechanism is stepwise, the ^{13}C-kinetic isotope effect will be smaller for the deuterated substrate. Coversely, if the reaction is concerted and the chemical steps are completely rate determining, the ^{13}C-effect should be unaffected by deuteration. The mechanism was found to change from stepwise to concerted on changing $NADP^+$ to acetylpyridine-$NADP^+$. This is a good lesson that mechanisms can change with changing reagents or with different enzymes.

3. Secondary isotope effects

These result from bond cleavage between atoms *adjacent* to the isotopically substituted atom. Secondary isotope effects are caused by a change in the electronic hybridization of the bond linking the isotope, rather than by cleavage of the bond. One example of this in enzymatic reactions is the substitution of deuterium or tritium for hydrogen on the C-1 carbon of substrates for lysozyme (Chapter 15). An oxocarbenium ion is formed in the reaction (equation 2.83), and the C-1 carbon changes from sp^3 to sp^2. Model reactions give a value of 1.14 for k_H/k_D, compared with the 1.11 found for the enzymatic reaction.[60]

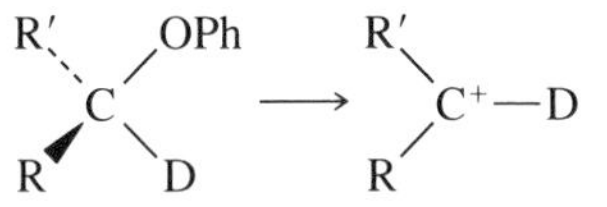

$$\underset{R}{\overset{R'}{\underset{\diagdown}{C}}}\overset{OPh}{\diagup}D \longrightarrow \underset{R}{\overset{R'}{\underset{\diagdown}{C^+}}}-D \tag{2.83}$$

4. Solvent isotope effects

These are found from comparing the rates of a reaction in H_2O and D_2O. They are usually a result of proton transfers between electronegative atoms *accompanying* the bond making and breaking steps between the heavier atoms in reactions such as the following:

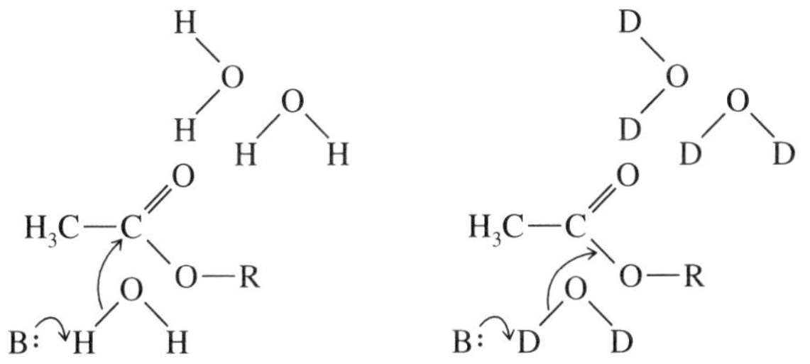

$$\tag{2.84}$$

Solvent isotope effects differ in origin from those generated by the cleavage of a C—H bond, and their cause is not fully understood. Proton transfers between electronegative atoms are very rapid compared with normal bond making and breaking steps, and the proton probably sits at the bottom of a potential energy well during the reaction. There is no loss of zero-point energy such as occurs in the transfer of a proton from carbon, where the actual C—H bond breaking step is slow. Solvation of the reagents and secondary isotope effects caused by the exchange of deuterium for hydrogen in the reagents may make contributions to the solvent isotope effect.

Solvent isotope effects are a useful diagnostic tool in simple chemical reactions, although they can be variable. For example, it is found that general-base-catalyzed reactions such as equation 2.84 have a k_H/k_D of about 2, whereas the nucleophilic attack on an ester has a k_H/k_D of about 1. The isotope effects in enzymatic reactions are more difficult to analyze because the protein has so many protons that may exchange with deuterons from D_2O.[61] Also there may be slight changes in the structure of the protein on the change of solvent.

H. Summary of classical factors of enzyme catalysis

Many of the important lessons of this chapter are summed up in Figure 2.11. It is not so much that enzymatic reactions are particularly fast as that uncatalyzed reactions are very slow. An enzyme is able to speed up a reaction considerably by removing some of the unfavorable features of the uncatalyzed reaction and the

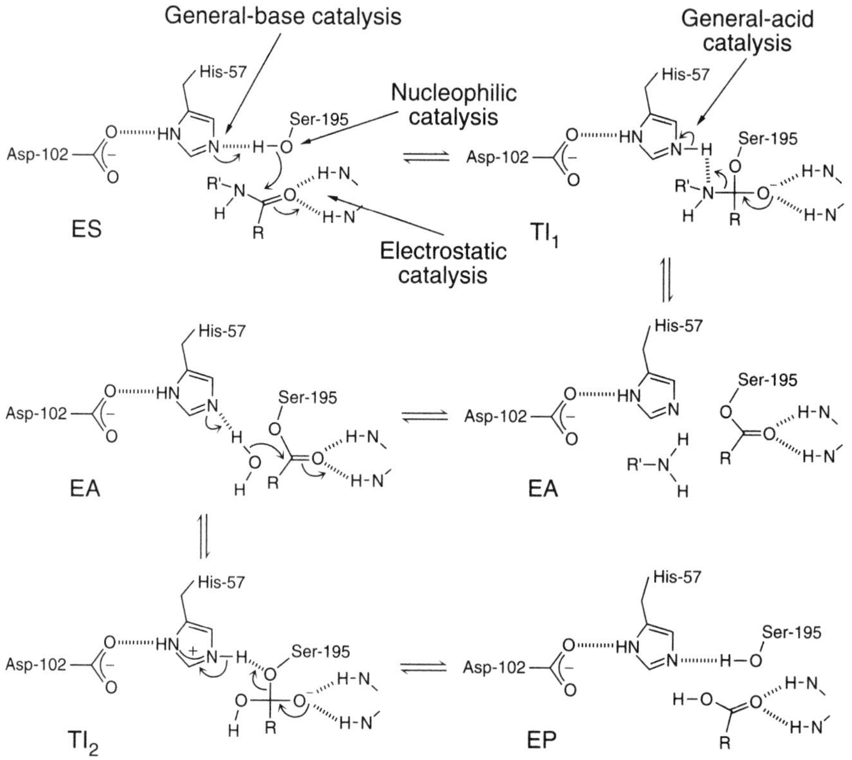

Figure 2.11 Illustration of the classical chemical factors of catalysis in the mechanism of chymotrypsin. The O of the $>C=O$ group of the substrate in the enzyme-substrate complex (ES) is held between two $\delta+$ charged HN< groups of the peptide backbone of the protein. The C of the $>C=O$ is held in the correct position for attack by the γ-OH of Ser-195. His-57 acts as general-base catalyst to activate the γ-OH group as a nucleophile. A tetrahedral adduct (TI_1) is formed in which the negative charge on the $>C-O^-$ substrate is stabilized by the backbone $>NH$ groups. The interaction between Asp-102 and His-57 increases to stabilize the positive charge on it. The R'NH— group is activated as a leaving group by His-57 shuttling its proton to convert it into $R'NH_2$, which is expelled to form the acylenzyme (EA). Deacylation then occurs by water being activated as a nucleophile by His-57, to generate the second tetrahedral adduct (TI_2) and regenerate the enzyme via the enzyme-product complex (EP). The pK_a of ~ 7 of His-57 makes it optimally active as both a general-base and general-acid catalyst at pH 7.

unfavorable aspects of simple catalysis in solution. It is tempting to estimate the individual components of catalysis, but we confront another important element of enzymatic catalysis that is not usually found in simple reactions in solution: the utilization of the binding energy between enzyme and substrate. This can dominate catalysis. The role of binding energy is discussed in Chapter 12, as is the importance of subsites that are distant from the seat of reaction. The individual components of catalysis are inextricably linked to the interplay of binding energy and chemical catalysis. Dissection of the structure and activity of the tyrosyl-tRNA synthetase by protein engineering (Chapter 15) has shown how contributions to catalysis are delocalized over the whole active site. Some of the concepts in this chapter that were speculative when first written, such as "electronic strain" in active sites, have been verified by protein engineering (Chapter 17, section E).

References

1. H. Pelzer and E. Wigner, *Z. Phys. Chem.* **B15**, 445 (1932).
2. H. Eyring, *Chem. Rev.* **17**, 65 (1935).
3. M. G. Evans and M. Polanyi, *Trans. Faraday Soc.* **31**, 875 (1935).
4. A. A. Frost and R. G. Pearson, *Kinetics and mechanism*, Wiley (1961).
5. G. S. Hammond, *J. Am. Chem. Soc.* **77**, 334 (1955).
6. A. Warshel, *Computer modeling of chemical reactions in enzymes and in solutions*, Wiley (1991).
7. J. N. Brønsted and K. Pedersen, *Z. Phys. Chem.* **A108**, 185 (1923).
8. T. H. Fife, *Accts. Chem. Res.* **5**, 264 (1972).
9. G. A. Craze and A. J. Kirby, *J. Chem. Soc. Perk. II* **61** (1974).
10. R. F. Atkinson and T. C. Bruice, *J. Am. Chem. Soc.* **96**, 819 (1974).
11. F. G. Bordwell and W. J. Boyle, Jr., *J. Am. Chem. Soc.* **94**, 3907 (1972).
12. A. R. Fersht and A. J. Kirby, *J. Am. Chem. Soc.* **89**, 4853, 4857 (1967); **90**, 5818, 5826 (1968).
13. T. St. Pierre and W. P. Jencks, *J. Am. Chem. Soc.* **90**, 3817 (1968).
14. Extrapolated from the data of E. Gaetjens and H. Morawetz, *J. Am. Chem. Soc.* **82**, 5328 (1960), and V. Gold, D. G. Oakenfull, and T. Riley, *J. Chem. Soc.* **1968B**, 515 (1968).
15. A. R. Fersht and A. J. Kirby, *J. Am. Chem. Soc.* **90**, 5833 (1968).
16. T. Higuchi, L. Eberson, and J. D. Macrae, *J. Am. Chem. Soc.* **89**, 3001 (1967).
17. W. P. Jencks, F. Barley, R. Barnett, and M. Gilchrist, *J. Am. Chem. Soc.* **88**, 4464 (1966).
18. M. I. Page and W. P. Jencks, *Proc. Natl. Acad. Sci. USA* **68**, 1678 (1971).
19. J. D. Dunitz, *Science* **264**, 670 (1994).
20. B. Tidor and M. Karplus, *J. Molec. Biol.* **238**, 405 (1994).
21. A. V. Finkelstein and J. Janin, *Protein Engineering* **3**, 1 (1989).
22. D. R. Storm and D. E. Koshland, Jr., *Proc. Natl. Acad. Sci. USA* **66**, 445 (1970); *J. Am. Chem. Soc.* **94**, 5805 (1972).
23. T. C. Bruice, A. Brown, and D. C. Harris, *Proc. Natl. Acad. Sci. USA* **68**, 658 (1971).
24. W. P. Jencks and M. I. Page, *Biochem. Biophys. Res. Commun.* **57**, 887 (1974).

25. G. Náray-Szabo, M. Fuxreiter and A. Warshel, *Computational approaches to biochemical reactivity*, G. Náray-Szabo and A. Warshel, eds., Kluwer Academic Publishers, Holland, 237–293 (1997).
26. M. F. Perutz, *Proc. R. Soc.* **B167**, 448 (1967).
27. C. A. Vernon, *Proc. R. Soc.* **B167**, 389 (1967).
28. A. Warshel, *Proc. Natl. Acad. Sci. USA* **75**, 5250 (1978).
29. A. Warshel, *Biochemistry* **20**, 3167 (1981).
30. W. G. J. Hol, L. M. Halie, and C. Sander, *Nature, Lond.* **294**, 532 (1981).
31. N. Strater, W. N. Lipscomb, T. Klabunde, and B. Krebs. *Angewandte Chemie* (*International Edition*) **35**, 2024 (1996).
32. D. A. Buckingham, C. E. Davis, D. M. Foster, and A. M. Sargeson, *J. Am. Chem. Soc.* **92**, 5571 (1970).
33. D. A. Buckingham, J. MacB. Harrowfield, and A. M. Sargeson, *J. Am. Chem. Soc.* **96**, 1726 (1974).
34. E. Chaffee, T. P. Dasgupta, and G. M. Harris, *J. Am. Chem. Soc.* **95**, 4169 (1973).
35. D. A. Palmer and G. M. Harris, *Inorg. Chem.* **13**, 965 (1974).
36. D. A. Buckingham and L. M. Engelhardt, *J. Am. Chem. Soc.* **97**, 5915 (1975).
37. J. E. Coleman, *Prog. Bioorg. Chem.* **1**, 159 (1971), and references therein.
38. P. Woolley, *Nature, Lond.* **258**, 677 (1975).
39. K. K. Kannan, M. Petef, K. Fridborg, H. Cid-Dresdener, and S. Lovgren, *FEBS Lett.* **73**, 115 (1977).
40. D. A. Buckingham, F. R. Keene, and A. M. Sargeson, *J. Am. Chem. Soc.* **96**, 4981 (1974).
41. T. H. Finlay, J. Valinsky, K. Sato, and R. H. Abeles, *J. Biol. Chem.* **247**, 4197 (1974).
42. B. T. Golding and L. Radom, *J. Am. Chem. Soc.* **98**, 6331 (1976).
43. G. A. Hamilton and F. H. Westheimer, *J. Am. Chem. Soc.* **81**, 6332 (1959).
44. S. G. Warren, B. Zerner, and F. H. Westheimer, *Biochemistry* **5**, 817 (1966).
45. A. R. Fersht and W. P. Jencks, *J. Am. Chem. Soc.* **92**, 5442 (1970).
46. A. R. Fersht, *J. Am. Chem. Soc.* **93**, 3504 (1971).
47. R. G. Pearson, *J. Chem. Educ.* **45**, 103, 643 (1968).
48. W. P. Jencks and J. Carriuolo, *J. Am. Chem. Soc.* **82**, 1778 (1960).
49. W. P. Jencks and M. Gilchrist, *J. Am. Chem. Soc.* **90**, 2622 (1968).
50. B. A. Connolly and F. Eckstein, *J. Biol. Chem.* **256**, 9450 (1981).
51. S. L. Huang and M. D. Tsai, *Biochemistry* **21**, 1530 (1982).
52. J. S. Thomsen, *Phys. Rev.* **91**, 1263 (1953).
53. R. M. Krupka, H. Kaplan, and K. J. Laidler, *Trans. Faraday Soc.* **62**, 2754 (1966).
54. C. G. Swain, E. C. Stivers, J. F. Reuwer, Jr., and L. J. Staod, *J. Am. Chem. Soc.* **80**, 5885 (1958).
55. W. H. Saunders Jr., *J. Am Chem. Soc.* **107**, 164 (1985).
56. Y. Cha, C. J. Murray, and J. P. Klinman, *Science* **243**, 1325 (1989).
57. B. J. Bahnson and J. P. Klinman. *Methods In Enzymology* **249**, 373 (1995).
58. B. J. Bahnson, T. D. Colby, J. K. Chin, B. M. Goldstein, and J. P. Klinman, *Proc. Natl Acad. Sci.* (*USA*) **94**, 12797 (1997).
59. W. A. Edens, J. L. Urbauer, and W. W. Cleland *Biochemistry* **36**, 1141 (1997).
60. L. E. H. Smith, L. H. Mohr, and M. A. Raftery, *J. Am. Chem. Soc.* **95**, 7497 (1973).
61. A. J. Kresge, *J. Am. Chem. Soc.* **95**, 3065 (1972).

The Basic Equations of Enzyme Kinetics

3

A. Steady state kinetics

The concept of the steady state is used widely in dynamic systems. It generally refers to the situation in which the value of a particular quantity is constant—is "in a steady state"—because its rate of formation is balanced by its rate of destruction. For example, the population of a country is in a steady state when the birth and immigration rates equal those of death and emigration. Similarly, the concentration of a metabolite in a cell is at a steady state level when it is being produced as rapidly as it is being degraded. In enzyme kinetics, the concept is applied to the concentrations of enzyme-bound intermediates. When an enzyme is mixed with a large excess of substrate, there is an initial period, known as the *pre–steady state*, during which the concentrations of these intermediates build up to their steady state levels. Once the intermediates reach their steady state concentrations, the reaction rate changes relatively slowly with time. It is during this steady state period that the rates of enzymatic reactions are traditionally measured. The steady state is an approximation, since the substrate is gradually depleted during the course of an experiment. But, provided that the rate measurements are restricted to a short time interval over which the concentration of the substrate does not greatly change, it is a very good approximation.

Although the use of pre–steady state kinetics is undoubtedly superior as a means of analyzing the chemical mechanisms of enzyme catalysis (Chapters 4 and 7), steady state kinetics is more important for the understanding of metabolism, since it measures the catalytic activity of an enzyme in the steady state conditions in the cell.

1. The experimental basis: The Michaelis-Menten equation[1]

In deriving the following kinetic expressions, we assume that the concentration of the enzyme is negligible compared with that of the substrate. Except in the procedures for the rapid reaction measurements described in the next chapter, this is generally true in practice because of the high catalytic efficiency of enzymes. We also assume that what is being measured is the *initial rate v* of formation of products (or depletion of substrates) — that is, the rate of formation of the first few percent of the products — so that they have not significantly accumulated and the substrates have not been appreciably depleted. Under these conditions, changes in reagent concentrations are generally linear with time.

It is found experimentally in most cases that v is directly proportional to the concentration of enzyme, $[E]_0$. However, v generally follows *saturation kinetics* with respect to the concentration of substrate, $[S]$, in the following way (Figure 3.1). At sufficiently low $[S]$, v increases linearly with $[S]$. But as $[S]$ is increased, this relationship begins to break down and v increases less rapidly than $[S]$ until, at sufficiently high or *saturating* $[S]$, v tends toward a *limiting* value termed V_{max}. This is expressed quantitatively in the Michaelis-Menten equation, the basic equation of enzyme kinetics:

$$v = \frac{[E]_0[S]k_{cat}}{K_M + [S]} \tag{3.1}$$

where

$$k_{cat}[E]_0 = V_{max} \tag{3.2}$$

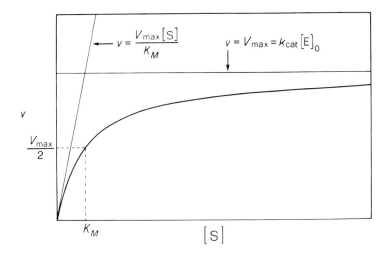

Figure 3.1 Reaction rate v plotted against substrate concentration $[S]$ for a reaction obeying Michaelis-Menten (or saturation) kinetics.

The concentration of substrate at which $v = \frac{1}{2}V_{max}$ is termed K_M, the Michaelis constant. Note that at low [S], where $[S] \ll K_M$,

$$v = \frac{k_{cat}}{K_M} [E]_0 [S] \tag{3.3}$$

2. Interpretation of the kinetic phenomena for single-substrate reactions: The Michaelis-Menten mechanism

In 1913, L. Michaelis and M. L. Menten developed the theories of earlier workers and proposed the following scheme:

$$E + S \underset{}{\overset{K_S}{\rightleftharpoons}} ES \xrightarrow{k_{cat}} E + P \tag{3.4}$$

The catalytic reaction is divided into two processes. The enzyme and the substrate first combine to give an enzyme–substrate complex, ES. This step is assumed to be rapid and reversible with no chemical changes taking place; the enzyme and the substrate are held together by noncovalent interactions. The chemical processes then occur in a second step with a first-order rate constant k_{cat} (the turnover number). The rate equations are solved in the following manner.

From equation 3.4,

$$\frac{[E][S]}{[ES]} = K_S \tag{3.5}$$

and

$$v = k_{cat} [ES] \tag{3.6}$$

Also, the total enzyme concentration, $[E]_0$, and that of the free enzyme, $[E]$, are related by

$$[E] = [E]_0 - [ES] \tag{3.7}$$

Thus

$$[ES] = \frac{[E]_0 [S]}{K_S + [S]} \tag{3.8}$$

and

$$v = \frac{[E]_0 [S] k_{cat}}{K_S + [S]} \tag{3.9}$$

This is identical to equation 3.1, where K_M is equal to the dissociation constant of the enzyme–substrate complex, K_S.

The concept of the enzyme–substrate complex is the foundation stone of enzyme kinetics and our understanding of the mechanism of enzyme catalysis. In honor of its introducer, this noncovalently bound complex is often termed the *Michaelis* complex.

There is, of course, an enzyme–product complex, EP, through which the reverse reaction proceeds. We assume in these analyses that the dissociation of EP is fast and so can be ignored in the forward reaction. The initial-rate assumption allows us to ignore the accumulation of the EP complex and the reverse reaction, since [P] is always very low.

3. Extensions and modifications of the Michaelis-Menten mechanism

A distinction must be drawn between the equation and the mechanism proposed by Michaelis and Menten. Their equation holds for many mechanisms, but their mechanism is not always appropriate.

The Michaelis-Menten mechanism assumes that the enzyme–substrate complex is in thermodynamic equilibrium with free enzyme and substrate. This is true only if, in the following scheme, $k_2 \ll k_{-1}$:

$$\text{E} + \text{S} \underset{k_{-1}}{\overset{k_1}{\rightleftharpoons}} \text{ES} \overset{k_2}{\longrightarrow} \text{E} + \text{P} \tag{3.10}$$

The case in which k_2 is comparable to k_{-1} was first analyzed by G. E. Briggs and J. B. S. Haldane in 1925.

a. Briggs-Haldane kinetics:[2] $K_M > K_S$

The solution of scheme 3.10 is somewhat more complicated than the solution of the Michaelis-Menten scheme: the steady state approximation is applied to the concentration of ES. That is, if the reaction rate measured is approximately constant over the time interval concerned, then [ES] is also constant:

$$\frac{d[\text{ES}]}{dt} = 0 = k_1[\text{E}][\text{S}] - k_2[\text{ES}] - k_{-1}[\text{ES}] \tag{3.11}$$

Substituting equation 3.7 gives

$$[\text{ES}] = \frac{[\text{E}]_0[\text{S}]}{[\text{S}] + (k_2 + k_{-1})/k_1} \tag{3.12}$$

and, since $v = k_2[\text{ES}]$,

$$v = \frac{[\text{E}]_0[\text{S}]k_2}{[\text{S}] + (k_2 + k_{-1})/k_1} \tag{3.13}$$

This is identical to the Michaelis-Menten equation (3.1), where now

$$K_M = \frac{k_2 + k_{-1}}{k_1} \tag{3.14}$$

Since K_S for the dissociation of [ES] is equal to k_{-1}/k_1, we have

$$K_M = K_S + \frac{k_2}{k_1} \tag{3.15}$$

Of course, when $k_{-1} \gg k_2$, equation 3.14 simplifies to $K_M = K_S$.

b.   Intermediates occurring after ES: $K_M < K_S$

The Michaelis-Menten scheme may be extended to cover a variety of cases in which additional intermediates, covalently or noncovalently bound, occur on the reaction pathway. It is found in all examples that the Michaelis-Menten equation still applies, although K_M and k_{cat} are now combinations of various rate and equilibrium constants. K_M is always less than or equal to K_S in these cases. Suppose that, as for example in the following scheme, there are several intermediates and the final catalytic step is slow:

$$E + S \underset{K_S}{\rightleftharpoons} ES \underset{K}{\rightleftharpoons} ES' \underset{K'}{\rightleftharpoons} ES'' \underset{slow}{\xrightarrow{k_4}} E + P \tag{3.16}$$

where $[ES'] = K[ES]$ and $[ES''] = K'[ES']$. Then

$$K_M = \frac{K_S}{1 + K + KK'} \tag{3.17}$$

and

$$k_{cat} = \frac{k_4 KK'}{1 + K + KK'} \tag{3.18}$$

The chymotrypsin-catalyzed hydrolysis of esters and amides proceeds through scheme 3.19:

$$E + S \underset{K_S}{\rightleftharpoons} ES \underset{P_1}{\xrightarrow{k_2}} EAc \xrightarrow{k_3} E \tag{3.19}$$

where EAc is an "acylenzyme." Applying the steady state assumption to [EAc], it may be shown that

$$v = [E]_0[S] \left\{ \frac{k_2 k_3/(k_2 + k_3)}{K_S k_3/(k_2 + k_3) + [S]} \right\} \tag{3.20}$$

This is a Michaelis-Menten equation in which

$$K_M = K_S \frac{k_3}{k_2 + k_3} \tag{3.21}$$

and

$$k_{cat} = \frac{k_2 k_3}{k_2 + k_3} \tag{3.22}$$

or, alternatively,

$$\frac{1}{k_{cat}} = \frac{1}{k_2} + \frac{1}{k_3} \tag{3.23}$$

(The physical basis of this reciprocal relationship is discussed in section K.) In more complex reactions, the generalization that $K_M < K_S$ can break down.

c.   All three mechanisms occur in practice

In the Briggs-Maldane mechanism, when k_2 is much greater than k_{-1}, k_{cat}/K_M is equal to k_1, the rate constant for the association of enzyme and substrate. It is shown in Chapter 4 that association rate constants should be on the order of $10^8 \ s^{-1} \ M^{-1}$. This leads to a diagnostic test for the Briggs-Haldane mechanism: the value of k_{cat}/K_M is about 10^7 to $10^8 \ s^{-1} \ M^{-1}$. Catalase, acetylcholinesterase, carbonic anhydrase, crotonase, fumarase, and triosephosphate isomerase all exhibit Briggs-Haldane kinetics by this criterion (see Chapter 4, Table 4.4).

It is extremely common for intermediates to occur after the initial enzyme–substrate complex, as in equation 3.19. However, it is often found for physiological substrates that these intermediates do not accumulate and that the slow step in equation 3.19 is k_2. (A theoretical reason for this is discussed in Chapter 12, where examples are given.) Under these conditions, K_M is equal to K_S, the dissociation constant, and the original Michaelis-Menten mechanism is obeyed to all intents and purposes. The opposite occurs in many laboratory experiments. The enzyme kineticist often uses synthetic, highly reactive substrates to assay enzymes, and covalent intermediates frequently accumulate.

B.   The significance of the Michaelis–Menten parameters

1.   The meaning of k_{cat}: The catalytic constant

In the simple Michaelis-Menten mechanism in which there is only one enzyme–substrate complex and all binding steps are fast, k_{cat} is simply the first-order rate constant for the chemical conversion of the ES complex to the EP complex. For more complicated reactions, k_{cat} is a function of all the first-order rate constants, and it cannot be assigned to any particular process except when simplifying features occur. For example, in the Briggs-Haldane mechanism, when the dissociation of the EP complex is fast, k_{cat} is equal to k_2 (equation 3.10). But if dissociation of the EP complex is slow, the rate constant for this process contributes to k_{cat}, and, in the extreme case in which EP dissociation is far slower than the chemical steps, k_{cat} will be equal to the dissociation rate constant. In the example of equation 3.19, k_{cat} was seen to be a function of k_2 and k_3. But if one of these constants is much smaller than the other, it becomes equal to k_{cat}. For example, if $k_3 \ll k_2$, then, from equation 3.22, $k_{cat} = k_3$. An extension of this is that k_{cat} cannot be greater than any first-order rate constant on the forward reaction pathway.[3] It thus sets a *lower* limit on the chemical rate constants.

The constant k_{cat} is often called the turnover number of the enzyme because it represents the maximum number of substrate molecules converted to products per active site per unit time, or the number of times the enzyme "turns over" per unit time.

Rule: The k_{cat} is a first-order rate constant that refers to the properties and reactions of the enzyme–substrate, enzyme–intermediate, and enzyme–product complexes.

2. The meaning of K_M: Real and apparent equilibrium constants

Although it is only for the simple Michaelis-Menten mechanism or in similar cases that $K_M = K_S$, the true dissociation constant of the enzyme–substrate complex, K_M may be treated for *some* purposes as an *apparent* dissociation constant. For example, the concentration of free enzyme in solution may be calculated from the relationship

$$\frac{[E][S]}{\Sigma [ES]} = K_M \tag{3.24}$$

where $\Sigma [ES]$ is the sum of *all* the bound enzyme species.*

---

* It is possible to devise kinetic curiosities that give Michaelis-Menten kinetics without the enzyme being saturated with the substrate. For example, in the following scheme—where the active form of the enzyme reacts with the substrate in a second-order reaction to give the products and an inactive form of the enzyme, E', which slowly reverts to the active form—apparent saturation kinetics are followed with $k_{cat} = k_2$ and $K_M = k_2/k_1$. Equation 3.24 applies to this example if E' is treated as a "bound" form of the enzyme:

$$E + S \xrightarrow{k_1} E' + P$$
$$\downarrow k_2$$
$$E$$

The concept of *apparent* values is very useful, and it appears in other phenomena, such as in pK_a values (section L3). Quite often, a pK_a value does not represent the microscopic ionization of a particular group but is a combination of this value and various equilibrium constants between different conformational states of the molecule. The result is an *apparent* pK_a which may be handled titrimetrically as a simple pK_a. This simple-minded approach must not be taken too far, and, when one is considering the effects of temperature, pH, etc., on an apparent K_M, one must realize that the rate-constant components of this term are also affected. The same applies to k_{cat} values. The literature contains examples in which breaks in the temperature dependence of k_{cat} have been interpreted as indicative of conformational changes in the enzyme, when, in fact, they are due to a different temperature dependence of the individual rate constants in k_{cat}, e.g., k_2 and k_3 in equation 3.22.

An illustration of the way that K_M is a measure of the amount of enzyme that is bound in any form whatsoever to the substrate is given by the following mechanism (cf. the chymotrypsin mechanism, equation 3.19):

$$E + S \underset{}{\overset{K_S}{\rightleftharpoons}} ES \xrightarrow{k_2} ES' \xrightarrow{k_3} E + P \tag{3.25}$$

Application of the steady state approximation to [ES'] gives

$$[ES'] = [ES]\frac{k_2}{k_3} \tag{3.26}$$

When k_2 is much greater than k_3, [ES'] is much greater than [ES], so that ES' makes a more important contribution to K_M than does ES and is the predominant enzyme-bound species. Without solving the equations for the reaction, we can say intuitively that K_M must be smaller than K_S by a factor of about k_3/k_2; i.e.,

$$K_M \approx K_S \frac{k_3}{k_2} \tag{3.27}$$

In all cases, K_M is the substrate concentration at which $v = V_{max}/2$.

Rule: The K_M is an apparent dissociation constant that may be treated as the overall dissociation constant of all enzyme-bound species.

3. The meaning of k_{cat}/K_M: The specificity constant

It was pointed out earlier that the reaction rate for low substrate concentrations is given by $v = (k_{cat}/K_M)[E]_0[S]$ (equation 3.3); that is, k_{cat}/K_M is an *apparent* second-order rate constant. It is not a true microscopic rate constant except in the extreme case in which the rate-determining step in the reaction is the encounter of enzyme and substrate.

The importance of k_{cat}/K_M is that it relates the reaction rate to the concentration of free, rather than total, enzyme. This is readily seen from equation 3.3, mentioned above, since at low substrate concentrations the enzyme is largely unbound and $[E] \approx [E]_0$. At such concentrations, the reaction rate is thus given by

$$v = [E][S]\frac{k_{cat}}{K_M}$$

It will be shown later (equation 3.41) that this result holds at *any* substrate concentration. It will also be shown later (equation 3.44) that k_{cat}/K_M determines the specificity for competing substrates. For this reason, k_{cat}/K_M is sometimes referred to as the "specificity constant."

The value of k_{cat}/K_M cannot be greater than that of any second-order rate constant on the forward reaction pathway.[3] It thus sets a *lower* limit on the rate constant for the association of enzyme and substrate.

Rule: The k_{cat}/K_M is an apparent second-order rate constant that refers to the properties and the reactions of the free enzyme and free substrate.

C. Graphical representation of data

It is very useful to transform the Michaelis-Menten equation into a linear form for analyzing data graphically and detecting deviations from the ideal behavior. One of the best known methods is the double-reciprocal or Lineweaver-Burk plot. Inverting both sides of equation 3.1 and substituting equation 3.2 gives the Lineweaver-Burk plot:[4]

$$\frac{1}{v} = \frac{1}{V_{max}} + \frac{K_M}{V_{max}[S]} \tag{3.28}$$

Plotting $1/v$ against $1/[S]$ (Figure 3.2) gives an intercept of $1/V_{max}$ on the y axis

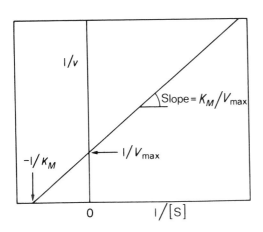

Figure 3.2 The Lineweaver-Burk plot.

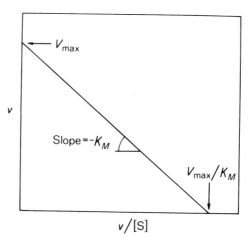

Figure 3.3 The Eadie-Hofstee plot.

as $1/[S]$ tends toward zero, and of $1/[S] = -1/K_M$ on the x axis. The slope of the line is K_M/V_{max}.

Another common plot is that of G. S. Eadie and B. H. J. Hofstee (equation 3.29).[5,6] Equations 3.1 and 3.2 may be rearranged to give

$$v = V_{max} - \frac{K_M v}{[S]} \tag{3.29}$$

Plotting v against $v/[S]$ (Figure 3.3) gives an intercept of V_{max} on the y axis as $v/[S]$ tends toward zero. The slope of the line is equal to $-K_M$. The intercept on the x axis is at $v/[S] = V_{max}/K_M$.

The Lineweaver-Burk plot has the disadvantage of compressing the data points at high substrate concentrations into a small region and emphasizing the points at lower concentrations. It does have the advantage that the values of v for a given value of $[S]$ are easy to read from it.

The Eadie plot does not compress the higher values, but the values of v against $[S]$ are more difficult to determine rapidly from it. The Eadie plot is considered more accurate and generally superior.[7,8]

D. Inhibition

As well as being irreversibly inactivated by heat or chemical reagents, enzymes may be *reversibly* inhibited by the noncovalent binding of inhibitors. There are four main types of inhibition.

1. Competitive inhibition

If an inhibitor I binds reversibly to the active site of the enzyme and prevents S binding and *vice versa*, I and S compete for the active site and I is said to be a *competitive* inhibitor. In the case of the simple Michaelis-Menten mechanism (equation 3.4, where $K_M = K_S$), an additional equilibrium must be considered, i.e.,

$$\begin{array}{c} E \underset{}{\overset{S, K_M}{\rightleftharpoons}} ES \xrightarrow{k_{cat}} E + P \\ {\scriptstyle I, K_I} \updownarrow \\ EI \end{array} \qquad (3.30)$$

where $K_I = [E][I]/[EI]$.

Solving the equilibrium and rate equations by using

$$[E]_0 = [ES] + [EI] + [E] \qquad (3.31)$$

gives

$$v = \frac{[E]_0[S]k_{cat}}{[S] + K_M(1 + [I]/K_I)} \qquad (3.32)$$

K_M is apparently increased by a factor of $(1 + [I]/K_I)$. This equation holds for all mechanisms obeying the Michaelis-Menten equation. Competitive inhibition affects K_M only and not V_{max}, since infinitely high concentrations of S displace I from the enzyme.

2. Noncompetitive, uncompetitive, and mixed inhibition

Different inhibition patterns occur if I and S bind simultaneously to the enzyme instead of competing for the same binding site (Figure 3.4):

$$\begin{array}{ccc} E & \underset{}{\overset{S, K'_M}{\rightleftharpoons}} & ES \xrightarrow{k_{cat}} \\ {\scriptstyle I, K_I} \updownarrow & & \updownarrow {\scriptstyle I, K'_I} \\ EI & \underset{}{\overset{S, K_M}{\rightleftharpoons}} & ESI \xrightarrow{k'} \end{array} \qquad (3.33)$$

It may be shown from the Michaelis-Menten mechanism—in the simplifying case in which the dissociation constant of S from EIS is the same as that from ES (i.e., $K_M = K'_M$) but in which ESI does not react (i.e., $k' = 0$)—that

$$v = \frac{[E]_0[S]k_{cat}/(1 + [I]/K_I)}{[S] + K_M} \qquad (3.34)$$

This is termed *noncompetitive inhibition:* K_M is unaffected, but k_{cat} is lowered by

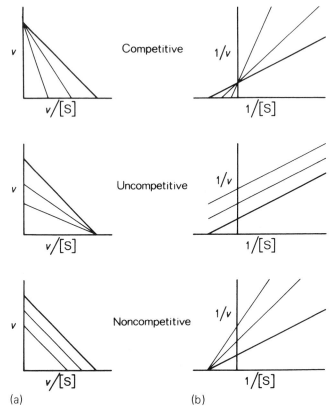

Figure 3.4 (a) Eadie-Hofstee and (b) Lineweaver-Burk plots of different types of inhibition. The heavier line in each plot shows the reaction in the absence of inhibitor; the lighter lines are for the reaction in the presence of inhibitor.

a factor of $(1 + [I]/K_1)$. More commonly, the dissociation constant of S from EIS is different than that from ES. In this case, both K_M and k_{cat} are altered and the inhibition is termed *mixed*. A further type of inhibition, *uncompetitive*, occurs when I binds to ES but not to E (Figure 3.5).

E. Nonproductive binding

In some reactions, a substrate binds in an alternative unreactive mode at the active site of the enzyme, in competition with the productive mode of binding:

$$
\begin{array}{c}
\quad \overset{k_2}{}\; ES \xrightarrow{k_2} E + P \\
{}^{K_S}\nearrow \\
E \\
{}_{K'_S}\searrow \\
\quad ES'
\end{array}
\qquad (3.35)
$$

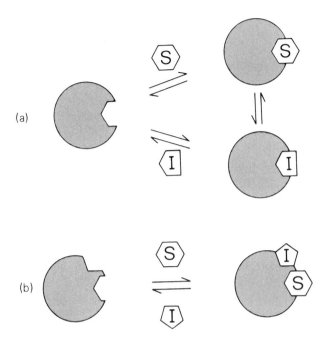

Figure 3.5 (a) Competitive inhibition: inhibitor and substrate compete for the same binding site. For example, indole, phenol, and benzene bind in the binding pocket of chymotrypsin and inhibit the hydrolysis of derivatives of tryptophan, tyrosine, and phenylalanine. (b) Noncompetitive inhibition: inhibitor and substrate bind simultaneously to the enzyme. An example is the inhibition of fructose 1,6-diphosphatase by AMP. This type of inhibition is very common with multisubstrate enzymes. A rare example of uncompetitive inhibition of a single-substrate enzyme is the inhibition of alkaline phosphatase by L-phenylalanine. This enzyme is composed of two identical subunits, so presumably the phenylalanine binds at one site and the substrate at the other. [From N. K. Ghosh and W. H. Fishman, *J. Biol. Chem.* **241**, 2516 (1966); see also M. Caswell and M. Caplow, *Biochemistry* **19**, 2907 (1980).]

This is known as nonproductive binding. The effect of such binding on the Michaelis-Menten mechanism is to lower both the k_{cat} and the K_M. The k_{cat} is lowered since, at saturation, only a fraction of the substrate is bound productively. The K_M is lower than the K_S because the existence of additional binding modes must lead to apparently tighter binding.

Solving equation 3.35 by the usual procedures gives

$$v = \frac{[E]_0[S]k_2}{K_S + [S](1 + K_S/K_S')} \tag{3.36}$$

Comparison with the Michaelis-Menten equation (3.1) gives

$$k_{cat} = \frac{k_2}{1 + K_S/K_S'} \tag{3.37}$$

and

$$K_M = \frac{K_S}{1 + K_S/K_S'}$$ (3.38)

It should be noted that

$$\frac{k_{cat}}{K_M} = \frac{k_2}{K_S}$$ (3.39)

That is, k_{cat}/K_M is unaffected by the presence of the additional binding mode, since k_{cat} and K_M are altered in a compensating manner. For example, if the non-productive site binds a thousand times more strongly than the productive site, K_M will be a thousand times lower than K_S, but since only one molecule in a thousand is productively bound, k_{cat} is a thousand times lower than k_2 (strictly speaking, the factor is 1001).

F. $k_{cat}/K_M = k_2/K_S$

The result of equation 3.39 for nonproductive binding is quite general. It applies to cases in which intermediates occur on the reaction pathway as well as in the nonproductive modes. For example, in equation 3.19 for the action of chymotrypsin on esters with accumulation of an acylenzyme, it is seen from the ratios of equations 3.21 and 3.22 that $k_{cat}/K_M = k_2/K_S$. This relationship clearly breaks down for the Briggs-Haldane mechanism in which the enzyme–substrate complex is not in thermodynamic equilibrium with the free enzyme and substrates. It should be borne in mind that K_M might be a complex function when there are several enzyme-bound intermediates in rapid equilibrium, as in equation 3.16. Here k_{cat}/K_M is a function of all the bound species.

G. Competing substrates

1. An alternative formulation of the Michaelis-Menten equation

Suppose that two substrates compete for the active site of the enzyme:

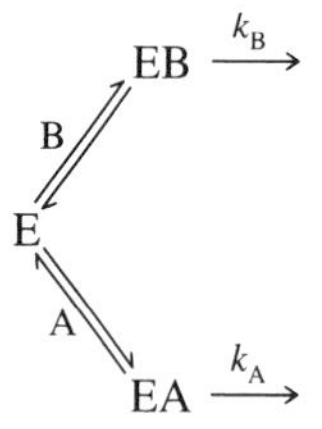

$$
\begin{array}{c}
\quad\quad EB \xrightarrow{k_B} \\
B \nearrow\!\!\!\nwarrow \\
E \\
A \searrow\!\!\!\swarrow \\
\quad\quad EA \xrightarrow{k_A}
\end{array}
$$ (3.40)

The reaction rates may be calculated by the usual steady state or Michaelis-

Menten assumptions. However, there is an alternative approach for rapidly calculating the ratio of the reaction rates. Substitution of equation 3.24 into the Michaelis-Menten equation (3.1) gives

$$v = \frac{k_{cat}}{K_M} [E][S] \tag{3.41}$$

where [E] is the concentration of free or unbound enzyme. This is a useful equation since it is based on [E] rather than $[E]_0$. Several important relationships may be inferred directly from this equation without the need for a detailed mechanistic analysis, as shown below.

2. Specificity for competing substrates

If two substrates A and B compete for the enzyme, then

$$-\frac{d[A]}{dt} = v_A = \left(\frac{k_{cat}}{K_M}\right)_A [E][A] \tag{3.42}$$

and

$$-\frac{d[B]}{dt} = v_B = \left(\frac{k_{cat}}{K_M}\right)_B [E][B] \tag{3.43}$$

which give

$$\frac{v_A}{v_B} = \frac{(k_{cat}/K_M)_A[A]}{(k_{cat}/K_M)_B[B]} \tag{3.44}$$

The important conclusion is that specificity, in the sense of discrimination between two competing substrates, is determined by the ratios of k_{cat}/K_M and not by K_M alone. Since k_{cat}/K_M is unaffected by nonproductive binding (section E) and by the accumulation of intermediates (section F), these phenomena do not affect specificity (see Chapter 13). Note that equation 3.44 holds at all concentrations of substrates.

H. Reversibility: The Haldane equation

1. Equilibria in solution

$$S \underset{k_r}{\overset{k_f}{\rightleftharpoons}} P \tag{3.45}$$

$$K_{eq} = \frac{[P]}{[S]} = \frac{k_f}{k_r} \tag{3.46}$$

An enzyme cannot alter the equilibrium constant between the free-solution concentrations of S and P. This places constraints on the relative values of k_{cat}/K_M for the forward and reverse reactions. Specifically, since the rates of formation of P and S are equal at equilibrium, application of equation 3.44 gives

$$\left(\frac{k_{cat}}{K_M}\right)_S [E][S] = \left(\frac{k_{cat}}{K_M}\right)_P [E][P] \tag{3.47}$$

so that

$$\frac{(k_{cat}/K_M)_S}{(k_{cat}/K_M)_P} = K_{eq} \tag{3.48}$$

This relationship is known as the Haldane equation, after J. B. S. Haldane.[9]

2. Equilibria on the enzyme surface (internal equilibria)

The Haldane equation does not relate the equilibrium constant between ES and EP to that between S and P in solution. The equilibrium constant for the enzyme-bound reagents is often very different from that in solution for several reasons:

1. *Strain.* The geometry of the active site may be such that, for example, P is bound more tightly than S. The equilibrium on the enzyme surface will favor P more than the equilibrium in solution.
2. *Nonproductive binding.* If the enzyme has binding modes for S other than the catalytically productive mode, these will favor [ES] in the equilibrium.
3. *Entropy.* In the case where the product is two separate molecules,

$$S \rightleftharpoons P + P' \tag{3.49}$$

the equilibrium constant in solution has a term reflecting the favorable gain in entropy on formation of two molecules from one. However, if both P and P' are bound on the enzyme surface,

$$ES \rightleftharpoons EPP' \tag{3.50}$$

the relevant equilibrium constant will not have this entropic contribution.

The tyrosyl-tRNA synthetase alters the equilibrium constant for the formation of tyrosyl-adenylate by a factor of 10^7 by a "strain" mechanism[10] (Chapter 15, section I).

I. Breakdown of the Michaelis–Menten equation

Apart from essentially trivial reasons such as an experimental inability to measure initial rates, there are two main reasons for the failure of the Michaelis-Menten equation.

The first possibility is substrate inhibition. A second molecule of substrate binds to give an ES_2 complex that is catalytically inactive. If, in a simple Michaelis-Menten mechanism, the second dissociation constant is K'_S, then

$$v = \frac{[E]_0[S]k_{cat}}{K_S + [S] + [S]^2/K'_S} \tag{3.51}$$

At low concentrations of S, the rate is given by $v = [E]_0[S]k_{cat}/K_S$, as usual. But as [S] increases, there is first a maximum value of v followed by a decrease.

The second possibility is substrate activation: an ES_2 complex is formed that is more active than ES.

J. Multisubstrate systems

We have dealt so far with enzymes that react with a single substrate only. The majority of enzymes, however, involve two substrates. The dehydrogenases, for example, bind both NAD^+ and the substrate that is to be oxidized. Many of the principles developed for the single-substrate systems may be extended to multisubstrate systems. However, the general solution of the equations for such systems is complicated and well beyond the scope of this book. Many books devoted almost solely to the detailed analysis of the steady state kinetics of multisubstrate systems have been published, and the reader is referred to these for advanced study.[11-14] The excellent short accounts by W. W. Cleland[15] and K. Dalziel[16] are highly recommended.

From the point of view of this book, the most important experimental observation is that most reactions obey Michaelis-Menten kinetics when the concentration of one substrate is held constant and the other is varied. Furthermore, in practice, only a limited range of mechanisms is commonly observed. In this section we shall just list some common pathways and give a glossary of terms.

Reactions in which all the substrates bind to the enzyme before the first product is formed are called *sequential*. Reactions in which one or more products are released before all the substrates are added are called *ping-pong*. Sequential mechanisms are called *ordered* if the substrates combine with the enzyme and the products dissociate in an obligatory order. A *random* mechanism implies no obligatory order of combination or release. The term *rapid equilibrium* is applied when the chemical steps are slower than those for the binding of reagents. Some examples follow.

1. The random sequential mechanism

$$(3.52)$$

The complex EAB is called a *ternary* or *central* complex.

2. The ordered mechanism

$$E \xrightleftharpoons{A} EA \xrightleftharpoons{B} EAB \longrightarrow \qquad (3.53)$$

Ordered mechanisms often occur in the reactions of the NAD^+-linked dehydrogenases, with the coenzyme binding first. The molecular explanation for this is that the binding of the dinucleotide causes a conformational change that increases the affinity of the enzyme for the other substrate (see Chapter 16).

3. The Theorell-Chance mechanism

The Theorell-Chance mechanism is an ordered mechanism in which the ternary complex does not accumulate under the reaction conditions, as is found for horse liver alcohol dehydrogenase:

$$E \xrightleftharpoons{A} EA \xrightarrow{B \quad P} EQ \qquad (3.54)$$

(P is one product and Q the other—acetaldehyde and NADH, respectively, for the liver alcohol dehydrogenase.)

4. The ping-pong (or substituted-enzyme or double-displacement) mechanism

The following type of reaction, in which the enzyme reacts with one substrate to give a covalently modified enzyme and releases one product, and then reacts with the second substrate, gives rise to the characteristic family of Lineweaver-Burk plots illustrated in Figure 3.6:

$$E + A \rightleftharpoons E \cdot A \rightleftharpoons E-P + Q \qquad (3.55)$$

$$E-P + B \rightleftharpoons E-P \cdot B \longrightarrow E + P-B \qquad (3.56)$$

A ping-pong reaction occurs, for example, when a phosphate-transferring enzyme, such as phosphoglycerate mutase, is phosphorylated by one substrate to form a phosphorylenzyme (E—P in equation 3.55), which then transfers the

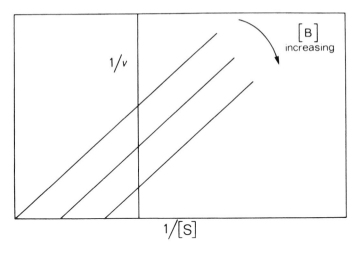

Figure 3.6 The characteristic parallel-line reciprocal plots of ping-pong kinetics. As the concentration of the second substrate in the sequence increases, V_{max} increases, as does the K_M for the first substrate. V_{max}/K_M, the reciprocal of the slope of the plot, remains constant (see equation 3.68).

phosphoryl group to a second substrate (equation 3.56). In the case of phospho-glycerate mutase (equation 3.57), N is the imidazole side chain of a histidine

$$E\text{—}N + ROPO_3^{2-} \rightleftharpoons E\text{—}N{\cdot}ROPO_3^{2-} \underset{ROH}{\searrow} E\text{—}N\text{—}PO_3^{2-} \overset{R'OH}{\searrow}$$

$$E\text{—}N + R'OPO_3^{2-} \quad (3.57)$$

residue. Another example occurs in the transfer of an acyl group from acetyl–coenzyme A to sulfanilimide or to another amine in a reaction catalyzed by acetyl–coenzyme A : arylamine acetyltransferase (equation 3.58). This reac-

$$E\text{—}SH + CH_3COS\text{-}CoA \rightleftharpoons E\text{—}SH{\cdot}CH_3COS\text{-}CoA \underset{HS\text{-}CoA}{\searrow}$$

$$E\text{—}S\text{—}COCH_3 \overset{RNH_2}{\searrow} E\text{—}SH + RNHCOCH_3 \quad (3.58)$$

tion almost certainly involves the formation of an acylthioenzyme in which the —SH of a cysteine residue is acylated.[17] The rate equation for reactions 3.57 and 3.58 is given later (3.66):

In many ways, ping-pong kinetics is the most mechanistically informative of all the types of steady state kinetics, since information is given about the occur-rence of a covalent intermediate. The finding of ping-pong kinetics is often used as evidence for such an intermediate, but because other kinetic pathways can give rise to the characteristic parallel double-reciprocal plots of Figure 3.6, the evidence must always be treated with caution and confirming data should be sought.

Steady state kinetics may be used to distinguish between the various mechanisms mentioned above. Under the appropriate conditions, their application can determine the order of addition of substrates and the order of release of products from the enzyme during the reaction. For this reason, the term "mechanism" when used in steady state kinetics often refers just to the sequence of substrate addition and product release.

K.   Useful kinetic shortcuts

We end steady state kinetics with two ideas that should provide insight into why many rate equations have their particular mathematical forms. These ideas point to useful short cuts for quickly noting the effects of additional intermediates on mechanisms, and even for solving certain complicated mechanisms by inspection instead of by analyzing the full steady state rate equations.

1.   Calculation of net rate constants[18]

It is possible to reduce the rate constants for a series of reactions as in equation 3.59

$$A \underset{k_{-1}}{\overset{k_1}{\rightleftharpoons}} B \underset{k_{-2}}{\overset{k_2}{\rightleftharpoons}} C \underset{k_{-3}}{\overset{k_3}{\rightleftharpoons}} D \underset{k_{-4}}{\overset{k_4}{\rightleftharpoons}} E \overset{k_5}{\longrightarrow} F \tag{3.59}$$

to a single net rate constant, or to a series of single rate constants, by just considering a net rate constant for the flux going through each step. To illustrate this, we consider the simpler reaction

$$X \underset{k_{-1}}{\overset{k_1}{\rightleftharpoons}} Y \overset{k_2}{\longrightarrow} Z \tag{3.60}$$

The rate of X going to Z via Y is given by the rate of X going to Y($= k_1[X]$) times the probability of Y going to Z rather than reverting to X [i.e., $k_2/(k_{-1} + k_2)$]. The net rate constant for X $\rightarrow$ Y, k_1', is thus given by

$$k_1' = \frac{k_1 k_2}{k_{-1} + k_2} \tag{3.61}$$

The same treatment can be applied to equation 3.59, starting from the irreversible step on the right-hand side and progressively working to the left. For example, the net rate constant for D $\rightarrow$ E, k_4', equals $k_4 k_5/(k_{-4} + k_5)$, as in equation 3.61. The net rate constant for C $\rightarrow$ D, k_3', is calculated by analogy with equation 3.61 to be $k_3 k_4'/(k_{-3} + k_4')$. This is continued sequentially to give eventually the net rate constant for A $\rightarrow$ B, i.e., $k_1' = k_1 k_2'/(k_{-1} + k_2')$.

This procedure provides a very simple means of calculating partitioning ratios where there is a branch point in a pathway. For example, suppose that in equation 3.59, A may also give directly a product P with rate constant k_P as well as giving F via B, C, D, and E:

$$(3.62)$$

The partitioning is simply calculated by using the net rate constant k_1' in equation 3.62. The rate of formation of F relative to that of P is thus equal to k_1'/k_P. Cleland has shown how the net rate constant method may be applied to more complex problems.[18]

2. Use of transit times instead of rate constants

The value of V_{max}, k_{cat}, or any rate constant for a series of sequential reactions may be derived by considering the *time* taken for each step, as follows. The dimensions of v are moles per second. The dimensions of $1/v$ are seconds per mole, and $1/v$ is the time taken for 1 mol of reagents to give products. Similarly, the reciprocal of k_{cat} (i.e., $[E]_0/V_{max}$) has the dimensions of seconds and is the time taken for one molecule of reagent to travel the whole reaction pathway in the steady state at saturating [S]. The reciprocal rate constant may be considered a *transit time*.

For a series of reactions as in equation 3.63, the reciprocal of the rate constant for any individual step is its transit time.

$$E{\cdot}P_1 \xrightarrow{k_1} E{\cdot}P_2 \xrightarrow{k_2} E{\cdot}P_3 \xrightarrow{k_3} E{\cdot}P_4 \xrightarrow{k_4} \cdots \xrightarrow{k_{n-1}} E{\cdot}P_n \qquad (3.63)$$

The total time for one molecule to be converted from P_1 to P_n, $1/k$, is given by the sum of the transit times for each step. That is,

$$\frac{1}{k} = \frac{1}{k_1} + \frac{1}{k_2} + \frac{1}{k_3} + \frac{1}{k_4} + \cdots + \frac{1}{k_{n-1}} \qquad (3.64)$$

where $1/k = 1/k_{cat} = [E]_0/V_{max}$ for saturating conditions. This is precisely the relationship derived earlier for k_{cat} in the acylenzyme mechanism for chymotrypsin (equation 3.23), and it gives the physical reason for the reciprocal relationship between k_{cat} and the first-order rate constants on the pathway: the reciprocals of the rate constants, i.e., the transit times, are additive, so that the time taken for a molecule to traverse the whole reaction pathway is the sum of the times taken for each step. For concentrations of S below saturating, the transit time for the whole reaction is $[E]_0/v$. (The transit times are analogous to resistances in electrical circuits, and may be summed in the same way: for a series of reactions, the residence times are additive; for parallel reaction pathways, the reciprocals of the residence times are additive.) This is the physical reason why the rate laws for steady state mechanisms are usually written in terms of $[E]_0/v$.

As an example, we consider the Briggs-Haldane mechanism:

$$E + A \underset{k_{-1}}{\overset{k_1}{\rightleftharpoons}} EA \xrightarrow{k_2} E + P \tag{3.65}$$

The binding step is reduced to the net rate constant

$$\frac{k_1[A]k_2}{k_{-1} + k_2}$$

The total transit time is given by

$$\frac{[E]_0}{v} = \frac{k_{-1} + k_2}{k_1[A]k_2} + \frac{1}{k_2} \tag{3.66}$$

Equation 3.66 is, in fact, the Lineweaver-Burk double-reciprocal plot (equation 3.28) in slight disguise: $[E]_0$ has been moved to the left-hand side, and from equation 3.14, $K_M = (k_{-1} + k_2)/k_1$.

The ping-pong mechanisms of equations 3.57 and 3.58, for example, may similarly be solved by inspection. Restating those equations as

$$E + A \underset{k_{-1}}{\overset{k_1}{\rightleftharpoons}} EA \xrightarrow{k_2} E{-}P \xrightarrow{k_3[B]} E + P{-}B \tag{3.67}$$

The transit times are summed to give

$$\frac{[E]_0}{v} = \frac{k_{-1} + k_2}{k_1[A]k_2} + \frac{1}{k_2} + \frac{1}{k_3[B]} \tag{3.68}$$

The total transit time is longer because of the third step. If the reciprocals of equation 3.68 are taken, a Michaelis-Menten-type equation is generated.

This method is very useful for solving the kinetics of polymerization reactions (e.g., DNA polymerization, Chapter 14) by inspection: the time taken to synthesize a polymer is the sum of the transit times for the addition of each monomer. For example, suppose that each step in equation 3.63 is a Michaelis-Menten process of

$$E{\cdot}P_x \underset{k_{-y}}{\overset{k_y[Y]}{\rightleftharpoons}} E{\cdot}P_x{\cdot}Y \xrightarrow{k_x} E{\cdot}P_{x+1} \tag{3.69}$$

Then, the step may be reduced to a single net rate constant k'_x, as above, given by

$$k'_x = \frac{k_y[Y]k_x}{k_{-y} + k_x} \tag{3.70}$$

and the transit times may be summed to give $[E]_0/v$.

L. Thermodynamic cycles

Thermodynamic functions such as Gibbs free energy G, enthalpy H, and entropy S, are functions of state. This means that they depend only on the state of the system being considered and not on how that system came into being. Changes in functions of state (e.g., ΔG, ΔH, and ΔS) between two states depend only on the initial and final states and not on the route between them. It is allowed, therefore, to set up notional schemes that contain hypothetical intermediates or ignore real ones when analyzing functions of state. *A corollary of this is that one cannot determine the order of events in a kinetic scheme from measurements at equilibrium.* Thermodynamic cycles provide some of the most useful manipulations in analyzing thermodynamics. These are easily understood from a series of examples.

1. Basic thermodynamic cycles

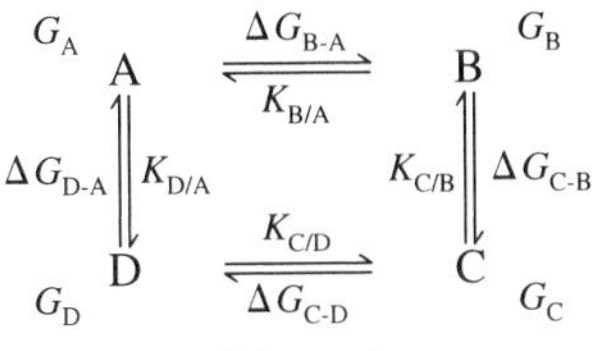

Scheme 1

Consider four species A, B, C, and D in equilibrium as in scheme 1. The individual equilibria are defined by $[B]/[A] = K_{B/A}$; $[C]/[B] = K_{C/B}$; $[C]/[D] = K_{C/D}$; and $[D]/[A] = K_{D/A}$. The difference in free energy between A and B, ΔG_{B-A} ($= G_B - G_A$) is related to $K_{B/A}$ by $\Delta G_{B-A} = -RT \ln K_{B/A}$; and so on.

The relationships among all the equilibrium constants and among all the free energies may be solved by algebra. For example, $[C]/[A] = ([B]/[A])([C]/[B])$, and thus $[C]/[A] = K_{B/A} \cdot K_{C/B}$. Also, $[C]/[A] = ([D]/[A])([C]/[D])$, and so $[C]/[A] = K_{D/A} \cdot K_{C/D}$. as well.

Thus,

$$K_{B/A} \cdot K_{C/B} = K_{D/A} \cdot K_{C/D} \tag{3.71}$$

And so,

$$\Delta G_{B-A} + \Delta G_{C-B} = \Delta G_{D-A} + \Delta G_{C-D} \tag{3.72}$$

The above procedure is simple, but it can be time-consuming for more complex examples. One can write down the correct relationships between the equilibrium constants or the free energies by inspection as follows. The overall equilibrium constant between, say, A and C must be the same for any route. Therefore, the products of the equilibrium constant for one route from A to C must be the same as for another route. And so, $K_{B/A} \cdot K_{C/B} = K_{D/A} \cdot K_{C/D}$ (i.e., equation 3.71). The difference in free energy between A and, say, D is similarly independent of route. Thus, $\Delta G_{C-A} = \Delta G_{B-A} + \Delta G_{C-B} = \Delta G_{D-A} + \Delta G_{C-D}$ (i.e., equation 3.72).

Two important corollaries from equation 3.71 that recur in many of the examples below are as follows.

1. When three of the four equilibrium constants are known, the fourth is automatically defined and so cannot vary independently. For example, if $K_{B/A}$, $K_{C/D}$, and $K_{D/A}$ are determined, then $K_{C/B} = (K_{D/A} \cdot K_{C/D})/K_{B/A}$.
2. The ratio of equilibrium constants of the vertical sides of the cycle is always equal to that of the constants of the horizontal sides. That is:

$$\frac{K_{B/A}}{K_{C/D}} = \frac{K_{D/A}}{K_{C/B}} \tag{3.73}$$

2. Two ligands or substrates binding to an enzyme

Suppose that two ligands A and B can bind to a protein E (scheme 2). The equilibria are defined by four equilibrium constants: the dissociation constant of A from E.A, $K_A = [E][A]/[E.A]$; of B from E.B, $K_B = [E][B]/[E.B]$; of A from E.A.B, $K'_A = [E.B][A]/[E.A.B]$; and B from E.A.B, $K'_B = [E.A][B]/[E.A.B]$.

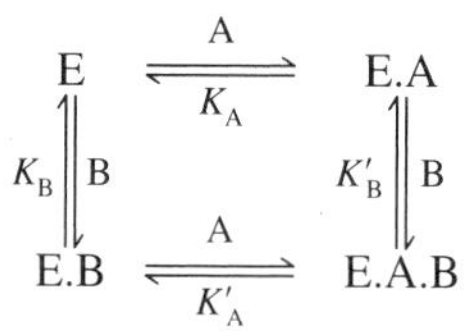

Scheme 2

Either from inspection or by solving the equations,

$$K_A \cdot K'_B = K'_A \cdot K_B \tag{3.74}$$

Note that if the binding of A is unaffected by the binding of B, so that $K_A = K'_A$, then K_B must equal K'_B. If the binding of B perturbs the binding of A such that $K_A \neq K'_A$, then from equation 3.74:

$$\frac{K_A}{K'_A} = \frac{K_B}{K'_B} \tag{3.75}$$

That is, the binding of B is perturbed by the binding of A by the same factor. There is a "coupling" energy, $\Delta\Delta G_{int}$ between the binding of the two that is defined by

$$\Delta\Delta G_{int} = -RT \ln \frac{K_A}{K'_A} = -RT \ln \frac{K_B}{K'_B} \tag{3.76}$$

A related example to scheme 2 is the binding of a substrate to an enzyme that has an ionizing group. One of the ligands in scheme 2 would be the substrate and the other would be the proton of the ionizing group. This is analyzed in Chapter 5, section B1. If the dissociation constant of the substrate varies with pH, then it is inevitable that the pK_a of the ionizing group is correspondingly perturbed.

Interesting binding isotherms are found when A and B are the same molecules and the binding of one affects the binding of the other. This is discussed at length in Chapter 10.

3. Linked ionization and equilibria: Microscopic and macroscopic constants

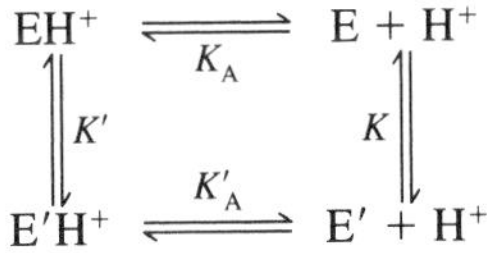

$$
\begin{array}{ccc}
EH^+ & \xrightarrow{\quad K_A \quad} & E + H^+ \\
\downarrow{\scriptstyle K'} & & \downarrow{\scriptstyle K} \\
E'H^+ & \xrightarrow{\quad K'_A \quad} & E' + H^+
\end{array}
$$

Scheme 3

Suppose that an enzyme E has an alternative conformation E' and that both take part in ionization equilibria. This happens with α-chymotrypsin on deprotonation of Ile-16, which is protonated in the active conformation. The equilibrium constants in scheme 3 are defined by $K = [E']/[E]$; $K' = [E'H^+]/[EH^+]$; $K_A = [H^+][E]/[EH^+]$; and $K'_A = [H^+][E']/[E'H^+]$. The apparent equilibrium constant between E' and E must consider all ionic forms and is defined by

$$K_{app} = \frac{[E'] + [E'H^+]}{[E] + [EH^+]} \tag{3.77}$$

Substituting the ionization constants into this equation, we get

$$K_{app} = \frac{[E'] + [E'][H^+]/K'_A}{[E] + [E][H^+]/K_A} \qquad (3.78)$$

This reduces to

$$K_{app} = \frac{[E'](1 + [H^+]/K'_A)}{[E](1 + [H^+]/K_A)} \qquad (3.79)$$

Substituting $K = [E']/[E]$ gives

$$K_{app} = \frac{K(1 + [H^+]/K'_A)}{1 + [H^+]/K_A} \qquad (3.80)$$

The value of K_{app} is pH dependent. It is seen in Chapter 5 (equation 5.8) that equation 3.80 is the equation for the ionization of an acid of ionization constant K_A. At high pH where $[H^+] \ll K_A$ or K'_A, $K_{app(highpH)} = K$. At low pH where $[H^+] \gg K_A$ or K'_A, $K_{app(lowpH)} = KK_A/K'_A$. Applying the rules for cyclic equilibria above, $K_{app(highpH)} = K'$. The values of K_{app} at extremes of pH should be obvious from examination of scheme 3.

These linked equilibria can give an anomalous pK_a for the ionizing group. The apparent ionization constant, $K_{A(app)}$ is defined by

$$K_{A(app)} = \frac{([E'] + [E])[H^+]}{[E'H^+] + [EH^+]} \qquad (3.81)$$

This is so because if one titrates the protein as a function of pH, either directly by the addition of base or by following any of its physical properties as a function of pH, then the contributions of both protonated states and of both unprotonated states are measured. Substituting the equations for equilibria into equation (3.81) gives

$$K_{A(app)} = \frac{(K[E] + [E])[H^+]}{K^+[EH^+] + [EH^+]} \qquad (3.82)$$

On substituting $K_A = [H^+][E]/[EH^+]$ and rearranging, this reduces to

$$K_{A(app)} = \frac{K_A(1 + K)}{1 + K^+} \qquad (3.83)$$

That is, the apparent ionization constant is perturbed by a factor of $(1 + K)/(1 + K^+)$. This is significant when K differs from K' and at least one of them is large. For chymotrypsin, K is small (< 1) and K' is large ($\gg 1$), so that $K_{A(app)} \ll K_A$. The physical picture is that as the pH is raised, the effective equilibrium is between EH^+ and E'. Thus, two processes occur on dissociation of EH^+ — the loss of the proton followed by a change in equilibrium — and so ionization is a composite function of equilibrium and ionization constants. The equilibrium constants K, K^+, K_A and K'_A are termed *microscopic* equilibrium constants $K_{A(app)}$ is variously termed a macroscopic, group, or apparent ionization constant. Note that the four microscopic equilibrium constants are related by $K \cdot K_A = K' \cdot K'_A$.

4. Hypothetical steps: Mutations[19]

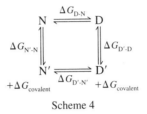

Scheme 4

In Chapter 17, we discuss the effects of mutation of side chains on protein stability. A protein that has a native structure N and a denatured state D is converted into N' and D' by substitution of one of its amino acid residues. D and N differ only in their noncovalent interactions because none of the covalent bonds are altered on denaturation, as are D' and N'. We can measure the free energies of denaturation directly (ΔG_{D-N} and $\Delta G_{D'-N'}$) and draw a cycle (scheme 4).

$$\Delta G_{D-N} - \Delta G_{D'-N'} = \Delta G_{N'-N} - \Delta G_{D'-D} + \Delta G_{covalent} - \Delta G_{covalent} \tag{3.84}$$

The change in the energy of the covalent interactions ($\Delta G_{covalent}$) is the same for N to N' and D to D', and so

$$\Delta G_{D-N} - \Delta G_{D'-N'} = \Delta G_{N'-N} - \Delta G_{D'-D} \tag{3.85}$$

for the noncovalent energy changes. Theoreticians can calculate the "alchemical" changes $\Delta G_{N'-N}$ and $\Delta G_{D'-D}$ so that the cycle can be filled in to compare theory and experiment. There is an important message in equation 3.85: the measured changes in energy on mutation are the differences between the changes in energy of the folded protein and are relative to the changes in the denatured state.

5. Double mutant cycles[20,21]

We can detect whether two side chains, —X and —Y, in a protein E interact by mutating them singly and then together to generate the family of mutants,

E-XY (wild type), E-X, E-Y, and E. We draw a cycle (scheme 5)

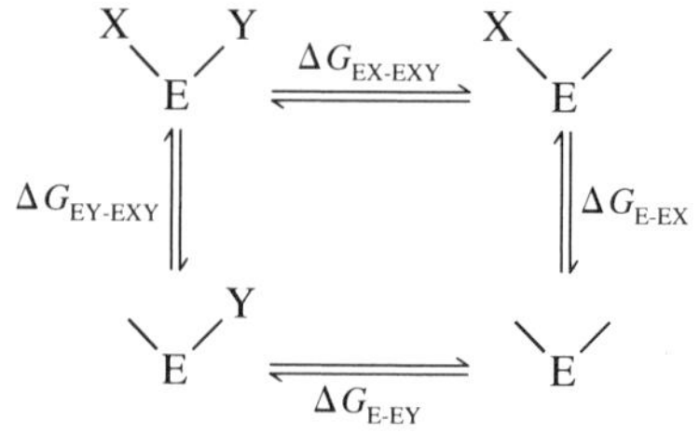

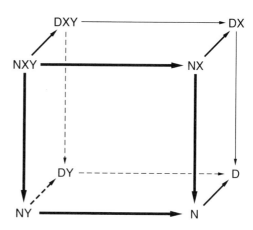

Scheme 5

in which changes in free energy for the mutations are denoted by $\Delta G_{EX\text{-}EXY}$ $(= \Delta G_{D\text{-}N(E\text{-}XY)} - \Delta G_{D\text{-}N(E\text{-}X)})$; $\Delta G_{E\text{-}EY}$ $(= \Delta G_{D\text{-}N(E\text{-}Y)} - \Delta G_{D\text{-}N(E)})$; $\Delta G_{EY\text{-}EXY}$ $(= \Delta G_{D\text{-}N(E\text{-}XY)} - \Delta G_{D\text{-}N(E\text{-}Y)})$; and $\Delta G_{E\text{-}EX}$ $(= \Delta G_{D\text{-}N(E\text{-}X)} - \Delta G_{D\text{-}N(E)})$. If the X and Y do not interact directly or indirectly, then the mutation of E-XY to E-Y should cause the same change in energy as the mutation of E-X to E (i.e., $\Delta G_{EY\text{-}EXY} = \Delta G_{E\text{-}EX}$). If X and Y interact favorably, then the favorable energy of interaction is lost on the mutation of E-XY to E-Y, but is not lost on E-X to E (i.e., $\Delta G_{EY\text{-}EXY} > \Delta G_{E\text{-}EX}$).

The changes in energy are formally equivalent to those in scheme 2 and equation 3.76. We can define a coupling energy $\Delta\Delta G_{int}$ by

$$\Delta\Delta G_{int} = \Delta G_{EY\text{-}EXY} - \Delta G_{E\text{-}EX} = \Delta G_{EX\text{-}EXY} - \Delta G_{E\text{-}EY} \qquad (3.86)$$

The cycle in scheme 5 is deceptive; it really should contain four separate cycles, each containing an equilibrium between the native and denatured states, as in scheme 4. These cycles may be drawn as a three-dimensional cube (Figure 3.7),

Figure 3.7 A double-mutant cycle cube.

which is both prettier and more informative. We can either go through the algebra and solve the equations or just look at the cube and write down by inspection:

$$\Delta\Delta G_{int} = (\Delta\Delta G_{int})_N - (\Delta\Delta G_{int})_D \tag{3.87}$$

That is, the measured coupling energy between X and Y is equal to their coupling energy in the native structure, $(\Delta\Delta G_{int})_N$, minus that in the denatured state, $(\Delta\Delta G_{int})_D$. Because residues do not usually interact much in random denatured states, $(\Delta\Delta G_{int})_D = 0$. Thus, $\Delta\Delta G_{int}$ can be a direct measure of interactions in the native structure, unlike the changes in energy of single mutations (equation 3.85).

References

1. L. Michaelis and M. L. Menten, *Biochem. Z.* **49**, 333 (1913).
2. G. E. Briggs and J. B. S. Haldane, *Biochem. J.* **19**, 338 (1925).
3. L. Peller and R. A. Alberty, *J. Am. Chem. Soc.* **81**, 5907 (1959).
4. H. Lineweaver and D. Burk, *J. Am. Chem. Soc.* **56**, 658 (1934).
5. G. S. Eadie, *J. Biol. Chem.* **146**, 85 (1942).
6. B. H. J. Hofstee, *Nature, Lond.* **184**, 1296 (1959).
7. J. E. Dowd and D. S. Riggs, *J. Biol. Chem.* **249**, 863 (1965).
8. G. L. Atkins and I. A. Nimmo, *Biochem. J.* **149**, 775 (1975).
9. J. B. S. Haldane, *Enzymes*, Longmans, Green and Co. (1930). M.I.T. Press (1965).
10. T. N. C. Wells, C. Ho, and A. R. Fersht, *Biochemistry* **25**, 6603 (1986).
11. H. J. Fromm, *Initial rate enzyme kinetics*, Springer (1975).
12. J. T.-F. Wong, *Kinetics of enzyme mechanisms*, Academic Press (1975).
13. I. H. Segel, *Enzyme kinetics*, Wiley (1975).
14. A. Cornish-Bowden, *Principles of enzyme kinetics*, Butterworth (1975).
15. W. W. Cleland, *The Enzymes* **2**, 1 (1970).
16. K. Dalziel, *The Enzymes* **10**, 2 (1975).
17. W. P. Jencks, M. Gresser, M. S. Valenzuela, and F. C. Huneeus, *J. Biol. Chem.* **247**, 3756 (1972).
18. W. W. Cleland, *Biochemistry* **14**, 3220 (1974).
19. A. R. Fersht, A. Matouschek, and L. Serrano, *J. Molec. Biol.* **224**, 771 (1992).
20. P. J. Carter, G. Winter, A. J. Wilkinson, and A. R. Fersht, *Cell* **38**, 835 (1984).
21. A. Horovitz, *Folding and Design* **1**, R121 (1996).

Measurement and Magnitude of Individual Rate Constants

4

Methods for measurement: An introduction to pre – steady state kinetics

Steady state kinetic measurements on an enzyme usually give only two pieces of kinetic data, the K_M value, which may or may not be the dissociation constant of the enzyme-substrate complex, and the k_{cat} value, which may be a microscopic rate constant but may also be a combination of the rate constants for several steps. The kineticist does have a few tricks that may be used on occasion to detect intermediates and even measure individual rate constants, but these are not general and depend on mechanistic interpretations. (Some examples of these methods will be discussed in Chapter 7.) In order to measure the rate constants of the individual steps on the reaction pathway and detect transient intermediates, it is necessary to measure the rate of approach to the steady state. It is during the time period in which the steady state is set up that the individual rate constants may be observed.

Since values of k_{cat} lie between 1 and 10^7 s^{-1}, measurements must be made in a time range of 1 to 10^{-7} s. This requires either techniques for rapidly mixing and then observing the enzyme and substrate, or totally new methods. Also, since the events that are to be observed occur on the enzyme itself, the enzyme must be available in substrate quantities. The development of apparatus for measuring these rapid reactions and of techniques for isolating large quantities of pure proteins has revolutionized enzyme and protein folding kinetics.

In sections A to C, we shall discuss four types of techniques. The first is *rapid mixing*. This is extremely useful since it is possible to mix two solutions in a fraction of a millisecond, and the majority of enzyme turnover numbers are less than 1000 s^{-1}. Rapid mixing techniques are now standard laboratory practice because of their ease and their wide range of application.

A. Rapid mixing and sampling techniques

1. The continuous-flow method

In 1923, H. Hartridge and F. J. W. Roughton introduced the continuous-flow method to solution kinetics in order to study the combination of deoxyhemoglobin with ligands.[1] The principle of the method is illustrated in Figure 4.1. Two syringes are connected by a mixing chamber to a flow tube. One syringe is filled with enzyme, the other is filled with substrate, and the two are compressed at a constant rate. The two solutions mix thoroughly in the mixing chamber, pass down the flow tube, and "age." At a constant flow rate the age of the solution is linearly proportional to the distance down the flow tube and the flow rate; e.g., if the flow rate is 10 m s^{-1}, then 1 cm from the mixing chamber the solution is 1 ms old, 10 cm from the chamber it is 10 ms old, and so on. The flow rate of the liquid must be kept above a critical velocity in order to ensure "turbulent flow." Below this value, which is about 2 m s^{-1} for a tube with a 1-mm diameter, the flow may be laminar, the liquid at the center traveling faster than that near the wall. This places an upper time limit on the apparatus for a particular length of tube. The "dead time" is the shortest time after mixing at which the reaction can be observed.

Continuous flow has recently had a revival in studying protein folding, where dead times as low as $10 \ \mu s$ have been achieved.[2] A particularly economical and modern design is illustrated in Figure 4.2.[3] The apparatus, designed by H. Roder, has a dead time of $45 \ \mu s$ and maximum times up to 1 ms or so. The flow cell is 10 mm long by $250 \ \mu m$ in cross section. The liquids to be mixed are sprayed from two capillaries over a ball of platinum, which causes sufficient turbulence for rapid mixing. The whole of the length of the cell is illuminated, and the fluoresced (or transmitted) light is simultaneously collected over the whole length by a CCD detector that resolves the timecourse into 1035 points.

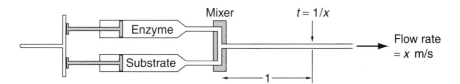

Figure 4.1 Continuous-flow apparatus.

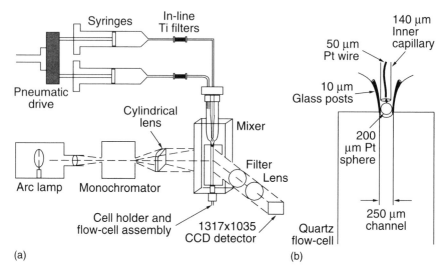

Figure 4.2 (a) Continuous-flow machine with 45-μs dead time. (b) Exploded view of the mixing chamber. Turbulent flow of the two liquids over the platinum sphere gives improved mixing. [Courtesy of M. C. Ramachrandra Shastry and H. Roder.]

2. The stopped-flow method

The stopped-flow method was introduced by Roughton in 1934[4] and greatly improved by B. Chance some six years later.[5] The principle is illustrated in Figure 4.3.[6] In contrast to the setup in the continuous-flow system, the two driving syringes are compressed to express about 50 to 200 μL from each, and then they are mechanically stopped. Suppose that there is an observation point 1 cm from

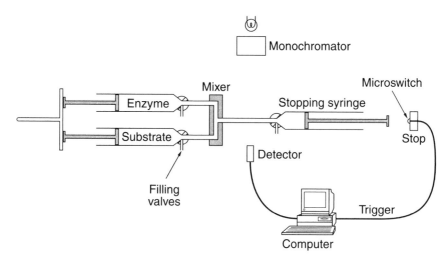

Figure 4.3 Stopped-flow apparatus.

the mixing chamber. If the flow rate is 10 m s^{-1} during the period of compression, during this *continuous-flow* period the detector sees a solution that is 1 ms old. When the flow is stopped, the solution ages normally with time and the detector sees the events occurring after 1 ms.

The stopped-flow method is a routine laboratory tool, whereas the continuous-flow apparatus is used in a few specialized cases only. The stopped-flow technique requires only 100 to 400 μL of solution or less for the complete time course of a reaction; the dead time is as low as 0.5 ms or so; and observations may be extended to several minutes. Stopped flow does, however, require a rapid detection and recording system.

3. Rapid quenching techniques

Instead of using a photomultiplier or other detector in the flow systems, the solutions may be quenched by, say, the addition of an acid, such as trichloroacetic acid, and the reaction products directly analyzed by chromatographic or other techniques.

a. The quenched-flow technique

The simplest form of the method is to submerge the end of the observation tube of a continuous-flow apparatus in a beaker of acid. A somewhat more sophisticated version is illustrated in Figure 4.4. A third syringe mixes the quenching acid with the reagent solutions via a second mixing chamber. Such an apparatus

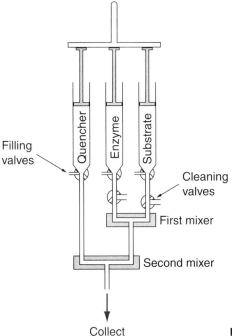

Figure 4.4 Quenched-flow apparatus.

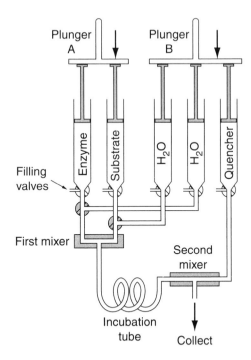

Figure 4.5 Pulsed quenched-flow apparatus.

may have a dead time of only 4 or 5 ms. However, the maximum practical reaction time that may be measured using small volumes of reagents is about 100 to 150 ms; otherwise, excessively long reaction tubes are required.

b. The pulsed quenched-flow technique

The time range of the quenched-flow technique may be extended by a procedure similar to that of stopped flow.[7] As illustrated in Figure 4.5 the enzyme and the substrate (25 μL of each) are first mixed and driven into an incubation tube by a plunger actuated by compressed air. After the desired time interval—15 ms or longer—a second plunger is actuated to drive the incubated mixture with a pulse of distilled water into a second mixer, where it is quenched. Quenched flow is a useful technique for studying transient intermediates in protein folding by ^{1}H/^{2}H-exchange measurements (Chapter 18, section F4).

B. Flash photolysis

The time involved in mixing places a limit on the dead time of flow techniques. The only way to increase the time resolution is to cut out the mixing by using a premixed solution of reagents that can be perturbed in some way to allow a measurable reaction to occur. A classic method from physical chemistry is flash photolysis, in which a particular bond in a reagent is cleaved by a pulse of light so that reactive intermediates are formed. This method was introduced in 1959 by Q. H. Gibson[8] for studying the dissociation of CO from carbonmonoxy-hemoglobin and carbonmonoxy-myoglobin, and the subsequent recombination with CO and other ligands. The procedure has been refined for these compounds by using mode-locked dye lasers to study dissociation on the femtosecond time scale, and recombination on the picosecond time scale.[9]

Unlike rapid mixing and sampling methods, flash photolysis cannot be generally applied to reactions because a suitable target for the flash is rarely available. However, flash photolysis has been adapted as a general procedure for initiating reactions that require ATP or GTP.[10] A derivative of ATP termed caged ATP (or GTP) (equation 4.1) may be photodissociated by a flash at 347 nm with a half-time of milliseconds to generate ATP.[11] As caged ATP is generally unreactive, this provides a general means of studying ATP-utilizing reactions, e.g., muscle contraction,[11] on the millisecond time scale.

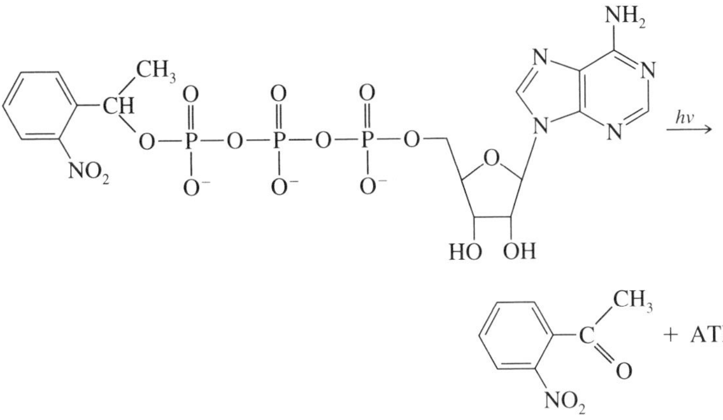

$$(4.1)$$

Flash photolysis has been used for initiating rapid reactions in the crystalline state, which can be followed by the von Laue method of data collection and analysis (Chapter 1, section F1c).[12]

C. Relaxation methods

1. Temperature jump

An alternative method of overcoming the time delay of mixing is to use a relaxation method. An equilibrium mixture of reagents is preincubated and the equilibrium is perturbed by an external influence. The rate of return, or relaxation, to equilibrium is then measured. The most common procedure for this is temperature jump (Figure 4.6).[13] A solution is incubated in an absorbance or fluorescence cell and its temperature is raised through 5 to 10°C in less than a microsecond by the discharge of a capacitor (or, in more recent developments, in 10 to 100 ns by the discharge of an infrared laser). If the equilibrium involves an

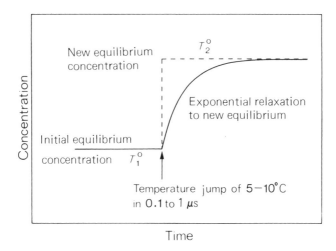

Figure 4.6 Illustration of temperature jump.

enthalpy change, the equilibrium position will change. The system will proceed to its new equilibrium position via a series of *relaxation times*, τ ($\equiv$ the reciprocal of the rate constant).

Clearly, this method cannot be applied to systems in which there are irreversible chemical processes. It is most suitable for situations involving simple ligand binding (such as NAD^+ with a dehydrogenase), inhibitor binding, or conformational changes in the protein. There have been some attempts to combine the temperature-jump with the stopped-flow method.

2. Nuclear magnetic resonance

NMR is a technique that does not perturb the chemistry of a system but that can measure the rate constants in an equilibrium mixture. The accessible time is usually limited; depending upon both the isotope involved and various physical processes, the time range can be slower than that for stopped flow, or one or two orders of magnitude faster.

NMR has been used for measuring the *dissociation* rate constants of enzyme–inhibitor complexes from *exchange broadening*. This phenomenon was discussed in Chapter 1, section G2d, where its role in quantifying the rotation of amino acid side chains between two states was described. The method may be used in an analogous way for measuring the rate of exchange between enzyme-bound and free inhibitor. The line-width analysis method is limited to a narrow range of rate constants, and, in practice, it cannot be applied to complex reaction pathways. Exchange broadening has been used with success in following protein folding on a submillisecond time scale (Chapter 18, section A).[14]

D. Analysis of pre–steady state and relaxation kinetics

Some of the fundamental differences between steady state and pre–steady state enzyme kinetics will become apparent later. Pre–steady state kinetics is concerned with the detection and analysis of transient enzyme-bound species as they arise and decay during the early phase of reaction. The processes generally observed are just the first-order rates of change from one enzyme species to another. Their time courses are generally exponential curves, unlike the linear plots of steady state kinetics. During these processes, the enzyme undergoes a single turnover in contrast to the multiple recycling of the steady state.

It is relatively straightforward to solve the differential equations for the time dependence of the transients in simple cases. However, it is important to understand the physical meaning of why a particular case gives rise to a particular form of solution. In this section we will concentrate on an intuitive approach to this understanding. Once a feel for the subject has been developed, algebraic mistakes will not be made and some complex kinetic schemes may be solved by inspection.

1. Simple exponentials

a. Irreversible reactions

Suppose that a compound A transforms into B with a first-order rate constant k_f, and the reaction proceeds to completion:

$$A \xrightarrow{k_f} B \tag{4.2}$$

Then

$$\frac{d[B]}{dt} = k_f[A] \tag{4.3}$$

and

$$\frac{d[A]}{dt} = -k_f[A] \tag{4.4}$$

This is solved by integration to give

$$[A]_t = [A]_0 \exp(-k_f t) \tag{4.5}$$

where $[A]_0$ is the initial concentration of A. Since

$$[A]_t + [B]_t = [A]_0 \tag{4.6}$$

it is also true that

$$[B]_t = [A]_0 \{1 - \exp(-k_f t)\} \tag{4.7}$$

Both [A] and [B] follow simple exponentials.

It should be noted that the half-life of the reaction, $t_{1/2}$, where [A] = [B] = $[A]_0/2$, is given by

$$\exp(-k_f t) = \tfrac{1}{2} \tag{4.8}$$

That is,

$$t_{1/2} = \frac{0.6931}{k_f} = 0.6931\tau \tag{4.9}$$

b. The method of initial rates

When discussing more complex examples we shall use an adaptation of the *method of initial rates* to illustrate the physical meaning of some of the expressions. This method is often used in experiments that are too slow to follow over a complete time course, or in which there are complicating side reactions.

The initial rate v_0 of equation 4.2 is given by

$$v_0 = k_f[A]_0 \tag{4.10}$$

The value of k_f is determined by dividing v_0 by the expected change in reagent concentrations:

$$k_f = \frac{v_0}{\Delta[A]_0} \tag{4.11}$$

(Note that for an irreversible reaction, $\Delta[A]_0 = [A]_0$.)

c. Reversible reactions

In this case, A does not completely transform to B, but there is an equilibrium concentration of A:

$$A \underset{k_r}{\overset{k_f}{\rightleftarrows}} B \tag{4.12}$$

and

$$K_{eq} = \frac{[B]}{[A]} = \frac{k_f}{k_r}$$ (4.13)

Here,

$$\frac{d[A]}{dt} = -k_f[A] + k_r[B]$$ (4.14)

Substitution of equation 4.6 gives

$$\frac{d[A]}{dt} = -k_f[A] + k_r([A]_0 - [A])$$ (4.15)

This equation may be integrated by separating the variables and multiplying each side by an exponential factor:

$$\frac{d[A]}{dt} + [A](k_f + k_r) = k_r[A]_0$$ (4.16)

$$\frac{d[A]}{dt} \exp{(k_f + k_r)t} + [A](k_f + k_r) \exp{(k_f + k_r)t} = k_r[A]_0 \exp{(k_f + k_r)t}$$ (4.17)

$$\therefore \frac{d}{dt} \{[A] \exp{(k_f + k_r)t}\} = k_r[A]_0 \exp{(k_f + k_r)t}$$

$$\therefore [A] \exp{(k_f + k_r)t} = \frac{k_r}{k_f + k_r} [A]_0 \exp{(k_f + k_r)t} + \text{constant}$$ (4.18)

Using the boundary conditions that at $t = 0$, $[A] = [A]_0$, and at $t = \infty$, $[A]_{eq}$, the equilibrium concentration of $[A]$, is given by

$$[A]_{eq} = \frac{[A]_0 k_r}{k_f + k_r}$$ (4.19)

(from equations 4.6 and 4.13), the solution is

$$[A]_t = \frac{[A]_0}{k_f + k_r} \{k_f \exp{[-(k_f + k_r)t]} + k_r\}$$ (4.20)

The expression for the time course (equation 4.20) may be divided into various factors. There is first the exponential term with the rate constant, or in terms of relaxation kinetics, the reciprocal relaxation time $1/\tau$ given by

$$\frac{1}{\tau} = k_f + k_r \tag{4.21}$$

Second, there is an *amplitude* factor given by

$$\frac{k_f}{k_f + k_r} \tag{4.22}$$

Now suppose that the reaction is started from the other direction, with B initially present, but not A. Then

$$[B]_t = \frac{[B]_0}{k_f + k_r} \{k_r \exp\left[-(k_f + k_r)t\right] + k_f\} \tag{4.23}$$

This expression has the same relaxation time as equation 4.20, but a different amplitude factor.

The first important point to be noted is that the rate constant for the approach to equilibrium is greater than either of the individual first-order rate constants, k_f and k_r, and is equal to their sum. The reason why this is so is readily understood by applying the principle of initial rates. The initial velocity in the reversible reaction (4.12) is the same as in the irreversible one (4.2), but the former reaction does not have to proceed so far. For example, before B accumulates, only A is present, so again

$$v_0 = k_f[A]_0$$

—however, the total change [A] is given by

$$\Delta[A]_0 = [A]_0 - [A]_{eq} \tag{4.24}$$

Substitution of equation 4.19 gives

$$\Delta[A]_0 = \frac{[A]_0 \, k_f}{k_f + k_r} \tag{4.25}$$

$$\therefore \frac{1}{\tau} = \frac{v_0}{\Delta[A]_0} = k_f + k_r \tag{4.26}$$

The second point to be noted is that k_f and k_r cannot be assigned without a knowledge of the amplitude factor. This basic symmetry in the relaxation times occurs in many cases, and, in general, the rate constants for unimolecular reactions cannot be assigned unless the concentrations of A and B at equilibrium may be determined. It will be seen later that when the reactions are not unimolecular but pseudo-unimolecular because of the presence of a second reagent, the relaxation time will have a concentration dependence that removes this ambiguity.

2. Association of enzyme and substrate

$$[E] + [S] \rightleftharpoons [ES] \tag{4.27}$$

If $[S] \gg [E]$, the reaction is effectively first-order since the concentration of S is hardly affected by the reaction. If the second-order rate constant for the association is k_{on} and that for dissociation is k_{off}, then the system reduces to

$$[E] \underset{k_{off}}{\overset{k_{on}[S]}{\rightleftharpoons}} [ES] \tag{4.28}$$

The relaxation time for this reaction is, from equation 4.21,

$$\frac{1}{\tau} = k_{off} + k_{on}[S] \tag{4.29}$$

Two points should be noted: (1) Because the rate constants are *pseudo-unimolecular*, there is a concentration dependence, so k_{on} and k_{off} may be resolved without the amplitude factor. (2) There is a lower limit to $1/\tau$; that is, $1/\tau$ cannot be less than k_{off}. This sets a limit on the measurement of these rate constants. A good stopped-flow spectrophotometer can cope only with rate constants of 1000 s^{-1} or less, and many enzyme-substrate dissociation constants are faster than this.

A favorable example is given in Figure 4.7. The dissociation rate constant of tyrosine from its complex with tyrosyl-tRNA synthetase is low, so the association and dissociation rate constants can be measured by stopped flow. (Note that sometimes a two-step process may appear to be a single-step reaction: see section 6).

Where there is no subsequent turnover of a substrate, such as occurs on the omission of a cosubstrate in a multisubstrate reaction, or on inhibitor binding, the temperature-jump technique is generally the most useful tool for the determination of these constants.

3. Consecutive reactions

a. Irreversible reactions

The simplest case of consecutive reactions is

$$A \xrightarrow{k_1} B \xrightarrow{k_2} C \tag{4.30}$$

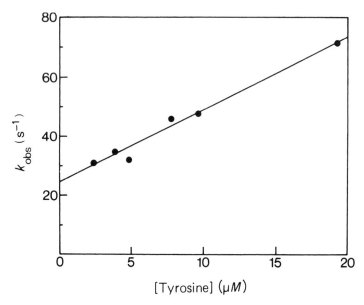

Figure 4.7 Binding of tyrosine to the tyrosyl-tRNA synthetase from *Bacillus stearothermophilus*. [From A. R. Fersht, R. S. Mulvey, and G. L. E. Koch, *Biochemistry* **14**, 13 (1975).]

This is solved by simply using the conservation equation and the integration procedures above to give equations 4.31.

$$[A] = [A]_0 \exp(-k_1 t)$$

$$[B] = \frac{[A]_0 k_1}{k_2 - k_1} [\exp(-k_1 t) - \exp(-k_2 t)]$$

$$[C] = [A]_0 \left\{ 1 + \frac{1}{k_1 - k_2} [k_2 \exp(-k_1 t) - k_1 \exp(-k_2 t)] \right\}$$

(4.31)

B is a transient intermediate that appears and then disappears (Figure 4.8). If $k_1 \gg k_2$, it is formed with rate constant k_1 and then slowly decomposes with rate constant k_2. However, if $k_2 \gg k_1$, [B] reaches a steady state level with rate constant k_2 and decays slowly with rate constant k_1. The apparently paradoxical situation is that the intermediate appears to be formed with its decomposition rate constant and to decompose with its formation rate constant! This is readily understood on the initial-rate treatment. When $k_1 \ll k_2$, [B] reaches a steady state level given by

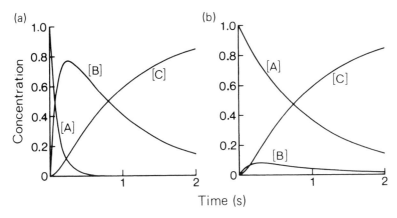

Figure 4.8 Plots of the concentrations of A, B, and C in the reaction $A \rightarrow B \rightarrow C$ (equation 4.30). (a) $k_1 = 10\ \text{s}^{-1}$, $k_2 = 1\ \text{s}^{-1}$. (b) $k_1 = 1\ \text{s}^{-1}$, $k_2 = 10\ \text{s}^{-1}$. Note that (1) the progress curves for C are identical; (2) the *shapes* of the two curves for B are identical—they differ only in amplitude; and (3) [A] decreases 10 times more rapidly in (a). Thus, unless [A] is monitored, the two examples cannot be distinguished on the basis of measured rate constants only.

$$\frac{d[\text{B}]}{dt} = 0 = k_1[\text{A}] - k_2[\text{B}] \tag{4.32}$$

The initial concentration of B when the steady state is set up is given by

$$[\text{B}]_{\text{SS}} \approx \frac{k_1}{k_2}[\text{A}]_0 \quad (\text{i.e.,} \ll [\text{A}]_0) \tag{4.33}$$

and

$$v_0 = k_1[\text{A}]_0 \tag{4.34}$$

$$\therefore \frac{1}{\tau} = \frac{v_0}{[\text{B}]_{\text{SS}}} = k_2 \tag{4.35}$$

In this latter case, where $k_2 > k_1$, B is at a very low concentration; in the former case, where $k_1 > k_2$, it accumulates. The two cases can be resolved if [B] can be monitored in absolute terms.

This type of kinetic situation sometimes occurs in protein renaturation experiments, in which the kinetics are often monitored by fluorescence changes. The biphasic traces cannot be resolved in such circumstances unless the quantum yield of the transient intermediate is known, so that its absolute concentration can be determined.

An example of the application of these equations is found in Chapter 7, section D. The aminoacyl-tRNA synthetase that specifically esterifies the tRNA molecule that accepts valine, $tRNA^{Val}$, "corrects" the error when it mistakenly forms an aminoacyl adenylate with threonine by the following scheme:

$$E \cdot Thr\text{—}AMP \cdot tRNA \xrightarrow[AMP]{transfer} E \cdot Thr\text{-}tRNA \xrightarrow{hydrolysis} E + Thr + tRNA \tag{4.36}$$

It will be seen that the rate of disappearance of E·Thr-AMP·tRNA was directly measured by the formation of AMP, that the intermediate E·Thr-tRNA was directly measured, and that the second step (hydrolysis) was measured directly and independently by isolating the mischarged tRNA and adding it to the enzyme.

b. Quasireversible reactions: Steady state
A more common situation in enzyme kinetics is the following:

$$E \xrightarrow[+P_1]{k_1[S]} ES' \xrightarrow{k_2} E + P_2 \tag{4.37}$$

For example, chymotrypsin reacts with p-nitrophenyl acetate (AcONp) according to the above scheme (when $[AcONp] \ll K_S$ for the first step) to give an intermediate acylenzyme, EAc:

$$E \xrightarrow[HONp]{k_1[AcONp]} EAc \xrightarrow[AcOH]{H_2O} E \tag{4.38}$$

(where $k_1 = k_{cat}/K_M$ for this step).

Since the acylenzyme is continuously being formed and turning over, its concentration is in the steady state (provided that $[AcONp] \gg [E]$). The steady state concentration of the acylenzyme is given by

$$\frac{d}{dt}[EAc] = 0 = k_1[AcONp][E] - k_2[EAc]_{SS} \tag{4.39}$$

Now, since

$$[E] + [EAc] = [E]_0 \tag{4.40}$$

equation 4.39 can be written as

$$0 = k_1[AcONp]([E]_0 - [EAc]_{SS}) - k_2[EAc]_{SS} \tag{4.41}$$

or

$$[EAc]_{SS} = \frac{k_1[AcONp][E]_0}{k_2 + k_1[AcONp]} \tag{4.42}$$

Applying the initial-rate treatment, we have

$$v_0 = k_1[E]_0[AcONp] \tag{4.43}$$

$$\frac{1}{\tau} = \frac{v_0}{[EAc]_{SS}} \tag{4.44}$$

$$= k_2 + k_1[AcONp] \tag{4.45}$$

Just as in the case of reversible reactions, the intermediate is formed with a rate constant that is greater than the rate constant for the transformation of the preceding intermediate.

The analytical solution for the rate is

$$[HONp] = [E]_0 \left(\frac{k_1'}{k_1' + k_2} \right)$$

$$\times \left(\frac{k_1'}{k_1' + k_2} \{1 - \exp[-(k_1' + k_2)t]\} + k_2 t \right) \tag{4.46}$$

where $k_1' = k_1[AcONp]$. If the rules of saturation kinetics are observed for the acylation step, k_1' is of the form $k_{cat}[S]/(K_M + [S])$. There is an initial exponential phase, which dies out after t is about 5 times greater than τ, and a linear term that eventually predominates.

c. Consecutive reversible reactions

The general solution for these reactions is given in section 6. We shall deal here with cases in which one step is fast compared with the other. Under these circumstances the relaxation times are on different time scales and do not "mix" with each other.

1. First step fast (pre-equilibrium).

$$E \underset{k_{-1}}{\overset{k_1[S]}{\rightleftharpoons}} ES \underset{k_{-2}}{\overset{k_2}{\rightleftharpoons}} ES' \tag{4.47}$$

This example may be readily solved by inspection if two simple rules are applied: (1) There must be two sets of relaxation times, since there are two sets of reactions involved. (2) Because the reactions occur on different time scales, they may be dealt with separately.

The first relaxation time is for the binding step. By analogy with equation 4.29, this is given by

$$\frac{1}{\tau_1} = k_{-1} + k_1[S] \qquad (4.48)$$

The second relaxation time is for the slow step. This is an example of a reversible reaction, and, by analogy with equation 4.21, the reciprocal relaxation time is given by the sum of the forward and the reverse rate constants for the step. However, the effective forward rate constant for this step is given by k_2 multiplied by the fraction of the enzyme that is in the form of ES; i.e.,

$$\frac{1}{\tau_2} = k_{-2} + \frac{k_2[S]}{[S] + K_S} \qquad (4.49)$$

where

$$K_S = \frac{k_{-1}}{k_1} \qquad (4.50)$$

2. *Second step fast.* Reaction 4.47 could involve a substrate-induced conformational change in the enzyme, where ES′ is just a different conformational state, or, alternatively, it could involve the accumulation of an intermediate on the pathway. The following reaction illustrates the displacement of an equilibrium between two conformational states of an enzyme caused by the binding of a substrate to one form only:

$$E' \underset{k_{-1}}{\overset{k_1}{\rightleftharpoons}} E \underset{\text{fast}}{\overset{S, K_S}{\rightleftharpoons}} ES \qquad (4.51)$$

This situation is found for the binding of ligands to chymotrypsin, which exists in two conformational states. Only one of these states binds aromatic substrates. It may be shown from the formal analysis to be given in section 6 that

$$\frac{1}{\tau_2} = k_1 + k_{-1}\left(\frac{K_S}{[S] + K_S}\right) \qquad (4.52)$$

Case 2 may be distinguished from case 1 in that $1/\tau_2$ *decreases* with increasing [S]. This may be understood by analogy with the examples of irreversible and reversible reactions (equations 4.2 and 4.12). Clearly, when [S] is very high the reaction is essentially irreversible, since E' is transformed completely to ES and so $1/\tau_2$ tends toward k_1. Similarly, as [S] tends toward zero there is very little ES, and $1/\tau_2$ tends toward $k_1 + k_{-1}$. Hence the concentration dependence.

4. Parallel reactions

Parallel reactions are said to arise when a compound undergoes two or more reactions simultaneously. Enzymatic reactions often occur in parallel, when an activated intermediate may react with several competing acceptors:

$$
\begin{array}{c}
\quad\; B \\
k_B \nearrow \\
A \\
k_C \searrow \\
\quad\; C
\end{array}
\tag{4.53}
$$

The kinetic equations are easily solved by integration, but it is instructive to solve them intuitively. It is obvious that [A] decreases with a rate constant that is the sum of k_B and k_C, and also that B and C are formed in the ratio of the rate constants. Since the rates of formation of B and C depend on [A], B and C must each be formed with a rate constant that is the same as for the disappearance of A. Therefore, we have

$$
[A] = [A]_0 \exp\left[-(k_B + k_C)t\right]
\tag{4.54}
$$

$$
[B] = \frac{[A]_0 k_B}{k_B + k_C} \{1 - \exp\left[-(k_B + k_C)t\right]\}
\tag{4.55}
$$

$$
[C] = \frac{[A]_0 k_C}{k_B + k_C} \{1 - \exp\left[-(k_B + k_C)t\right]\}
\tag{4.56}
$$

The situation is similar to that for reversible reactions (equation 4.12), in that the relaxation time is composed of the sum of those for two reactions.

Chapter 7 includes examples of parallel reactions, e.g., the attack of various nucleophiles on acylchymotrypsins, measured by steady state and pre–steady state kinetics.

5. Derivation of equations for temperature jump

As an illustration, consider the association of an enzyme and a substrate in a one-step reaction:

$$
E + S \underset{k_{-1}}{\overset{k_1}{\rightleftharpoons}} ES
\tag{4.57}
$$

Suppose that because of the change in temperature, the equilibrium moves to a new position, so that

$$[E] = [E]_{eq} + e \qquad (4.58)$$

$$[S] = [S]_{eq} + s \qquad (4.59)$$

and

$$[ES] = [ES]_{eq} + es \qquad (4.60)$$

where $[E]_{eq}$, $[S]_{eq}$, and $[ES]_{eq}$ are the equilibrium concentrations at the new temperature. Then

$$\frac{d[ES]}{dt} = k_1([E]_{eq} + e)([S]_{eq} + s) - k_{-1}([ES]_{eq} + es) \qquad (4.61)$$

$$= k_1[E]_{eq}[S]_{eq} - k_{-1}[ES]_{eq}$$
$$+ k_1([E]_{eq}s + [S]_{eq}e + e \cdot s) - k_{-1}es \qquad (4.62)$$

Equation 4.62 may be simplified, since the first two terms on the righthand side cancel out (they are equal at equilibrium). Also, because the reagents are conserved, $e = s = -es$. And since $[ES]_{eq}$ is a constant, $d[ES]/dt = des/dt$. Therefore,

$$-\frac{des}{dt} = k_1([E]es + [S]es + e \cdot s) + k_{-1}es \qquad (4.63)$$

Now, if the perturbation from equilibrium is small, the second-order term $e \cdot s$ may be ignored. Equation 4.63 may then be integrated to give the relaxation time:

$$\frac{1}{\tau} = k_1([E] + [S]) + k_{-1} \qquad (4.64)$$

If the equilibrium is perturbed only slightly, the return to equilibrium is always a first-order process, even though the reagents may be present at similar concentrations.

6. A general solution of two-step consecutive reversible reactions

The solution of the following equation involves simultaneous linear differential equations:

$$A \underset{k_{-1}}{\overset{k_1}{\rightleftarrows}} B \underset{k_{-2}}{\overset{k_2}{\rightleftarrows}} C \qquad (4.65)$$

Two relaxation times are obtained:

$$\frac{1}{\tau_1} = \frac{p + q}{2} \tag{4.66}$$

$$\frac{1}{\tau_2} = \frac{p - q}{2} \tag{4.67}$$

where

$$p = k_1 + k_{-1} + k_2 + k_{-2}$$

and $\hspace{10cm}$ (4.68)

$$q = [p^2 - 4(k_1 k_2 + k_{-1} k_{-2} + k_1 k_{-2})]^{1/2}$$

These basic equations may be manipulated to cover many cases. A useful trick is to express the rate constants as the sums and products of the relaxation times:

$$\frac{1}{\tau_1} + \frac{1}{\tau_2} = k_1 + k_{-1} + k_2 + k_{-2} \tag{4.69}$$

$$\frac{1}{\tau_1 \tau_2} = k_1 k_2 + k_{-1} k_{-2} + k_1 k_{-2} \tag{4.70}$$

The equations are easy to solve if a concentration dependence is involved. For example, if the sequence is the pseudo-first-order series

$$E \underset{k_{-1}}{\overset{k_1'[S]}{\rightleftharpoons}} ES \underset{k_{-2}}{\overset{k_2}{\rightleftharpoons}} ES' \tag{4.71}$$

then $k_1'[S]$ may be substituted for k_1 in equations 4.69 and 4.70. Also, by analogy with equation 4.64 for temperature jump, $k_1'([S] + [E])$ may be substituted for k_1 in a relaxation experiment.

The equations may be simplified if one of the relaxation times is much faster than the other. For example, if in equation 4.71 the first step is fast, $1/\tau_2$ and $k_2 + k_{-2}$ may be ignored in 4.69. The value of τ_2 may then be obtained by substituting 4.69 and 4.70. In the case of a temperature-jump experiment, this gives

$$\frac{1}{\tau_1} = k_1'([E] + [S]) + k_{-1} \tag{4.72}$$

$$\frac{1}{\tau_2} = k_{-2} + \frac{k_2([E] + [S])}{k_{-1}/k_1' + [E] + [S]} \tag{4.73}$$

The same manipulations may be performed for

$$E \underset{k_{-1}}{\overset{k_1}{\rightleftarrows}} E' \underset{k_{-2}}{\overset{k_2'[S]}{\rightleftarrows}} E'S \tag{4.74}$$

where the first step is slow, to give

$$\frac{1}{\tau_1} = k_{-2} + k_2'([E'] + [S]) \tag{4.75}$$

$$\frac{1}{\tau_2} = k_1 + \frac{k_{-1}(k_{-2}/k_2')}{k_{-2}/k_2' + [E'] + [S]} \tag{4.76}$$

Two practical points should be noted. The kinetic mechanisms in equations 4.71 and 4.74 may be distinguished by the concentration dependence of $1/\tau_2$. For 4.71 this increases with increasing [S]; for 4.74 it decreases. But there are situations that are difficult to resolve. For example, in equation 4.74, if $[E'] \gg [E]$ there will be a burst of formation of ES' with relaxation time τ_1, followed by a small increase at relaxation time τ_2 as E converts to E'. The concentration dependence of τ_2 will be small, since $k_1 \gg k_{-1}$ for $[E'] \gg [E]$. This can be mistaken for the scheme in equation 4.71, where only a little ES' is formed. In this case also, the concentration dependence of $1/\tau_2$ is small, because $k_{-2} \gg k_2$. In both cases the amplitudes of the changes will often be small and the rate constants difficult to measure accurately.

A more common situation that leads to difficulties is the two-step combination of an enzyme and substrate, as in equation 4.71, where the dissociation constant for the first step, k_{-1}/k_1', is high. If measurements are made only in the region where $k_{-1}/k_1' > [E] + [S]$, equation 4.73 reduces to*

$$\frac{1}{\tau_2} = k_{-2} + k_1'\frac{k_2}{k_{-1}}([E] + [S]) \tag{4.77}$$

This has the form of a simple one-step association of an enzyme and a substrate, as in equation 4.64, and may mistakenly be interpreted as this. In such a case, the association rate constant would appear to be $k_1'(k_2/k_{-1})$, a value lower than the true rate constant of k_1'. Some of the low values to be shown in Table 4.1 are undoubtedly caused by this. Measurements should always be extended to high substrate concentrations to search for a leveling off of rate as predicted by equation 4.73 (Figure 4.9).

---

* Equation 4.73 was derived on the assumption that $k_{-1} \gg k_2$. If k_2 is appreciable, equation 4.77 should be modified to

$$\frac{1}{\tau_2} = k_{-2} + \frac{k_1'k_2}{k_{-1} + k_2}([E] + [S])$$

Table 4.1 *Association and dissociation rate constants for macromolecular interactions*

Protein	Ligand	k_1 $(s^{-1} M^{-1})$	k_{-1} (s^{-1})	Ref.
Protein–nucleic acids				
Phenylalanyl-tRNA synthetase	tRNA^{Phe}	1.6×10^8	27	1
Seryl-tRNA synthetase	tRNA^{Ser}	2.1×10^8	11	2
Tyrosyl-tRNA synthetase	tRNA^{Tyr}	2.2×10^8	1.5	3
		1.4×10^8	53	
Protein-protein				
Trypsin	Basic pancreatic	1.1×10^6	6.6×10^{-8}	4
Anhydrotrypsin	trypsin inhibitor	7.7×10^5	8.5×10^{-8}	4
Trypsin	Pancreatic	6.8×10^6	2.2×10^{-4}	4
Anhydrotrypsin	secretory trypsin inhibitor	4×10^6	1.4×10^{-3}	4
Insulin	Insulin	1.2×10^8	1.5×10^4	5
β-Lactoglobulin	β-Lactoglobulin	4.7×10^4	2.1	5
α-Chymotrypsin	α-Chymotrypsin	3.7×10^3	0.68	5
Barnase	Barstar	$>5 \times 10^{9\,a}$	8.0×10^{-6}	6

[a] Varies greatly with ionic strength; see text.

1 G. Krauss, D. Römer, D. Riesner, and G. Maass, *FEBS Letts.* **30**, 6 (1973).
2 A. Pingoud, D. Riesner, D. Boehme, and G. Maass, *FEBS Letts.* **30**, 1 (1973).
3 A. Pingoud, D. Boehme, D. Riesner, R. Kownatski, and G. Maass, *Eur. J. Biochem.* **56**, 617 (1975).
4 J.-P. Vincent, M. Peron-Renner, J. Pudles, and M. Lazdunski, *Biochemistry* **13**, 4205 (1974).
5 R. Koren and G. G. Hammes, *Biochemistry* **15**, 1165 (1976).
6 G. Schreiber and A. R. Fersht, *Nature Structural Biology* **3**, 427 (1996).

7. Experimental application of pre–steady state kinetics

Later, we shall discuss several examples of the successful application of transient kinetics to the solution of enzyme mechanisms (Chapter 7) and to protein folding (Chapters 18 and 19). Here, we briefly describe some of the strategies and tactics used by the kineticist to initiate a transient kinetic study. On many occasions, steady state kinetics and other studies have set kineticists a well-defined and specific question to answer. At other times, they just wish to study a particular system to gather information. In both cases there is no substitute for

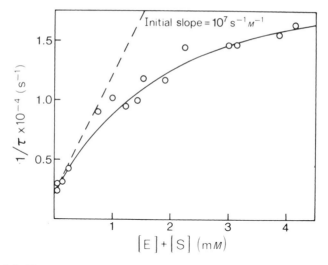

Figure 4.9 The two-step binding of (*N*-acetylglucosamine)$_2$ to lysozyme. Solutions of enzyme (0.03 to 0.2 m*M*) and ligand (0.02 to 4.1 m*M*) were temperature-jumped from 29 to 38°C at pH 6. Experiments at low substrate concentration are in the linear region of the curve and give an apparent second-order rate constant of 10^7 s^{-1} *M*$^{-1}$ for binding. However, measurements at higher concentrations reveal that the rate reaches a plateau, indicating a two-step process. [From E. Holler, J. A. Rupley, and G. P. Hess, *Biochem. Biophys. Res. Commun.* **37**, 423 (1969).]

imagination and insight in designing the incisive experiment. But there is a systematic approach that can be used.

The analysis of a relaxation time may be divided into two basic steps:

1. The relaxation time must be assigned to a specific physical event.
2. This physical event must be fitted into the overall kinetic mechanism of the process under study.

Let us now consider two examples. The first is the binding of a ligand to a protein under noncatalytic conditions. Two likely physical events are the initial binding step and a ligand-induced conformational change. The first experiment is the measurement of the concentration dependence and the number of relaxation times in order to determine the number of intermediate states and the rate constants for their interconversion. Under ideal circumstances, the number of relaxation times will be equal to the number of steps in the reaction. Even if only one relaxation time is found, its concentration dependence may indicate a two-step process by being nonlinear (e.g., equations 4.73 and 4.76). Additional information may be obtained by using more than one physical probe, for example fluorescence and absorbance, and by studying both the ligand and the protein, since some steps may show up in one of these and not in the other. Further physical processes may occur, such as proton release or uptake during the reaction, or a change in the state of aggregation of the protein. The former provides an additional convenient probe since changes of pH may be measured by a chromophoric pH indicator. Aggregation complicates the kinetics, but may be detected and measured to provide additional information. Relaxation techniques are often more powerful than flow methods for these simple reactions because they can measure faster processes. However, there are times when stopped flow is more useful: For example, processes that are too slow for detection by temperature jump may be measured by stopped flow; and also, certain experiments, such as those to determine the effect of a large change in pH, can be performed only by mixing (although small changes in pH can be made in a temperature-jump experiment by using a buffer whose pK_a is temperature-dependent).

The second example is the reaction mechanism of an enzyme under catalytic conditions. In addition to determination of the binding steps and conformational changes described above, it is even more important to measure the bond making and breaking steps and to detect the chemical intermediates in the reaction. The chemical steps are usually best studied by the methods that directly measure the concentration of chemical species—for example, stopped-flow spectrophotometry and quenched flow. Indirect measurements of chemical steps, such as a change in protein fluorescence, must somehow be assigned. The ideal situation for study is the accumulation of an intermediate that may be detected and measured. In general, as many different probes as possible should be used in order to confirm existing information and add further details. Good examples of the application of such methods are to be found in Chapter 7.

E. The absolute concentration of enzymes

1. Active-site titration and the magnitudes of "bursts"

The calculation of rate constants from steady state kinetics and the determination of binding stoichiometries requires a knowledge of the concentration of active sites in the enzyme. It is not sufficient to calculate this specific concentration value from the relative molecular mass of the protein and its concentration, since isolated enzymes are not always 100% pure. This problem has been overcome by the introduction of the technique of active-site titration, a combination of steady state and pre–steady state kinetics whereby the concentration of active enzyme is related to an initial burst of product formation. This type of situation occurs when an enzyme-bound intermediate accumulates during the reaction. The first mole of substrate rapidly reacts with the enzyme to form stoichiometric amounts of the enzyme-bound intermediate and product, but then the subsequent reaction is slow since it depends on the slow breakdown of the intermediate to release free enzyme.

$$E + S \xrightarrow{\ k_1'\ } EI \xrightarrow{\ k_2\ } E + P_2 \qquad\qquad (4.78)$$
$$\phantom{E + S \xrightarrow{\ k_1'\ } EI}\ + P_1$$

Clearly, if in equation 4.78 k_1' is very fast and k_2 is negligibly slow, the release of P_1 is easily measured and related to the concentration of enzyme. However, in practice, k_2 is generally not negligible, so that there is an initial burst of formation of P_1 followed by a progressive increase as the intermediate turns over. The mathematics of this situation was described previously (equations 4.37 to 4.46). It was shown that the overall release of products is linear with time after an initial transient. From equation 4.46 it can be seen that the linear portion extrapolates back to a burst, π, given by

$$\pi = [E]_0 \left(\frac{k_1'}{k_1' + k_2} \right)^2 \tag{4.79}$$

(Figures 4.10 and 4.11). It should be noted that the burst depends on a "squared" relationship with the rate constants. If the ratio $k_1':k_2$ is high, the squared term is close to 1, so that the burst is equal to the enzyme concentration. If this condition does not hold, the concentration will be underestimated unless both rate constants are measured and substituted into equation 4.79.

2. The dependence of the burst on substrate concentration

In equation 4.78, the term k_1' is the apparent first-order rate constant for the formation of the intermediate under the particular reaction conditions. In general, this will follow the Michaelis-Menten equation, that is,

$$k_1' = \frac{k_{cat}[S]}{[S] + K_M} \tag{4.80}$$

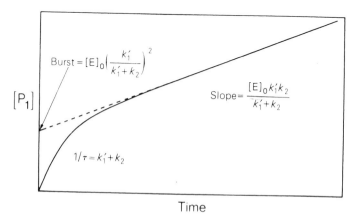

Figure 4.10 The principle of active-site titration.

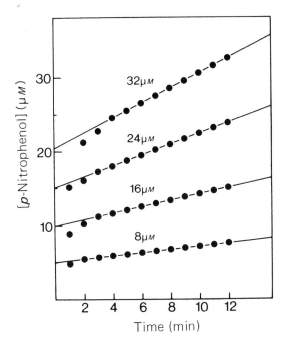

Figure 4.11 The "original" active-site titration experiment. The indicated concentrations of chymotrypsin were mixed with p-nitrophenyl ethyl carbonate.

$$EtOCO_2 - \langle\!\!\!\!\!\!\!\bigcirc\!\!\!\!\!\!\!\rangle - NO_2$$

The acylenzyme E—O—COOEt is rapidly formed but hydrolyzes slowly. Note that about 0.63 mol of p-nitrophenol is released per mole of enzyme in the "burst." Either the enzyme is only 63% pure (active), or the rate constant for the formation of the acylenzyme is not sufficiently greater than that for deacylation for the acylenzyme to accumulate fully. [From B. S. Hartley and B. A. Kilby, *Biochem. J.* **56**, 288 (1954).]

where k_{cat} and k_M refer to the first step. At sufficiently low concentrations of S, there will be no burst, but, provided that k_{cat} is greater than k_2, one will occur at higher concentrations. Substituting equation 4.80 into 4.79 gives

$$\frac{1}{\sqrt{\pi}} = \frac{1}{\sqrt{[E]_0}}\left(1 + \frac{k_2}{k_{cat}} + \frac{K_M k_2}{[S]k_{cat}}\right) \qquad (4.81)$$

If k_{cat} is much greater than k_2, equation 4.81 may be used to extrapolate the burst from measurements at various substrate concentrations. It is obvious that when k_2 is not negligible, care must be taken not to underestimate the concentration of the enzyme.

3. Active-site titration versus rate assay

Active-site titration is not always applicable, since it requires the accumulation of an intermediate in the reaction. The more usual procedure is to determine the concentration of an enzyme from a rate assay. This has the disadvantage that it does not give the absolute concentration of the enzyme unless it has been calibrated against an active-site titration. Further, rate measurements are sensitive to the reaction conditions. Whereas these may be controlled with some precision in a particular laboratory, they often vary from laboratory to laboratory. Active-site titration suffers from the disadvantage that several milligrams of enzyme are required for a spectrophotometric assay, but this quantity may be reduced a thousand times by using radioactive methods. With its relative insensitivity to precise reaction conditions and its yield of absolute values for the concentrations of enzyme solutions, active-site titration has been a most important factor in providing highly reproducible data and making possible the comparison of rate constants from steady state and pre–steady state kinetics. (The rate constants of pre–steady state kinetics generally involve exponential processes that do not depend on the concentration of enzyme, whereas steady state rates are typically directly proportional to the concentration.)

PART 2

The magnitude of rate constants for enzymatic processes

A. Upper limits on rate constants[15]

1. Association and dissociation

A simple way of analyzing the rate constants of chemical reactions is the *collision theory* of reaction kinetics. The rate constant for a bimolecular reaction is considered to be composed of the product of three terms: the frequency of collisions, Z; a steric factor, p, to allow for the fraction of the molecules that are in the correct orientation; and an activation energy term to allow for the fraction of the molecules that are sufficiently thermally activated to react. That is,

$$k_2 = Zp \exp\left(-\frac{E_{act}}{RT}\right) \tag{4.82}$$

The maximum value for the bimolecular rate constant occurs when the activation energy E_{act} is zero and the steric factor is 1. The rate is then said to be *diffusion-controlled*, and it is equal to the encounter frequency of the molecules. Assuming that the reacting molecules are uncharged spheres of radius r_A and r_B, the encounter frequency may be calculated as

$$Z = \left(\frac{2RT}{3000\eta} \right) \frac{(r_A + r_B)^2}{r_A r_B} \tag{4.83}$$

(where η is the viscosity). For two molecules of the same radius in water at 25°C, the encounter frequency is equal to $7 \times 10^9 \text{ s}^{-1} M^{-1}$. It should be noted that two large molecules collide at exactly the same rate as two small ones. This is because the increase of target area exactly compensates for the slower diffusion of the larger molecules. However, the rate of encounter of a small molecule with a large molecule is higher than this value because of the combination of the large target area of the latter and the high mobility of the former. More sophisticated calculations, allowing for the possibility of favorable electrostatic interactions (at one extreme) and an unfavorable geometry for a small molecule hitting a particular target area of a larger one (at the other extreme), give a range of 10^9 to $10^{11} \text{ s}^{-1} M^{-1}$ for the encounter frequency. A similar treatment gives a range of 10^9 to 10^{12} s^{-1} for the upper limit on the *dissociation* rate constants of bimolecular complexes. Many of the second-order rate constants that do not involve the proton or hydroxide ion are found experimentally to be about $10^9 \text{ s}^{-1} M^{-1}$.

The rate constants for the association of proteins with one another and with other macromolecules are profoundly influenced by the geometry of the interaction and by electrostatic factors. Only a small part of each protein may be involved in the formation of a protein-protein complex, which imposes a bad steric factor on the reaction. Accordingly, protein-protein association rate constants may be as low as $10^4 \text{ s}^{-1} M^{-1}$ (Table 4.1). But there is very fast association at $> 5 \times 10^9 \text{ s}^{-1} M^{-1}$ at low ionic strength for proteins that have complementary charged surfaces, such as barnase with its polypeptide inhibitor barstar (whose properties are discussed in Chapter 19), thrombin with its polypeptide inhibitor hirudin, and ferricytochrome c with ferrocytochrome b_5.

The contribution of electrostatic interactions to fast association was analyzed by applying the classical Debye-Hückel theory of electrostatic interactions between ions to mutants of barnase and barstar whose ionic side chains had been altered by protein engineering (Chapter 14).[16] The association fits a two-step model that is probably general (equation 4.84).

$$A + B \underset{k_{-1}}{\overset{k_1}{\rightleftharpoons}} A.B \xrightarrow{k_2} C \tag{4.84}$$

There is first the formation of a loose encounter complex, followed by the precise docking. Electrostatic interactions affect the equilibrium constant for the formation of the encounter complex. These depend greatly on ionic strength, I; interactions are the highest at low ionic strength and are masked at high ionic strength. A simplified electrostatic model assuming two uniformly charged spheres predicts the rate constant for the association, k_{ass}, to vary with the charge on each component, z, and ionic strength according to the equation:

$$\log k_{ass} = (\log k_{ass})_{I=0} + \alpha \log f_{\pm}^{*} \tag{4.85}$$

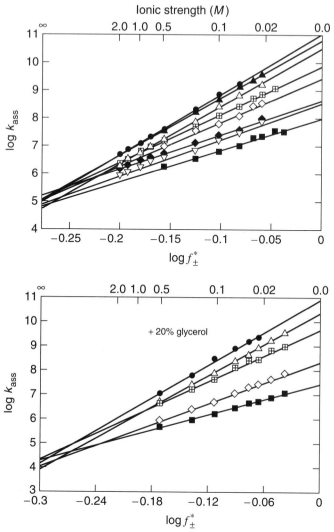

Figure 4.12 Rate constants for the association of barnase and barstar at different ionic strengths. The ionic strength I is varied by changing the concentration of NaCl. $\log f_{\pm}^{*} = A|z_A z_B|\sqrt{I}/(1 + Ba\sqrt{I})$. A and B are constants that depend only on the temperature and dielectric constant of the solution; a is the closest distance of approach of the ions. In this context, z_A and z_B are the charges on the Na^+ and Cl^- ions. The simple Debye-Hückel theory breaks down rapidly with increasing concentration for multivalent ions, and so simple linear plots are not found with $MgCl_2$ over such a wide concentration range. [From G. Schreiber and A. R Fersht, *Nature Structural Biology* **3**, 427 (1996).]

where $\alpha = -2z_A z_B/2.303$ and $f_{\pm}^{*}$ is the electrostatic component of the mean rational activity coefficient of the ions. $f_{\pm}^{*}$ varies with ionic strength. The plots of log k_{ass} against log $f_{\pm}^{*}$ in Figure 4.12 for a series of mutants of barnase and barstar of varying values of z_A and z_B all converge to a value of about 10^5 s^{-1} M^{-1} at infinite ionic strength, which should completely mask the electrostatic component. This implies that the basal association rate constant is 10^5 s^{-1} M^{-1}. The addition of 20% glycerol significantly increases the viscosity and so lowers the association rate constant by a factor of seven, as predicted from equation 4.83, which is a diagnostic test for diffusion-controlled reactions. It may also be derived that

$$(\log k_{ass})_{I=0} = (\log k_{ass})_0 + \alpha\beta \tag{4.86}$$

where $(\log k_{ass})_0$ is the rate constant in the absence of electrostatic interactions and β is a simple function of temperature. A plot of $(\log k_{ass})_{I=0}$ versus α for different mutants (Figure 4.13) also extrapolates to a value of 10^5 s^{-1} M^{-1} for the basal association rate constant in the absence of electrostatic effects. A value of 5×10^4 s^{-1} M^{-1} is similarly calculated for the basal association rate constant of hirudin and thrombin.

The data on barnase and barstar may be analyzed by simple[17] or computer-simulation methods.[18,19]

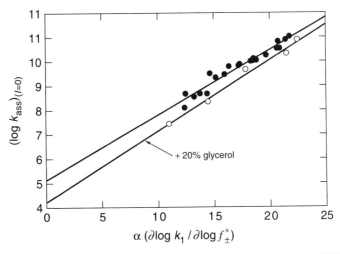

Figure 4.13 Plot of $(\log k_{ass})_{I=0}$, the rate constants for the association of different mutants at zero ionic strength, against α (equation 4.86). This extrapolates at $\alpha = 0$ to the basla association rate constant in the absence of electrostatic effects. [From G. Schreiber and A. R. Fersht, *Nature Structural Biology* **3**, 427 (1996).]

2. Chemical processes

The upper limit on the rate constant of any unimolecular or intramolecular reaction is the frequency of a molecular vibration, about 10^{12} to 10^{13} s^{-1}.

3. Proton transfers

Favorable proton transfers between electronegative atoms such as O, N, and S are extremely fast. The bimolecular rate constants are generally diffusion-controlled, being 10^{10} to 10^{11} s^{-1} M^{-1} (Table 4.2). For example, the rate constant for the transfer of a proton from H_3O^+ to imidazole, a favorable transfer since imidazole is a stronger base than H_2O, is 1.5×10^{10} s^{-1} M^{-1} (Table 4.3). The rate constant for the reverse reaction, the transfer of a proton from the imidazolium ion to water, may be calculated from the difference in their pK_a's by using the following equations:

$$\frac{[B][H^+]}{[BH^+]} = K_a \quad \frac{[A^-][H^+]}{[HA]} = K'_a \tag{4.87}$$

$$\frac{[B][HA]}{[BH^+][A^-]} = K_a/K'_a \tag{4.88}$$

$$[B] + [HA] \underset{k_{-1}}{\overset{k_1}{\rightleftharpoons}} [BH^+] + [A^-] \tag{4.89}$$

so that

$$\frac{k_{-1}}{k_1} = \frac{K_a}{K'_a} \tag{4.90}$$

Table 4.2 *Proton-transfer rate constants (25°C, s^{-1} M^{-1})[a]*

H$^+$ and	k (s^{-1} M^{-1})	OH$^-$ and	k (s^{-1} M^{-1})
OH$^-$	1.4×10^{11}		
Inorganic acid anions	$10^8 - 10^{11}$	Inorganic acids	$\sim 10^{10}$
Carboxylates	$\sim 5 \times 10^{10}$	Carboxylic acids	$\sim 10^{10}$
Phenolates	$\sim 5 \times 10^{10}$	Phenols	
Enolates	$\sim 5 \times 10^{10}$	Enols	$\sim 10^{10}$
Amines	10^{10}	Ammonium	$\sim 10^{10}$
		ions	$\sim 3 \times 10^{10}$
Carbanions	$<1 - \sim 10^{10}$	Carbon acids	$<1 - \sim 10^{10}$
		Phosphoric acids	$10^8 - 10^{11}$

[a] From M. Eigen, *Nobel Symp.* **5**, 245 (1967).

Table 4.3 *Proton-transfer rates involving imidazole* ($pK_a = 6.95$)[a]

Donor (DH$^+$)	pK_a	k (DH$^+ \rightarrow$ Im) (s^{-1} M^{-1})	k (ImH$^+ \rightarrow$ D) (s^{-1} M^{-1})
H$_3$O$^+$	-1.74	1.5×10^{10}	31
H$_2$O	15.74	45	2.3×10^{10}
CH$_3$CO$_2$H	4.76	1.2×10^9	7.7×10^6
HATP^{3-}	6.7	2×10^9	1×10^9
p-Nitrophenol	7.14	4.5×10^8	7.0×10^8
HP$_2$O$_7^{3-}$	8.45	1.1×10^8	3.6×10^9
Phenol	9.95	1×10^7	1×10^{10}
CO$_3^{2-}$	10.33	1.9×10^7	2×10^{10}
Glucose	12.3	1.6×10^5	2×10^{10}

[a] At 25°C, ionic strength = 0. The H$_2$O rates are calculated on [H$_2$O] = 55 M. [From M. Eigen and G. G. Hammes, *Adv. Enzymol.* **25**, 1 (1963).]

The rate constant for the transfer of a proton from the imidazolium ion ($pK_a = 6.95$) to water ([H$_2$O] = 55 M, $pK_a = -1.74$) is calculated from equation 4.87 to be 1.7×10^3 s^{-1}.

Proton transfers from carbon acids and to carbon bases are generally much slower. This is because the lower electronegativity of carbon requires that the negative charge on a carbon base be stabilized by electron delocalization. The consequent reorganization of structure and solvent may slow down the overall transfer rate.

It was once thought that the rate of equilibrium of the catalytic acid and basic groups on an enzyme with the solvent limited the rates of acid- and base-catalyzed reactions to turnover numbers of 10^3 s^{-1} or less. This is because the rate constants for the transfer of a proton from the imidazolium ion to water and from water to imidazole are about 2×10^3 s^{-1}. However, protons are transferred between imidazole or imidazolium ion and buffer species in solution with rate constants that are many times higher than this. For example, the rate constants with ATP, which has a pK_a similar to imidazole's, are about 10^9 s^{-1} M^{-1}, and the ATP concentration is about 2 mM in the cell. Similarly, several other metabolites that are present at millimolar concentrations have acidic and basic groups that allow catalytic groups on an enzyme to equilibrate with the solvent at 10^7 to 10^8 s^{-1} or faster. Enzyme turnover numbers are usually considerably lower than this, in the range of 10 to 10^3 s^{-1}, although carbonic anhydrase and catalase have turnover numbers of 10^6 and 4×10^7 s^{-1}, respectively.

B. Enzymatic rate constants and rate-determining processes

1. Association of enzymes and substrates

Calculations suggest that the diffusion-controlled encounter frequency of an enzyme and a substrate should be about 10^9 s^{-1} M^{-1}. The observed values in Table 4.4 tend to fall in the range of 10^6 to 10^8 s^{-1} M^{-1}. The faster ones are close to

Table 4.4 *Association and dissociation rate constants for enzyme–substrate interactions*

Enzyme	Substrate	k_1 (s^{-1} M^{-1})	k_{-1} (s^{-1})	Ref.
Protein-small ligands				
Catalase	H_2O_2	5×10^6		1
Catalase–H_2O_2	H_2O_2	1.5×10^7		1
Chymotrypsin	Proflavin	1.2×10^8	8.3×10^3	2
	Acetyl-L-tryptophan *p*-nitrophenyl ester	6×10^7	6×10^4	3
	Furylacryloyl-L-tryptophanamide	6.2×10^6		4
	Trifluorylacetyl-D-tryptophan	1.5×10^7		5
	Indole	1.9×10^7	5.8×10^3	6
Creatine kinase	ADP	2.2×10^7	1.8×10^4	7
	MgADP	5.3×10^6	5.1×10^3	
Glyceraldehyde 3-phosphate dehydrogenase	NAD$^+$	1.9×10^7	1×10^3	8
		1.4×10^6	210	
Lactate dehydrogenase (rabbit muscle)	NADH	$\sim 10^9$	$\sim 10^4$	9

1 B. Chance, in *Currents in biochemical research* (D. E. Green, Ed.), Wiley, p. 308 (1956).
2 U. Quast, J. Engel, H. Heumann, G. Krause, and E. Steffen, *Biochemistry* **13**, 2512 (1974).
3 M. Renard and A. R. Fersht, *Biochemistry* **12**, 4713 (1973).
4 G. P. Hess, J. McConn, E. Ku, and G. McConkey, *Phil. Trans. R. Soc.* **B257**, 89 (1970).
5 S. H. Smallcombe, B. Ault, and J. H. Richards, *J. Am. Chem. Soc.* **94**, 4585 (1972).
6 R. Maehler and J. R. Whitaker, *Biochemistry* **21**, 4621 (1982).
7 G. G. Hammes and J. K. Hurst, *Biochemistry* **8**, 1083 (1969).
8 K. Kirschner, M. Eigen, R. Bittman, and B. Voigt, *Proc. Natl. Acad. Sci. USA* **56**, 1661 (1966).
9 G. H. Czerlinski and G. Schreck, *J. Biol. Chem.* **239**, 913 (1964).

diffusion-controlled, but the slower ones are significantly lower than the limit. This may be partly due to desolvation requirements in some cases, or, as is more likely in others, to a two-step process that appears as a single step. For example, at low concentrations the binding of $(NAG)_2$ to lysozyme appears to occur at about 5×10^6 s^{-1} M^{-1}. But extension of the measurements to higher concentrations shows that the binding is a two-step process with an association rate constant of 4×10^7 s^{-1} M^{-1} at pH 4.4 and 31°C (see section D6 and Figure 4.9).[20]

Enzyme	Substrate	k_1 (s^{-1} M^{-1})	k_{-1} (s^{-1})	Ref.
Protein-small ligands				
Lactate	NADH	5.5×10^7	39	10
dehydrogenase (pig heart)	Oxamate	8.1×10^6	17	10
Liver alcohol dehydrogenase	NADH	2.5×10^7	9	11
Lysozyme	$(NAG)_2$	4×10^7	1×10^5	12, 13
Malate dehydrogenase	NADH	5×10^8	50	14
Pyruvate carboxylase– Mn^{2+}	Pyruvate	4.5×10^6	2.1×10^4	15
Ribonuclease	Uridine 3'-phosphate	7.8×10^7	1.1×10^4	16
	Uridine 2', 3'-cyclic phosphate	1×10^7	2×10^4	17
Tyrosyl-tRNA synthetase	Tyrosine	2.4×10^6	24	18

10 H. d'A. Heck, *J. Biol. Chem.* **244**, 4375 (1969).
11 J. D. Shore and H. Gutfreund, *Biochemistry* **9**, 4655 (1970).
12 E. Holler, J. A. Rupley, and G. P. Hess, *Biochem. Biophys. Res. Commun.* **37**, 423 (1969).
13 J. H. Baldo, S. E. Halford, S. L. Patt, and B. D. Sykes, *Biochemistry* **14**, 1893 (1975).
14 G. Czerlinski and G. Schreck, *Biochemistry* **3**, 89 (1964).
15 A. S. Mildvan and M. C. Scrutton, *Biochemistry* **6**, 2978 (1967).
16 G. G. Hammes and F. G. Walz, Jr., *J. Am. Chem. Soc.* **91**, 7179 (1969).
17 E. J. del Rosario and G. G. Hammes, *J. Am. Chem. Soc.* **92**, 1750 (1970).
18 A. R. Fersht, R. S. Mulvey, and G. L. E. Koch, *Biochemistry,* **14**, 13 (1975).

$$E + (NAG)_2 \underset{1.2 \times 10^5 \, s^{-1}}{\overset{4 \times 10^7 \, s^{-1} M^{-1}}{\rightleftharpoons}} E \cdot (NAG)_2 \underset{1.3 \times 10^3 \, s^{-1}}{\overset{1.7 \times 10^4 \, s^{-1}}{\rightleftharpoons}} E' \cdot (NAG)_2 \qquad (4.91)$$

2. Association can be rate-determining for k_{cat}/K_M

Table 4.5 shows that for some efficient enzymes, k_{cat}/K_M may be as high as 3×10^8 s^{-1} M^{-1}. In these cases, the rate-determining step for this parameter, which is the apparent second-order rate constant for the reaction of free enzyme with free substrate, is close to the diffusion-controlled encounter of the enzyme and the substrate. Briggs-Haldane kinetics holds for these enzymes (Chapter 3, section B3).

Table 4.5 *Enzymes for which k_{cat}/K_M is close to the diffusion-controlled association rate*

Enzyme	Substrate	k_{cat} (s^{-1})	k_M M	k_{cat}/K_M (s^{-1} M^{-1})	Ref.
Acetylcholin-esterase	Acetylcholine	1.4×10^4	9×10^{-5}	1.6×10^8	1
Carbonic	CO_2	1×10^6	0.012	8.3×10^7	2
anhydrase	HCO_3^-	4×10^5	0.026	1.5×10^7	3
Catalase	H_2O_2	4×10^7	1.1	4×10^7	4
Crotonase	Crotonyl-CoA	5.7×10^3	2×10^{-5}	2.8×10^8	5
Fumarase	Fumarate	800	5×10^{-6}	1.6×10^8	6
	Malate	900	2.5×10^{-5}	3.6×10^7	6
Triosephosphate isomerase	Glyceraldehyde 3-phosphate	4.3×10^3	4.7×10^{-4}	$2.4 \times 10^{8\,a}$	7
β-Lactamase	Benzylpenicillin	2.0×10^3	2×10^{-5}	1×10^8	8
Superoxide dismutase	Superoxide	—	—	7×10^9	9

[a] The observed value is 9.1×10^6 s^{-1} M^{-1}. The tabulated value is calculated on the basis of only 3.8% of the substrate being reactive, since 96.2% is hydrated under the conditions of the experiment.

1 T. I. Rosenberry, *Adv. Enzymol.* **43**, 103 (1975).
2 J. C. Kernohan, *Biochim. Biophys. Acta.* **81**, 346 (1964).
3 J. C. Kernohan, *Biochim. Biophys. Acta.* **96**, 304 (1965).
4 Y. Ogura, *Archs. Biochem. Biophys.* **57**, 288 (1955).
5 R. M. Waterson and R. L. Hill, *Fedn. Proc.* **30**, 1114 (1971).
6 J. W. Teipel, G. M. Hass, and R. L. Hill, *J. Biol. Chem.* **243**, 5684 (1968).
7 S. J. Putman, A. F. W. Coulson, I. R. T. Farley, B. Riddleston, and J. R. Knowles, *Biochem. J.* **129**, 301 (1972).
8 J. Fisher, J. G. Belasco, S. Khosla, and J. R. Knowles, *Biochemistry* **19**, 2985 (1980).
9 I. Bertini, S. Mangani, and M. S. Viezzoli, *Advances in Inorganic Chemistry* **45**, 127 (1998).

3. Dissociation of enzyme–substrate and enzyme–product complexes

Dissociation rate constants are much lower than the diffusion-controlled limit, since the forces responsible for the binding must be overcome in the dissociation step. In some cases, enzyme–substrate dissociation is slower than the subsequent chemical steps, and this gives rise to Briggs-Haldane kinetics.

4. Enzyme–product release can be rate-determining for k_{cat}

Product release is sometimes rate-determining at saturating substrate concentrations with some dehydrogenases. Examples of this are the dissociation of NADH from glyceraldehyde 3-phosphate dehydrogenase at high pH,[21] of NADH from horse liver alcohol dehydrogenase at low salt,[22,23] and of NADPH from glutamate dehydrogenase.[24] Note that the phrase product release is used, and not product dissociation. This is because the overall release of products can involve steps in addition to dissociation, such as conformational changes, and these may be the rate-determining steps rather than the dissociation itself (see the next section).

5. Conformational changes

There are many documented cases of substrate-induced conformational changes with rate constants in the range of 10 to 10^4 s^{-1}, and also instances in which discrepancies in rate constants indicate rate-determining protein isomerizations.[24] Isomerizations are often associated with slow steps; for example, the dissociation of NADH from some dehydrogenases involves a concomitant conformational change. However, there are few direct demonstrations that a conformational change is, by itself, rate-determining. An exception is in the reactions of triosephosphate isomerase. A loop of the protein acts as a "trapdoor" to encircle the substrate and product. The opening and closing of the trapdoor closely matches the values of k_{cat} for the reaction[25] (Chapter 1, section G2e).

It should be noted that the rate-determining step of a reaction changes with substrate concentration, since the rate is proportional to k_{cat} at saturating concentrations of substrate, and to k_{cat}/K_M at low concentrations. When a step is said to be rate-determining and the reaction conditions are not stated, the reaction is usually at saturating substrate concentrations.

References

1. H. Hartridge and F. J. W. Roughton, *Proc. R. Soc.* **A104**, 376 (1923).
2. C. K. Chan, Y. Hu, S. Takahashi, D. L. Rousseau, W. A. Eaton, and J. Hofrichter, *Proc. Natl. Acad. Sci. USA.* **94**, 1779 (1997).
3. M. C. Ramachandra Shastry and H. Roder, *Nature Structural Biology* **5**, 385 (1998).
4. F. J. W. Roughton, *Proc. R. Soc.* **B115**, 475 (1934).

5. B. Chance, *J. Franklin Inst.* **229**, 455, 613, 637 (1940).
6. Q. H. Gibson, *J. Physiol.* **117**, 49P (1952).
7. A. R. Fersht and R. Jakes, *Biochemistry* **14**, 3350 (1975).
8. Q. H. Gibson, *Progr. Biophys. Biophys. Chem.* **2**, 1 (1959).
9. J. L. Martin, A. Migus, C. Poyart, Y. Lecarpentier, A. Antonetti, and A. Orszag, *Biochem. Biophys. Res Commun.* **107**, 803 (1982).
10. J. H. Kaplan, B. Forbush III, and J. F. Hoffman, *Biochemistry* **17**, 1929 (1978).
11. Y. E. Goldman, M. G. Hibberd, J. A. McRay, and D. R. Trentham, *Nature, Lond.* **300**, 701 (1982).
12. I. Schlichting and R. S. Goody, *Triggering methods in crystallographic enzyme kinetics*, in *Macromolecular Crystallography*, C. W. Carter and R. M. Sweet, eds., Academic Press, 467 (1997).
13. G. Czerlinski and M. Eigen, *Z. Electrochem.* **63**, 652 (1959).
14. G. S. Huang and T. G. Oas, *Proc. Natl. Acad. Sci. USA.* **92**, 6878 (1995).
15. M. Eigen and G. G. Hammes, *Adv. Enzymol.* **25**, 1 (1963).
16. G. Schreiber and A. R. Fersht, *Nature Structural Biology* **3**, 427 (1996).
17. J. Janin, *Proteins: Structure, Function and Genetics* **28**, 153 (1997).
18. R. R. Gabdoulline and R. C. Wade, *METHODS: A Companion to Methods in Enzymology* **14**, 329 (1998).
19. M. Vijaykoumar, K. Y. Wong, G. Schreiber, A. R. Fersht, A. Szabo, and H. X. Zhou, *J. Molec. Biol.* **278**, 1015 (1998).
20. E. Holler, J. A. Rupley, and G. P. Hess, *Biochem. Biophys. Res. Commun.* **37**, 423 (1969).
21. D. R. Trentham, *Biochem. J.* **122**, 71 (1971).
22. H. Theorell and B. Chance, *Acta Chem. Scand.,* **5**, 1127 (1951).
23. J. D. Shore and H. Gutfreund, *Biochemistry* **9**, 4655 (1970).
24. A. di Franco, *Eur. J. Biochem.* **45**, 407 (1974).
25. J. C. Williams and A. E. McDermott, *Biochemistry* **34**, 8309 (1995).

The pH Dependence of Enzyme Catalysis

5

T he activities of many enzymes vary with pH in the same way that simple acids and bases ionize. This is not surprising, since, as we saw in Chapter 1, the active sites generally contain important acidic or basic groups (Table 5.1). It is to be expected that if only one protonic form of the acid or base is catalytically active, the catalysis will somehow depend on the concentration of the active form. In this chapter we shall see that k_{cat}, K_M, and k_{cat}/K_M are affected in different ways by the ionizations of the enzyme and enzyme–substrate complex.

A. Ionization of simple acids and bases: The basic equations

It is usual to discuss the ionization of a base B in terms of its conjugate acid BH^+ in order to use the same set of equations for both acids and bases. The ionization constant K_a is defined by

$$K_a = \frac{[B][H^+]}{[BH^+]} \tag{5.1}$$

Or for an acid HA and its conjugate base A^-,

$$K_a = \frac{[A^-][H^+]}{[HA]} \tag{5.2}$$

The pK_a is defined by

Table 5.1	pK_a's of ionizing groups[a]	
		pK_a
Group	Model compounds (small peptides)	Usual range in proteins
Amino acid α-CO_2H	3.6	
Asp (CO_2H)	4.0	2–5.5
Glu (CO_2H)	4.5	
His (imidazole)	6.4	5–8
Amino acid α-NH_2	7.8	~8
Lys (ϵ-NH_2)	10.4	~10
Arg (guanidine)	~12	—
Tyr (OH)	9.7	9–12
Cys (SH)	9.1	8–11
Phosphates	1.3, 6.5	—

[a] Data mainly from C. Tanford, *Adv. Protein Chem.* **17**, 69 (1962); C. Tanford and R. Roxby, *Biochemistry* **11**, 2192 (1972); Z. Shaked, R. P. Szajewski, and G. M. Whitesides, *Biochemistry* **19**, 4156 (1980).

$$pK_a = -\log K_a \tag{5.3}$$

Equations 5.1 and 5.3 (or 5.2) may be rearranged to give the Henderson-Hasselbalch equation:

$$pH = pK_a + \log \frac{[B]}{[BH^+]} \tag{5.4}$$

It is readily seen from this equation that the pK_a of an acid or a base is the pH of half neutralization when the concentrations of B and BH^+ are equal.

The variation of the concentrations of HA and A^- with the proton concentration is found from rearranging equation 5.2 as

$$[HA] = \frac{[A]_0[H^+]}{K_a + [H^+]} \tag{5.5}$$

and

$$[A^-] = \frac{[A]_0 K_a}{K_a + [H^+]} \tag{5.6}$$

where $[A]_0 = [HA] + [A^-]$.

Suppose that there is some quantity L (an absorption coefficient, a rate constant, etc.) such that the corresponding property of a solution (the absorbance, the reaction rate, etc.) is the product of this quantity and the concentration. If the value of L for the molecule HA is L_{HA}, and the value for the molecule A^- is L_{A^-}, then the observed value of the property at a particular pH, $L_H[A]_0$, is given by $L_H[A]_0 = L_{HA}[HA] + L_{A^-}[A^-]$. That is:

$$L_H[A]_0 = \frac{L_{HA}[A]_0[H^+]}{K_a + [H^+]} + \frac{L_{A^-}[A]_0 K_a}{K_a + [H^+]} \qquad (5.7)$$

so that

$$L_H = \frac{L_{HA}[H^+] + L_A K_a}{K_a + [H^+]} \qquad (5.8)$$

An example of equation 5.8 in which L is an equilibrium constant K is plotted in Figure 5.1. A further example is shown in Figure 5.2, where L is a rate constant k

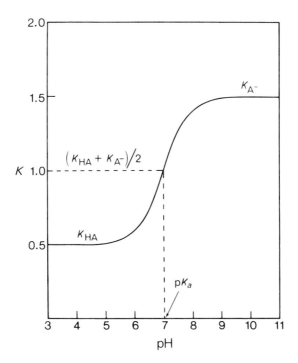

Figure 5.1 Plot of the pH dependence of an arbitrary constant K that has the value of K_{HA} for the acidic form and K_A for the basic form of an acid of pK_a 7.

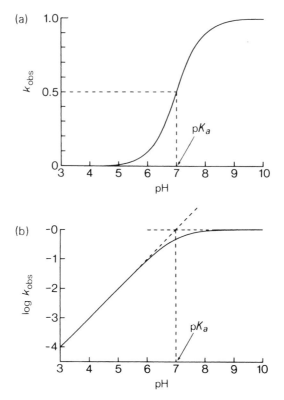

Figure 5.2 Plots of (a) k_{obs} and (b) log k_{obs} against pH for a reaction whose rate depends on the basic form of an acid of pK_a 7.

for a reaction that depends on the basic form of an acid (i.e., $L_{HA} = 0$). This is plotted in two ways. Note in the logarithmic plot that there is a linear decrease with decreasing pH at low pH. The point of intersection of the two linear regions is the pK_a.

A *doubly ionizing* system such as

$$H_2A \overset{K_1}{\rightleftharpoons} HA^- + H^+ \overset{K_2}{\rightleftharpoons} A^{2-} + 2H^+ \tag{5.9}$$

may be analyzed to give

$$L_H = \frac{[H^+]^2 L_{H_2A} + [H^+]K_1 L_{HA^-} + K_1 K_2 L_{A^{2-}}}{K_1 K_2 + [H^+]K_1 + [H^+]^2} \tag{5.10}$$

It is frequently found in enzyme reactions that activity depends on two groups, one in the acidic form and one in the basic. This is equivalent to the situation in equation 5.9, where rate varies as $[HA^-]$, leading to equation 5.10, in which $L_{H_2A} = L_{A^{2-}} = 0$. Equation 5.10 is then the equation of a bell-shaped curve with its maximum at $pH = (pK_{a1} + pK_{a2})/2$.

1. Extraction of pK_a's by inspection of equations

The crucial point in equations 5.8 and 5.10 is that the points of inflection in the plots of L_H against pH—i.e., the pK_a's—are determined by the *denominator* of the fraction. The numerator determines the amplitudes of the functions. The apparent pK_a's of a complex kinetic equation may be found by rearranging the equation to the form of equation 5.8 (for a singly ionizing system) or to that of equation 5.10 (for a doubly ionizing system), and then comparing the denominators. This may be illustrated by considering the pH dependence of $1/L_H$ instead of L_H in equation 5.8; i.e.,

$$\frac{1}{L_H} = \frac{K_a + [H^+]}{L_{HA}[H^+] + L_{A^-}K_a} \tag{5.11}$$

Rearranging equation 5.11 to be of the same form as 5.8 gives

$$\frac{1}{L_H} = \frac{K_a/L_{HA} + [H^+]/L_{HA}}{K_a L_{A^-}/L_{HA} + [H^+]} \tag{5.12}$$

We can say directly from examining the denominator of equation 5.12 that the plot of $1/L_H$ against pH will have an apparent ionization constant, K_{app}, given by

$$K_{app} = K_a \frac{L_{A^-}}{L_{HA}} \tag{5.13}$$

B. The effect of ionizations of groups in enzymes on kinetics

Although enzymes contain a multitude of ionizing groups, it is usually found that plots of rate against pH take the form of simple single or double ionization curves. This is because the only ionizations that are of importance are those of groups that are directly involved in catalysis at the active site, or those of groups elsewhere that are responsible for maintaining the active conformation of the enzyme.

We shall now analyze some of the simple examples, beginning with the Michaelis-Menten mechanism. We can make four simplifying assumptions that may break down in some circumstances but that often hold in practice:

1. The groups act as perfectly titrating acids or bases (this is generally a fair approximation).

2. Only one ionic form of the enzyme is active (this is usually true).
3. All intermediates are in protonic equilibrium; i.e., proton transfers are faster than chemical steps (this is generally true—see Chapter 4).
4. The rate-determining step does not change with pH (this may break down, with interesting consequences).

1. The simple theory: The Michaelis-Menten mechanism

$$
\begin{array}{ccc}
\text{E} & \underset{}{\overset{K_S}{\rightleftharpoons}} \text{ES} & \xrightarrow[\text{slow}]{k_{cat}} \\
K_E \updownarrow & \updownarrow K_{ES} & \\
\text{HE} & \underset{K'_S}{\rightleftharpoons} \text{HES} &
\end{array}
\tag{5.14}
$$

In equation 5.14, the ionization constant of the free enzyme is K_E; that of the enzyme–substrate complex is K_{ES}; and the dissociation constants of HES and ES are K'_S and K_S, respectively.

All four equilibrium constants cannot vary independently, due to the cyclic nature of the equilibria (see Chapter 3, section L). Once three are fixed, the fourth is defined by

$$
K_E K'_S = K_{ES} K_S
\tag{5.15}
$$

Equation 5.15 may be derived by multiplying the various dissociation and ionization constants or simply by inspection, since the process HES → ES → E must give the same energy change as HES → HE → E.

The two important conclusions from equation 5.15 are:
1. If $K_E = K_{ES}$, then $K_S = K'_S$, and there is no pH dependence for binding.
2. If $K_E \neq K_{ES}$, i.e., if the pK_a is perturbed on binding, then

$$
K_S = K'_S \frac{K_E}{K_{ES}}
\tag{5.16}
$$

and the binding of S must of necessity be pH-dependent.

2. The pH dependence of k_{cat}, k_{cat}/K_M, and $1/K_M$[1,2]

The pH dependence of v is obtained by expressing the concentration of ES in terms of $[E]_0$ to give, after the necessary algebra,

$$
v_H = \frac{k_{cat}[E]_0[S]}{K_S + [S](1 + [H^+]/K_{ES}) + K_S[H^+]/K_E}
\tag{5.17}
$$

The pH dependence of k_{cat} is derived from equation 5.17 when [S] is much greater than K_S; i.e.,

$$(V_{max})_H = [E]_0(k_{cat})_H = \frac{k_{cat}[E]_0 K_{ES}}{K_{ES} + [H^+]} \tag{5.18}$$

Comparison of the denominator of equation 5.18 with that of 5.8 shows that the pH dependence of V_{max} or k_{cat} follows the ionization constant of the enzyme–substrate complex, K_{ES}.

The apparent value of K_M at each pH may be found by rearranging equation 5.17 to the form of the basic Michaelis-Menten equation (equation 3.1):

$$(K_M)_H = \frac{K_S K_{ES} + [H^+]K_S K_{ES}/K_E}{K_{ES} + [H^+]} \tag{5.19}$$

K_M also follows the ionization of the enzyme–substrate complex.

The pH dependence of k_{cat}/K_M is given by the variation of v at low values of [S] (or, alternatively, by the ratios of equations 5.18 and 5.19). Simplifying equation 5.17 by putting [S] close to zero and rearranging gives

$$\left(\frac{k_{cat}}{K_M}\right)_H = \frac{(k_{cat}/K_S)K_E}{K_E + [H^+]} \tag{5.20}$$

The value of k_{cat}/K_M (and v for [S] much less than K_M) follows the ionization of the free enzyme.

3. A simple rule for the prediction and assignment of pK_a's

The above results are part of the general rule that "the plot of the equilibrium constant K or the rate constant k for the process X $\rightarrow$ Y as a function of pH follows the ionization constants of X." Let us now apply this rule to scheme 5.14, bearing in mind that the substrate may also have ionizing groups:

1. *The pH dependence of* k_{cat}. The process concerned is

$$\text{ES} \xrightarrow{k_{cat}} \text{E} + \text{P} \tag{5.21}$$

The value of k_{cat} follows the pK_a of the enzyme–substrate complex.

2. *The pH dependence of* K_M. The process concerned is

$$\text{ES} \xrightarrow{K_M} \text{E} + \text{S} \tag{5.22}$$

The pH dependence of K_M follows the ionization of the enzyme–substrate complex.

3. *The pH dependence of $1/K_M$.** The process concerned is

$$E + S \xrightarrow{1/K_M} ES \tag{5.23}$$

The pH dependence of $1/K_M$ follows the ionizations in the free enzyme and the free substrate.

4. *The pH dependence of k_{cat}/K_M.* The process concerned is

$$E + S \xrightarrow{k_{cat}/K_M} E + P \tag{5.24}$$

The pH dependence of k_{cat}/K_M follows the ionizations in the free enzyme and the free substrate.

C. Modifications and breakdown of the simple theory

The simple theory outlined above has to be modified to account for the pH dependence of the catalytic parameters in mechanisms more complicated than the basic Michaelis-Menten.

1. Modifications due to additional intermediates

a. Intermediates on the reaction pathway[3]

$$E + S \underset{}{\overset{K_S}{\rightleftharpoons}} ES \xrightarrow{k_2} EA \xrightarrow{k_3} E + P \tag{5.25}$$

The presence of additional intermediates does not affect the pH dependence of k_{cat}/K_M or $1/K_M$, since they represent changes from the free enzyme and free substrate only. The pH dependence of these still gives the pK_a's of the free enzyme and free substrate. But the pH dependence of k_{cat} and K_M now concerns changes from the intermediate (EA) as well as from the enzyme–substrate complex. If the intermediate EA in equation 5.25 is the major enzyme-bound species, the pH dependence of k_{cat} and K_M will give the pK_a's of EA. If both ES and EA accumulate, the pH profiles give pK_a's that are the weighted means of those of ES and EA.[4]

---

* Some find it confusing that plotting the inverse of a function, e.g., $1/K_M$ instead of K_M, gives a different pK_a. The mathematical reason for this was given in equations 5.11 to 5.13. The pH dependence of K_M contains the information for determining the pK_a's of E, S, and ES, but they are manifested in different ways in different plots. An alternative procedure for analyzing a plot of log K_M vs. pH has been given by M. Dixon.[1]

b. Nonproductive binding modes[5]

When a substrate binds in a nonproductive mode as well as in the productive mode, the pH dependence of k_{cat} and K_M may give an apparent pK_a for the catalytically important group at the active site; such a pK_a is far from the real value in the productive complex. This happens if the ratio of productive to nonproductive binding changes on ionization of the group. Suppose that the activity of the enzyme is dependent on a group being in the basic form at high pH. If, say, the substrate is bound with more being in the productive mode at low pH than at high pH, then as the pH decreases through the pK_a of the catalytic group, the decrease in rate as the group becomes protonated will be partially compensated for by an increase in productive binding. This has the effect of lowering the apparent pK_a controlling the pH dependence of k_{cat} from the value in the productive complex. The pH dependence of K_M is affected in an identical manner.

$$
\begin{array}{ccc}
\text{HE} & \overset{K_a}{\rightleftharpoons} & \text{E} \\[4pt]
K'_S \big\updownarrow & & \big\updownarrow K_S \\[6pt]
\text{HES} & \overset{K'_a}{\rightleftharpoons} \text{ES} & \xrightarrow[\text{slow}]{k_2} \text{E} + \text{P} \\[4pt]
K' \big\updownarrow & & \big\updownarrow K \\[6pt]
\text{HES}' & \overset{K''_a}{\rightleftharpoons} & \text{ES}'
\end{array}
\tag{5.26}
$$

The algebraic solution of such a situation is shown in scheme 5.26. HES and ES are the productively bound complexes, and K' and K are the equilibrium constants between these and the nonproductively bound complexes HES′ and ES′ as defined in the scheme. Solving the rate equation by the usual means gives

$$
k_{cat} = \frac{k_2}{K(1 + [H^+]/K''_a) + 1 + [H^+]/K'_a}
\tag{5.27}
$$

$$
K_M = \frac{K_S(1 + [H^+]/K_a)}{K(1 + [H^+]/K''_a) + 1 + [H^+]/K'_a}
\tag{5.28}
$$

Through rearranging equations 5.27 and 5.28 to the form of equation 5.8 and inspecting the denominator, we can obtain an observed pK_a given by

$$
pK_{a(obs)} = pK'_a - \log \frac{1 + K}{1 + K'}
\tag{5.29}
$$

Nonproductive binding modes do not affect the pH dependence of k_{cat}/K_M or $1/K_M$, for the reasons discussed in the previous case.

2. Breakdown of the simple rules: Briggs-Haldane kinetics and change of rate-determining step with pH: Kinetic pK_a's[4,6-8]

$$E + S \underset{k_{-1}}{\overset{k_1}{\rightleftharpoons}} ES \overset{k_2}{\longrightarrow} E + P \tag{5.30}$$

In the extreme case of Briggs-Haldane kinetics (equation 5.30), the rate constant for the chemical step is larger than that for the dissociation of the enzyme–substrate complex. In this case, k_{cat}/K_M is equal to k_1, the association constant of the enzyme and substrate. Suppose again that the catalytic activity of the enzyme depends on a group being in its basic form. At high pH, k_2 is faster than k_{-1}. But as the pH is lowered, k_2 decreases while the base becomes protonated, until k_2 is slower than k_{-1}. The apparent pK_a to which this leads in the pH profile of k_{cat}/K_M is lower than the pK_a of the important base (Figure 5.3).[7] It may be

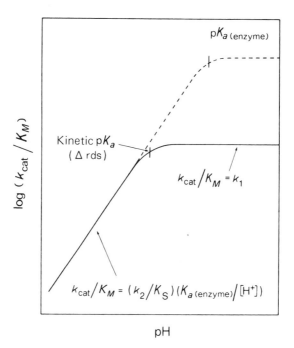

Figure 5.3 Illustration of the "kinetic pK_a" caused by a change in rate-determining step with pH. At low pH, k_{cat}/K_M falls off as

$$\left(\frac{k_2}{k_{-1}/k_1}\right)\left(\frac{K_a}{K_a + [H^+]}\right)$$

—which can be simplified because $K_S = k_{-1}/k_1$ and $[H^+] \gg K_a$. At high pH, k_{cat}/K_M levels off at k_1, the second-order constant for the association of enzyme and substrate.

shown that if the ionization constant of the group on the enzyme is K_a, the apparent ionization constant is given by

$$K_{app} = \frac{K_a(k_2 + k_{-1})}{k_{-1}} \tag{5.31}$$

K_{app} is termed a *kinetic* pK_a, since it does not represent a real ionization but is composed of the ratios of rate constants that are not for proton transfers. Kinetic pK_a's occur whenever there is a change of rate-determining step with pH.

3. An experimental distinction between kinetic and equilibrium pK_a's[7,8]

A pK_a that is composed of the combination of equilibrium constants and the individual pK_a's of titrating groups, such as that for the nonproductive binding scheme (equation 5.29), titrates as a real pK_a (see also Chapter 3, section L3). If one was to add acid or base to the enzyme–substrate complexes in scheme 5.26, the catalytic group would titrate according to the pK_a given by equation 5.29. Similarly, if one was to measure the fraction of the base in the ionized form by some spectroscopic technique, it would be found to ionize with the pK_a of equation 5.29. This does not happen with kinetic pK_a's. In the case of the change of rate-determining step with pH for k_{cat}/K_M in the Briggs-Haldane mechanism, although the pK_a from the kinetics will be that given by equation 5.31, direct measurement of the titration of the catalytic group will give its true ionization constant K_a.

4. Microscopic and macroscopic pK_a's

When an acid exists in different forms, such as HES and HES′ in scheme 5.26, the pK_a in each of the forms is termed a *microscopic* pK_a. The pK_a with which the system is found to titrate, e.g., $pK_{a(obs)}$ in equation 5.29, is called variously a *macroscopic*, an *apparent*, or a *group* pK_a. To all intents and purposes it is a real pK_a, unlike a kinetic pK_a.

D. The influence of surface charge on pK_a's of groups in enzymes

The surface of an enzyme contains many polar groups. Chymotrypsinogen, for example, has on its surface 4 arginine and 14 lysine residues, which are positively charged, and 7 aspartate and 5 glutamate residues, which are negatively charged. These provide an ionic atmosphere, or electrostatic field, which may stabilize or destabilize buried or partly buried ionic groups.

Early estimates of the effects of surface charge used chemical modification to change radically the surface charge of the protein to be either highly positive or highly negative. Relatively low changes were observed in the pK_a of the active site histidine of chymotrypsin because the effects of the surface charges were masked by a cloud of counter-ions (see Chapter 4, Part 2, section A1). The precision weapon of protein engineering was used to modify single charges on the surface of subtilisin, which revealed significant effects of distant charged residues on the pK_a of its active site His-64[9,10] (Figure 5.4). The carboxylates of Asp-99 and Glu-156 are 12.2 and 14.6 Å, respectively, distant from the imidazole of His-64. These stabilize the protonated form of His-64. Mutation of both negatively charged residues to positively charged lysine residues decreases the pK_a of His-64 by a whole unit at low ionic strength. The value of k_{cat}/K_M for the hydrolysis of the negatively charged substrate succinyl-Ala-Ala-Pro-Phe-p-nitroanilide increases with the increased positive charge on the enzyme. Similar systematic modifications of charged residues in subtilisin and barnase have been used to map out the effective dielectric of the proteins and the effects of ionic strength to benchmark theoretical methods for their calculation[11] (see Chapter 11).

The effects of mutating surface charge are mimicked by changes of pH. Below pH 5, the carboxylates become protonated, whereas above pH 9 the ammonium groups of lysine residues deprotonate, causing them to lose their positive charge. The changes in protonation state cause perturbations in the titration curves of the enzyme at extremes of pH. These effects are most marked at low

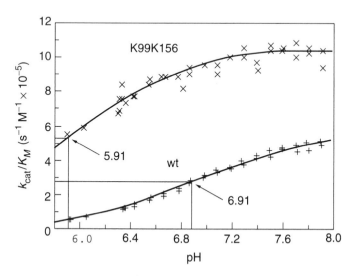

Figure 5.4 Plot of the pH dependence of k_{cat}/K_M for the hydrolysis of succinyl-Ala-Ala-Pro-Phe-p-nitroanilide by wild-type subtilisin and the double mutant with Asp-99 → Lys and Glu-156 → Lys. [From A. J. Russell and A. R. Fersht, *Nature* **328**, 496 (1987).]

ionic strengths of 0.2 M or less, but they may be depressed by high ionic strengths to become negligible at 1 M. Fortunately, many titration curves are measured between pH 5 and pH 9, where few surface groups titrate, and excellent results are obtained at ionic strengths of 0.1 M.[8,12]

As well as perturbing the pK_a of the catalytic acid or base, the large change of surface charge may alter the rate constants for the chemical steps. The change of surface charge is analogous to the change of ionic strength in nonenzymatic ionic reactions, and the two effects are analogous to secondary and primary salt effects in physical organic chemistry.

E. Graphical representation of data

Suppose that one of the kinetic quantities, such as k_{cat}, depends on the enzyme being in the acidic form. Then the rate will depend upon the ionization constant K_a and the pH according to

$$(k_{cat})_H = \frac{k_{cat}[H^+]}{K_a + [H^+]} \tag{5.32}$$

where (k_{cat}) is the observed value at the particular $[H^+]$ (equation 5.8). Equation 5.32 is in the same form as the Michaelis-Menten equation, so K_a may be found by plots analogous to the Lineweaver-Burk or Eadie plots (equations 3.28 and 3.29). For example, because

$$(k_{cat})_H = k_{cat} - \frac{K_a(k_{cat})_H}{[H^+]} \tag{5.33}$$

k_{cat} and K_a may be obtained by plotting $(k_{cat})_H$ against $(k_{cat})_H/[H^+]$. Similarly, if the rate depends on the enzyme being in the basic form, it may be shown that

$$(k_{cat})_H = k_{cat} - \frac{(k_{cat})_H[H^+]}{K_a} \tag{5.34}$$

Here $(k_{cat})_H$ must be plotted against $(k_{cat})_H[H^+]$ (see Figure 5.5).

For more complicated examples, in which the kinetic quantity does not fall to zero at either high or low pH (as is often found in plots of association or dissociation constants against pH), the procedure is modified by plotting the difference between the observed value at the particular pH and one of the limiting values at the extremes of pH. Modern practice is to fit data directly to equation 5.32, as described in Chapter 6, section E.

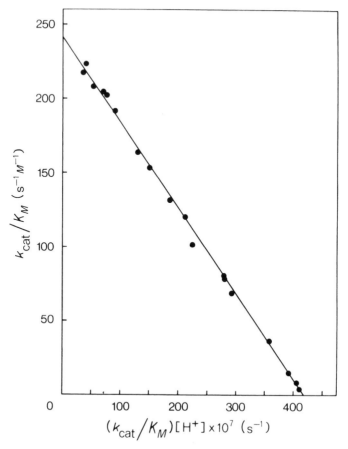

Figure 5.5 Illustration of the use of equation 5.34 for determining the pK_a of the active site of an enzyme. This example is the hydrolysis of acetyl-L-tyrosine *p*-acetylanilide by α-chymotrypsin at 25°C. Note that k_{cat}/K_M rather than k_{cat} is used here. [From A. R Fersht and M. Renard, *Biochemistry* **13**, 1416 (1974).]

F. Illustrative examples and experimental evidence

The most-studied enzyme in this context is chymotrypsin. Besides being well characterized in both its structure and its catalytic mechanism, it has the advantage of a very broad specificity. Substrates may be chosen to obey the simple Michaelis-Menten mechanism, to accumulate intermediates, to show nonproductive binding, and to exhibit Briggs-Haldane kinetics with a change of rate-determining step with pH.

The pH dependence of k_{cat}/K_M for the hydrolysis of substrates follows a bell-shaped curve with a maximum at pH 7.8 and pK_a's of 6.8 and 8.8 for α-chymotrypsin, and a maximum at pH 7.9 and pK_a's of 6.8 and 9.1 for δ-chymotrypsin. The pK_a of 6.8 represents the ionization of the catalytically important base at the active site, whereas the high-pH ionization is due to the α-amino group of Ile-16, which holds the enzyme in a catalytically active conformation. This conformationally important ionization does not affect k_{cat}—which usually gives a sigmoid curve of pK_a 6 to 7—but it does cause an increase of K_M at high pH.

1. The pK_a of the active site of chymotrypsin

a. The free enzyme

The theory predicts that unless there is a change of rate-determining step with pH, the pH dependence of k_{cat}/K_M for all non-ionizing substrates should give the same pK_a: that for the free enzyme. With one exception, this is found (Table 5.2). At 25°C and ionic strength 0.1 M, the pK_a of the active site is 6.80 ± 0.03. The most accurate data available fit very precisely the theoretical ionization curves between pH 5 and 8, after allowance has been made for the fraction of the enzyme in the inactive conformation. The relationship holds for amides with which no intermediate accumulates and the Michaelis-Menten mechanism holds, and also for esters with which the acylenzyme accumulates.

b. Acetyl-L-tryptophan p-nitrophenyl ester: Briggs-Haldane kinetics with change of rate-determining step with pH

The one exception is the pH dependence of k_{cat}/K_M for the hydrolysis of acetyl-L-tryptophan p-nitrophenyl ester. This gives an apparent pK_a of 6.50 for the free

Table 5.2 Some pK_a's for the active site of chymotrypsin

| | pK_a | | |
Substrate	k_{cat}/K_M	k_{cat}	Ref.
Acetyl-L-phenylalanine alaninamide[a]	6.80	6.6	1
Formyl-L-phenylalanine semicarbazide[a]	6.84	6.32	1
Acetyl-L-tyrosine p-acetylanilide[a]	6.77	—	1
p-Nitrophenyl acetate[a]	6.85	—	2
Acetyl-L-phenylalanine ethyl ester[a]	6.8	6.85	3
Acetyl-L-tyrosine p-acetylanilide[b]	6.83	—	4
Acetyl-L-tryptophan p-nitrophenyl ester[b]	6.50	6.9	4

[a] α-Chymotrypsin, 25°C, ionic strength 0.1 M.
[b] δ-Chymotrypsin, 25°C, ionic strength 0.95 M.
1 A. R. Fersht and M. Renard, *Biochemistry* **13**, 1416 (1974).
2 M. L. Bender, G. E. Clement, F. J. Kézdy, and H. d'A. Heck, *J. Am. Chem. Soc.* **86**, 3680 (1964).
3 B. R. Hammond and H. Gutfreund, *Biochem. J.* **61**, 187 (1955).
4 M. Renard and A. R. Fersht, *Biochemistry* **12**, 4713 (1973).

enzyme. The reason for the low value is explained in section C2. At high pH, the association of enzyme and substrate is partly rate-determining for k_{cat}/K_M. The limiting value of k_{cat}/K_M at high pH is 3×10^7 s^{-1} M^{-1}, a value close to that for the diffusion-controlled encounter of enzyme and substrate. However, at low pH, the chemical steps slow down and become rate-determining as the enzyme becomes protonated and less active. The rate constant for the association step may be calculated from equation 5.31 to be 6×10^7 s^{-1} M^{-1}. The data are consistent with scheme 5.35 and a pK_a of 6.8 for the active site.[7,13] This is a rare example in which steady state kinetics may be analyzed to give rate constants for several specific steps on the reaction pathway.

$$\text{Ac-Trp-ON-Ph} + \text{CT} \underset{6 \times 10^4 \text{s}^{-1}}{\overset{6 \times 10^7 \text{s}^{-1} M^{-1}}{\rightleftharpoons}} \text{Ac-Trp-ON-Ph} \cdot \text{CT}$$
$$\downarrow 7 \times 10^4 \text{s}^{-1}$$
$$\text{Ac-Trp-CT}$$
$$\downarrow 65 \text{ s}^{-1}$$
$$\text{Ac-Trp} + \text{CT}$$

(5.35)

(CT = chymotrypsin)

c. The enzyme–substrate complex

The pK_a of the enzyme–substrate complex is not constant like that of the free enzyme, since the binding of substrates perturbs the pK_a of the active site. The pK_a values of k_{cat} for the hydrolysis of amides ranges from 6 to 7. In consequence, the value of K_M increases at low pH according to equation 5.16.

The precise determination of the pK_a for the enzyme–substrate complex from the pH dependence of K_M is difficult for two reasons. First, the variation of K_M is relatively small, and highly accurate data are required. Secondly, two plateau regions must be determined, one at low pH as well as one at high pH. This means that measurements are required over a wider range of pH than for the determination of k_{cat} or k_{cat}/K_M, and are thus more susceptible to perturbations from the ionizations of other groups.

d. The ionization at high pH[14]

The pK_a at high pH is caused by a substrate-independent conformational change in the enzyme, a change that may be monitored directly by physical techniques such as optical rotation and fluorescence yield. Kinetic measurements give the same pK_a as that found by these methods.

G. Direct titration of groups in enzymes

Several methods have been developed for the direct titration of some ionizing side chains in proteins. One of the major problems is in identifying the desired group among all the similar ones.

1. The effect of D_2O on pH/pD and pK_a's

Many of the spectroscopic methods use D_2O as the solvent rather than H_2O, in order to separate the required signals from those coming from the solvent. It has been found experimentally that the glass electrode gives a lower reading in D_2O by 0.4 units:

$$pH = pD + 0.4 \qquad\qquad (5.36)$$

In mixtures of H_2O and D_2O, the glass electrode reading is related to pH by

$$pH = pD = 0.3139\alpha + 0.0854\alpha^2 \qquad\qquad (5.37)$$

where α is the atom fraction of deuterium, [D]/([D] + [H]).[15] However, pK_a's are higher in D_2O and in D_2O/H_2O mixtures than in H_2O. The increase in pK_a sometimes balances the decreased reading on the glass electrode, so that the measured pK_a in D_2O is often assumed to be the true pK_a in water. The effect of the solvent on ionization constants is somewhat variable, though, and the simplification can lead to errors of a few tenths of a pH unit.

2. Methods

a. Nuclear magnetic resonance

NMR is a most powerful method for determining pK_a values of residues in proteins because it can detect individual signals and their titration with pH.

^{1}H NMR has been particularly useful in determining the pK_a's of histidines in proteins. The signals from the C-2 and C-4 protons are further downfield than the bulk of the resonances of the protein, and are resolvable in D_2O solutions. The chemical shift changes on protonation, so the histidines may be readily titrated. Where there is more than one histidine it is difficult to assign the individual pK_a's. In the case of ribonuclease, chemical modification, selective deuteration, and a knowledge of the crystal structure have allowed the pK_a's of all four histidines to be assigned.[16,17]

The resonances of protons in hydrogen bonds may be shifted downfield to such an extent that they may be observed in H_2O solutions. The proton between Asp-102 and His-57 in chymotrypsinogen, chymotrypsin, and other serine proteases has been located and its resonance found to titrate with a pK_a of 7.5[18] (although the pK_a is for the dissociation of the proton on the other nitrogen of the imidazole ring).

^{15}N NMR has also been used to measure the pK_a value of the active-site histidine of the serine proteases,[19] confirming that it is the histidine and not the buried aspartate that ionizes with a pK_a of about 7.

^{13}C NMR may be used to measure the pK_a's of lysine and aspartate residues. ^{1}H NMR has been used to follow the ionization of tyrosine residues.[20]

^{31}P NMR is a most valuable technique. Not only can it be used to titrate the pK_a values of phosphates, but it can also be used to measure the ionization of known phosphate residues *in vivo*, such as those of 2,3-diphosphoglycerate, and thus to determine intracellular pH.[21]

b.   Infrared spectroscopy[22]

Carboxyl groups absorb at about 1710 cm^{-1}, and carboxylates at about 1570 cm^{-1}. Infrared difference spectra in D_2O solutions have been used to measure the pK_a's of abnormal carboxyl groups in α-lactoglobulin (7.5) and lysozyme (2.0, 6.5).

c.   Ultraviolet difference spectroscopy[23]

Ultraviolet difference spectra have frequently been used to measure the ionization of the phenolic hydroxyl of tyrosines. The sulfydryls of cysteines and the imidazoles of histidines are also amenable to difference spectroscopy.

d.   Fluorescence[23,24]

Fluorescence is useful when the ionization of a group perturbs the spectrum of a neighboring tryptophan, the major fluorescent species in proteins, or causes a conformational change that perturbs the fluorescence of the protein as a whole. Tyrosines may be titrated in the absence of tryptophans (which fluoresce more strongly).

e.   Difference titration[25]

The direct titration of a protein with acid or base usually gives an uninterpretable ionization curve because of the overlapping titrations of the many groups. However, one or two ionizations may be isolated by comparing the titration curve with that of the enzyme after a particular group has been modified. An example from classical protein chemistry is the specific esterification of Asp-52 of lysozyme with triethoxonium fluoroborate. The difference titration between this modified form and the native protein gives not only the pK_a of Asp-52 but also the effect of this ionization on the other important acid at the active site, Glu-35. Nowadays, such ionizing groups are replaced using protein engineering (Chapter 14).

f.   Denaturation difference titration[26]

Practically speaking, it is not possible to determine by titration whether or not a buried group in a protein, such as Asp-102 in chymotrypsin and chymotrypsinogen, is ionized. Accurate titration is possible only between pH 3 and 11 because of the high background concentrations of hydroxyl ions and protons outside this range. For this reason it is not possible to detect a missing ionization in the overall titration curve if the protein has any other abnormally titrating groups. Chymotrypsin, for example, has three abnormally low titrating carboxyls which are still ionized below pH 3. However, these groups titrate normally when the protein is denatured, and the number of carboxyls titrating in the denatured protein is easily determined. It was shown that Asp-102 is ionized in the high-pH forms of chymotrypsin and chymotrypsinogen by measuring the proton uptake on denaturation and adding this to the number of carboxyls known to be ionized in the denatured protein. In order to improve the accuracy, the majority of the surface carboxyls were converted to amides to lower the number of ionizing groups.

g. Chemical modification

The rate of inhibition of enzymes by irreversible inhibitors has been used in the same way as normal kinetics to give the pK_a's of the free enzyme and the enzyme–inhibitor complex. The pK_a's of other residues have been measured from the pH dependence of their reaction with chemical reagents. For example, since the basic forms of amines react with acetic anhydride[27] or dinitrofluorobenzene,[28–30] their extent of ionization is given from the relative rates of reaction as a function of pH. This has been used to measure the pK_a's of amino groups in proteins by modifying them with radioactive reagents, digesting the protein, separating the peptides, and measuring their specific radioactivities. In this way the pK_a's of several groups may be determined simultaneously, as has been done for elastase and chymotrypsin.

A useful variation of this technique is to measure the rate of exchange of tritium from tritiated water with the C-2 proton of the imidazole ring of a histidine.[31,32] The rate constant for the exchange depends on the state of ionization of the imidazole, and is faster for the unprotonated form. This procedure is a useful adjunct to NMR experiments, which also measure the pK_a's of histidine residues.

These experiments are tedious to perform, but the assignments of pK_a are unambiguous.

H. The effect of temperature, polarity of solvent, and ionic strength on pK_a's of groups in enzymes and in solution

Ions are stabilized by a polar solvent. The electrostatic dipoles of the solvent directly interact with the electrical charges of the ions, and the dielectric constant decreases the tendency of the ions to reassociate. The ionization of a neutral acid, as in equation 5.38, is depressed by the addition of a solvent of low polarity to an aqueous solution (Table 5.3).

$$HA = A^- + H^+ \tag{5.38}$$

On the other hand, the ionization of a cationic acid (equation 5.39) is insensitive to solvent polarity, since there is no change of charge in the equilibrium.

$$BH^+ = B + H^+ \tag{5.39}$$

Table 5.3 *Effect of organic solvents on pK_a's at 25°C[a]*

Wt % dioxane	Acetic acid	Tris[b]	Benzoylarginine	Glycine $-CO_2H$	Glycine $-NH_3^+$
0	4.76	8.0	3.34	2.35	9.78
20	5.29	8.0	—	2.63	9.91
45	6.31	8.0	—	3.11	10.2
50	—	8.0	4.59	—	—
70	8.34	8.0	4.60	3.96	11.3

[a] From H. S. Harned and B. B. Owen, *The physical chemistry of electrolytic solutions,* Reinhold, pp. 755–56 (1958); T. Inagami and J. M. Sturtevant, *Biochim. Biophys. Acta* **38**, 64 (1960).
[b] $(HOCH_2)_3CNH_2$.

The procedure of using the effect of solvent polarity on the pK_a of a group in an enzyme to tell whether it is a cationic or a neutral acid could be unreliable. A partly buried acid is shielded from the full effects of the solvent, and the electrostatic interactions with the protein may be more important. For example, the pK_a of the acylenzyme of benzoylarginine and trypsin is almost invariant in 0 to 50% dioxane/water and increases only slightly in 88% dioxane/water.[33] This suggests that the ionizing group is cationic, which is consistent with it being the imidazole moiety of His-57. (The pK_a of acetic acid increases by some 6 units under those conditions.) On the other hand, although increasing ionic strength decreases the pK_a's of carboxylic acids in solution, the pK_a's of Asp-52 and Glu-35 in lysozyme are increased by increasing ionic strength because of an effect on the surface charge of the protein.[25]

Imidazole groups in solution have enthalpies of ionization of about 30 kJ/mol (7 kcal/mol), whereas carboxylic acids have negligible enthalpies of ionization. But the changes in the enthalpy of the solvating water molecules also make important contributions to these values, and so the solution values cannot be extrapolated to partly buried groups in proteins.

I. Highly perturbed pK_a's in enzymes

Many amine bases and carboxylic acids in proteins titrate with anomalously high or low pK_a's (Table 5.4). The reasons are quite straightforward, and depend on the microenvironment. If a carboxyl group is in a region of relatively low polarity, its pK_a will be raised, since the anionic form is destabilized. Alternatively, if the carboxylate ion forms a salt bridge with an ammonium ion, it will be stabi-

Table 5.4 *Some highly perturbed pK_a's of groups in proteins[a]*

Enzyme	Residue	pK_a
Lysozyme	Glu-35	6.5
Lysozyme–glycolchitin complex	Glu-35	~8.2
Acetoacetate decarboxylase	Lys (ϵ-NH$_2$)	5.9
Chymotrypsin	Ile-16 (α-NH$_2$)	10.0
α-Lactoglobulin	CO$_2$H	7.5
Papain	His-159	3.4

[a] From text references 25, 25, 6, 11, 20, and 34, respectively.

lized by the positive charge so it will be more acidic. Conversely, if an amino group is buried in a nonpolar region, like the lysine in the active site of acetoacetate decarboxylase, protonation is inhibited and the pK_a is lowered. An ammonium ion in a salt bridge, such as Ile-16 in chymotrypsin, is stabilized by the negative charge on the carboxylate ion. Deprotonation is inhibited and the pK_a is raised.

References

1. M. Dixon, *Biochem. J.* **55**, 161 (1953).
2. R. A. Alberty and V. Massey, *Biochim. Biophys. Acta* **13**, 347 (1954).
3. M. L. Bender, G. E. Clement, F. J. Kézdy, and H. d'A. Heck, *J. Am. Chem. Soc.* **86**, 3680 (1964).
4. A. R. Fersht and Y. Requena, *J. Am. Chem. Soc.* **93**, 7079 (1971).
5. J. Fastrez and A. R. Fersht, *Biochemistry* **12**, 1067 (1973).
6. D. E. Schmidt, Jr., and F. H. Westheimer, *Biochemistry* **10**, 1249 (1971).
7. M. Renard and A. R. Fersht, *Biochemistry* **12**, 4713 (1973).
8. A. R. Fersht and M. Renard, *Biochemistry* **13**, 1416 (1974).
9. P. G. Thomas, A. J. Russell, and A. R. Fersht, *Nature, Lond.* **318**, 375 (1985).
10. A. J. Russell and A. R. Fersht, *Nature, Lond.* **328**, 496 (1987).
11. R. Loewenthal, J. Sancho, T. Reinikainen, and A. R. Fersht, *J. Molec. Biol.* **232**, 574 (1993).
12. F. J. Kézdy, G. E. Clement, and M. L. Bender, *J. Am. Chem. Soc.* **86**, 3690 (1964).
13. A. C. Brower and J. F. Kirsch, *Biochemistry* **21**, 1302 (1982).
14. A. R. Fersht, *J. Molec. Biol.* **64**, 497 (1972), and references therein.
15. L. Pentz and E. R. Thornton, *J. Am. Chem. Soc.* **89**, 6931 (1967).
16. J. L. Markley, *Biochemistry* **14**, 3546 (1975).
17. S. M. Dudkin, M. Ya. Karpeisky, V. G. Sakharovskii, and G. I. Yakovlev, *Dokl. Akad. Nauk SSSR* **221**, 740 (1975).
18. G. Robillard and R. G. Shulman, *J. Molec. Biol.* **86**, 519 (1974).

19. W. W. Bachovchin, R. Kaiser, J. H. Richards, and J. D. Roberts, *Proc. Natl. Acad. Sci. USA* **78**, 7323 (1981).
20. S. Karplus, G. H. Snyder, and B. D. Sykes, *Biochemistry* **12**, 1323 (1973).
21. R. B. Moon and J. H. Richards, *J. Biol. Chem.* **248**, 7276 (1973).
22. S. N. Timasheff and J. A. Rupley, *Archs. Biochem. Biophys.* **150**, 318 (1972).
23. S. N. Timasheff, *The Enzymes* **2**, 371 (1970).
24. R. W. Cowgill, *Biochim. Biophys. Acta* **94**, 81 (1965).
25. S. M. Parsons and M. A. Raftery, *Biochemistry* **11**, 1623, 1630, 1633 (1972).
26. A. R. Fersht and J. Sperling, *J. Molec. Biol.* **74**, 137 (1973).
27. H. Kaplan, K. J. Stephenson, and B. S. Hartley, *Biochem. J.* **124**, 289 (1971).
28. A. L. Murdock, K. L. Grist, and C. H. W. Hirs, *Archs. Biochem. Biophys.* **114**, 375 (1966).
29. R. J. Hill and R. W. Davis, *J. Biol. Chem.* **242**, 2005 (1967).
30. W. H. Cruickshank and H. Kaplan, *Biochem. Biophys. Res. Commun.* **46**, 2134 (1972).
31. H. Matsuo, M. Ohe, F. Sakiyama, and K. Narita, *J. Biochem., Tokyo* **72**, 1057 (1972).
32. M. Ohe, H. Matsuo, F. Sakiyama, and K. Narita, *J. Biochem., Tokyo* **75**, 1197 (1974).
33. T. Inagami and J. M. Sturtevant, *Biochim. Biophys. Acta* **38**, 64 (1960).
34. F. A. Johnson, S. D. Lewis, and J. A. Shafer, *Biochemistry* **20**, 44 (1981).

Practical Methods for Kinetics and Equilibria

6

T he basic requirement for any kinetic or equilibrium study is the availability of a convenient assay for monitoring the concentrations of reagents. In this chapter, we introduce the more common methods, emphasizing their practical uses and limitations rather than giving an in-depth or exhaustive survey, and we outline how binding and kinetic data are analyzed.

A.　Spectrometry and methods for kinetics

1.　Spectrophotometry

Many substances absorb light in the ultraviolet or visible regions of the spectrum. If the intensity of the light shining onto a solution of the compound is I_0 and that transmitted through is I, the absorbance A of the solution is defined by

$$A = \log \frac{I_0}{I} \tag{6.1}$$

The absorbance usually follows *Beer's law*,

$$A = \epsilon c l \tag{6.2}$$

where ϵ is the *absorption* or *extinction coefficient* of the compound, c is its concentration (usually in units of molarity, mol/liter), and l is the pathlength (in cm) of the light through the solution. A compound may be assayed by measuring A if ϵ is known.

Spectrophotometry is particularly useful with naturally occurring chromophores. For example, the rates of many dehydrogenases may be measured from the rate of appearance of NADH at 340 nm ($\epsilon = 6.23 \times 10^3\ M^{-1}\ cm^{-1}$), because NAD^+ does not absorb at this wavelength. Otherwise artificial substrates may be used, such as p-nitrophenyl esters with esterases, since the p-nitrophenolate ion absorbs at 400 nm ($\epsilon = 1.8 \times 10^4\ M^{-1}\ cm^{-1}$).

The sensitivity of the method depends on the extinction coefficient involved; for $\epsilon = 10^4$, the lower limit of detectability is about 0.5 nmol using a conventional spectrophotometer requiring at least 0.5 mL of solution and an absorbance of 0.01.

a. Some possible errors

The usual source of error is the breakdown of Beer's law. This may happen in several ways: the chromophore aggregates or forms micelles at high concentrations with a change in absorption coefficient; at high background absorbances, so much light of the correct wavelength is absorbed that light of other wavelengths leaking through the monochromator becomes significant ("stray light"); the bandwidth of the monochromator is set so wide that wavelengths other than those absorbed by the chromophore are transmitted to the sample; the solutions are turbid and scatter light.

2. Spectrofluorimetry[1]

Some compounds absorb light and then re-emit it at a longer wavelength. This is known as fluorescence. The efficiency of the process is termed the quantum yield, q, which is equal to the number of quanta emitted per number absorbed (the ratio is always less than 1). Natural fluorophores include NADH, which absorbs at 340 nm and emits at about 460 nm. The major fluorophore in proteins is tryptophan, absorbing at 275 to 295 nm and re-emitting at 330 to 340 nm. Tyrosine fluoresces weakly in the same region. Many synthetic substrates for esterases, phosphatases, sulfatases, and glycosidases are based on 4-methylumbelliferone, a highly fluorescent derivative of phenol.

Fluorimetry is theoretically about 100 times more sensitive than spectrophotometry for the detection of low concentrations. Fluorescence is measured at right angles to the exciting beam against a dark background. A drift of 5% in the excitation intensity leads to only a 5% change in the signal. Absorbance measurements involve detecting a small change in the transmitted light; i.e., a small change against a high background. Any fluctuation in intensity is magnified: a 5% change is equivalent to an absorbance of 0.02.

Although the intensity of the fluoresced light is proportional to the intensity of the exciting beam, it is not possible to compensate indefinitely for decreasing concentration by increasing the excitation intensity. This is because the solvent scatters light—via Rayleigh scattering at the same wavelength as the excitation, and Raman scattering at a longer wavelength—and this eventually swamps the emitted light. Also, the compound that is being excited will be destroyed by photolysis at sufficiently high intensities.

a. Some properties of fluorescence

Fluorescence occurs by a photon being absorbed by a compound to give an excited state that decays by re-emission of a photon. (The lifetime of the excited state is about 10 ns.) Decay may also take place by a collision with another molecule, such as an iodide ion, or by a transfer of energy to another group in the molecule. The fluorescence is then said to be *quenched*. The quenching of the fluorescence of tryptophans of a protein on the binding of ligands is a useful way of measuring the extent of binding.

Fluorescence may also be *enhanced*. Sometimes a compound has a low quantum yield in aqueous solution but a higher one in nonpolar media. The dyes toluidinyl- and anilinyl-naphthalene sulfonic acid fluoresce very weakly in water, but strongly when they are bound in the hydrophobic pockets of proteins. Interestingly enough, if they are bound next to a tryptophan residue, they may be excited by light that is absorbed by the tryptophan at 275 to 295 nm and whose energy is transferred to them. Tryptophan and NADH fluoresce relatively weakly in water, and their fluorescence may be enhanced in the nonpolar regions of proteins.

b. Some possible errors

Errors usually arise through some form of quenching, such as the inadvertent addition of a substance such as potassium iodide, or more commonly by *concentration quenching*. This occurs when the fluorophore or ligand absorbs significantly through a Beer's law effect and reduces the intensity of the exciting or emitted light. For example, if the solution has an absorbance of A, then the average intensity of the exciting light in the cell is derived from equation 6.1:

$$I_{ave} = I_0 \times 10^{-\frac{1}{2}A} \tag{6.3}$$

The true fluorescence may be calculated from the above equation, from more sophisticated correction formulae, or from standard curves.[1]

3. Circular dichroism[2,3]

Circular dichroism (CD) is a property of optically active molecules. Plane-polarized light can be considered to be composed of two vectorial components that are circularly polarized; one is right-handed (R) and polarized in a clockwise direction; the other is left-handed (L) and polarized in an anticlock direction. The two components may not be equally absorbed when plane polarized light is shone through an optically active sample at a wavelength at which the light is absorbed. Thus, there are different extinction coefficients for the right- and left-handed circularly polarized vectors, ϵ_R and ϵ_L, and the phenomenon is called circular dichroism. Circular dichroism is intimately linked with optical rotation but has largely superseded it as a technique for examining proteins. The importance of CD is that the polypeptide backbone is optically active in the far ultraviolet (170–250 nm), and different secondary structures produce characteristic spectra. Further, some natural chromophores sometimes exhibit idiosyncratic CD behavior when placed in a heterogeneous environment in a folded protein. Tryptophan and tyrosine, for example, can give CD signals in the near ultraviolet ($\sim 270-300$ nm). CD has a sensitivity of detection in the near ultraviolet similar to that of absorbance measurement. In the far ultraviolet, it is comparable to fluorescence.

The extinction (absorption) coefficient was defined from equation 6.2:

$$\epsilon = \frac{A}{cl} \tag{6.4}$$

The CD is the difference in extinction (absorption) coefficient, $\Delta\epsilon$, between the right- and left-handed polarized light:

$$\Delta\epsilon = \epsilon_L - \epsilon_R \tag{6.5}$$

or, from equation 6.4:

$$\Delta\epsilon = \frac{(A_L - A_R)}{cl} \tag{6.6}$$

where A_L is the absorbance of the left-handed polarized light and A_R that of the right-handed polarized light. $A_L - A_R$ tends to be in the region of 0.005 to 0.1% of A_L or A_R. $\Delta\epsilon$ has units of cm M^{-1} or 10^3 cm^2 mol^{-1}. Many CD instruments give their output in ellipticity, Θ, measured in degrees, where

$$\Theta = 33(A_L - A_R) \tag{6.7}$$

Θ is often converted into the molar ellipticity $[\Theta]$ by the formula

$$[\Theta] = \frac{100\Theta}{cl} \tag{6.8}$$

so that $[\Theta]$ has the dimensions of deg cm^2 dmol^{-1} (d = deci). From equations 6.6 to 6.8,

$$[\Theta] = 3300\Delta\epsilon \tag{6.9}$$

Another term that is frequently used is $[\Theta]_{MRW}$, where MRW stands for mean residue weight.

$$[\Theta]_{MRW} = \frac{[\Theta]}{N} \tag{6.10}$$

where N is the number of residues in the protein.

The principal wavelengths at which the peptide bond absorbs light are around 222 nm, 208–210 nm, and 191–193 nm. The principal types of secondary structure have some characteristic features (Figure 6.1). The α helix is characterized by a strong negative CD signal at 222 and 208 nm and a strong positive CD signal at 192 nm. Signals from β sheet are weaker and more easily obscured.

a. Some possible errors

Other chromophores can absorb in the peptide region and obscure results. Tyr can give bands at 194 and 224 nm, the imidazole of histidine at low pH at 222 nm, tryptophan at 218 and 196 nm, and the disulfide bonds of cystine residues at 250 nm.

4. Automated spectrophotometric and spectrofluorimetric procedures

The spectroscopic assays are simplest and most accurate when the products have a spectrum that is different from that of the reagents, so that the products may be observed directly. Assays in which aliquots of the reaction mixture are developed with a reagent to give a characteristic color, such as amino acids with ninhydrin, may be automated by using a proportionating pump that mixes the reagents in the desired ratios and pumps them through a flow cell in a spectrophotometer or fluorimeter. Besides being less tedious for routine work, these procedures are much more reproducible and accurate than the conventional approach.

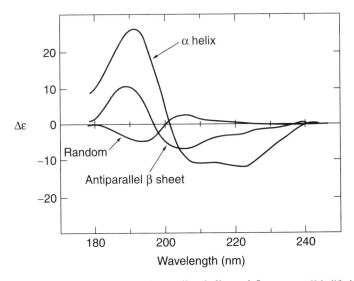

Figure 6.1 Ideal CD spectra of random coil, α helix, and β structure. [Modified from L. A. Compton and W. C. Johnson, Jr., *Anal. Biochem.* **155**, 155 (1986).]

5. Coupled assays

Some reactions that do not give chromophoric or fluorescent products may be coupled with another enzymatic reaction that does. Many of these coupled reactions are based on the formation or the disappearance of NADH. The formation of pyruvate may be linked with the conversion of NADH to NAD^+ by lactate dehydrogenase:

$$\text{Pyruvate} + \text{NADH} \underset{}{\overset{\text{lactate dehydrogenase}}{\rightleftharpoons}} \text{lactate} + NAD^+ \tag{6.11}$$

The formation of ATP may be coupled to the above reaction by the use of pyruvate kinase or another enzyme. For example, phosphofructokinase may be assayed by the following scheme:

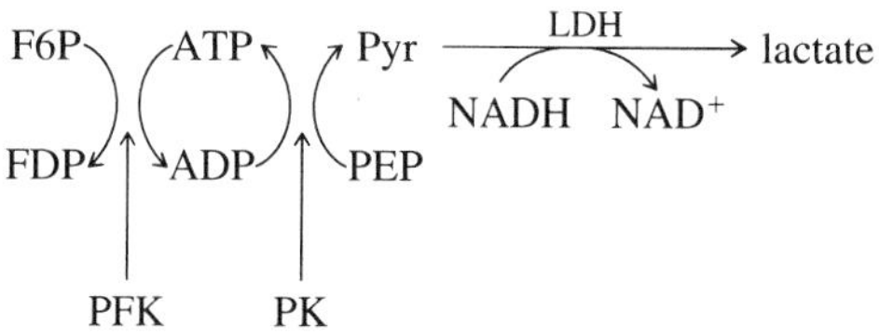

(F6P = fructose 6-phosphate, FDP = fructose 1,6-diphosphate, Pyr = pyruvate, PEP = phosphoenolpyruvate, PFK = phosphofructokinase, PK = pyruvate kinase, LDH = lactate dehydrogenase)

Another assay for phosphofructokinase involves converting the fructose 1,6-diphosphate to dihydroxyacetone phosphate and glyceraldehyde 3-phosphate with aldolase, equilibrating the triosephosphates with triosephosphate isomerase, and then measuring the production of NADH on the oxidation of the glyceraldehyde phosphate by glyceraldehyde 3-phosphate dehydrogenase.

6. Automatic titration of acid or base

A hydrolytic reaction that releases acid may be followed by titration with base. This is best done automatically by use of the pH-stat. A glass electrode registers the pH of the solution, which is kept constant by the automatic addition of base from a syringe controlled by an electronic circuit. Reaction volumes as low as 1 mL may be used, and the limit of detectability is about 50 nmol (5 to 10 μL of base at 5×10^{-3} to 1×10^{-2} M). The usual source of error with this apparatus is the buffering effect of dissolved CO_2.

7. Radioactive procedures

The most sensitive assay methods available involve the use of radioactively labeled substrates and reaction volumes of 20 to 100 μL.

Radioactivity is measured either in curies (Ci) (1 Ci is 2.22×10^{12} decompositions per minute) or in becquerels (Bq) (1 Bq is 1 decomposition per second).

In practice, radioactive decay is measured in terms of counts per minute by using a scintillation counter with the common isotopes ^{3}H, ^{14}C, ^{35}S, and ^{32}P. These isotopes emit β radiation (electrons) when they decay. The radiation may be monitored by using a scintillant, which converts the radiation into light quanta that are registered as "counts" by a photomultiplier. The low energy emission of ^{3}H is counted with an efficiency of between 15 and 40%, and the higher energy emissions from ^{14}C, ^{35}S, and ^{32}P with an efficiency of about 80%. Because the energy of the ^{3}H emission is so different from the others, it may be counted in their presence in a "double labeling" experiment by monitoring different regions of the energy spectrum separately (Table 6.1).

The energy of the emission from ^{32}P is so high that it may be monitored in the absence of scintillant due to the Čerenkov effect. On passing through water or a polyethylene scintillation vial, the electrons move so rapidly that they spontaneously emit photons, which may be detected at about 40% efficiency by using the counter at an open-window setting. This technique has been used to count ^{32}P and ^{14}C when both are present, by first monitoring the Čerenkov radiation and then counting again after the addition of scintillant.

The sensitivity of detection depends on the specific activity of the compound. Some examples are given in Table 6.2. Detection is possible for as low as 10^{-18} to 10^{-17} mol, which is some 6 or 7 orders of magnitude below the lower limit of spectrophotometry.

a. Some possible errors

1. Quenching. The emission from tritium is so weak that it is readily absorbed by a filter paper or chromatographic strip on which it is adsorbed, or by precipitates. Fewer problems occur with ^{14}C, but the position of the energy

Table 6.1 *Some common radioactive isotopes*

Isotope	Half-life	Specific activity of 100% isotopic abudance (Ci/mol)	Type of emission	Maximum energy of emission (MeV)
^{14}C	5730 yr	62.4	β	0.156
^{3}H	12.35 yr	2.9×10^4	β	0.0186
^{35}S	87.4 days	1.49×10^6	β	0.167
^{32}P	14.3 days	9.13×10^6	β	1.709
^{125}I	60 days	2.18×10^6	γ	—
^{131}I	8.06 days	1.62×10^7	β, γ	0.247–0.806
^{75}Se	120 days	1.09×10^6	γ	—

Table 6.2 *Sensitivity of radioactive methods*

Substrate	Specific activity (Ci/mol)	Limit of detectability[a] (pmol)
[^{14}C]methionine (90% enriched)	280	0.1
[CH$_3$-^{3}H]methionine	8×10^4	1×10^{-3}
[^{35}S]methionine	1×10^6	3×10^{-5}
[γ-^{32}P]ATP	1×10^7	3×10^{-6}

[a] Based on 50 counts per minute as the minimum measurable.

maximum may be lowered on rare occasions. Colored materials or charcoal may absorb the light emitted from the scintillant.

2. *^{3}H transfer.* Tritium is sometimes transferred from substrates to solvents or proteins. For this reason and more importantly for that above, the use of ^{14}C is preferable to ^{3}H, although the higher specific activities and relative cheapness of ^{3}H-labeled compounds often more than compensate for the disadvantages.

3. *Miscellaneous.* The addition of water to scintillants often lowers the efficiency of counting, so it should always be added in constant amounts. Sometimes the addition of base causes artifacts. Strip lighting can cause some scintillants to phosphoresce, and the effect can last several minutes.

b. Separation of reaction products

Clearly, in order to assay a reaction, it is necessary to separate the products from the starting materials to measure the radioactivity. Chromatography and high-voltage electrophoresis are very useful since the supporting paper may be cut into strips and the *relative* activities of the various regions accurately measured. Much more convenient, though, is the use of a filter pad to adsorb or trap the desired compound selectively. For example, labeled ATP may be adsorbed onto charcoal and collected on a glass fiber disk, or adsorbed onto a disk impregnated with charcoal, and the remaining reagents washed away. Filter disks of diethylammoniumethyl-cellulose have been used to adsorb anionic reagents selectively, and disks of carboxymethyl-cellulose to trap cations. Proteins and any strongly bound ligand often adsorb to nitrocellulose filters. Covalently labeled proteins or polynucleotides may be precipitated with acid and collected on glass fiber filters if a heavy precipitate is formed, or on nitrocellulose if the precipitate is light. (These disks do not lower the efficiency of counting of ^{14}C or ^{32}P.) Double-stranded DNA adsorbs to glass fiber filters; single-stranded DNA adsorbs to nitrocellulose.

8. Label-free optical detection[4]

A ligand is anchored to a polymer (e.g., dextran) that is immobilized on a solid surface (gold in the case of a commercial Biacore™ instrument) or silica. On the binding of a protein to the ligand, the refractive index changes from the value of 1.33 in water to up to 1.4. The change in refractive index can be measured by microrefractometry or by reflectometry. The strength of the signal depends on the mass bound. The sensitivity is $\sim$ pg/mm^2. A solution of protein is flowed over the ligand, and the rate of binding and the equilibrium fraction bound are directly determined. Rate constants for dissociation may be measured by washing off the protein. One advantage of the procedure is that it does not depend on any particular spectroscopic probe but is absolutely general. Another is that it can be used under "dirty" conditions; for example, for monitoring the binding of one antibody to its antigen in a mixture of components. The disadvantage is that the measurements are made on a solid surface. This causes less problems for equilibria than for kinetics. The author has seen rate constants for the binding of ligands that are clearly too low by two orders of magnitude because the rate constants for the heterogeneous reaction may be considerably slower than those in the homogeneous reaction.

B. Plotting kinetic data

Kinetic and equilibrium data are now generally analyzed by computer fitting. But learning to fit data manually is a valuable investment of time that enables one to understand what is going on and where errors occur.

1. Exponentials

a. Single

$$A \xrightarrow{k} B \tag{6.12}$$

In the first-order reaction of equation 6.12, the concentration of B at time t, $[B]_t$, is related to the final concentration of B, $[B]_\infty$ (the endpoint), by

$$[B]_t = [B]_\infty \{1 - \exp(-kt)\} \tag{6.13}$$

(cf. equation 4.7). The value of k is usually obtained from the following semilogarithmic plot of $[B]_\infty - [B]_t$ against t:

$$\ln([B]_\infty - [B]_t) = \ln[B]_\infty - kt \tag{6.14}$$

It is important to determine the endpoint $[B]_\infty$ accurately, since errors here cause serious errors in the derived rate constant.

b. The Guggenheim method

In cases in which the endpoint cannot be determined, the Guggenheim method may be used.[5] The differences between pairs of readings at t and $t + \Delta t$ (where Δt is a constant time that must be at least 2 or 3 times the half-time) are plotted against t in the semilogarithmic plot, since it may be shown that

$$\ln ([B]_{t+\Delta t} - [B]_t) = \text{constant} - kt \tag{6.15}$$

c. Consecutive exponentials

A series of exponentials of the form

$$[B] = X\{\exp(-k_1 t) - \exp(-k_2 t)\} \tag{6.16}$$

(cf. equation 4.31) are relatively straightforward to solve if one of the rate constants is more than 5 or 10 times faster than the other. The slower process is plotted as a simple semilogarithmic plot by using the data from the tail end of the curve after the first process has died out (that is, after 5 to 10 half-lives of the first process have occurred). The data may then be fed back into equation 6.16 to give the faster rate constant. This is often done graphically, as shown in Figure 6.2. If the rate constants are not separated by this factor, it is simpler to try to change their ratio by a change in the reaction conditions than to use a least squares method.

2. Second-order reactions

$$A + B \xrightarrow{k_2} C \tag{6.17}$$

Second-order kinetics are best dealt with by converting them to pseudo-first-order kinetics by using one of the reagents in large excess over the other. If $[A] \gg [B]$ in equation 6.17, $[A]$ changes little during the reaction and

$$\frac{d[C]}{dt} = -\frac{d[B]}{dt} = k_2[A]_0[B] \tag{6.18}$$

The disappearance of B and the appearance of C follow exponential first-order kinetics with a pseudo-first-order rate constant of $k_2[A]_0$. Plotting a series of such reactions with varying $[A]_0$, the initial concentration of A, gives k_2.

The analytical solution for the reaction when $[A]_0$ is similar to $[B]_0$ is

$$\frac{\ln ([A]_0[B]/[B]_0[A])}{[B]_0 - [A]_0} = k_2 t \tag{6.19}$$

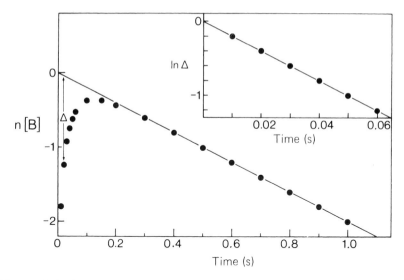

Figure 6.2 Graphical method of analyzing the kinetics of a reaction obeying equation 6.16. The logarithm of [B] is plotted against time. The rate constant for the slower process is obtained from the slope of the linear region after the faster process has died out. The rate constant for the faster process is obtained by plotting the logarithm of Δ (the difference between the value of [B] at a particular time and the value of [B] extrapolated back from the linear portion of the plot) against time for the earlier points. The rate constants for this example are 20 and 2 s^{-1}, respectively.

—an equation that is tedious to plot. However, the equation is greatly simplified when $[A]_0 = [B]_0$. The analytical solution becomes

$$\frac{1}{[A]_0 - [C]} - \frac{1}{[A]_0} = k_2 t \qquad (6.20)$$

This simple equation holds to a good approximation even when the initial concentrations are not exactly equal. In this case, the average of $[A]_0$ and $[B]_0$ should be used in equation 6.20 instead of $[A]_0$.

3. Michaelis-Menten kinetics

Although it has become fashionable to fit data for a complete progress curve directly to the integrated form of the Michaelis-Menten equation by using a computer, the most satisfactory method of determining k_{cat}, k_{cat}/K_M, and K_M is to use the classic approach of measuring initial rates (i.e., the first 5% or less of the reaction) and making a plot such as that of Eadie and Hofstee (equation 3.29). The following multiples of the K_M are a good range of substrate concentrations: 8; 4; 2; 1; 0.5; 0.25; 0.125.

C. Determination of protein – ligand dissociation constants[6]

1. Kinetics

As we discussed in Chapter 3, the K_M for an enzymatic reaction is not always equal to the dissociation constant of the enzyme–substrate complex, but may be lower or higher depending on whether or not intermediates accumulate or Briggs-Haldane kinetics hold. Enzyme–substrate dissociation constants cannot be derived from steady state kinetics unless mechanistic assumptions are made or there is corroborative evidence. Pre–steady state kinetics are more powerful, since the chemical steps may often be separated from those for binding.

The *dissociation constants of competitive inhibitors* are readily determined from inhibition studies by using equation 3.32. This equation holds whether or not the K_M for the substrate is a true dissociation constant. The inhibition must first be shown to be competitive by determining the apparent K_M for the substrate at different concentrations of the inhibitor, and calculating the K_1 from the apparent K_M and equation 3.32. Significant changes in K_M are obtained only at relatively high values of the inhibitor concentration.

2. Equilibrium dialysis

This is a method of directly measuring the concentrations of free and enzyme-bound ligand. A solution of the enzyme and ligand is separated from a solution of the ligand by a semipermeable membrane across which only the small ligand may equilibrate (Figure 6.3). After equilibration, a sample from the chamber containing protein gives the sum of the concentrations of free and bound ligand, whereas a sample from the other chamber gives the concentration of free ligand. Measurements may be made by using radioactively labeled ligands in chamber volumes of only 20 μL and sampling triplicate aliquots of 5 μL. However, since the apparatus requires at least 1 to 2 hours of equilibration with even the most porous membranes, the method cannot be used with unstable ligands or enzymes. *Nonequilibrium dialysis* has been used to make rapid measurements.[7] In this technique, the *rate* of diffusion of the ligand across the membrane from the

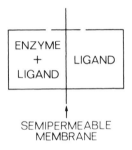

Figure 6.3 The principle of equilibrium dialysis.

side containing enzyme is measured as a function of concentration; binding slows down the rate.

3. Equilibrium gel filtration[8]

Certain gels, such as the commercial products Sephadex, Sephacryl, and Biogel, are made with pores that are large enough to be occupied by small ligands but not by proteins. If a chromatography column is packed with one of these gels and a solution of protein and ligand is applied, the protein travels faster through the column than the ligand does, since the ligand has to pass through the volume of solution surrounding the beads of gel and in the pores, whereas the protein has only to travel through the surrounding water. In equilibrium gel filtration, such a column is equilibrated with a solution of the ligand, and a sample of the enzyme in the equilibrating buffer is applied to the column. As the protein travels through the column it drags any bound ligand with it at the same flow rate. The result is that a peak of ligand travels through the column in the position of the protein, and a trough follows at the position normally occupied by a small ligand. The area under the peak should be the same as that in the trough, and should be equal to the amount of bound ligand.

One advantage of this method is that the enzyme is in contact with any particular substrate molecule for a short time only, and so can be used in cases in which the enzyme slowly hydrolyzes or chemically transforms the ligand. Another advantage is that some available gels are able to distinguish between the size of one polymer and another, so that, for example, the binding of a tRNA (M_r = 25 000) to an aminoacyl-tRNA synthetase (M_r = 100 000) may be measured.

The method may be scaled down. A convenient size for the binding of small ligands is a 1-mL tuberculin syringe packed with Sephadex G-25. A sample volume of 100 μL is applied, and individual drops (about 35 to 40 μL) are collected in siliconized tubes. The volume of each drop is measured by drawing it into a syringe. However, where the gel has to distinguish between a protein and a large ligand, much larger columns have to be used (a typical example being a 0.7- by 18-cm column for the tRNA/aminoacyl-tRNA synthetase).

a. Some possible errors

The criteria for an accurate experiment are that (1) the base line must be constant, (2) there must be a region of base line between the peak and the trough to show that the ligand and the protein have sufficiently different mobilities, and (3) the area of the peak must equal that of the trough. (Since it is very difficult to make up an enzyme–ligand solution that is exactly the same as the solution of the ligand in the equilibrating buffer, it is advisable to assay an aliquot of the enzyme solution before it is added to the column, in order to make any necessary correction to the area of the trough.)

Artifacts can occur due to the specific retardation of a ligand. ATP, ADP, and AMP, for example, bind to Sephadex and are retarded. The faster mobility of

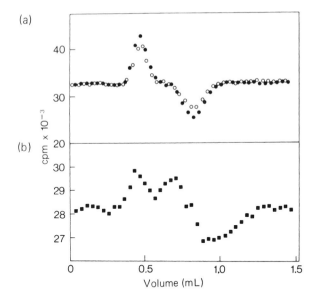

Figure 6.4 Experimental curves for equilibrium gel filtration. (a) The 1-mL tuberculin syringe was incubated in 109-μM[^{14}C]valine (○) or 109-μM[^{14}C]valine and 4-mM ATP (●). Then 100 μL of a solution of 26-μM valyl-tRNA synthetase was added to the same solution. Stoichiometries of 0.8 and 1.1, respectively, were found for the binding of the amino acid. Note the return to baseline between the peak and the trough—the mark of a good equilibrium gel filtration experiment. (b) An artifact-induced double peak obtained from the binding of [γ-^{32}P]ATP and valine to the enzyme. Some of the labeled ATP hydrolyzed to [^{32}P]orthophosphate, which traveled down the column faster than the [γ-^{32}P]ATP did.

phosphate and pyrophosphate produced in reactions in which [γ-^{32}P]ATP is hydrolyzed causes additional peaks and troughs (Figure 6.4).

4. Ultracentrifugation[9]

The binding of a small polymer to a larger one, such as the binding of tRNA to an aminoacyl-tRNA synthetase, cannot be determined by equilibrium dialysis, and also requires relatively large volumes for equilibrium gel filtration. In this case, the binding may be measured on a 100- to 200-μL scale by using the analytical ultracentrifuge. The cell is filled with a mixture of, say, the tRNA and aminoacyl-tRNA synthetase, and the absorbances of the bound tRNA and the free tRNA are directly measured by the ultraviolet optics during sedimentation. The higher-molecular-weight complex of the enzyme and the tRNA sediments faster than the free tRNA, and there is a sharp moving boundary of absorbance as the complex moves down the cell (Figure 6.5).

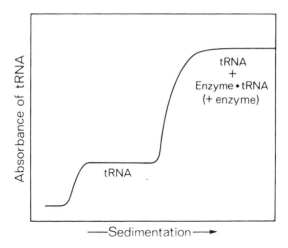

Figure 6.5 Sedimentation of a mixture of tRNA and aminoacyl-tRNA synthetase. The complex of tRNA and enzyme sediments faster than the free tRNA. The free enzyme and its complex with tRNA migrate together since they are in rapid equilibrium, the tRNA exchanging between the two.

Analytical ultracentrifugation is by far the best method for determining the state of oligomerization of small peptides.

5.   Filter assays[10]

Many proteins are adsorbed (along with any slowly dissociating bound ligands) on nitrocellulose filters, while the free ligands are not retained. In special cases, this provides a very economical procedure for assaying binding. In particular, much useful data on the binding of nucleic acids to proteins has resulted. Care must be taken to avoid possible errors. Binding is often less than 100% efficient, and it may be variable from batch to batch of filters. The filters also saturate at fairly low concentrations of protein.

6.   Spectroscopic methods

Except in NMR experiments, in which the concentrations of free and bound ligand may often be individually measured, spectroscopic methods do not usually give a direct measurement of the number of bound ligand molecules. On the binding of the ligand there is indeed a change in spectroscopic signal that is related to the fraction of the protein that is binding. But without additional evidence, the number of ligand molecules binding cannot be known. In these spectroscopic methods, the binding of a ligand to an enzyme typically causes either a change in the fluorescence of the protein, or a change in the fluorescence or the optical spectrum of the ligand. The usual procedure is to add increasing amounts of the ligand to a relatively dilute solution of the protein, and to plot the change in the spectroscopic signal by one of the procedures given in the following section. A variation of the procedure for ligands that are not chromophores is to measure their competitive inhibition of the binding of chromophoric ligands.

7. Stoichiometric titration

If the dissociation constant of the complex is sufficiently low, it may be possible to determine the number of equivalents of ligand that are required to give the maximum spectral change that occurs when all the binding sites are occupied. For example, increasing amounts of the ligand are added to a solution in which the concentration of the protein is at least 10 times higher than the dissociation constant. The results are plotted as in Figure 6.6 to give the stoichiometry.

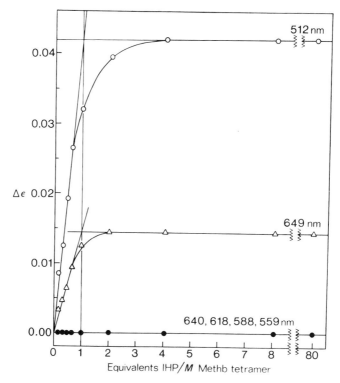

Figure 6.6 Spectrophotometric titration of the binding of inositol hexaphosphate (IHP) to methemoglobin (Methb). The complex has an increased absorbance at 512 and 649 nm, and no increases at 640, 618, 588, and 599 nm. The concentration of methemoglobin (20 μM) is about 14 times higher than the dissociation constant of 1.4 μM for the complex. The intersection of the slope of the increase in absorbance with the maximum value gives the stoichiometry (1, in this case). Note that this simple procedure cannot be used if the protein is not initially present at such a high concentration relative to the dissociation constant, since the assumption is that all the added ligand is bound to the protein for the early additions.

8. Microcalorimetry[11]

The heat of binding of a ligand to a protein may be measured using isothermal titration calorimetry (ITC), and this can be used to construct binding isotherms.[12,13] Aliquots of ligand are automatically added to a stirred sample of the protein, and the heat change is measured from the amount of electrical current required to keep the temperature constant with respect to a reference cell. The amount of protein required depends on the enthalpy of binding. About 1.4 mL of 10 μM protein for a change of 80 kJ/mol (20 kcal/mol) is adequate for an accurate determination. State-of-the-art equipment, which was just released in 1998, is five times more sensitive. The minimum concentration of the protein required for an accurate measurement determines the type of information that can be obtained. If the concentration is much higher than the dissociation constant of the ligand, then the stoichiometry and enthalpy of binding, but not the dissociation constant, may be determined with high precision. If the minimum concentration is similar to the dissociation constant, then the constant may measured as well. If the minimum concentration is lower than the dissociation constant, then just the constant and the enthalpy change may be measured. The principal advantages of ITC are that the enthalpy is determined directly, the specific heat can be measured, and no optical signals are required.

The use of differential scanning microcalorimetry for measuring the thermal denaturation of proteins is described in Chapter 17, section Ale. Typically, 0.5–1 mg of protein in 1 mL of buffer, or 0.1–0.2 mg in 0.5 mL with the most sensitive apparatus, is required for an accurate determination of the enthalpy of denaturation. The thermodynamics of dissociation of a reversibly bound ligand may be calculated from its effects on the denaturation curve of a protein.[14] The binding of ligands always raises the apparent T_m (temperature at 50% denaturation) of a protein because of the law of mass action: the ligand does not bind to the denatured state of the protein, and so binding displaces the denaturation equilibrium toward the native state.

D. Plotting binding data

1. The single binding site

The binding of a ligand to a single site on a protein is described by the following equations:

$$PL \xrightleftharpoons{K_L} P + L \tag{6.21}$$

$$K_L = \frac{[P][L]}{[PL]} \tag{6.22}$$

where [P] and [L] are the concentrations of the unbound protein and ligand.

In terms of the total protein concentration $[P]_0$,

$$[PL] = \frac{[P]_0[L]}{[L] + K_L} \tag{6.23}$$

Equation 6.23 is in the same form as the Michaelis-Menten equation, and may be manipulated in the same way. A good strategy in plotting the data is to use the equivalent of the Eadie plot:

$$[PL] = [P]_0 - K_L \frac{[PL]}{[L]} \tag{6.24}$$

A plot of $[PL]$ against $[PL]/[L]$ gives K_L.

Equation 6.24 cannot be used directly with spectroscopic data since $[PL]$ is not known. However, because $[PL]$ is usually directly proportional to the change in the spectroscopic signal being observed, we have

$$\Delta F = \Delta F_{max} - K_L \frac{\Delta F}{[L]} \tag{6.25}$$

where ΔF is the change in spectroscopic signal when $[L]$ is added to the protein solution. A plot of ΔF against $\Delta F/[L]$ gives K_L and ΔF_{max}, the change in signal when all the protein is converted into complex.

2. Multiple binding sites

a. Identical

If there are n identical noninteracting sites on the protein, equation 6.24 may be modified to the *Scatchard plot*.

$$v = n - K_L \frac{v}{[L]} \tag{6.26}$$

where v is the number of moles of ligand bound per mole of protein. The stoichiometry n and K_L are obtained from the plot of v against $v/[L]$.

b. Nonidentical

If there are two classes of sites, one weak and the other strong, the Scatchard plot will be biphasic and composed of the sum of two different Scatchard plots. The determination of the values of K_L from such plots is satisfactory only when they differ by at least a factor of 10.

E.   Computer fitting of data

There are many commercial programs available for fitting data to theoretical curves. They are extremely powerful but very dangerous. Used properly, they fit data directly to nonlinear curves without the need for transformation of data to linear equations that can distort the statistics, and they allow data to be fitted to complex equations that could not be solved by hand. But a computer just gives the best fit to the particular equation, $y = f(x)$, that you choose, and your data might not follow that equation.

1. Always check for deviation from the theoretical curves by using the computer to plot the *residuals* of y against x. (That is $y_{calculated} - y_{observed}$ versus x.) These should be randomly distributed around the x-axis. Any systematic or periodic deviation is evidence that the data do not fit the chosen equation. Where possible, also plot the data using a linear transformation to see if there is random or systematic deviation from this.

2. Always have a sufficient number of data points. Use a time window that focuses on the time range of interest—not the interval from the Big Bang to the current age of the universe.

3. The curve fitting programs cope better with fewer variables in the equations. Try to reduce the number of variables. For example, suppose you have to fit a multiphasic curve to three exponentials that are moderately separated in time. There are seven unknowns: three rate constants; three amplitudes; and an endpoint. If the slowest phase is sufficiently separated from the second, first fit the tail of the slowest phase to a single exponential. Then fit the whole curve to a triple exponential equation in which the rate constant and the amplitude that were derived for the third phase are used as constants. Use a time window that focuses on the first two phases and not the whole time course. Similarly, if the first phase is much faster than the second and third, fit the tail of the process to two exponentials. Then fit the fast time region to a triple exponential in which the last two phases have fixed rate constants and amplitudes.

F.   Statistics, errors of observation, and accuracy

Computer programs give an estimated error for fitting a curve (usually it is a standard error; see below). When fitting experimental data for a single progress curve for, say, a first-order reaction, this error is a measure of the fit of the data to that experimental curve. It is not a real measure of how close that measurement should be to the true rate constant. It is usually a considerable underestimate of the true error, which can be obtained only by making a series of measurements and analyzing them statistically. The following simple summary of statistical quantities is useful when analyzing data.

1. Normal or Gaussian distribution

The distribution of errors of measurement is usually analyzed according to the Gaussian or normal distribution. This applies to sampling a population that is subject to a random distribution. The normal distribution follows the equation

$$y = \frac{1}{\sigma\sqrt{2\pi}} e^{-(x-\bar{x})^2/2\sigma^2} \tag{6.27}$$

The quantity σ is called the *standard deviation* of the population and is a measure of the spread or dispersion of the values about the mean, $\bar{x}$ (see Table 6.3). The smaller the value of σ, the narrower the spread. σ is calculated from

$$\sigma = \left[\frac{1}{n} \sum_{i=1}^{n} (x_i - \bar{x})^2 \right]^{1/2} \tag{6.28}$$

2. Errors in sampling

Suppose one takes n measurements of a quantity and the measurements are subject to random errors. The value of the mean so measured follows a normal distribution about the true mean. Most calculators give a standard deviation calculated from the formula

$$\sigma_{\text{calc}} = \left[\frac{1}{n-1} \sum_{i=1}^{n} (x_i - \bar{x})^2 \right]^{1/2} \tag{6.29}$$

(which is equation 6.28 modified slightly). But the important quantity for the calculation of accuracy is the *standard error of the mean*, α, where

$$\alpha = \frac{\sigma_{\text{calc}}}{n^{1/2}} \tag{6.30}$$

The relationship between distribution of the measured value of $\bar{x}$ and α follows exactly the same distribution as in Table 6.3. Thus, the probability of the true

Table 6.3 *Probability of x being within various limits of its mean value ($\bar{x}$)*

Range of values	Probability of x within this range
$\bar{x} \pm 0.6745\sigma$	0.5
$\bar{x} \pm \sigma$	0.683
$\bar{x} \pm 2\sigma$	0.954
$\bar{x} \pm 3\sigma$	0.997

mean being between $\bar{x} \pm \alpha$ is 0.683, between $\bar{x} \pm 2\alpha$ is 0.954, and between $\bar{x} \pm 3\alpha$ is 0.997. Note that the value of α itself has a standard error that depends on n: $\alpha_\alpha = \alpha/(2n)^{1/2}$.

3. Combining errors of measurement

a. Sum or difference

Suppose we have a quantity y that is the sum or difference of two other quantities, a and b, and each has a standard error of α_a or α_b. Then the standard error of y is given by:

$$\alpha_y = (\alpha_a^2 + \alpha_b^2)^{1/2} \tag{6.31}$$

This may be extended to any series of a, b, c, etc.:

$$\alpha_y = (\alpha_a^2 + \alpha_b^2 + \alpha_c^2 \cdots)^{1/2} \tag{6.32}$$

b. Product or quotient

Suppose $y = ab$; then

$$\alpha_y = (b\alpha_a^2 + a\alpha_b^2)^{1/2} \tag{6.33}$$

For more complex multiplications, it is easier to calculate the error as a fraction. For example, if $y = abc$, then

$$\frac{\alpha_y}{y} = \left[\left(\frac{\alpha_a}{a} \right)^2 + \left(\frac{\alpha_b}{b} \right)^2 + \left(\frac{\alpha_c}{c} \right)^2 \right]^{1/2} \tag{6.34}$$

Equation 6.34 may be manipulated for a power law, $y = a^n$, to give

$$\frac{\alpha_y}{y} = \frac{n\alpha_a}{a} \tag{6.35}$$

or for a quotient, if $y = a/bc$, to give

$$\frac{\alpha_y}{y} = \left[\left(\frac{\alpha_a}{a} \right)^2 + \left(\frac{\alpha_b}{b} \right)^2 + \left(\frac{\alpha_c}{c} \right)^2 \right]^{1/2} \tag{6.36}$$

(which is the same as equation 6.34).

c. Logarithm

From elementary calculus, small changes, δ, in a natural logarithm of a quantity y and in y are related by $\delta \ln y = \delta y/y$. Consequently,

$$\alpha_{\ln y} = \frac{\alpha_y}{y} \tag{6.37}$$

Since log (to base 10) and ln are related by ln y = 2.303 log y,

$$\alpha_{\log y} = \frac{\alpha_y/y}{2.303} \tag{6.38}$$

d. Weighting data

Measurements that have a low standard error are clearly more significant than those with a large standard error. Each measurement should be weighted proportionally to the reciprocal of the square of its standard error. Suppose there are n measurements of a value and each value x_i has a standard error α_i; then the weighted mean value is given by

$$\bar{x} = \frac{\sum x_i/\alpha_i^2}{\sum 1/\alpha_i^2} \tag{6.39}$$

The standard error of the weighted mean is given by

$$\alpha = \left[\frac{1}{(n-1)\sum(1/\alpha_i^2)} \sum \frac{(x_i - \bar{x})^2}{\alpha_i^2} \right]^{1/2} \tag{6.40}$$

4. Poisson distribution

The statistics of processes such as radioactive decay and emission of light that produce a flux of particles or *distributive* polymerase enzymes that add residues at random to growing polymer chains obey the Poisson distribution (see Chapter 14). The number of particles measured per unit time or the number of residues added to a particular chain varies about the mean value $\bar{x}$ according to equation 6.41.

$$p(k) = \frac{\bar{x}^k}{k!} e^{\bar{x}} \tag{6.41}$$

where $p(k)$ is the probability of k particles being emitted or residues added. The Gaussian and Poisson distributions approximate each other when there are a large number of particles involved in the Poisson.

The standard deviation of the Poisson distribution is equal to $\bar{x}^{1/2}$. This has profound consequences in designing experiments involving radioactivity and light intensity.

5. Signal to noise in absorbance, circular dichroism, fluorescence, and radioactive counting

Processes such as radioactive decay and emission of light produce a flux of particles. The number of particles measured per unit time varies about the mean value n according to the Poisson distribution, with the standard deviation equal to $\sqrt{n}$. Thus, there is an inherent noise that is proportional to $\sqrt{n}$ in measuring n because there are fluctuations in its apparent strength. Suppose the noise, $N = \alpha \sqrt{n}$. Then the ratio of noise to signal, (N/S) is given by

$$\frac{N}{S} = \frac{\alpha}{\sqrt{n}} \tag{6.42}$$

Thus, the greater the value of n, the lower the (N/S) ratio. For processes such as radioactive decay and fluorescence, the greater the specific activity or the greater the intensity of the exciting light, the lower the error or noise in the measurements. Similarly, the longer the period of counting, the more the precision. Rapid reaction experiments where optical signals are collected over shorter times are inherently noisy because the shorter the time range, the smaller the number of particles counted. Accuracy improves according to the square root of the time of collection or the time constant of the electronic filter that is used to smooth the signal. Absorbance and CD measurements respond in a more complex way because of the logarithmic relationship of absorbance to the intensity of transmitted light: there is a balance between the advantage of having a high absorbance and the loss of signal. According to Beer's law (equation 6.1), $A = \log (I_0/I)$, where A is the absorbance, I_0 the light intensity entering the solution, and I the transmitted light. Thus,

$$I = I_0 10^{-A} \tag{6.43}$$

From equation 6.42, the ratio of the noise in transmitted intensity, δI, to I is given by

$$\frac{\delta I}{I} = \pm \frac{\alpha 10^{A/2}}{\sqrt{I_0}} \tag{6.44}$$

The relationship between fluctuations in A (δA) and those in I may be derived from

$$\frac{dA}{dI} = \frac{d(\log (I_0/I))}{dI} = \frac{-1}{2.303I} \tag{6.45}$$

so that $\delta A = \delta I/2.303I$. The ratio of signal to noise in A, (N/S)$_A$, is given by $\delta A/A$. Substituting equation 6.45 into 6.44 and dividing through by A gives

$$\left(\frac{N}{S}\right)_A = \pm \frac{\delta A}{A} = \pm \frac{\alpha 10^{A/2}}{2.303A/\sqrt{I_0}} \tag{6.46}$$

The minimum is found by differentiating (N/S)$_A$ with respect to A. The absorbance where the value of (N/S)$_A$ is at a minimum is given by

$$A_{\min} = \frac{2}{2.303} = 0.87$$

(6.47)

The optimal absorbance to use in absorbance or CD measurements for best ratio of signal to noise is 0.87.

Appendix: Measurement of protein concentration

For many purposes, it is not necessary to know the absolute concentration of a protein with precision. For example, equation 6.25 for the determination of a protein–ligand dissociation constant is independent of the protein concentration, provided the ligand is in vast excess. For many measurements on enzyme activity, only the concentration of active enzyme is necessary, which can be measured by active site titration, as described in Chapter 2, section E. For some purposes, such as the measurement of stoichiometry by equation 6.26, or especially for calorimetric measurements, the concentrations must be measured with high precision. The routine procedure for determining protein concentration is from optical absorbance, either from a known extinction coefficient at 280 nm or from the binding of a dye such as Coomassie Blue (the "Bradford" assay).[15] The extinction coefficient is determined simply and accurately from the values for the constituent individual amino acids that absorb at 280 nm tabulated by Gill and von Hippel.[16] The extinction coefficients do depend slightly on their environment within the native protein ($\pm 5-10\%$), and so the denatured protein is used initially as the standard, and the native protein is calibrated against this. The relevant extinction coefficients at 280 nm in 6-M guanidinium chloride are: tryptophan, 5690, tyrosine, 1280; and cysteine, 120 $M^{-1}cm^{-1}$. The absorbance in the Coomassie Blue assay can vary by a factor of about two, depending on the sequence of the protein, and so the assay must be calibrated. The simplest and most readily available procedure for independently determining the protein concentration is quantitative amino acid analysis.

The absorbance at 280 nm can be affected by light scattering, which increases sharply with decreasing wavelength. There is always a small amount of scattering from the inherent size of the protein, which can be corrected by measuring the absorbance at different wavelengths.[17] But oligomerization can cause very large effects. It is always advisable to scan the spectrum of the protein rather than just measure the absorbance at 280 nm, to detect not only the increasing scattering with decreasing wavelength but also other artifacts.

References

1. M. R. Eflink, *Meth. Enzymol.* **278**, 221 (1997).
2. R. Woody, *Meth. Enzymol.* **246**, 34 (1995).
3. G. Fasman, *Circular dichroism and the conformational analysis of biomolecules*, Plenum Press (1996).

4. P. Schuck, *Curr. Opin. Biotech.* **8**, 498 (1997).
5. E. A. Guggenheim, *Phil. Mag.* **2**, 538 (1926).
6. G. Weber, *Protein interactions*, Chapman and Hall (1992).
7. P. England and G. Hervé, *Biochemistry* **31**, 9725 (1992).
8. J. P. Hummel and W. J. Dreyer, *Biochim. Biophys. Acta* **63**, 530 (1962).
9. J. Behlke, *Europ. Biophys. J. Biophys. Lett.* **25**, 5 (1997).
10. M. Yarus and P. Berg, *Anal. Biochem.* **35**, 450 (1970).
11. A. Cooper and C. M. Johnson, in *Methods in molecular biology, Vol. 22: Microscopy, optical spectroscopy, and macroscopic techniques*, C. Jones, B. Mulloy, and A. H. Thomas, eds., Humana Press, 109–124; 137–150 (1994).
12. T. Wiseman, S. Williston, J. F. Brandts, and L. N. Lin, *Anal. Biochem.* **179**, 131 (1989).
13. D. R. Bundle and B. W. Sigurskjold, *Meth. Enzymol.* **247**, 288 (1994).
14. J. C. Martínez, M. El Harrous, V. V. Filiminov, P. L. Mateo, and A. R. Fersht, *Biochemistry* **33**, 3919 (1994).
15. M. M. Bradford, *Anal. Biochem.* **72**, 248 (1976).
16. S. C. Gill and P. H. von Hippel, *Anal. Biochem.* **182**, 319 (1989).
17. A. F. Drake, in *Methods in molecular biology, Vol. 22: Microscopy, optical spectroscopy, and macroscopic techniques,* C. Jones, B. Mulloy, and A. H. Thomas, eds. Humana Press, 173–182 (1994).

Detection of Intermediates in Enzymatic Reactions

7

The mechanism of an enzymatic reaction is ultimately defined when all the intermediates, complexes, and conformational states of the enzyme are characterized and the rate constants for their interconversion are determined. The task of the kineticist in this elucidation is to detect the number and sequence of these intermediates and processes, define their approximate nature (that is, whether covalent intermediates are formed or conformational changes occur), measure the rate constants, and, from studying pH dependence, search for the participation of acidic and basic groups. The chemist seeks to identify the chemical nature of the intermediates, by what chemical paths they form and decay, and the types of catalysis that are involved. These results can then be combined with those from x-ray diffraction and NMR studies and calculations by theoretical chemists to give a complete description of the mechanism.

We shall now discuss some of the techniques that have been used to detect intermediates and delineate reaction pathways, using some well-known enzymes as examples.

A. Pre–steady state versus steady state kinetics

It is often said that kinetics can never prove mechanisms but can only rule out alternatives. Although this is certainly true of steady state kinetics, in which the only measurements made are those of the rate of appearance of products or disappearance of reagents, it is not true of pre–steady state kinetics. If the intermediates on a reaction pathway are directly observed and their rates of formation and decay are measured, kinetics can prove a particular mechanism. This is the basic strength of pre–steady state kinetics: the enzyme is used in *substrate* quantities and the events on the enzyme are directly observed. There may, of course, be intermediates that remain undetected because they do not accumulate or give rise to spectral signals, or simply because they are beyond the time scale of the measurements, but an overall reaction pathway may be proved in a scientifically acceptable manner by pre–steady state kinetics. The basic weakness of steady state kinetics is that the evidence is always ambiguous. No direct information is obtained about the number of intermediates, so the minimum number is always assumed. This is not to say that pre–steady state kinetics should be performed to the exclusion of steady state kinetics, but rather that a combination of the two approaches should be used. Once pre–steady state kinetics has given information about the intermediates on the pathway, steady state kinetics becomes much more powerful.

There are certain practical advantages of pre–steady state kinetics. Processes that are essentially very simple may be measured, among them the stoichiometry of a burst process, the rate constant for the transfer of an enzyme-bound intermediate to a second substrate, and a ligand-induced conformational change. Also, the first-order processes that are usually measured are independent of enzyme concentration, unlike the rate constants of steady state measurements. Although very high concentrations of enzymes are required for the rapid reaction measurements, these are usually close to those that occur *in vivo*. Furthermore, the high enzyme concentrations are usually similar to those used for making direct measurements on the physical state of the protein, so that data may be obtained for the state of aggregation, etc., under the reaction conditions.

Pre–steady state kinetics involves direct measurements, and direct measurements are always preferable, especially considering the tendency of enzymes to "misbehave" (section B5).

1. Detection of intermediates: What is "proof"?

Much of the following discussion centers on determining the chemical pathways of enzymatic reactions. This usually requires the detection of the chemical intermediates involved, since they give *positive* evidence. We shall consider that an intermediate is "proved" to be on a reaction pathway if the following criteria are satisfied:

1. The intermediate is isolated and characterized.
2. The intermediate is formed sufficiently rapidly to be on the reaction pathway.
3. The intermediate reacts sufficiently rapidly to be on the reaction pathway.

These criteria require that pre–steady state kinetics be used at some stage in order to measure the relevant formation and decomposition rate constants of the intermediate. But the rapid reaction measurements are not sufficient by themselves, since the rate constants must be shown to be consistent with the activity of the enzyme under steady state conditions. Hence the power, and the necessity, of combining the two approaches.

It must always be borne in mind that an intermediate that has been isolated could be the result of a rearrangement of another intermediate, and might not itself be on the reaction pathway. This is why criteria 2 and 3 are needed. It is also possible that a genuine intermediate is isolated but that during the experimental work-up the enzyme takes on a different, low-activity conformation. In this case, criteria 2 and 3 are not met, even though the chemical nature of the intermediate is correct.

B. Chymotrypsin: Detection of intermediates by stopped-flow spectrophotometry, steady state kinetics, and product partitioning

The currently accepted mechanism for the hydrolysis of amides and esters catalyzed by the archetypal serine protease chymotrypsin involves the initial formation of a Michaelis complex followed by the acylation of Ser-195 to give an acylenzyme (Chapter 1) (equation 7.1). Much of the kinetic work with the enzyme has been directed toward detecting the acylenzyme. This work can be used to illustrate the available methods that are based on pre–steady state and steady state kinetics. The acylenzyme accumulates in the hydrolysis of activated or specific ester substrates ($k_2 > k_3$), so that the detection is relatively straightforward. Accumulation does not occur with the physiologically relevant peptides ($k_2 < k_3$), and detection is difficult.

$$E + RCO-X \underset{}{\overset{K_S}{\rightleftharpoons}} RCO-X \cdot E \xrightarrow{k_2} RCO-E \xrightarrow{k_3} RCO_2H + E \qquad (7.1)$$
$$+ XH$$

1. Detection of intermediates from a "burst" of product release

In 1954, B. S. Hartley and B. A. Kilby[1] examined the reaction of substrate quantities of chymotrypsin with excess p-nitrophenyl acetate or p-nitrophenyl ethyl carbonate. They noted that the release of p-nitrophenol did not extrapolate back to zero but instead involved an initial "burst," equal in magnitude to the concentration of the enzyme (Chapter 4, Figure 4.10). They postulated that initially the ester rapidly acylated the enzyme in a mole-to-mole ratio, and that the subsequent turnover of the substrate involved the relatively slow hydrolysis of the acylenzyme as the rate-determining step. This was later verified by the stopped-flow experiments described in section B2.

Such burst experiments have since been performed on many other enzymes. But, bursts may be due to effects other than the accumulation of intermediates, and artifacts can occur. Some examples are the following:

1. The enzyme is converted to a less active conformational state on combination with the first mole of substrate.
2. The dissociation of the product is rate-determining.
3. There is severe product inhibition.

It is not a trivial matter to eliminate possibilities 1 and 2. But this has been done for chymotrypsin in the following series of stopped-flow experiments.

2. Proof of formation of an intermediate from pre–steady state kinetics under single-turnover conditions

The strategy is to measure the rate constants k_2 and k_3 of the acylenzyme mechanism (equation 7.1) and to show that each of these is either greater than or equal to the value of k_{cat} for the overall reaction in the steady state (i.e., apply rules 2 and 3 of section A1). This requires: (1) choosing a substrate (e.g., an ester of phenylalanine, tyrosine, or tryptophan) that leads to accumulation of the acylenzyme, (2) choosing reaction conditions under which the acylation and deacylation steps may be studied separately, and (3) finding an assay that is convenient for use in pre-steady state kinetics. The experiments chosen here illustrate stopped-flow spectrophotometry and chromophoric procedures.

a. Measurement of the rate constant for acylation, k_2

The step k_2 is isolated by mixing an excess of ester substrate with the enzyme. Under this condition, the acylenzyme is formed; it accumulates and then remains at a constant concentration over an extended period of time, as long as there is enough substrate present to ensure that the enzyme remains acylated (Chapter 4, equations 4.37 to 4.46). The process of the E·S complex giving the acylenzyme is thus isolated. The rate constant for this step is said to be measured under *single-turnover* conditions, since a single turnover of the enzyme from the E·S to the acylenzyme is measured, and not the recycling of enzyme as in the steady state.

Three different chromophoric procedures may be used. The first two depend on the synthesis of chromophoric substrates. The third utilizes an independent probe that can be applied to a wide range of substrates.

1. Chromophoric leaving group.[2-4] The original work on *p*-nitrophenyl acetate has been extended by synthesizing *p*-nitrophenyl esters of specific acyl groups, such as acetyl-L-phenylalanine, -tyrosine, and -tryptophan. The rate of acylation of the enzyme is determined from the rate of appearance of the nitrophenol or nitrophenolate ion, which absorbs at a different wavelength from the parent ester.

Acetyl-L-phenylalanine *p*-nitrophenyl ester

When the ester is mixed with the enzyme, there is a rapid exponential phase followed by a linear increase in the absorbance due to the nitrophenol. The rate constant for acylation and the dissociation constant of the enzyme-substrate complex may be calculated from the concentration dependence of the rate constant for the exponential phases (Chapter 4, equation 4.46). (The rate constant of the linear portion gives the deacylation rate, but this is a steady state measurement.) Unfortunately, nitrophenyl esters are often so reactive that the acylation rate is too fast for stopped-flow measurement.

2.   *Chromophoric acyl group.*[4,5] The spectrum of the furylacryloyl group depends on the polarity of the surrounding medium, and also on the nature of the moiety to which it is attached. The spectrum of furylacryloyl-L-tyrosine ethyl ester changes slightly when it is bound to chymotrypsin. There are also further changes on formation of the acylenzyme and on the subsequent hydrolysis. The rate constants for acylation and deacylation and the dissociation constant of the Michaelis complex may be measured by the appropriate experiments.

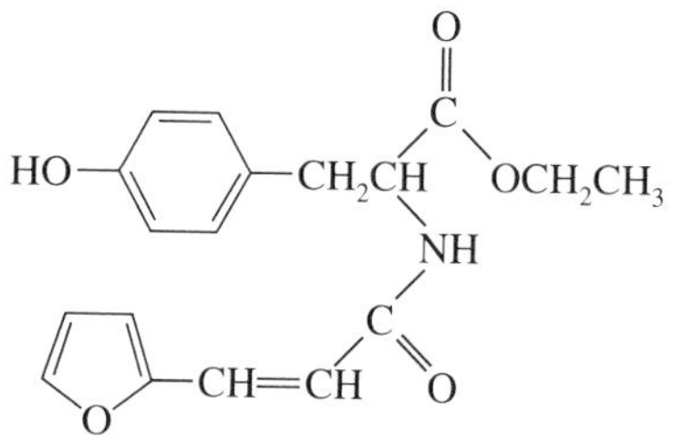

Furylacryloyl-L-tyrosine ethyl ester

When the ester is mixed with the enzyme, there is an initial change in absorbance that is due to the formation of the Michaelis complex. The rate constant for this is beyond the time scale of stopped flow, but the magnitude of the change can be used to calculate the dissociation constant. The absorbance then changes exponentially as the acylenzyme accumulates. There are further changes in the spectrum of the furylacryloyl group as the ester is gradually hydrolyzed to the free acid.

3.   *Chromophoric inhibitor displacement.*[6,7] The spectrum of the dye proflavin changes significantly with solvent polarity. It is a competitive inhibitor of chymotrypsin, trypsin, and thrombin, and it undergoes a large increase in absorbance at 465 nm ($\Delta\epsilon \approx 2 \times 10^4 \, M^{-1} \, cm^{-1}$) on binding (Figure 7.1).

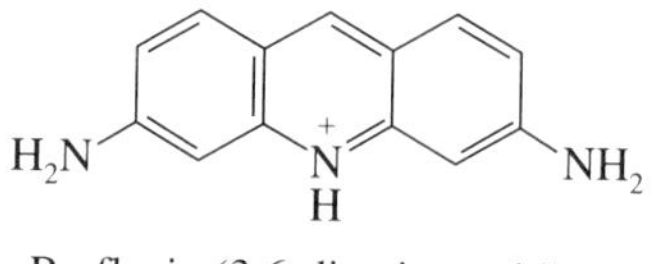

Proflavin (3,6-diaminoacridine)

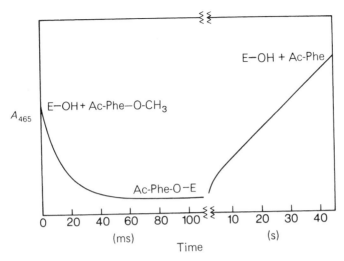

Figure 7.1 The proflavin displacement method. A solution of chymotrypsin ($10~\mu M$), proflavin ($50~\mu M$), and Ac-Phe-OCH$_3$ ($2~mM$) is mixed at pH 6 and 25°C in a stopped-flow spectrophotometer. The substrate-binding step is too fast to be observed. The rapid exponential decrease in absorbance at 465 nm is caused by the displacement of proflavin from the enzyme on formation of the acylenzyme. The slow increase in absorbance is due to the depletion of the substrate and the consequent decrease in the steady state concentration of the acylenzyme. [From A. Himoe, K. G. Brandt, R. J. Desa, and G. P. Hess, *J. Biol. Chem.* **244**, 3483 (1969).]

When an ester such as acetyl-L-phenylalanine ethyl ester is mixed with a solution of chymotrypsin and proflavin, the following events occur. There is a rapid displacement of some of the proflavin from the active site as the substrate combines with the enzyme, leading to a decrease in A_{465}. (This is complete in the dead time of the apparatus.) Then, as the acylenzyme is formed, the binding equilibrium between the ester and the dye is displaced, leading to the displacement of all the proflavin. The absorbance remains constant until the ester is depleted and the acylenzyme disappears. The dissociation constant of the enzyme-substrate complex may be calculated from the magnitude of the initial rapid displacement, whereas the rate constant for acylation may be obtained from the exponential second phase.

Use of the proflavin displacement method is far more convenient than use of the furylacryloyl group, since no special substrates have to be synthesized and one readily available compound can be used with all substrates. In general, it is better not to use modified substrates: not only are they chemically inconvenient to synthesize, but they are always open to criticism on the grounds that the results could be artifacts.

b.   Measurement of the rate constant for deacylation, k_3

The thematic approach to isolating the deacylation step is to generate the acylenzyme *in situ* in the stopped-flow spectrophotometer by mixing a substrate that acylates very rapidly with an excess or *stoichiometric* amount of the enzyme. The acylenzyme is formed in a rapid step that consumes all the substrate. This is then followed by relatively slow hydrolysis under single-turnover conditions. For example, acetyl-L-phenylalanine *p*-nitrophenyl ester may be mixed with chymotrypsin in a stopped-flow spectrophotometer in which the enzyme is acylated in the dead time. The subsequent deacylation may be monitored by the binding of proflavin to the free enzyme as it is produced in the reaction.[8]

There are also nonthematic methods that allow the formation of acylenzymes under conditions where they are stable, so that they can be stored in a syringe in a stopped-flow spectrophotometer. For example, it is possible to synthesize certain nonspecific acylenzymes and store them at low pH.[9–12] When they are restored to high pH, they are found to deacylate at the rate expected from the steady state kinetics. This approach has been extended to cover specific acylenzymes. When acyl-L-tryptophan derivatives are incubated with chymotrypsin at pH 3 to 4, the acylenzyme accumulates. The solution may then be "pH-jumped" by mixing it with a concentrated high-pH buffer in the stopped-flow spectrophotometer.[13,14] The deacylation rate has been measured by the proflavin displacement method and by using furylacrylolyl compounds.

c.   Characterization of the intermediate

The preceding experiments prove that there is an intermediate on the reaction pathway: in each case, the measured rate constants for the formation and decay of the intermediate are at least as high as the value of k_{cat} for the hydrolysis of the ester in the steady state. They do not, however, prove what the intermediate is. The evidence for covalent modification of Ser-195 of the enzyme stems from the early experiments on the irreversible inhibition of the enzyme by organophosphates such as diisopropyl fluorophosphate: the inhibited protein was subjected to partial hydrolysis, and the peptide containing the phosphate ester was isolated and shown to be esterified on Ser-195.[15,16] The ultimate characterization of acylenzymes has come from x-ray diffraction studies of nonspecific acylenzymes at low pH, where they are stable (e.g., indolylacryloyl-chymotrypsin),[17] and of specific acylenzymes at subzero temperatures and at low pH.[18] When stable solutions of acylenzymes are restored to conditions under which they are unstable, they are found to react at the required rate. These experiments thus prove that the acylenzyme does occur on the reaction pathway. They do not rule out, however, the possibility that there are further intermediates. For example, they do not rule out an initial acylation on His-57 followed by rapid intramolecular transfer. Evidence concerning this and any other hypothetical intermediates must come from additional kinetic experiments and examination of the crystal structure of the enzyme.

3. Detection of the acylenzyme in the hydrolysis of esters by steady state kinetics and partitioning experiments

In the last section we saw that stopped-flow kinetics can detect intermediates that *accumulate*. Detection of these intermediates by steady state kinetics is of necessity indirect and relies on inference. Proof depends ultimately on relating the results to the direct observations of the pre–steady state kinetics. But steady state kinetics can also detect intermediates that do not accumulate, and, by extrapolation from the cases in which accumulation occurs, can *prove* their existence and nature.

Detection of intermediates by steady state kinetics depends on:

1. The accumulation of an intermediate that is able to react either with an acceptor whose concentration may be varied, or, preferably, with several different acceptors.

2. The generation of a common intermediate E—R by a series of different substrates all containing the structure R. This intermediate must be able to react with different acceptors.

The hydrolysis of esters (and amides) by chymotrypsin satisfies these criteria. The hydrolysis of, say, acetyl-L-tryptophan *p*-nitrophenyl ester forms an acylenzyme that reacts with various amines such as hydroxylamine, alaninamide, hydrazine, etc., and also with alcohols such as methanol, to give the hydroxamic acid, dipeptide, hydrazide, and methyl ester, respectively, of acetyl-L-tryptophan. The same acylenzyme is generated in the hydrolysis of the phenyl, methyl, ethyl, etc., esters of the amino acid (and also during the hydrolysis of amides).

The kinetic consequences of the common intermediate can be used to diagnose its presence.

a. The rate-determining breakdown of a common intermediate implies a common value of V_{max} or k_{cat}

If several different substrates generate the same intermediate and if its breakdown is rate-determining, then they should all hydrolyze with the same value of k_{cat}:

$$\begin{matrix} \text{RCOX} \\ \text{RCOY} \\ \text{RCOZ} \end{matrix} \Bigg\rangle \longrightarrow \text{RCO—E} \xrightarrow{\text{slow}} \text{RCO}_2\text{H} + \text{E} \qquad (7.2)$$

This has been found for many series of ester substrates of chymotrypsin (Table 7.1) since the original study of H. Gutfreund and B. R. Hammond in 1959.[19] For weakly activated esters, the value of k_{cat} decreases to below that of k_3 because k_2 becomes partly rate-determining (equation 7.1). With amides, k_{cat} is very low and k_2 is completely rate-determining. The steady state analysis of k_{cat} in relation to k_2 and k_3 was presented in Chapter 3 [equation 3.22, where $k_{cat} = k_2 k_3/(k_2 + k_3)$]. Also given there was the relationship between K_M and K_S

Table 7.1 *Comparison of values of k_{cat} for the hydrolysis of substrates by α-chymotrypsin at pH 7.0 and 26°C*

Derivative	k_{cat} (s^{-1})	K_M (mM)	Rate-determining step	Ref.
N-Acetyl-L-tryptophan derivatives				
Amide	0.026	7.3	Acylation	1
Ethyl ester	27	0.1	Deacylation	1
Methyl ester	28	0.1	Deacylation	1
p-Nitrophenyl ester	30	0.002	Deacylation	1
N-Acetyl-L-phenylalanine derivatives				
Amide	0.039	37	Acylation	1
Ethyl ester	63	0.09	Deacylation	1
Methyl ester	58	0.15	Deacylation	1
p-Nitrophenyl ester	77	0.02	Deacylation	1
N-Benzoylglycine derivatives				
Ethyl ester	0.1	2.3	Mainly acylation	2
Methyl ester	0.14	2.4	Mainly acylation	2
Isopropyl ester	0.05	2.3	Acylation	2
Isobutyl ester	0.17	2.4	Mainly acylation	2
Choline ester	0.43	1.2	Deacylation	2
4-Pyridinemethyl ester	0.51	0.092	Deacylation	2
p-Methoxyphenyl ester	0.61	0.1	Deacylation	3
Phenyl ester	0.54	0.14	Deacylation	3
p-Nitrophenyl ester	0.54	0.03	Deacylation	3

1 B. Zerner, R. P. M. Bond, and M. L. Bender, *J. Am. Chem. Soc.* **86**, 3674 (1964).
2 R. M. Epand and I. B. Wilson, *J. Biol. Chem.* **238**, 1718 (1963).
3 A. Williams, *Biochemistry.* **9**, 3383 (1970). (The data are corrected to pH 7.0 from pH 6.91, assuming a pK_a of 6.8.)

[equation 3.21, where $K_M = K_S k_3/(k_2 + k_3)$]. This latter relationship is shown very clearly in Table 7.1 in the reactions of the derivatives of acetyl-L-tryptophan and acetyl-L-phenylalanine. For $k_2 \ll k_3$— i.e., the situation for the amide substrates—$K_M = K_S$. But for the ester substrates, $k_2 > k_3$ and so $K_M \simeq K_S k_3/k_2$. Since k_3 is the same for the derivatives of any one acylamino acid, K_M should decrease with increasing k_2, as found. The enhanced reactivity of the activated *p*-nitrophenyl ester is manifested in a low value for K_M. The trend of K_M values decreasing with increasing reactivity of the leaving group is seen also for the hydrolysis of benzoylglycine esters.

The occurrence of a common value of k_{cat} in the reaction of a series of substrates is not sufficient evidence for the accumulation of a common covalent intermediate whose breakdown is rate-determining. The value of k_{cat} is constant for the hydrolysis of a wide range of phosphate esters by alkaline phosphatase.[20,21] This was once interpreted as evidence for the rate-determining hydrolysis of a phosphorylenzyme. But it now seems likely that, at alkaline pH, dephosphorylation is rapid[22] and there is a rate-determining dissociation of inorganic phosphate from the enzyme–product complex $(E \cdot P_i)$.[23,24] The common value of k_{cat} is caused by a common intermediate, but it is a noncovalent one.

b. Partitioning of the intermediate between competing acceptors

If an intermediate that may react with different acceptors is generated, two procedures may be used for its detection.

The first involves determining product ratios. For example, the hydrolysis of a series of esters of hippuric acid by chymotrypsin in solutions containing hydroxylamine leads to the formation of the free hippuric acid and hippurylhydroxamic acid in a constant ratio (Table 7.2):

$$E + RCOOR' \longrightarrow RCO{-}E \begin{array}{c} \xrightarrow{NH_2OH} RCONHOH \\ \xrightarrow{H_2O} RCO_2H \end{array} \qquad (7.3)$$

The nonenzymatic hydrolysis under the same conditions leads to variable product ratios.[25] This is good evidence for a common intermediate.

The second procedure involves measuring the rates of formation of the products for a particular substrate at varying acceptor concentrations. This gives information on the rate-determining step of the reaction as well as detecting the intermediate. Suppose that the rate-determining step is the formation of the acylenzyme. Then, since the acceptor reacts with the acylenzyme after the

Table 7.2 *Product ratios in the hydrolysis of N-benzoylglycine esters by α-chymotrypsin in 0.1-M hydroxylamine*[a]

	Hydroxylaminolysis/hydrolysis	
Ester	Enzymatic, pH 6.6–6.8	Nonenzymatic, pH 12
Methyl	0.37	0.99
Isopropyl	0.38	0.29
Homocholine	0.37	1.73
4-Pyridinemethyl	0.37	3.03

[a] From R. M. Epand and I. B. Wilson, *J. Biol. Chem.* **238**, 1718 (1963); **240**, 1104 (1965).

rate-determining step, it cannot increase the rate of destruction of the ester. The overall formation rates will be as in Figure 7.2. If the rate-determining step is the hydrolysis of the acylenzyme, the acceptor increases the rate of its reaction and hence increases the overall reaction rate. The product formation rates will be as in Figure 7.3.

The steady state kinetics for partition may be calculated from

$$E + RCOOR' \underset{K_S}{\rightleftharpoons} E \cdot RCOOR' \xrightarrow[R'OH]{k_2} RCO-E \begin{array}{c} \xrightarrow{k_3'[H_2O]} RCO_2H \\ \xrightarrow{k_4[N]} RCO-N \end{array} \tag{7.4}$$

The following expressions for k_{cat} and K_M may be derived by the usual procedures for the reactions of chymotrypsin.[26] The kinetics are simplified in this example because the acceptor N does not bind to the enzyme:

$$K_M = K_S \frac{k_3'[H_2O] + k_4[N]}{k_2 + k_3'[H_2O] + k_4[N]} \tag{7.5}$$

For the formation of RCO_2H,

$$k_{cat} = \frac{k_2 k_3'[H_2O]}{k_2 + k_3'[H_2O] + k_4[N]} \tag{7.6}$$

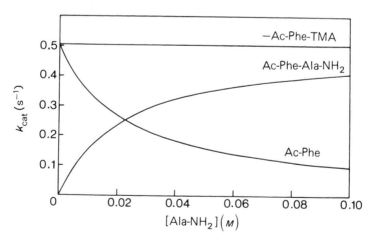

Figure 7.2 The chymotrypsin-catalyzed hydrolysis and transacylation reactions of acetylphenylalanine p-trimethylammoniumanilide (Ac-Phe-TMA) in the presence of various concentrations of Ala-NH$_2$. The values of k_{cat} for the depletion of Ac-Phe-TMA and the production of Ac-Phe and Ac-Phe-Ala-NH$_2$ are calculated from equation 7.3 by using $k_2 = 0.504$ s^{-1}, $k_3'[H_2O] = 144$ s^{-1}, and $k_4 = 6340$ s^{-1} M^{-1}. [From J. Fastrez and A. R. Fersht, *Biochemistry* **12**, 2025 (1973).]

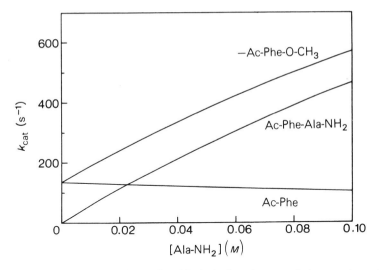

Figure 7.3 The chymotrypsin-catalyzed hydrolysis and transacylation reactions of Ac-Phe-OCH$_3$ in the presence of various concentrations of Ala-NH$_2$. The values of k_{cat} for the depletion of Ac-Phe-OCH$_3$ and the production of Ac-Phe and Ac-Phe-Ala-NH$_2$ are calculated from equation 7.3 by using $k_2 = 2200$ s^{-1}, $k_3'[H_2O] = 144$ s^{-1}, and $k_4 = 6340$ s^{-1} M^{-1}. [From J. Fastrez and A. R. Fersht, *Biochemistry* **12**, 2025 (1973).]

For the formation of RCO-N,

$$k_{cat} = \frac{k_2 k_4[N]}{k_2 + k_3'[H_2O] + k_4[N]} \tag{7.7}$$

Note that k_{cat}/K_M for the rate of disappearance of RCO$_2$R′ is equal to k_2/K_S, and is obtained by dividing the sum of the two values of k_{cat} from equations 7.6 and 7.7 by the K_M. It may be recalled from Chapter 3, section F, that this is always true for such a series of sequential reactions following an equilibrium binding step. Thus another criterion for the detection of an intermediate is provided. If the reaction of the nucleophiles involves the direct attack on the Michaelis complex, as in equation 7.8, k_{cat}/K_M will be a function of the concentration of N.[8] For example, if

$$E + RCOOR' \underset{K_S}{\rightleftharpoons} E \cdot RCOOR' \begin{array}{c} \xrightarrow{k_2[H_2O]} RCO_2H \\ \xrightarrow{k_3[N]} RCO\text{-}N \end{array} \tag{7.8}$$

then, for the disappearance of ester,

$$\frac{k_{cat}}{K_M} = \frac{k_2[H_2O] + k_3[N]}{K_S} \tag{7.9}$$

(Complications will arise, of course, if the nucleophiles bind to the enzyme and both compete for a single site.)

The speeding up of the deacylation rate by adding nucleophiles has been used to give the rate constant k_2 for deacylation with substrates for which k_3 is normally rate-determining. 1,4-Butanediol is a sufficiently good nucleophile that moderate concentrations cause the deacylation rate to become faster than k_2.[27] Values of some rate constants obtained by this method are listed in Table 7.3.

Partition experiments provide a very powerful approach for the detection of intermediates.

Table 7.3 *Kinetic constants for the hydrolysis of N-acyl-L-amino acid esters by α-chymotrypsin at 25°C, pH 7.8, and ionic strength 0.1 M, determined by partitioning experiments*[a]

Acyl	Amino acid	Ester	k_{cat} (s^{-1})	K_M (mM)	k_2 (s^{-1})	k_3 (s^{-1})	k_S (mM)
Acetyl	Gly	OCH_3	0.109	862	0.49	0.14	3380
Acetyl	Gly	OC_2H_5	0.051	445	0.094	0.11	823
Benzoyl	Gly	OCH_3	0.31	4.24	0.42	1.17	5.78
Acetyl	But	OCH_3	1.41	66.7	8.81	1.68	417
Benzoyl	But	OCH_3	0.32	1.41	0.41	1.52	1.79
Benzoyl	Ala	OC_2H_5	0.069	5.97	0.069	0.6	5.97
Acetyl	Norval	OCH_3	5.08	14.3	35.6	5.93	100
Benzoyl	Norval	OCH_3	2.45	0.85	4.16	5.93	1.45
Acetyl	Val	OCH_3	0.173	87.7	0.98	0.21	500
Acetyl	Val	OC_2H_5	0.152	110	0.55	0.21	398
Acetyl	Val	i-OC_3H_7	0.096	177	0.178	0.21	327
Chloroacetyl	Val	OCH_3	0.127	43	0.32	0.21	108.8
Benzoyl	Val	OCH_3	0.064	4.17	0.09	0.22	5.84
Acetyl	Norleu	OCH_3	16.1	5.37	103	19.1	34.4
Acetyl	Phe	OCH_3	97.1	0.93	796	111	7.63
Acetyl	Phe	OC_2H_5	68.6	1.85	265	92.7	7.14
Acetylala(L)	Phe	OCH_3	57.3	0.296	176	85	0.909
Benzoyl	Phe	OCH_3	30.7	0.0349	45.8	91.6	0.0524
Acetyl	Tyr	OC_2H_5	192	0.663	5000	200	17.2
Benzoyl	Tyr	OCH_3	90.9	0.018	364	121	0.072
Benzoyl	Tyr	OC_2H_5	85.9	0.022	249	131	0.0638
Acetylleu (L)	Tyr	OCH_3	65.7	0.0192	158	113	0.0461
Furoyl	Tyr	OCH_3	50	0.417	66.7	200	0.56

[a] From I. V. Berezin, N. F. Kazanskaya, and A. A. Klyosov, *FEBS Lett.* **15**, 121 (1971).

Table 7.4 *Product ratios in the hydrolysis of substrates*
by δ-chymotrypsin[a]

Substrate	Acceptor:	Transacylation/hydrolysis (M^{-1})		
		Ala-NH$_2$	Gly-NH$_2$	H$_2$NNH$_2$
Ac-Phe-OCH$_3$		43	13	2.2
Ac-Phe—NH—⟨benzene ring⟩—$\overset{+}{N}(CH_3)_3$		45	11	1.8
Ac-Phe-Ala-NH$_2$		43	9	—

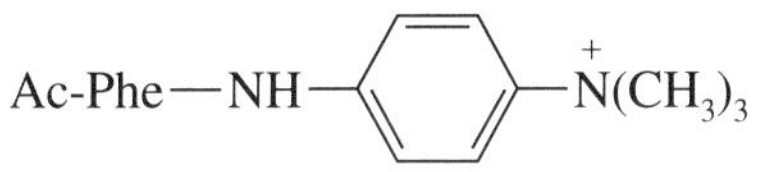

[a] At 25°C, pH 9.3. [From J. Fastrez and A. R. Fersht, *Biochemistry* **12**, 2025 (1973).]

4. Detection of the acylenzyme in the hydrolysis of amides and peptides[8]

The acylenzyme does not accumulate in the hydrolysis of amides, so detection is indirect and difficult. Fortunately, the direct detection of the acylenzyme in ester substrates can be used to provide a rigorous proof of the acylenzyme with amides.

The acylenzyme mechanism was proved for derivatives of acetyl-L-phenylalanine (Ac-Phe) as follows:

1. The hydrolysis of amides in the presence of acceptor nucleophiles gives the same product ratios as those found for the hydrolysis of the methyl ester (Ac-Phe-OCH$_3$) under the same conditions (Table 7.4). Furthermore, these product ratios are the same as those expected from direct rate measurements of the attack of the nucleophiles on Ac-Phe-chymotrypsin, generated *in situ* in the stopped-flow spectrophotometer (Table 7.5).

2. Under conditions where over 94% of the amide that is reacting in the presence of the acceptor nucleophiles forms Ac-Phe-nucleophile, there is no significant increase in the rate of disappearance of the amide. This is consistent with attack by the nucleophile after the rate-determining step—that is, after the formation of an intermediate, with at least 94% of the reaction going through this intermediate.

3. The final *proof* of the acylenzyme route comes from calculating the rate constant for the hydrolytic reaction from the rate constant for the reverse reaction (the synthesis of the substrate by the acylenzyme route) and the *Haldane equation* (Chapter 3, section H). It is found that amines will react with the acylenzyme to produce amides and peptides.

Table 7.5 *Rate constants for the attack of nucleophiles on Ac-Phe-δ-chymotrypsin[a]*

| | $k \, (s^{-1} M^{-1})$ | |
Nucleophile	Direct kinetic measurement	Calculated from product ratios[b]
Ala-NH$_2$	4800	6200
Gly-NH$_2$	1500	1600
H$_2$NNH$_2$	330	280
(H$_2$O	142 s^{-1})	

[a] At 25°C, pH 9.3. [From J. Fastrez and A. R. Fersht, *Biochemistry* **12**, 2025 (1973).]
[b] From Table 7.4.

Hence, by the principle of microscopic reversibility, the reverse reaction (the hydrolysis of peptides by the acylenzyme mechanism) must also occur. The question is whether or not this reaction is rapid enough to account for the observed hydrolysis rate. This can be answered by measuring $(k_{cat}/K_M)_S$ for the synthesis of a peptide by the acylenzyme route, and K_{eq} for the hydrolysis of the peptide; $(k_{cat}/K_M)_H$ for the hydrolytic reaction can then be calculated from the Haldane equation,

$$K_{eq} = \frac{(k_{cat}/K_M)_H}{(k_{cat}/K_M)_S}$$

The calculated value is close to the experimental value.

5. The validity of partitioning experiments and some possible experimental errors

Neither the occurrence of a constant value of V_{max} or a constant product ratio is sufficient proof of the presence of an intermediate. It was seen for alkaline phosphatase that a constant value for V_{max} is an artifact, and also that there is no *a priori* reason why the attack of acceptors on a Michaelis complex should not also give constant product ratios. In order for partitioning experiments to provide a satisfactory proof of the presence of an intermediate, they must be linked with rate measurements. When the rate measurements are restricted to steady state kinetics, the most favorable situation is when the intermediate accumulates. If the kinetics of equations 7.5 to 7.7 hold, it may be concluded beyond a reasonable doubt that an intermediate occurs. The ideal situation is a combination of partitioning experiments with pre–steady state studies, as described for chymotrypsin and amides.

Errors can arise when the enzymes "misbehave." Chymotrypsin is often treated as a solution of imidazole and serine. But proteins are quite sensitive to their environment; they often bind organic molecules and ions nonspecifically to alter their kinetic properties slightly. The first experimental rule is that reactions should be carried out, to the extent possible, at the same concentration of enzyme. Many proteins tend to aggregate somewhat, especially with increasing concentration, and this can cause changes in rate constants. The second rule is that product ratios should be determined by direct analysis of the products rather than by indirect measurements. For example, the rate of attack of Gly-NH$_2$ on the acylenzyme Bz-Tyr-chymotrypsin was once measured from the decrease in k_{cat} for the hydrolysis of Bz-Tyr-Gly-NH$_2$ on addition of Gly-NH$_2$. (The Gly-NH$_2$ inhibits the reaction, since it reacts with the acylenzyme to regenerate the Bz-Tyr-Gly-NH$_2$.) Unfortunately, it was subsequently found that amines bind to chymotrypsin, causing increases of up to 30% in the k_{cat}.[8] This increase is on the same order as the expected decreases due to the reversal of the reaction. In general, small changes of rate are not reliable in enzymatic reactions under circumstances that would give reliable results in chemical kinetics.

There are circumstances in which the simple rules for the partition of intermediates break down. If the acceptor nucleophile reacts with the acylenzyme before the leaving group has diffused away from the enzyme-bound intermediate, the partition ratio could depend on the nature of the leaving group (e.g., due to steric hindrance of attack, etc.). Also, the measurement of rate constants for the attack of the nucleophiles on the intermediate could be in slight error due to the nonspecific binding effects mentioned above.

C. Further examples of detection of intermediates by partition and kinetic experiments

1. Alkaline phosphatase

There is little doubt that a phosphorylenzyme is formed during the hydrolysis of phosphate esters by alkaline phosphatase.[28,29] The phosphorylenzyme is stable at low pH, and it may be isolated.[30,31] The phosphorylenzyme has been detected by gel electrophoresis.[28] Partition experiments using tris buffer as a phosphate acceptor give a constant product ratio (Table 7.6).[32,33] Earlier kinetic experiments have to be evaluated in light of the recent findings that the enzyme as isolated may contain tightly, although noncovalently, bound phosphate,[34] and that the rate-determining step in the reaction at high pH is the dissociation of the tightly bound phosphate[23,24] (although there is no unanimity on the kinetic points or on the stoichiometry of binding).[35,36] As mentioned earlier, it seems that the constant value of V_{max} for the hydrolysis of a wide series of phosphate esters (Table 7.7)[20,21] is the result of a slow, rate-determining dissociation of the enzyme-product complex, $E \cdot P_i$.

Table 7.6   Relative values of V_{max} for the hydrolysis of phosphate esters by alkaline phosphatase[a]

Phosphate	V_{max}	Phosphate	V_{max}
5'-AMP	1	dCTP	1.05
Pyrophosphate	1	Ribose 5-phosphate	0.7
3'-AMP	0.9	β-Glycerol phosphate	0.9
ApAp	0.6	Ethanolamine phosphate	0.7
ATP	1.05	Glucose 1-phosphate	0.8
dATP	1.05	Glucose 6-phosphate	0.9
dGTP	1.05	Histidinol phosphate	0.8
UDP	1.0	p-Nitrophenyl phosphate	1.0
5'-UMP	0.85		

[a] From L. A. Heppel, D. R. Harkness, and R. J. Hilmoe, *J. Biol. Chem.* **237**, 841 (1962).

Table 7.7   Product ratios in the hydrolysis of phosphate esters and phosphoramidates by alkaline phosphatase[a]

Phosphate	Transphosphorylation/ hydrolysis
Acceptor = 1-M tris, pH 8	
Phenyl	1.42
Cresyl	1.41
Chlorophenyl	1.38
p-tert-Butylphenyl	1.37
p-Nitrophenyl	1.37
o-Methoxy-p-methylphenyl	1.40
α-Naphthyl	1.40
β-Naphthyl	1.40
Acceptor = 2-M tris, pH 8.2	
p-Nitrophenyl	1.2
Phosphoramidates	1.1[b]

[a] From H. Barrett, R. Butler, and I. B. Wilson, *Biochemistry* **8**, 1042 (1969); S. Snyder and I. B. Wilson, *Biochemistry* **11**, 3220 (1972).
[b] Average for several phosphoramidates.

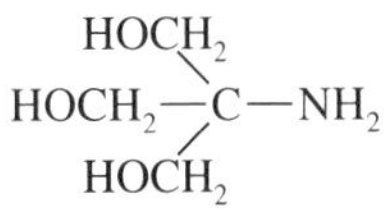

$$HOCH_2$$
$$HOCH_2-C-NH_2$$
$$HOCH_2$$

Tris
[*tris*-(hydroxymethyl)-aminomethane]

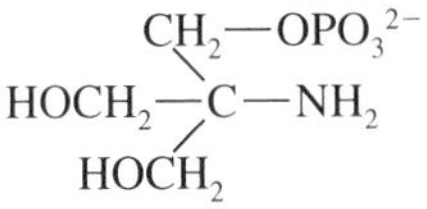

$$CH_2-OPO_3^{2-}$$
$$HOCH_2-C-NH_2$$
$$HOCH_2$$

O-Phosphoryl-tris

2. Acid phosphatase

Acid phosphatase gives a straightforward example of the accumulation of a phosphorylenzyme intermediate with rate-determining breakdown:

$$E-OH + RO\bar{P}O_3H \rightleftharpoons E-OH \cdot RO\bar{P}O_3H \xrightarrow{\text{fast}} E-O\bar{P}O_3H \xrightarrow{\text{slow}}$$
$$\searrow ROH$$

$$E-OH + H_2PO_4^- \quad (7.10)$$

A wide variety of esters is hydrolyzed with the same V_{max} (Table 7.8),[37] constant product ratios are found (Table 7.9); stopped-flow studies using *p*-nitrophenyl phosphate find a burst of 1 mol of *p*-nitrophenolate ion released per enzyme subunit; and the enzyme is covalently labeled by diisopropyl fluorophosphate.[38]

3. β-Galactosidase

Chapter 8 (section C3) presents stereochemical evidence that the hydrolysis of β-D-galactosides catalyzed by β-galactosidase (equation 7.11) involves two successive displacements on the C-1 carbon—i.e., involves an intermediate. Further evidence for an intermediate from partitioning experiments is presented in Table 7.10. There is constant partitioning between water and methanol.[39–41]

Table 7.8 *Hydrolysis of phosphate esters by prostatic acid phosphatase at pH 5 and 37°C[a]*

Phosphate ester	K_M (mM)	V_{max} (relative)
β-Glycerol phosphate	1.1	1
2'-AMP	0.28	1
Acetyl phosphate	0.17	1
3'-AMP	0.068	1
p-Nitrophenyl phosphate	0.034	1

[a] From G. S. Kilsheimer and B. Axelrod, *J. Biol. Chem.* **227**, 879 (1957).

Table 7.9 *Product ratios in the hydrolysis of phosphate esters by prostatic and phosphatase[a]*

Phosphate ester	Acceptor:	Transphosphorylation/hydrolysis	
		Ethanol	Ethanolamine
p-Nitrophenyl phosphate		0.29	0.044
Phenyl phosphate		0.26	0.044
3'-UMP		0.28	—
3'-AMP		0.30	0.046
β-Glycerol phosphate		0.28	0.041

[a] From W. Ostrowski and E. A. Barnard, *Biochemistry* **12**, 3893 (1973).

Examination of V_{max} suggests that formation of the intermediate is rate-determining for the weakly activated substrates since V_{max} is variable, but that hydrolysis of the intermediate is rate-determining for the highly activated dinitro compounds because the rate levels off. Consistent with this is the observation that the rate of disappearance of the 2,4-dinitro and 3,5-dinitro substrates is increased

Table 7.10 *Product ratios and relative values of V_{max} for the hydrolysis of β-galactosides by β-galactosidase at 25°C and pH 7.0–7.5[a]*

β-Galactoside	Methanolysis/ hydrolysis (M^{-1})	V_{max} (relative)	Rate-determining step
2,4-Dinitrophenyl	—	1.3	Degalactosylation
3,5-Dinitrophenyl	—	1.1	Degalactosylation
2,5-Dinitrophenyl	—	1.1	Degalactosylation
2-Nitrophenyl	1.97	1.0	—
3-Nitrophenyl	1.96	0.9	—
3-Chlorophenyl	2.08	0.5	Galactosylation
4-Nitrophenyl	1.99	0.2	Galactosylation
Phenyl	1.94	0.1	Galactosylation
4-Methoxyphenyl	2.14	0.1	Galactosylation
4-Chlorophenyl	2.13	0.02	Galactosylation
4-Bromophenyl	2.02	0.02	Galactosylation
Methyl	2.2	0.06	Galactosylation

[a] From T. M. Stokes and I. B. Wilson, *Biochemistry* **11**, 1061 (1972); M. L. Sinnott and O. M. Viratelle, *Biochem. J.* **133**, 81 (1973); M. L. Sinnott and I. J. Souchard, *Biochem J.* **133**, 89 (1973).

by added methanol, but not the rate of disappearance of the less reactive substrates (see Figures 7.2 and 7.3).

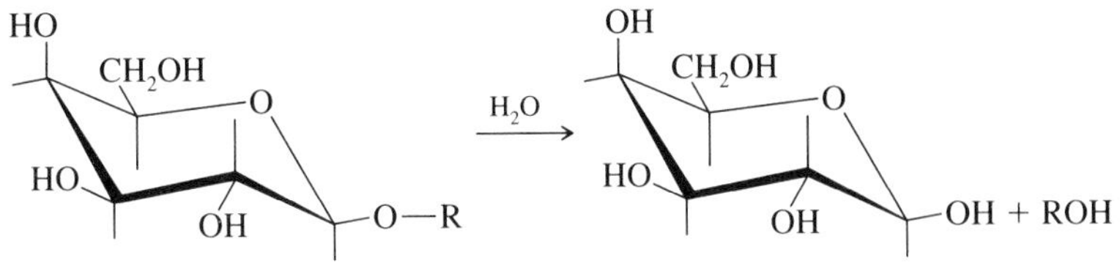

$$(7.11)$$

A change in rate-determining step is also indicated from $^{16}O/^{18}O$ kinetic isotope effects.[42] V_{max} for the hydrolysis of 4-nitrophenyl β-D-galactoside exhibits a $^{16}O/^{18}O$ effect of 1.022 (for the phenolic oxygen) compared with an expected value of 1.042 for complete C—O bond fission in the transition state. The 2,4-dinitrophenyl derivative for which degalactosylation is thought to be rate-determining has a negligible isotope effect on V_{max} (= 1.002). (Additional evidence suggests that the enzyme has an S_N2 reaction with the C-1 carbon to give a covalent intermediate for most substrates. The reaction of the activated dinitrophenyl derivative, however, possibly has a contribution from an S_N1 pathway.[42])

D. Aminoacyl–tRNA synthetases: Detection of intermediates by quenched flow, steady state kinetics, and isotope exchange

1. The reaction mechanism

The aminoacyl-tRNA synthetases catalyze the formation of aminoacyl-tRNA from the free amino acid (AA) and ATP:

$$AA + ATP + tRNA \xrightarrow{E} AA\text{-}tRNA + AMP + PP_i \qquad (7.12)$$

In the absence of tRNA, the enzymes will, with a few exceptions, activate amino acids to the attack of nucleophiles, and ATP to the attack of pyrophosphate.[43–46] This is done by forming a tightly bound complex with the aminoacyl adenylate, the mixed anhydride of the amino acid, and AMP. (The chemistry of activation is discussed in Chapter 2, section D2c.)

The activation to the attack of pyrophosphate is measured by the pyrophosphate exchange technique. The enzyme, the amino acid, and ATP are incubated with [^{32}P]-labeled pyrophosphate so that β,γ-labeled ATP is formed by the continuous recycling of the E-AA-AMP complex. The complex is formed as in equation 7.13, and the reaction is reversed by the attack of labeled pyrophosphate to generate labeled ATP. This process is repeated until the isotopic label is uniformly distributed among all the reagents.

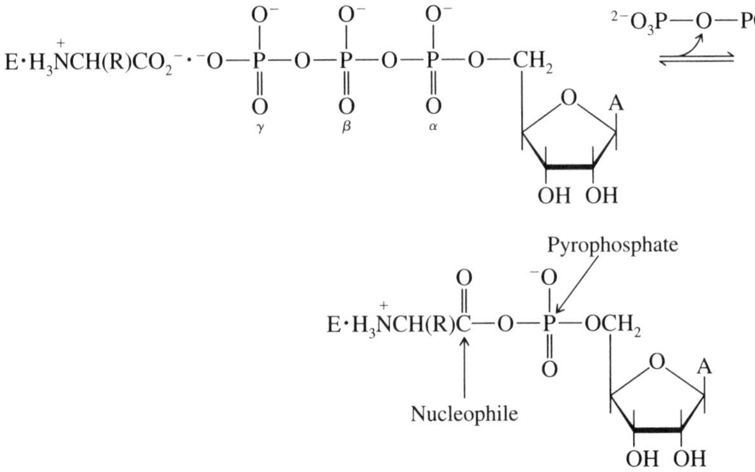

$$E \cdot \overset{+}{H_3}NCH(R)CO_2^{-} \cdot {}^{-}O - \overset{\overset{O^-}{\|}}{\underset{\underset{\gamma}{O}}{P}} - O - \overset{\overset{O^-}{\|}}{\underset{\underset{\beta}{O}}{P}} - O - \overset{\overset{O^-}{\|}}{\underset{\underset{\alpha}{O}}{P}} - O - CH_2 \quad\quad {}^{2-}O_3P - O - PO_3^{2-} \;\rightleftharpoons$$

Pyrophosphate

$$E \cdot \overset{+}{H_3}NCH(R)\overset{\overset{O}{\|}}{C} - O - \overset{\overset{-O}{\diagup}}{\underset{\underset{O}{\|}}{P}} - OCH_2 \qquad (7.13)$$

Nucleophile

The acylation of tRNA proceeds by the attack of a ribose hydroxyl of the terminal adenosine of the tRNA on the carbonyl group, as indicated in equations 7.13 and 7.14.

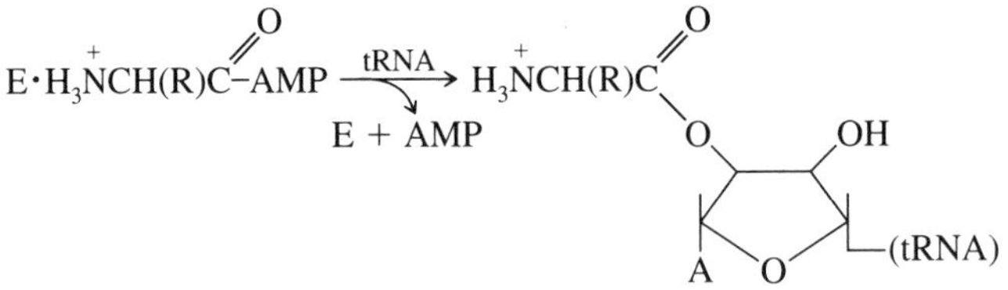

$$E \cdot \overset{+}{H_3}NCH(R)\overset{\overset{O}{\diagdown\!\!\!\diagup}}{C} - AMP \xrightarrow{tRNA} \overset{+}{H_3}NCH(R)\overset{\overset{O}{\diagdown\!\!\!\diagup}}{C} \qquad (7.14)$$

$$E + AMP$$

There is no doubt that the enzyme-bound aminoacyl adenylate is formed in the absence of tRNA. It may be isolated by chromatography and the free aminoacyl adenylate obtained by precipitation of the enzyme with acid.[47,48] Furthermore, the isolated complex will transfer its amino acid to tRNA. The crystal structure of the tyrosyl-tRNA synthetase bound to tyrosyl adenylate has been solved (Chapter 15, section B).

The following mechanism is derived logically from these observations:

$$E + ATP + AA \underset{PP_i}{\searrow} E \cdot AA\text{-}AMP \xrightarrow{tRNA} AA\text{-}tRNA + AMP + E \qquad (7.15)$$

Despite this, it seemed at one stage that not all the evidence was consistent with the aminoacyl adenylate pathway. An alternative mechanism appeared possible: in the presence of tRNA, perhaps an aminoacyl adenylate was *not* formed, and the reaction occurred instead by the simultaneous reaction of the tRNA, the amino acid, and ATP.[49] As is shown below, this mechanism is not correct, but at the time it was suggested it was a valid possibility. Furthermore, it made an important point; the finding of a partial reaction in the absence of one of the substrates (for example, the formation of the aminoacyl adenylate from the amino acid and ATP in the absence of tRNA) does not mean that the same reaction occurs in the presence of *all* the substrates. Indeed, there are examples in which such partial reactions have been found to be artifacts.

The aminoacyl adenylate pathway is proved very simply from three quenched-flow experiments by using the three criteria for proof: the intermediate is isolated; it is formed fast enough; and it reacts fast enough to be on the reaction pathway.[50] The following is found for the isoleucyl-tRNA synthetase (IRS):

1. When the preformed and isolated IRS·[^{14}C]Ile-AMP is mixed with tRNA in the pulsed quenched-flow apparatus (Figure 7.4), the first-order rate constant for the transfer of the [^{14}C]Ile to the tRNA is measured to be the same as the k_{cat} for the steady state aminoacylation of the tRNA under the same reaction conditions. The rate constant for the reaction of the intermediate is thus fast enough to be on the reaction pathway; furthermore, reaction of the intermediate appears to be the rate-determining step.

2. When IRS, isoleucine, tRNA, and [γ-^{32}P]ATP (labeled in the terminal phosphate) are mixed in the pulsed quenched-flow apparatus (Figure 7.5), there is a burst of release of labeled pyrophosphate before the steady state rate of aminoacylation of tRNA is reached. This means either that the aminoacyl adenylate is formed before the aminoacylation of tRNA, thus proving the

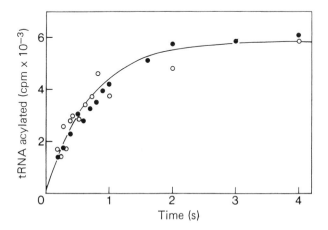

Figure 7.4   Transfer of [^{14}C]Ile from IRS·[14C]Ile-AMP to tRNA^{Ile} when the complex is mixed with excess tRNA in the pulsed quenched-flow apparatus.

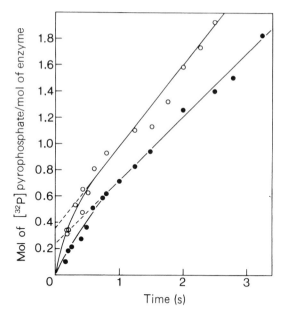

Figure 7.5 Release of [^{32}P]pyrophosphate when [^{32}P]ATP, isoleucine, tRNA, and enzyme are mixed in the pulsed quenched-flow apparatus. The extrapolated burst of product formation is below 1 mol per mol of enzyme because the concentrations of ATP are not saturating (Chapter 4, section D). Open circles (○) are for [ATP] = 2 × K_M; filled circles (●) are for [ATP] = 1 × K_M.

mechanism, *or* that there could be a pathway that involves the formation of aminoacyl-tRNA in a rapid process followed by a subsequent slow step, such as the dissociation of the IRS-Ile-tRNA complex:

$$E + AA + ATP + tRNA \xrightarrow{\text{fast}} E \cdot AA\text{-tRNA} \xrightarrow{\text{slow}} E + AA\text{-tRNA} \qquad (7.16)$$
$$AMP + PP_i$$

This reaction, which was thought to occur by many workers, is disproved by a third quenching experiment. Equation 7.16 predicts that there should be a burst of charging of tRNA, since 1 mol of enzyme-bound aminoacyl-tRNA is formed rapidly whereas the subsequent turnover is slow.

3. When IRS, [^{14}C]Ile, tRNA, and ATP are mixed in the quenched-flow apparatus (Figure 7.6), the initial rate of charging of tRNA extrapolates back through the origin without any indication of a burst of charging. The burst of pyrophosphate release is due to the formation of the aminoacyl adenylate before the transfer of the amino acid to tRNA.

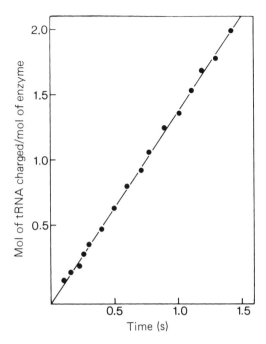

Figure 7.6 Initial rate of charging of tRNA^{Ile} when saturating concentrations of [^{14}C]Ile, ATP, and tRNA are mixed with the enzyme.

2.   The editing mechanism

Chapter 13 points out that during protein biosynthesis, the cell distinguishes between certain amino acids with an accuracy far greater than would be expected from their differences in structure. This sensitivity could be caused either by the specifically catalyzed hydrolysis of the aminoacyl adenylate complex of the "wrong" amino acid by the aminoacyl-tRNA synthetase, or by hydrolysis of the mischarged tRNA.[48] One example is the rejection of threonine by the valyl-tRNA synthetase (VRS).[51] This enzyme catalyzes the pyrophosphate exchange reaction in the presence of threonine, and also forms a stable VRS·Thr-AMP complex. In the presence of tRNA and threonine, the VRS acts as an ATP pyrophosphatase, hydrolyzing ATP to AMP and pyrophosphate, and does not catalyze the net formation of Thr-tRNA^{Val}. During this reaction there is the intermediate formation of VRS·Thr-AMP shown by the occurrence of the pyrophosphate exchange reaction.

Rapid quenching experiments show that the editing mechanism for the rejection of threonine involves the mischarging of tRNA followed by its rapid hydrolysis. The transiently mischarged tRNA may be trapped, isolated, and found to be hydrolyzed at the necessary rate. When the VRS·[14C]Thr-AMP complex is mixed with tRNA^{Val} in the quenched-flow apparatus, [^{14}C]Thr-tRNA^{Val} is

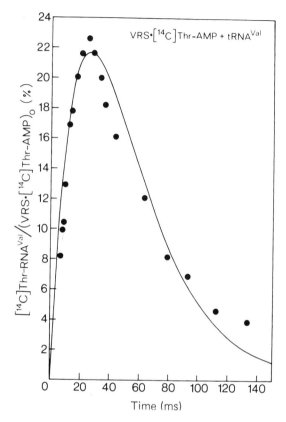

Figure 7.7 The transient formation of $[^{14}C]$Thr-tRNA^{Val} when VRS$\cdot$ $[^{14}C]$Thr-AMP is mixed with excess tRNA in the quenched-flow apparatus.

transiently formed (Figure 7.7). This may be isolated by rapidly quenching the reaction with phenol and precipitating the mischarged tRNA from the aqueous layer. The mischarged tRNA is hydrolyzed by the VRS with a rate constant of $40\ s^{-1}$ (Figure 7.8). The rate of transfer of the threonine from the VRS$\cdot$ Thr-AMP complex may be measured independently from the rate of liberation of AMP (using the VRS$\cdot$ Thr-$[^{32}P]$AMP compound: Figure 7.9). The solid curve in Figure 7.8 is calculated from the independently measured formation and hydrolysis rate constants given in equation 7.17. Kinetic data thus obtained are consistent with the levels of the mischarged tRNA in the steady state.[52]

$$\text{VRS}\cdot\text{Thr-AMP}\cdot\text{tRNA}^{Val} \xrightarrow{\ 36\ s^{-1}\ } \text{VRS}\cdot\text{Thr-tRNA}^{Val} + \text{AMP} \tag{7.17}$$
$$\downarrow 40\ s^{-1}$$
$$\text{VRS} + \text{Thr} + \text{tRNA}^{Val}$$

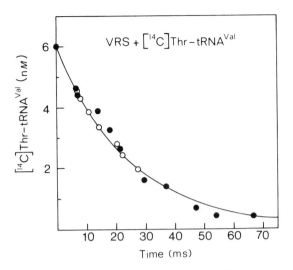

Figure 7.8 The VRS-catalyzed hydrolysis of mischarged [^{14}C]Thr-tRNA^{Val}.

One experimental point worth noting in regard to Figure 7.8 is that the rate constant for the deacylation of the mischarged tRNA can be measured by the pre–steady state kinetics even in the presence of a large fraction of uncharged tRNA. This is not easily done by steady state kinetics, because of the competi-

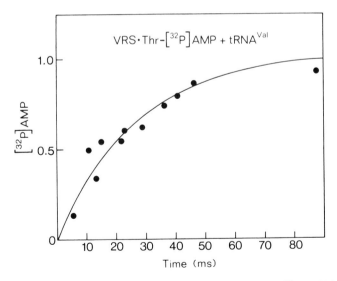

Figure 7.9 The rate of breakdown of Thr-AMP when VRS·Thr-[^{32}P]AMP is mixed with excess tRNA^{Val}.

tive inhibition by the uncharged material. But in the rapid quenching experiment, an excess of enzyme over total tRNA—i.e., a single turnover—is used.

E. Detection of conformational changes

The examples of intermediates discussed so far have been relatively clear cut, since real chemical changes that generate either covalent complexes with the enzyme or chemical products have been involved. A more difficult problem to work with is the detection and analysis of conformational changes of the enzyme on the binding of substrates and also on their interconversion to products. The approach to this problem was outlined in Chapter 4, section D7. A systematic analysis of the relaxation times on the binding of substrates or analogues is usually essential. In addition, since the property of the protein that is being monitored in the rapid reaction study is usually poorly defined (e.g., a change of tryptophan fluorescence), as much independent corroborative evidence as possible must be gathered and more than one probe used if available (Chapter 4, section D7). Because the nature of the problem is less concrete than for the detection of chemical intermediates, the examples are more different to analyze.

Frequently, the existence of a conformational change is inferred because rate constants measured for events in a reaction are inadequate to describe the overall scheme, or because they appear "suspicious." An example of the latter is illustrated in Chapter 4, Figure 4.9: the second-order rate constant measured for the binding of an inhibitor to lysozyme is too low for a diffusion-controlled reaction. Application of a full relaxation-time analysis to the binding revealed a second step that is presumably related to the small conformational change described in Chapter 1, section D3. Another example is illustrated in Figure 4.7, the binding of tyrosine to the tyrosyl-tRNA synthetase. The second-order rate constant is 2 to 3 orders of magnitude below that of diffusion control, so it is most likely that at least one conformational change is involved although not directly detected. Another step in the reaction of the enzyme, the formation of the E·Tyr-AMP complex from E·Tyr·ATP, probably involves a further conformational change: stopped-flow fluorescence has detected a large change in tryptophan fluorescence occurring with the same rate constant as that for the chemical step that was monitored independently by rapid quenching. The temporal coupling between the conformational change and the chemical step has not been shown by these studies. The kinetics just indicate a parallel change, but there could be distinct steps. For example, there could be a slow rate-determining conformational change followed by a rapid chemical step, or *vice versa*.

F. The Future

D. B. Northrop and F. B. Simpson have suggested that the major new development in the study of mechanism will be stopped-flow mass spectrometry.[53] Soft ionization techniques are routinely used for determining primary structures of peptides and proteins, and they are now rapidly being applied to noncovalent structures and protein-protein interactions. Northrop and Simpson propose interfacing mass spectrometry with a stopped-flow mixing device to study enzyme reactions during catalysis directly. Virtually all the reagents and products could be monitored simultaneously and directly. The transient-state measurements will provide the order of reaction events, and steady state measurements will provide the distribution of the different enzyme species. The combination of the transient and steady state measurements should allow the construction of the free energy diagram of the enzyme-catalyzed reaction, as has been laboriously done in the past by combining measurements from many different techniques, using many different probes.

Electrospray ionization mass spectrometry has recently been used to detect and measure the pre–steady state kinetics of formation of a glycosylenzyme in the reactions of a glycosidase, xylanase, which has a mechanism analogous to that of β-galactosidase (section C3).[54] The wild-type enzyme has glycosylation as the rate determining step so that the glycosylenzyme is present only at a low steady state concentration. But, a mutant (produced by protein engineering, Chapter 14) has rate determining deglycosylation so that the glycosylenzyme accumulates. Mass spectrometry detected the 5% accumulation of the intermediate for wild-type enzyme and could be used to measure easily the rate constant of $0.04\ s^{-1}$ for its formation with the mutant.[54]

References

1. B. S. Hartley and B. A. Kilby, *Biochem. J.* **56**, 288 (1954).
2. H. Gutfreund, *Discuss. Faraday Soc.* **20**, 167 (1955).
3. H. Gutfreund and J. M. Sturtevant, *Biochem. J.* **63**, 656 (1956).
4. A. Himoe, K. G. Brandt, R. J. DeSa, and G. P. Hess, *J. Biol. Chem.* **244**, 3483 (1969).
5. T. E. Barman and H. Gutfreund, *Biochem. J.* **101**, 411 (1966).
6. S. A. Bernhard and H. Gutfreund, *Proc. Natl. Acad. Sci. USA* **53**, 1238 (1965).
7. J. McConn, E. Ku, A. Himoe, K. G. Brandt, and G. P. Hess, *J. Biol. Chem.* **246**, 2918 (1971).
8. J. Fastrez and A. R. Fersht, *Biochemistry* **12**, 2025 (1973).
9. M. L. Bender, G. R. Schonbaum, and B. Zerner, *J. Am. Chem. Soc.* **84**, 2540 (1962).
10. M. Caplow and W. P. Jencks, *Biochemistry* **1**, 883 (1962).
11. S. A. Bernhard, S. J. Lau, and H. Noller, *Biochemistry* **4**, 1108 (1965).
12. J. De Jersey, D. T. Keough, J. K. Stoops, and B. Zerner, *Eur. J. Biochem.* **42**, 237 (1974).
13. C. G. Miller and M. L. Bender, *J. Am. Chem. Soc.* **90**, 6850 (1968).
14. A. R. Fersht, D. M. Blow, and J. Fastrez, *Biochemistry* **12**, 2035 (1973).
15. E. F. Jansen, M. D. Nutting, and A. K. Balls, *J. Biol. Chem.* **179**, 201 (1949).
16. N. K. Schaffer, S. C. May, and W. H. Summeson, *J. Biol. Chem.* **202**, 67 (1953).
17. R. Henderson, *J. Molec. Biol.* **54**, 341 (1970).
18. T. Alber, G. A. Petsko, and D. Tsernoglou, *Nature, Lond.* **263**, 297 (1976).

19. H. Gutfreund and B. R. Hammond, *Biochem. J.* **73**, 526 (1959).
20. A. Garen and C. Levinthal, *Biochim. Biophys. Acta* **38**, 470 (1960).
21. L. A. Heppel, D. R. Harkness, and R. J. Hilmoe, *J. Biol. Chem.* **237**, 841 (1962).
22. W. N. Aldridge, T. E. Barman, and H. Gutfreund, *Biochem. J.* **92**, 23C (1964).
23. W. E. Hull and B. D. Sykes, *Biochemistry* **15**, 1535 (1976).
24. W. E. Hull, S. E. Halford, H. Gutfreund, and B. D. Sykes, *Biochemistry* **15**, 1547 (1976).
25. R. M. Epand and I. B. Wilson, *J. Biol. Chem.* **238**, 1718 (1963); **240**, 1104 (1965).
26. M. L. Bender, G. E. Clement, C. R. Gunter, and F. J. Kézdy, *J. Am. Chem. Soc.* **86**, 3697 (1964).
27. I. V. Berezin, N. F. Kazanskaya, and A. A. Klyosov, *FEBS Lett.* **15**, 121 (1971).
28. M. Cocivera, J. McManaman, and I. B. Wilson, *Biochemistry* **19**, 2901 (1980).
29. M. Caswell and M. Caplow, *Biochemistry* **19**, 2907 (1980).
30. L. Engström, *Biochim. Biophys. Acta* **54**, 179 (1961); **56**, 606 (1962).
31. J. H. Schwartz and F. Lipmann, *Proc. Natl. Acad. Sci. USA* **47**, 1996 (1961).
32. H. Barrett, R. Butler, and I. B. Wilson, *Biochemistry* **8**, 1042 (1969).
33. S. L. Snyder and I. B. Wilson, *Biochemistry* **11**, 3220 (1972).
34. W. Bloch and M. J. Schlesinger, *J. Biol. Chem.* **248**, 5794 (1973).
35. J. F. Chlebowski, I. M. Armitage, P. P. Tusa, and J. E. Coleman, *J. Biol. Chem.* **251**, 1207 (1976).
36. D. Chappelet-Tordo, M. Iwatsubo, and M. Lazdundski, *Biochemistry* **13**, 3754 (1974).
37. G. S. Kilsheimer and B. Axelrod, *J. Biol. Chem.* **227**, 879 (1957).
38. W. Ostrowski and E. A. Barnard, *Biochemistry* **12**, 3893 (1973).
39. T. M. Stokes and I. B. Wilson, *Biochemistry* **11**, 1061 (1972).
40. M. L. Sinnott and O. M. Viratelle, *Biochem. J.* **133**, 81 (1973).
41. M. L. Sinnott and I. J. L. Souchard, *Biochem. J.* **133**, 89 (1973).
42. S. Rosenberg and J. F. Kirsch, *Biochemistry* **20**, 3189 (1981).
43. M. B. Hoagland, *Biochim. Biophys. Acta* **16**, 288 (1955).
44. P. Berg, *J. Biol. Chem.* **222**, 1025 (1956).
45. M. B. Hoagland, E. B. Keller, and P. C. Zamecnik, *J. Biol. Chem.* **218**, 345 (1956).
46. P. R. Schimmel and D. Söll, *Ann. Rev. Biochem.* **48**, 601 (1979).
47. A. Norris and P. Berg, *Proc. Natl. Acad. Sci USA* **52**, 330 (1964).
48. A. N. Baldwin and P. Berg, *J. Biol. Chem.* **241**, 839 (1966).
49. R. B. Loftfield, *Progr. Nucl. Acid Res. (& Mol. Biol.)* **12**, 87 (1972).
50. A. R. Fersht and M. M. Kaethner, *Biochemistry* **15**, 818 (1976).
51. A. R. Fersht and M. M. Kaethner, *Biochemistry* **15**, 3342 (1976).
52. A. R. Fersht and C. Dingwall, *Biochemistry* **18**, 1238 (1979).
53. D. B. Northrop and F. B. Simpson, *Bioorganic & Medicinal Chemistry* **5**, 641 (1997).
54. D. L. Zechel, L. Konerman, S. G. Withers, and D. J. Douglas, *Biochemistry* **37**, 7664 (1998).

Stereochemistry of Enzymatic Reactions

8

Stereospecificity is a hallmark of enzyme catalysis, so a knowledge of the basic principles of stereochemistry is essential for appreciating enzyme mechanisms. Stereochemical evidence can provide important information about the topology of enzyme–substrate complexes. In particular, the positions of catalytic groups on the enzyme relative to the substrate may often be indicated, as may be the conformation or configuration of a substrate or intermediate during the reaction. Further, comparison of the stereochemistry of the substrates and products may reveal the likelihood of intermediates during the reaction.

A. Optical activity and chirality

A compound is optically active, rotating the plane of polarization of plane-polarized light, if it is not superimposable on its mirror image, i.e., on its enantiomer. A simple diagnostic test for superimposability is to determine whether there is a plane or center of symmetry. The presence of such symmetry indicates a lack of activity; the absence indicates activity. As with many organic molecules of biological interest, the optical activity in the amino acids is caused by an asymmetric carbon atom with four different groups around it. This type of carbon is now called a *chiral* center (from Greek for hand). A carbon atom of the form $CR_2R'R''$ is called *prochiral*. Although it is not optically active because it is bound to two identical groups (and thus has a plane of symmetry, making it superimposable with its mirror image), it is *potentially* chiral since it can be made chiral by any operation that produces a difference between the two R groups. The idea that enzymes may be specific for only one enantiomer of a pair of optically active substrates is as old as the study of stereochemistry itself. L. Pasteur, that towering genius of chemistry and biochemistry, reported in 1858 a form of yeast that fermented dextrorotatory tartaric acid but not levorotatory. Early work on the proteases showed that derivatives of L-amino acids and not those of D-amino acids are hydrolyzed.

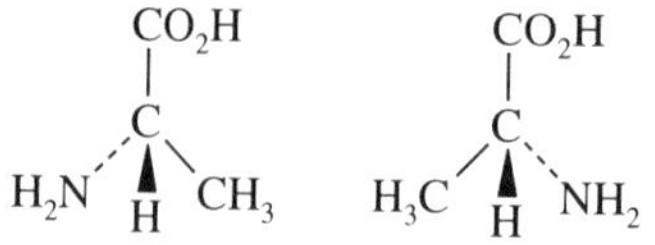

$$\text{(8.1)}$$

L-Alanine   D-Alanine

1. Notation[1,2]

The letters D and L, which are often used to denote configuration, have the drawback that they are not absolute but are relative to a reference compound. A more useful notation, denoting *absolute* configuration of a chiral center, is the RS convention. The groups around the chiral carbon are assigned an order of "priority" based on a series of rules that depend on atomic number and mass. The atom directly attached to the chiral carbon is considered first; the higher its atomic number, the higher its priority. For isotopes, the higher mass number has priority. For groups that have the same type of atom attached to the chiral carbon, the atomic numbers of the next atoms out are considered. This is best illustrated by the following list of the most commonly found groups: $-SH > -OR > -OH > -NHCOR > -NH_2 > -CO_2R > -CO_2H > -CHO > -CH_2OH > -C_6H_5 > -CH_3 > -T > -D > -H$. (Note: $-CHO$ has priority over $-CH_2OH$ because a $C{=}O$ carbon is counted as being bonded to *two* oxygen atoms.)

A chiral carbon is designated as being R or S as follows. The carbon is viewed from the direction opposite to the ligand of lowest priority. If the priority order of the remaining three ligands decreases in a clockwise direction, the absolute configuration of the molecule is said to be R (Latin, *rectus* = right). If it decreases in the counterclockwise direction it is said to be S (Latin, *sinister* = left. For example:

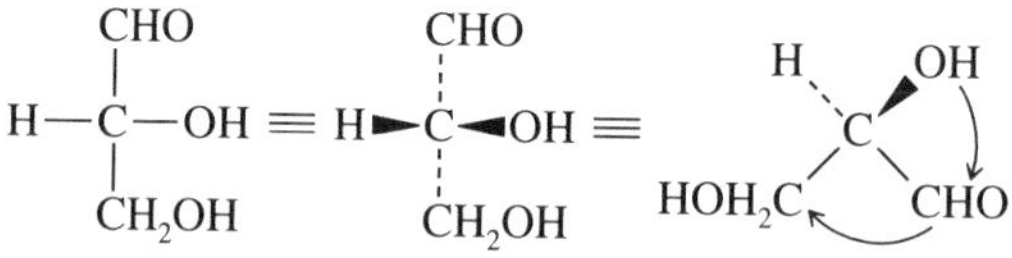

$$\text{(8.2)}$$

D-Glyceraldehyde = R-glyceraldehyde

$$\text{(8.3)}$$

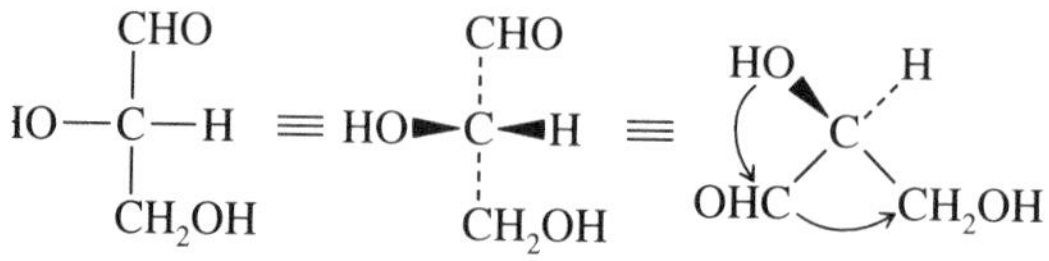

L-Glyceraldehyde = S-glyceraldehyde

(Note: The formulas for glyceraldehyde on the left-hand sides of equations 8.2 and 8.3 are written according to the Fischer projection notation. Bonds in the "east-west" direction come up out of the page and those running "north-south" go into the page, as represented in the middle formulas in the equations.)

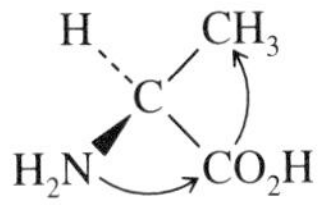

(8.4)

L-Alanine = S-alanine

If compounds contain different isotopes, e.g., hydrogen and deuterium, the priority rules are applied first to the atomic numbers. If this does not give an unambiguous assignment, the higher-mass-number isotope is given priority.

There is also a *prochirality* rule. For example, in ethanol, if we label the two protons by the subscripts a and b, and *arbitrarily* give H_a priority over H_b as in structure 8.5, H_a is said to be pro-R because of the clockwise order of priority. Conversely, H_b is pro-S. Note that if we repeat the treatment but give H_b priority over H_a, H_a is still found to be pro-R and H_b pro-S. Prochirality is thus absolute and does not depend on whether H_a or H_b is given priority.

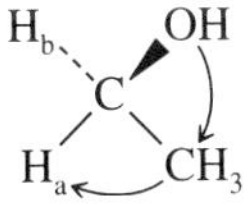

(8.5)

The two faces of a compound containing a trigonal carbon atom are described as *re* (rectus) and *si* (sinister) by a complicated set of rules. Two simple cases are illustrated by the faces presented to the reader by structures 8.6.

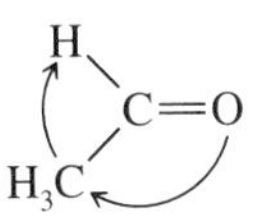

(8.6)

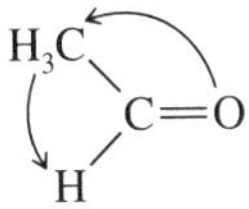

re   *si*

2. Differences between the stereochemistries of enzymatic and nonenzymatic reactions

The crucial difference between nonenzymatic and enzymatic reactions is that the former generally take place in a homogeneous solution, whereas the latter occur on the surface of a protein that is asymmetric. Because of this, an asymmetric enzyme is able to confer asymmetry on the reactions of symmetric substrates. For example, although the two hydrogen atoms in CH_2RR' are equivalent in simple chemical reactions, the equivalence may be lost when the compound binds to the asymmetric active site of an enzyme.[3] The attachment to the enzyme by R and R' in structure 8.7 causes the two hydrogen atoms to be exposed to different environments: H_a may be next to a catalytic base, whereas H_b may be in an inert position.

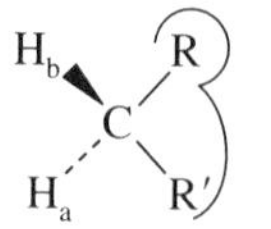

(8.7)

Another example of this is found in the reactions of carbonyl compounds and carbon-carbon double bonds. An optically active compound is formed if a reagent attacks just one side of a planar trigonal carbon. For example, if in equation 8.8 the nucleophile attacks acetaldehyde from the "front" side, the product on the left is formed, whereas attack from the "back" of the page gives the enantiomer on the right. (Note: Attack on a trigonal carbon always occurs perpendicular to the plane of the double bond.) In a simple chemical reaction in solution, there is an equal probability of attack at either face of the trigonal carbon in equation 8.8, so that a racemic mixture of 50% of each enantiomer is formed. But in an enzymatic reaction, attack may occur on one face only, because the substrate is firmly held at an asymmetric active site with only one face exposed to the attacking group.

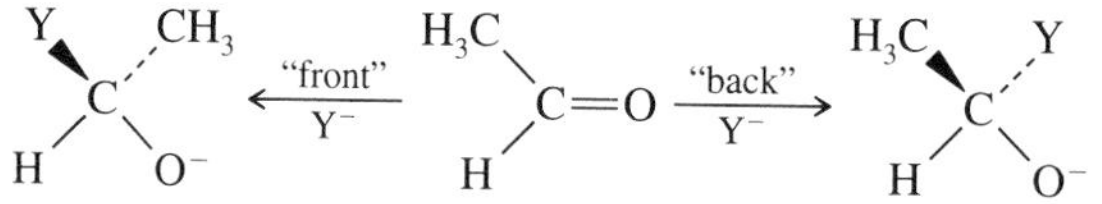

(8.8)

It was pointed out over 60 years ago that the recognition of a chiral (or, as subsequently realized, a prochiral) carbon by an enzyme implies that at least three of the groups surrounding the carbon atom must interact with the enzyme. This is the *multi-point attachment theory*.[4] If only two of the groups interact, the other two may be interchanged without affecting the binding of the substrate (structures 8.9).

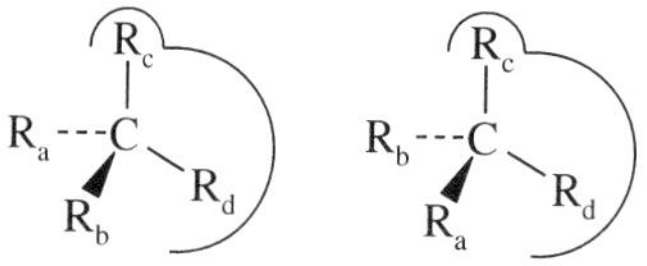

(8.9)

The structural basis of one of the classic examples of such stereospecificity, that of chymotrypsin for L-amino acid derivatives, is immediately obvious on examination of the crystal structure of the enzyme. D-Amino acid derivatives differ from those of L-amino acids by having the H atom and the side chain attached to the chiral carbon interchanged (structure 8.10). The D derivatives cannot bind because of steric hindrance between the side chain and the walls of enzyme around the position normally occupied by the H atom of L derivatives (Chapter 1).

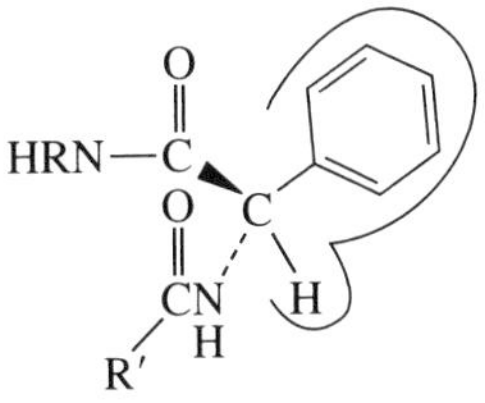

$$(8.10)$$

3. Conformation and configuration

The terms conformation and configuration, although often used interchangeably in biochemistry, have precise and different meanings: conformation refers to any one of a molecule's instantaneous orientations in space caused by free rotation about its single bonds; configuration refers to the geometry about a rigid or dissymmetric part of a molecule (e.g., about a double bond, a chiral center, or a single bond where there is steric hindrance to free rotation). In other words, a change of conformation requires just that single bonds rotate, whereas, in general, a change of configuration requires that covalent bonds be broken. The distinction between conformation and configuration becomes blurred in examples in which there is restricted rotation about single bonds: at low temperatures rotation is slow, so different conformational isomers are technically in different configurations; whereas at high temperatures the rate of rotation may be increased so that there is a conformational equilibration. It is important to preserve the distinction between the terms for the clear-cut examples.

B. Examples of stereospecific enzymatic reactions

1. NAD$^+$- and NADP$^+$-dependent oxidation and reduction

NAD$^+$ (structure 8.11, R = H) and NADP$^+$ (structure 8.11, R = PO$_3^-$) function as coenzymes in redox reactions by reversibly accepting hydrogen at the 4 position of the nicotinamide ring (equation 8.12).

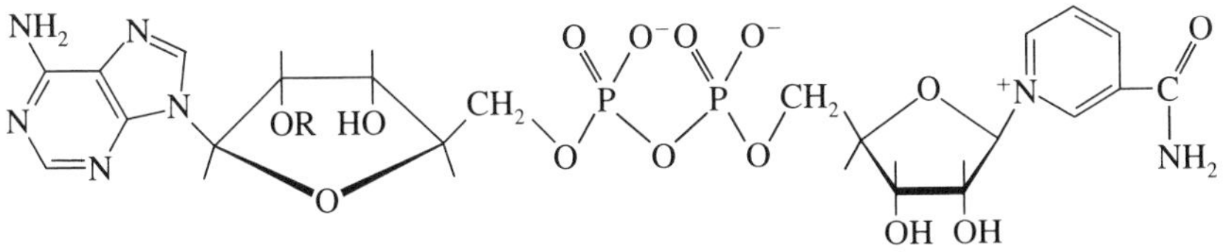

$$(8.11)$$

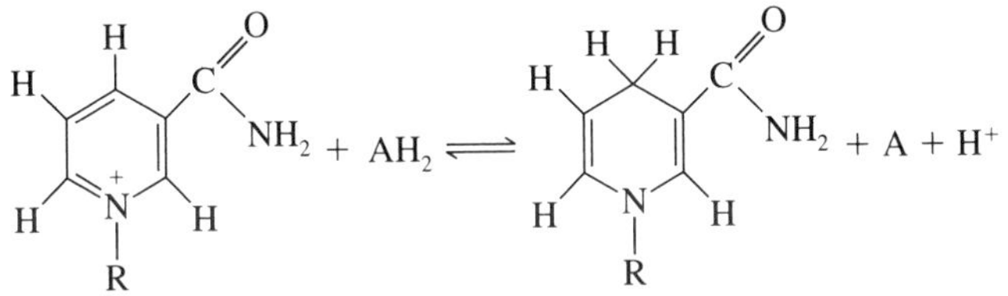

$$(8.12)$$

The 4 position in the dihydronicotinamide ring is prochiral. The faces of the nicotinamide ring and the C-4 protons of the dihydronicotinamide rings may be labeled according to the *RS* convention by giving the portion of the ring containing the $-CONH_2$ group priority over the other portion:

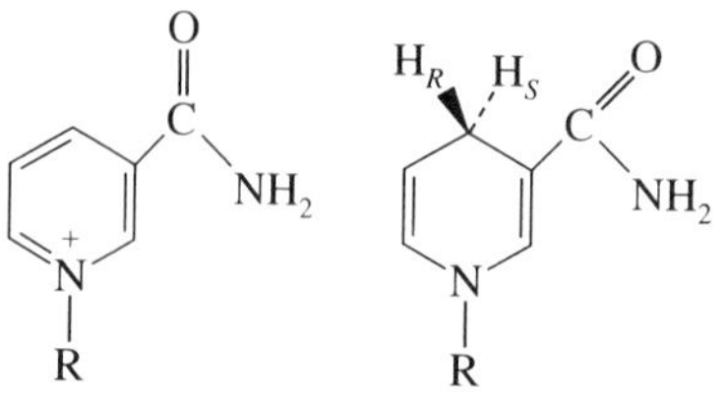

$$(8.13)$$

re

It was discovered in a historically important series of experiments that there is direct and stereospecific transfer between the substrate and NAD^+.[5,6] Yeast alcohol dehydrogenase transfers 1 mol of deuterium from CH_3CD_2OH to NAD^+. When the NADD formed with the enzyme and unlabeled acetaldehyde is incubated, *all* the deuterium is lost from the NADD and incorporated in the alcohol that is formed. The deuterium or hydrogen is transferred stereospecifically to one face of the NAD^+ and then transferred back from the same face. A nonspecific transfer to both faces would lead to a transfer of 50% of the deuterium back to the acetaldehyde (or considerably less than 50% because of the kinetic isotope effect slowing down deuterium transfer). Some dehydrogenases transfer to the same face as does the alcohol dehydrogenase (Class A dehydrogenases), others to the opposite face (Class B); these are listed in Chapter 16. It has subsequently been found that the Class A enzymes transfer to the *re* face of NAD^+ (or $NADP^+$) and use the pro-*R* hydrogen of NADH:

$$NAD^+ + CH_3CD_2OH \longrightarrow NADD + CH_3CDO + H^+ \qquad (8.14)$$

$$NADD + CH_3CHO + H^+ \longrightarrow NAD^+ + CH_3CDHOH \qquad (8.15)$$

The transfer to the aldehyde is also stereospecific. The alcohol formed is the *R* enantiomer. The *S* enantiomer (structure 8.16), formed from CH_3CDO and NADH, has a specific rotation of $-0.28 \pm 0.03°$.[7]

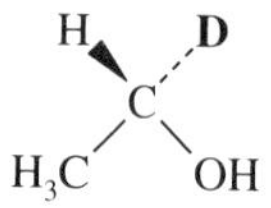

$$(8.16)$$

2. Stereochemistry of the fumarase-catalyzed hydration of fumarate

Fumarase catalyzes the addition of the elements of water across the double bond of fumarate to form malate (equation 8.17). An NMR analysis of the stereochemistry of the malic acid formed shows that the addition of D_2O is anti (D^+ adding to the *re* face at the top and OD^- adding to the *si* face at the bottom) rather than syn (both adding to the *re-re* face). Similarly, other enzymes in the citric acid cycle catalyze anti hydrations of double bonds.[8] It should be noted that, from the principle of microscopic reversibility, the dehydration reaction of malate must occur in a conformation in which the two carboxyl groups are anti, as are the elements of water that are to be eliminated. It will be seen below that this observation is critical in the analysis of the stereochemistry of the chiral methyl group.

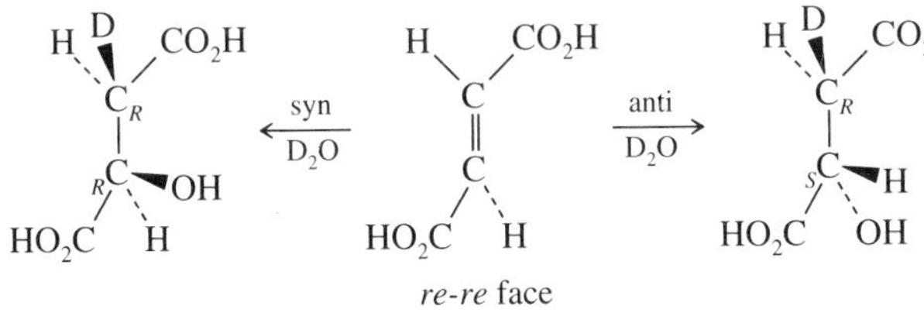

(8.17)

3. Demonstration that the enediol intermediate in aldose–ketose isomerase reactions is syn

As will be discussed in Chapter 16, the catalysis of a reaction by aldose–ketose isomerases involves an enediol intermediate in which the transferred proton (T in equation 8.18) remains on the *same* face of the intermediate. The stereochemistry of the products shows that the intermediate is syn rather than anti.[9]

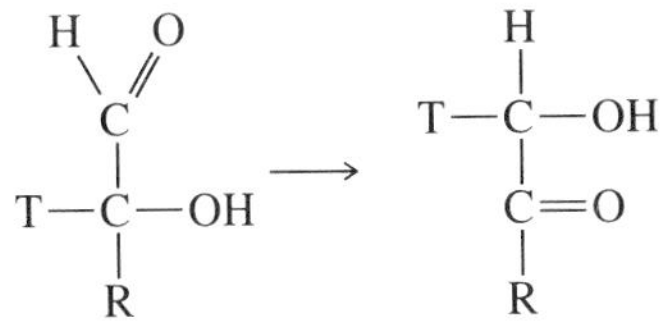

(8.18)

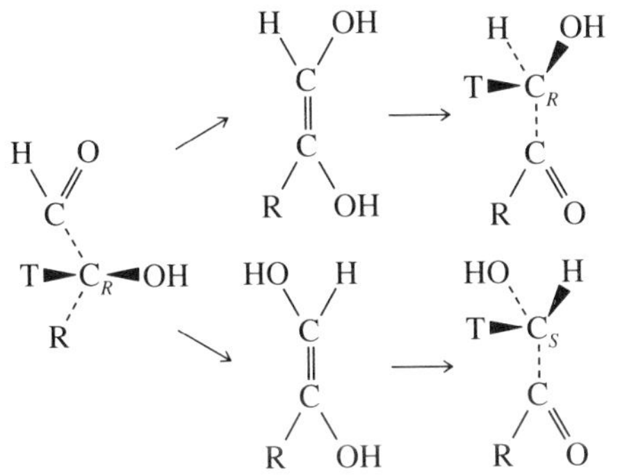

(8.19)

Aldoses that are R at C-2, as in equation 8.19, always form ketoses that are R at C-1. This implies the *syn*-enediol in the upper branch of 8.19 rather than the anti intermediate of the lower branch.

4. Use of locked substrates to determine the anomeric specificity of phosphofructokinase

Phosphofructokinase catalyzes the phosphorylation of fructose 6-phosphate to fructose 1,6-diphosphate (Chapter 10, section G):

$$\text{D-fructose 6-phosphate} + \text{ATP} \longrightarrow \text{D-fructose 1,6-diphosphate} + \text{ADP} \qquad (8.20)$$

In solution, the substrate exists as about 80% β anomer and 20% α anomer. The two forms rapidly equilibrate in solution via the open-chain keto form (equation 8.21).

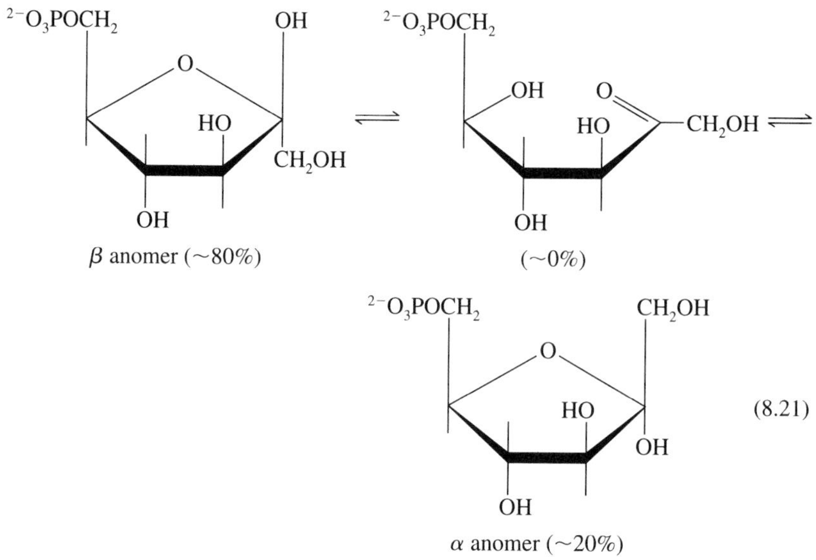

$$\tag{8.21}$$

The enzyme has been shown to be specific for the β form by rapid reaction measurements on a time scale faster than that for the interconversion of the anomers, and also by determination of the activity toward model substrates that are locked in either of the configurations. By using sufficient enzyme to phosphorylate all the active anomer of the substrate before the two forms can reequilibrate, it is found that 80% of the substrate reacts rapidly, and that the remaining 20% reacts at the rate constant for the anomerization. The kinetics were followed both by quenched flow using $[\gamma\text{-}^{32}\text{P}]\text{ATP}$[10] and by the coupled spectrophotometric assay of equation 6.4.[11] The other evidence comes from the steady state data on the following substrates:[12]

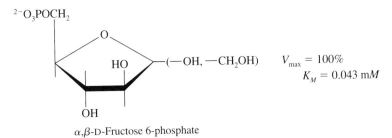

(—OH, —CH$_2$OH) $V_{max} = 100\%$
$K_M = 0.043$ mM

α,β-D-Fructose 6-phosphate

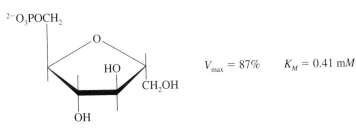

$V_{max} = 87\%$ $K_M = 0.41$ mM

2,5-Anhydro-D-mannitol 1-phosphate
(locked in β configuration)

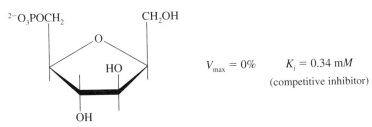

$V_{max} = 0\%$ $K_i = 0.34$ mM
(competitive inhibitor)

2,5-Anhydro-D-glucitol 1-phosphate
(locked in α configuration)

2,5-Anhydro-D-mannitol 1-phosphate is locked in a configuration that is equivalent to the β anomer of D-fructose: it lacks the 2-hydroxyl group and cannot undergo mutarotation to the equivalent of the α anomer because the ring cannot open. The glucitol derivative, on the other hand, is locked into the equivalent of the configuration of the α anomer. It is seen from the values of V_{max} and K_M (or K_i) that although both bind, only the β anomer is bound productively and is phosphorylated.

C. Detection of intermediates from retention or inversion of configuration at chiral centers

1. Stereochemistry of nucleophilic reactions

The reaction that usually begins most elementary courses on mechanistic organic chemistry is nucleophilic substitution at saturated carbon. There are two extreme forms of the mechanism. The biomolecular S_N2 reaction of a nucleophile (Y^-) and an alkyl halide leads to inversion of the chiral carbon (equation 8.22).

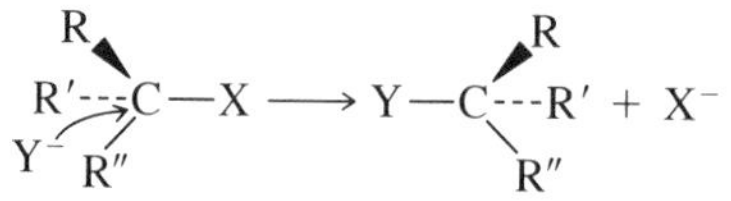

$$(8.22)$$

The unimolecular S_N1 reaction, on the other hand, generates a planar carbonium ion that reacts randomly at each face to generate a racemic product (equation 8.23).

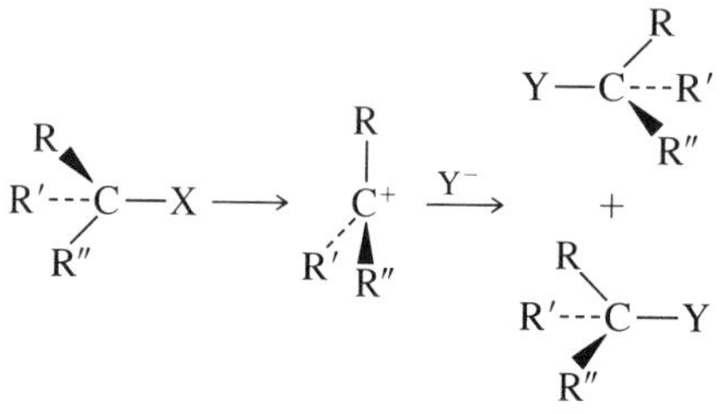

$$(8.23)$$

The inversion of configuration during an S_N2 reaction has led to certain criteria for the detection of intermediates in enzymatic reactions. For example, suppose that an enzyme catalyzes the nucleophilic reaction 8.22. Then a direct attack of Y^- on the alkyl halide will lead to inversion. But if the substrate first reacts with a nucleophilic group on the enzyme to give an intermediate that then reacts with Y^- (equation 8.24),

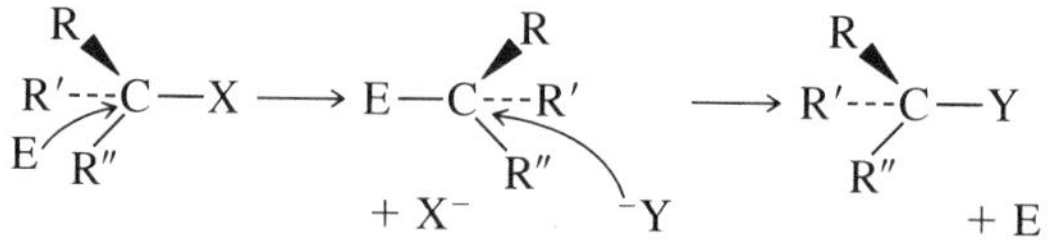

$$(8.24)$$

there will be two successive inversions, and hence an overall retention of configuration. Thus, retention in such reactions is generally taken as evidence for the presence of an intermediate on a reaction pathway, whereas inversion of configuration is interpreted as evidence for the direct reaction of one substrate with the other.[13] How valid are these and other stereochemical arguments?

2. The validity of stereochemical arguments

Stereochemical criteria by themselves have never solved a reaction mechanism. This is because there is the same basic philosophical problem in analyzing stereochemistry as there is in analyzing steady state kinetics (Chapter 7). Only the products and the starting materials are being directly observed; the information regarding the intermediates is indirect and inferential only. For example, although inversion of configuration can arise from a single nucleophilic displacement reaction, it can also arise from three successive displacements, or from five or from any *odd* number of successive displacements. Similarly, retention of configuration implies not just two successive displacement reactions but any *even* number. Proof, as discussed in Chapter 7, requires direct observation of intermediates. Stereochemical criteria do, however, place constraints on the range of possible mechanisms. Thus, as was true for steady state kinetics, *stereochemical evidence per se can never prove mechanisms but can only rule out alternative pathways.*

3. Intermediates in reactions of lysozyme and β-galactosidase

Lysozyme and β-galactosidase, which are both glycosidases, catalyze very similar reactions. Both enzymes are found to catalyze the alcoholysis of their polysaccharide substrates with retention of configuration at the C-1 carbon (equation 8.25).[14–17] This is consistent with the evidence presented in Chapter 7, section C3, that there is at least one (but probably only one) intermediate on the reaction pathway. However, kinetic isotope data are consistent with the interpretation that the intermediate in the reaction of β-galactosidase is covalent and that there are two successive S_N2 displacements, whereas the intermediate with lysozyme is a bound carbonium ion formed in an S_N1 reaction (Chapter 16). The carbonium ion, unlike an analogous one in solution, reacts stereospecifically on the enzyme. Thus, the stereochemical evidence by itself has given no indication of the *nature* of the intermediate.

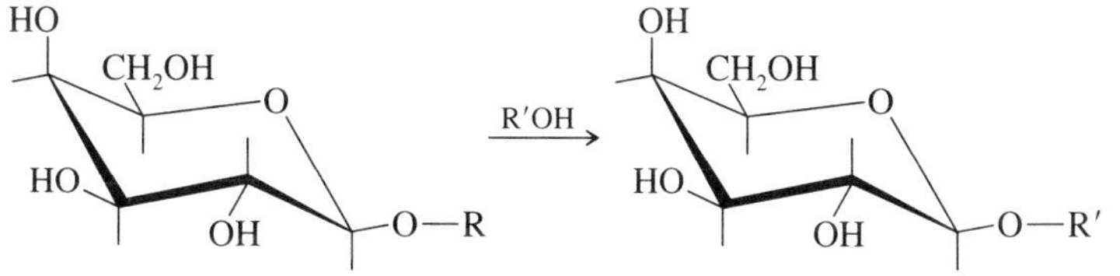

$$(8.25)$$

The above stereochemical experiments were relatively easy to perform because the natural substrates are chiral. We shall examine two areas in which clever chemistry was required to build chiral substrates.

D. The chiral methyl group

Many enzymatic reactions involve the conversion of methylene groups (CH_2XY or $CH_2=$) to methyl (CH_3-), and *vice versa*. The stereochemistry of these reactions may be studied by synthesizing a methyl group that is chiral. This is possible by using all three isotopes of hydrogen to form a classical asymmetric carbon atom (structure 8.26). The chemistry involved is most elegant, being based on a combination of organic and physical organic chemistry and enzymology.[18,19]

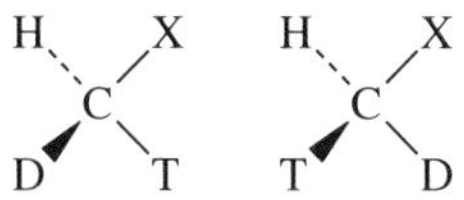

$$(8.26)$$

R-Methyl   *S*-Methyl

$(D = {}^2H; T = {}^3H)$

1. The fundamental difference between generating a chiral methyl group from a methylene group and converting a chiral methyl group into methylene

There is a crucial difference between generating a chiral methyl group and converting the same group to methylene; this distinction lies at the heart of much of the experimentation.[20] A methylene group CH_2XY is prochiral, and when it interacts with an enzyme by a three-point contact the two hydrogen atoms are non-equivalent. Their reactivity is thus governed by stereospecificity. A methyl group in CH_3X, on the other hand, rotates freely about the C—X bond, both in solution and when the group is bound to an enzyme (Chapter 1, section G2c). The reactivities of the hydrogen isotopes in CHDTX are thus governed by the *kinetic isotope effect* on the reaction. (Recall from Chapter 2, section G1, that, in general, H is transferred faster than D, which reacts faster than T: that is, $k_H > k_D > k_T$.) Accordingly, whereas a prochiral methylene compound may be transformed by a stereospecific reaction into a homogeneous chiral product, a chiral methyl group reacts to give a mixture of products, the distribution of which depends on the magnitude of the kinetic isotope effect. Further, because of the possibility that a chiral methyl group is gradually exchanging isotopes by recycling in a reversible reaction, it is important that either the reaction being studied is irreversible, or that the products are rapidly converted into stable form by a subsequent reaction so that the isotopes do not become scrambled by equilibration.

2. The chirality assay

a. The assay depends on radioactivity and not on optical measurements

It is clearly not practical to measure the optical activities of chiral methyl compounds because of the vast amounts of radioactive tritium that would have to be handled. Fortunately, as discussed in the next section, it is not necessary to produce substrates that are 100% [HDT]methyl. It is sufficient that practically all tritium-containing methyl groups also contain one deuterium, and that all, or nearly all, such methyl groups have the same chirality. There is thus a small amount of chiral, but also *radioactive*, methyl in a sea of nonradioactive material. By using a radioactivity assay on a mixture of unlabeled and labeled reagents, the only compounds that are measured are those that are labeled and hence chiral.

b. The assay uses the stereochemistry of known enzymatic reactions

R and *S* isomers of [HDT]acetic acid were synthesized by chemical and enzymatic methods that yield products of known stereochemistry.[18,19] The two isomers were then distinguished by using the following ingenious enzymatic assays. The acetic acid was first converted to acetyl–coenzyme A (by a reaction of the carboxyl group—and not the methyl—of acetic acid). The acetyl–coenzyme A was then condensed with glyoxylate to form malate in an essentially *irreversible* reaction catalyzed by malate synthase (equation 8.27). The crucial feature of this reaction is that it is subject to a normal kinetic isotope effect, so that more H than D

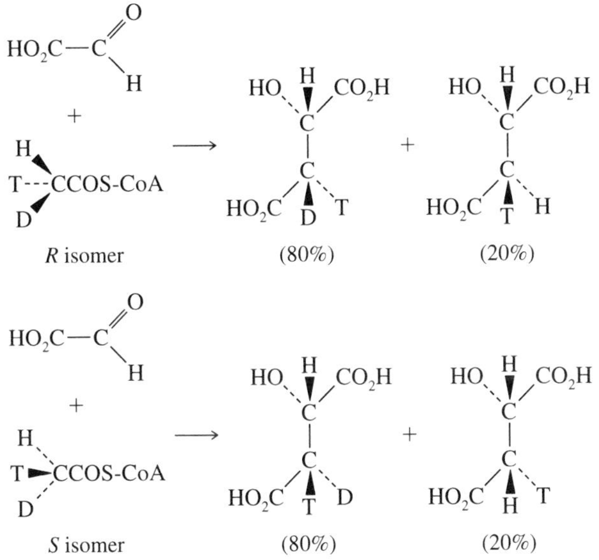

$$(8.27)$$

is lost on condensation. In fact, there is a k_H/k_D effect of 4 so that 80% of the radioactive product has lost H and 20% has lost D.[21] The stereochemistry may now be resolved by using the enzyme fumarase. The crucial feature of the fumarase reaction (equation 8.28) is that, as described previously (equation 8.17), the stereochemistry of dehydration of malate is controlled stereospecifically and is not influenced by a kinetic isotope effect: dehydration occurs from a malate conformation in which the two carboxyl groups are anti, as are the H and OH groups that are eliminated (equation 8.28).[8] It is found that the fumarate eventually formed from the R isomer of acetic acid gives a product containing 80% of the tritium and that the fumarate from the S isomer contains 20%, compared with a control reaction of TCH_2CO_2H, which gives a product containing 50%. This clearly provides an assay for the chirality of unknown acetyl groups: R should retain 80% of T, and S 20%.

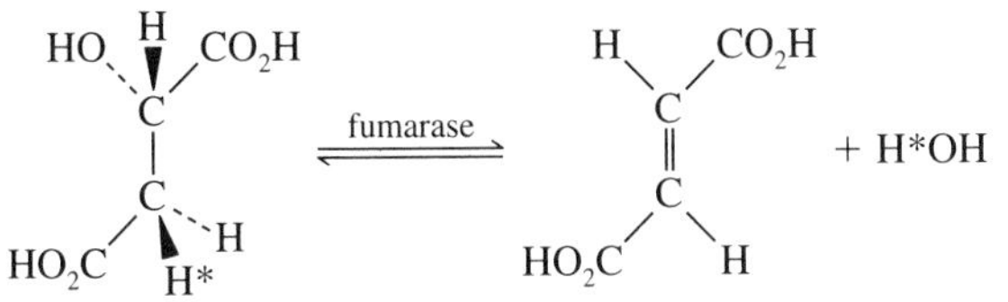

$$(8.28)$$

It should be noted that although the stereochemistry of the products of the malate synthase reaction is now known (see the next section), with the kinetic isotope effect taken into account, this information is not necessary for the chirality assay: all that is essential is that R and S isomers of [HDT]acetic acids give different yields of radioactive fumarate.[20]

3. Stereochemistry of the malate synthase reaction[18-21]

The stereospecificity of the malate synthase reaction was inferred from prior knowledge that the —$C(H)(OH)(CO_2H)$ moiety of malic acid is S and the specificity of fumarase is anti, and assuming a normal k_H/k_D effect of greater than 1. Given, then, that the malic acid has lost more H than D, inversion of the configuration of acetate will yield the products shown in equations 8.27, where the D-containing form is greater than 50% and the H-containing form less than 50%. Retention of configuration will give products with the opposite configurations (equations 8.29). Equations 8.29 predict that the fumarate formed from R-acetate in the condensation of glyoxylate with acetyl-coenzyme A will contain greater than 50% of the T if inversion has occurred, and less than 50% with retention. The observed figure of 80% is therefore consistent with inversion of configuration. Similarly, it is predicted, as found, that inversion of the configuration of the S isomer will give fumarate containing less than 50% of the T. It is seen that even though the rotation of a chiral methyl group is not constrained, stereochemical information may be derived, provided that there is a kinetic isotope effect. The absence of an effect would give 50% of each product in equations 8.27 and 8.29, and hence no information on stereospecificity.

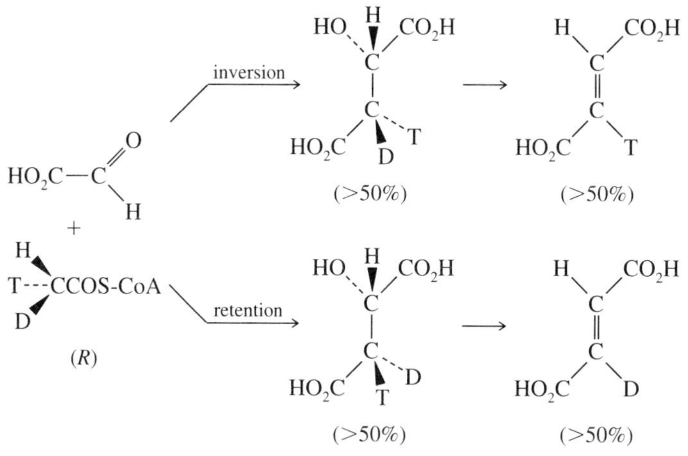

$$(8.29)$$

It has since been proved that malate synthase proceeds with inversion independently of any assumption about isotope effect. This has been done via a lyase system that splits malate to acetate[21]—a methylene-to-methyl conversion. By using methylene-labeled malate and generating chiral acetate, it could be shown unambiguously that this cleavage is an inversion. When the lyase reaction was applied to malate formed from chiral acetate by malate synthase, it was found that the acetate thus generated was of the same chirality as the starting material. This proved that there is also inversion in the reaction of malate synthase.

The stereospecificity of other methyl-methylene transformations has been assigned by converting the reaction products to acetyl–coenzyme A and using the malate synthase/fumarase assay. Similarly, the stereochemistry of transfer of chiral methyl groups between two acceptors has also been studied by converting the methyl group into chiral acetate. The stereochemistry of dozens of such reactions has been investigated and reviewed in detail.[22,23]

E. Chiral phosphate[24-27]

Phosphate esters play a central and ubiquitous role in biochemistry: the genetic information of all living organisms is stored in the phosphodiester polymers DNA and RNA (Chapter 14); the genetic information is translated into protein via RNA; the di- and triphosphates of nucleosides are important energy carriers, as are other phosphorylated compounds; groups are often activated by phosphorylation (Chapter 2); energy transduction (work and movement) is generally mediated via the hydrolysis of a nucleoside triphosphate (Chapter 10); phosphorylation of proteins and enzymes is typically an important part of control mechanisms (Chapter 10); phosphorylation of metabolites is often invoked to increase their solubility or prevent their passage across membranes. Phosphate transfer reactions are accordingly most common. The chemistry of phosphoryl transfer is far more complicated than that of acyl transfer. A very important piece of information in the elucidation of the mechanism of phosphoryl transfer is the stereochemistry.

1. A preview of phosphoryl transfer chemistry

Unlike carbon, which forms only stable tetravalent compounds, phosphorus forms stable trivalent, tetravalent, and pentavalent compounds. A phosphoryl transfer reaction such as

$$R'OPO_3^{2-} + ROH \longrightarrow ROPO_3^{2-} + R'OH \tag{8.30}$$

can proceed by either of two general types of mechanisms, analogous to the S_N1 and S_N2 mechanisms of carbon chemistry. In a *dissociative* mechanism, there is first the elimination of the leaving group to produce an unstable *metaphosphate* intermediate, and then the rapid addition of the nucleophile:

$$R'OPO_3^{2-} \longrightarrow R'O^- + PO_3^- \xrightarrow{ROH} ROPO_3^{2-} + R'OH \tag{8.31}$$

In an *associative* mechanism, there is first the addition of the nucleophile to give a pentacovalent intermediate, followed by the elimination of the leaving group. This mechanism is subdivided further. There is an *in-line* mechanism, in which the attacking nucleophile enters opposite the leaving group (equation 8.32),

$$RO-\underset{HO}{\overset{O}{P}}-OR' \longrightarrow RO-\underset{HO}{\overset{O^-}{\underset{O^-}{P}}}-OR' \longrightarrow RO-\underset{O^-OH}{\overset{O}{P}} + R'O^- \tag{8.32}$$

and there is an *adjacent* mechanism, in which the nucleophile enters on the same side as the leaving group (equation 8.33).

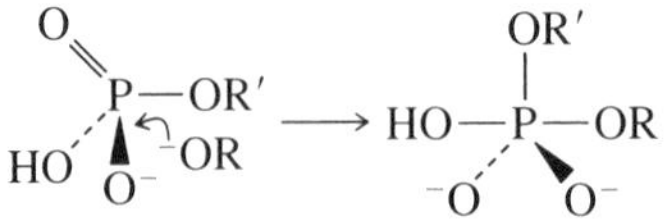

$$\text{(8.33)}$$

The adjacent mechanism involves an additional step that is a consequence of the symmetry of the intermediate. The pentacovalent intermediate is a trigonal bipyramid in which groups assume either of two topologically different positions, *equatorial* and *apical* (equations 8.34). In the in-line mechanism, the entering and leaving groups naturally take up apical positions. In the adjacent mechanism, however, the entering nucleophile is apical whereas the leaving group is equatorial. For the leaving group to be expelled, it must move to an apical position. (This is required by the principle of microscopic reversibility; groups enter apically and so must leave apically.) The movement occurs by a process termed *pseudorotation* (equation 8.34).

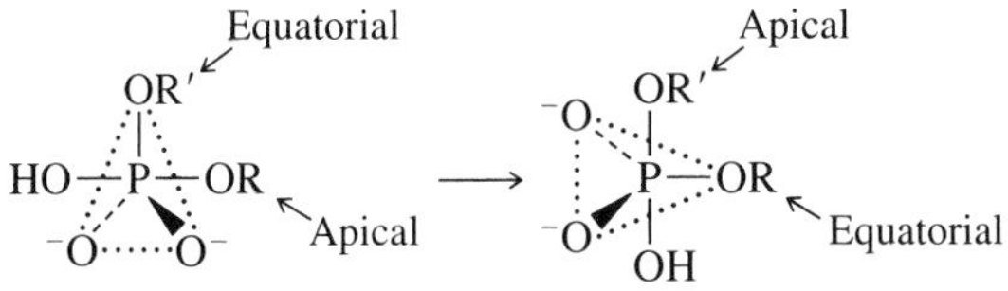

$$\text{(8.34)}$$

The stereochemical consequences of the three mechanisms in enzymatic reactions are illustrated in Figure 8.1. The phosphoryl group has the same tetrahedral geometry as saturated carbon compounds. A direct in-line associative reaction between two substrates leads to inversion (Figure 8.1a), as would an in-line dissociative mechanism (Figure 8.1b). Although a dissociative reaction in solution would generally lead to racemization, the stereochemistry in an enzymatic reaction is governed by the spatial arrangements of the substrates (cf. retention of configuration of the carbonium ion in lysozyme catalysis, section C3). The adjacent associative mechanism leads to retention of configuration (Figure 8.1c). In an enzymatic reaction there is the fourth possibility that a covalent intermediate is formed with a nucleophilic group on the enzyme (Figure 8.1d). The stereochemistry should obey the rules that normally apply when such intermediates occur: an odd number of intermediates with inversion at each step will lead to retention of configuration, whereas an even number will lead to inversion. *But*, the adjacent mechanism always leads to retention, since configuration is retained at each step.

2.   Chirality of phosphoryl derivatives

A phosphate diester of the form $ROPO_2OR'$ is prochiral, since one of the non-alkylated oxygen (^{16}O) atoms must be replaced to produce chirality. A monoester

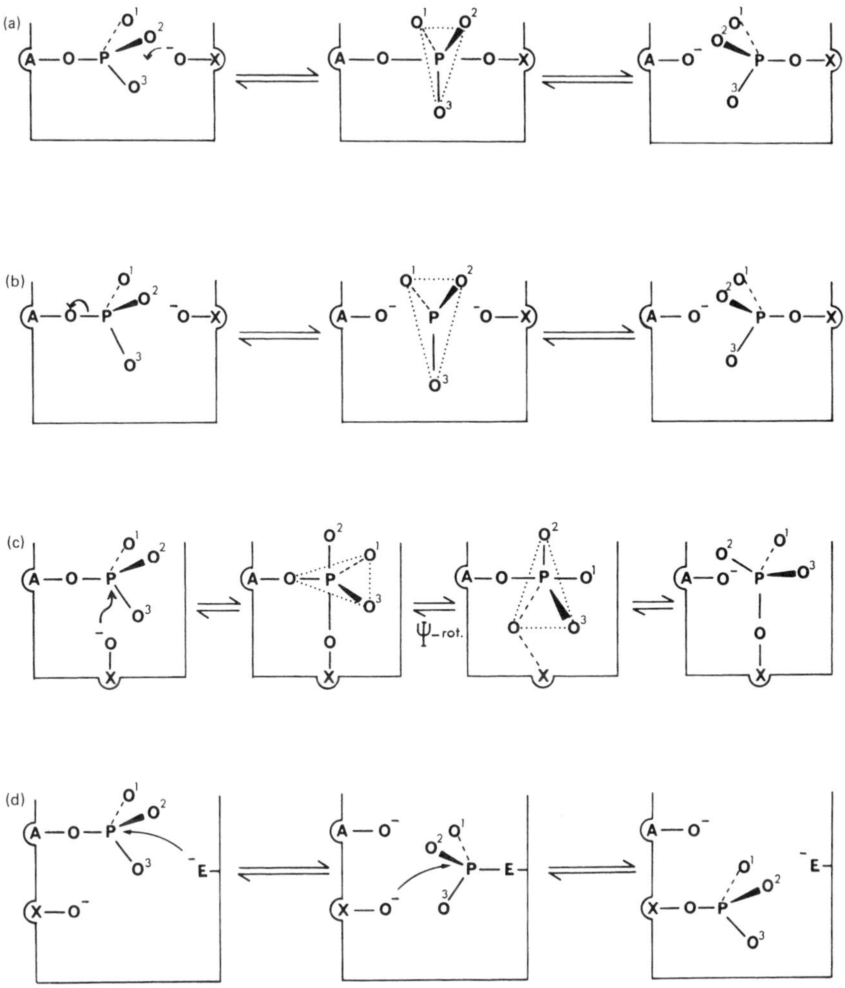

Figure 8.1 The four mechanisms of phosphoryl transfer to which phosphokinases are subject. The three peripheral oxygen atoms of the phosphoryl group are each labeled to illustrate the stereochemical consequences of the mechanisms. [From G. Lowe, P. M. Cullis, R. L. Jarvest, B. V. Potter, and B. S. Sproat, *Phil. Trans. R. Soc.* **B293**, 75 (1981).]

$ROPO_3^{2-}$ is *pro*-prochiral, since two oxygen atoms must be substituted to produce a chiral compound. Chiral phosphates have been synthesized *de novo* by using stereospecific chemical and enzymatic reactions with isotopic and/or atomic substitutions. For example, a chiral phosphorothioate may be synthesized from a prochiral phosphate by replacing an oxygen atom with a sulfur atom. Similarly, what would otherwise be a pro-prochiral phosphate has been synthesized as a chiral product by replacing one oxygen atom with sulfur and another with ^{17}O or ^{18}O, or by replacing two oxygen atoms with both isotopes to give $[^{16}O, {}^{17}O, {}^{18}O]$phosphate.

Assays of isotopic and chiral phosphoryl compounds have been greatly facilitated by ^{31}P NMR: substitution with ^{18}O causes a shift to higher field,[28,29] the magnitude of the isotope shift being related to the bond order;[30] ^{17}O, with its spin of $\frac{5}{2}$ and nuclear electronic quadrupole moment, rapidly relaxes the ^{31}P resonance and hence washes out its signal.[30–32] The methods will not be detailed in this chapter. Many of the assignments have involved some complex chemical and/or enzymatic derivatization.

The different degrees of substitution required to generate chirality have led to different methods for studying the stereochemistry of the various reactions. These reactions may be divided into four categories:

1. Prochiral substrate giving prochiral product (i.e., diester $\rightarrow$ diester).
2. Prochiral substrate giving pro-prochiral product (i.e., diester $\rightarrow$ monoester).
3. Pro-prochiral substrate giving pro-prochiral product (i.e., monoester $\rightarrow$ monoester).
4. Pro-prochiral substrate giving pro-pro-prochiral product (i.e., monoester $\rightarrow$ phosphate).

An example of each type of reaction follows.

3. Examples of chiral phosphoryl transfer

a. Prochiral substrate giving prochiral product

These reactions are the easiest to tackle, since they require only one phosphoryl oxygen to be substituted in both the substrate and the product. The classic example of this experiment is the first step in the hydrolysis of RNA catalyzed by bovine pancreatic ribonuclease. As discussed in detail in Chapter 16, ribonuclease catalyzes the hydrolysis of RNA by a two-step reaction in which a cyclic intermediate is formed. The stereochemistry of the first step (cyclization) (equation 8.35),

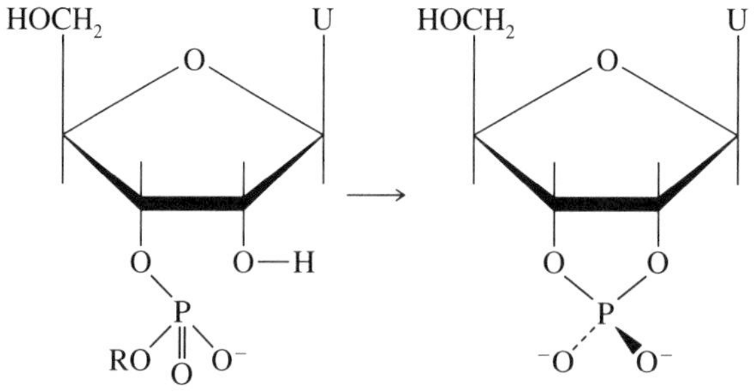

$$(8.35)$$

was solved by some powerful, elegant experiments with the chiral substrate uridine 2', 3'-cyclic phosphorothioate (structure 8.36).[33,34]

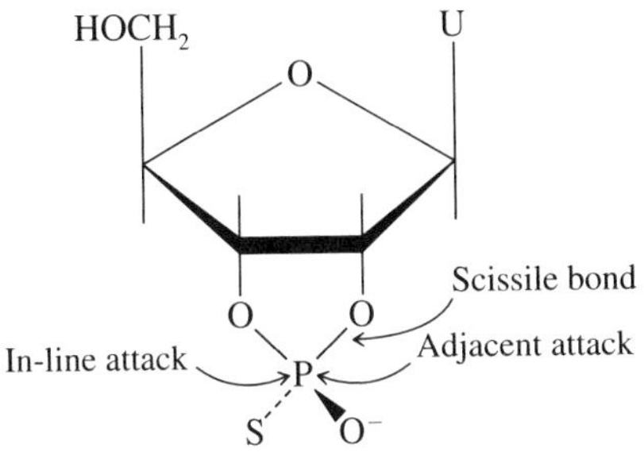

(8.36)

This compound was crystallized and its structure and absolute stereochemistry were determined by x-ray diffraction. Incubation with ribonuclease in aqueous methanol solution formed a methyl ester by the reverse of mechanism 8.35 (equation 8.37).

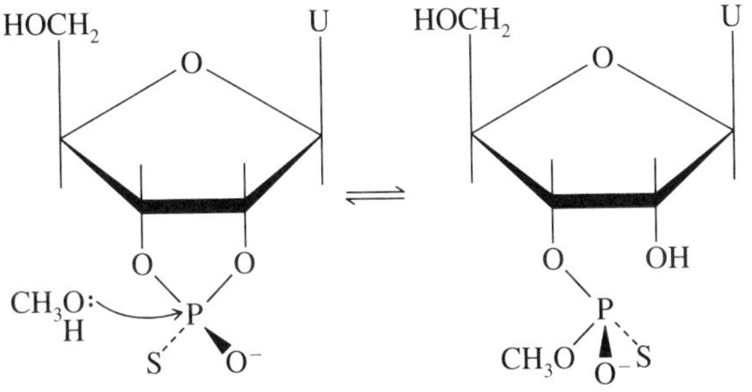

(8.37)

The methyl ester was crystallized and its absolute stereochemistry was determined by x-ray diffraction to be as in equation 8.37. This product corresponds to an in-line attack. When incubated with ribonuclease in aqueous solution, the methyl ester re-forms the original cyclic phosphorothioate (structure 8.36). This result is expected from the principle of microscopic reversibility, since the forward and reverse reactions must go through the same transition state. But it does show directly that the cyclization step involves an in-line attack: an adjacent attack of the ribose hydroxyl in the cyclization of the methyl ester as in the right-hand structure 8.38 would give the enantiomer of structure 8.36.

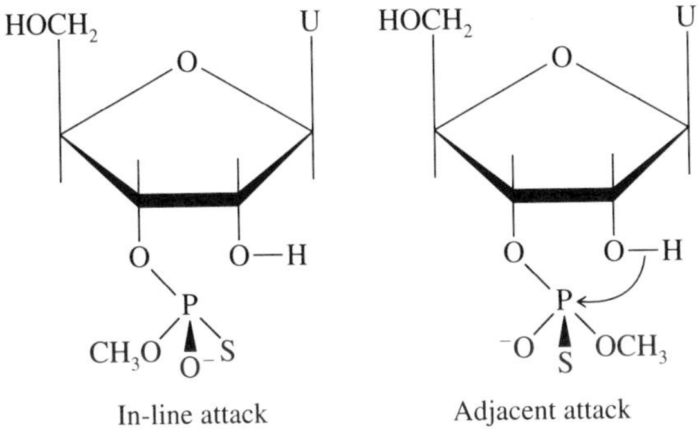

(8.38)

Later experiments involving similar reactions have used ^{31}P NMR on the products (or on chemical derivatives) to assign configuration.

b. Prochiral substrate giving pro-prochiral product

The second step of the ribonuclease reaction, ring opening (equation 8.39), is expected by analogy with the methanolysis step in the preceding section to be an in-line attack of water.

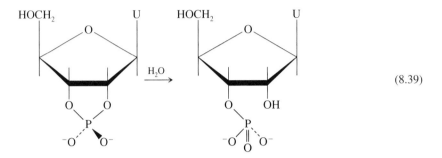

$$(8.39)$$

This was demonstrated directly, however, by using a technique generally applied to studying the reaction of prochiral phosphate giving pro-prochiral phosphate: the chiral phosphorothioate was hydrolyzed in $H_2^{18}O$ to give a chiral product whose configuration could be assigned (equation 8.40).[35]

$$(8.40)$$

c. Pro-prochiral substrate giving pro-prochiral product

These reactions involve the transfer of the phosphoryl moiety itself and require that two of the three phosphoryl oxygen atoms be tagged in the substrate. This can be achieved by synthesizing an ^{18}O-substituted phosphorothioate[36,37] or [^{16}O, ^{17}O, ^{18}O]phosphomonoester[38,39] substrate. An example of the latter is the demonstration that the hexokinase reaction (equation 8.41) proceeds with inversion of configuration.[25,40] The products were derivatized and their configurations were determined by NMR.[25]

$$\beta\text{-D-glucose} + \text{ATP} \rightleftharpoons \text{glucose 6-phosphate} + \text{ADP} \qquad (8.41)$$

d. Pro-prochiral substrate giving pro-pro-prochiral product

Experiments conducted so far for studying the stereochemistry of hydrolysis of phosphomonoesters to inorganic phosphate have used [^{16}O, ^{17}O, ^{18}O]thiophosphate.[41,42] For example, the hydrolysis of ATP catalyzed by the myosin ATPase (equation 8.42) has been shown to proceed with inversion of configuration.

$$\text{ADP}-^{18}\bullet-\overset{2}{\underset{\underset{^{16}O^-}{\overset{\|}{\overset{18}\bullet^-}}}{\overset{\displaystyle S}{\overset{\diagup}{P}}}}\quad+\ H_2^{17}\otimes\ \longrightarrow\ \text{ADP}\ +\ \underset{^{18}\bullet\ ^{16}O^-}{\overset{\displaystyle S}{\overset{\|}{P}}}\otimes^- \tag{8.42}$$

4. Positional isotope exchange

None of the experiments just discussed gives information on whether or not the reactions are associative or dissociative, since both mechanisms predict the same configurational changes. This aspect of the mechanism may be probed, however, by a different approach which is based on the procedure of *molecular isotope exchange*.[43] This may be illustrated by reactions of ATP. The dissociative mechanism is stepwise with prior formation of the metaphosphate intermediate. Thus, a kinase that catalyzes the transfer of the terminal phosphoryl group of ATP to an acceptor may, in the absence of the acceptor molecule, generate metaphosphate ion and ADP in a rapid and reversible reaction. If the β,γ-bridge oxygen is tagged as in equation 8.43,

$$\text{Ad}-\text{O}-\overset{\overset{\displaystyle O}{\|}}{\underset{\underset{\displaystyle O}{\|}}{P}}-\text{O}-\overset{\overset{\displaystyle O}{\|}}{\underset{\underset{\displaystyle O}{\|}}{P}}-\bullet-\overset{\overset{\displaystyle O}{\|}}{\underset{\underset{\displaystyle O}{\|}}{P}}-\text{O} \rightleftharpoons \left[\text{Ad}-\text{O}-\overset{\overset{\displaystyle O}{\|}}{\underset{\underset{\displaystyle O}{\|}}{P}}-\text{O}-\overset{\overset{\displaystyle O}{\|}}{\underset{\underset{\displaystyle O}{\|}}{P}}-\bullet\ +\ \overset{\overset{\displaystyle O\diagdown\ \diagup O}{}}{\underset{\underset{\displaystyle O}{\|}}{P}}\right]\rightleftharpoons$$

$$\left[\text{Ad}-\text{O}-\overset{\overset{\displaystyle O}{\|}}{\underset{\underset{\displaystyle O}{\|}}{P}}-\text{O}-\overset{\overset{\displaystyle \bullet}{\|}}{\underset{\underset{\displaystyle O}{\|}}{P}}-\text{O}\ +\ \overset{\overset{\displaystyle O\diagdown\ \diagup O}{}}{\underset{\underset{\displaystyle O}{\|}}{P}}\right] \rightleftharpoons \text{Ad}-\text{O}-\overset{\overset{\displaystyle O}{\|}}{\underset{\underset{\displaystyle O}{\|}}{P}}-\text{O}-\overset{\overset{\displaystyle \bullet}{\|}}{\underset{\underset{\displaystyle O}{\|}}{P}}-\text{O}-\overset{\overset{\displaystyle O}{\|}}{\underset{\underset{\displaystyle O}{\|}}{P}}-\text{O}$$

(Ad = adenosine)

$$\tag{8.43}$$

then because of the torsional symmetry of the phosphate group, the bridge oxygen may become scrambled.[44] There are several "may"s in this argument, so what is the status of data from such positional isotope exchange experiments? The rules of proof discussed at the beginning of Chapter 7 for the detection of intermediates apply equally well here. If scrambling occurs at a rate that is consistent with the value of k_{cat} for the transfer of the phosphoryl group to the acceptor in the steady state, this is very good evidence for the importance of the dissociative route. However, for the following reasons, the absence of scrambling does not necessarily rule out the dissociative mechanism. It is possible that:

1. There is no scrambling in the absence of the acceptor because it is required to cause a conformational change in the enzyme, or to exert an electrostatic effect on the reaction (i.e., the acceptor is a promoter).

2. The lifetime of the intermediate is very short compared with that for rotation of the phosphate, so that the intermediate reverts to substrate faster than it can rotate.

3. Rotation of the intermediate may be restricted because an oxyanion is bound by a group on the enzyme (e.g., via a bridging metal ion).

Positional isotope exchange experiments conducted on creatine kinase,[45] hexokinase,[46] and pyruvate kinase[47] have provided no evidence in favor of a dissociative mechanism. In each case, the data are consistent with a straightforward nucleophilic attack on phosphorus to generate a trigonal bipyramidal intermediate.[47]

5. A summary of the stereochemistry of enzymatic phosphoryl transfers

So far, all evidence is consistent with the interpretation that enzymatic reactions at phosphorus proceed with inversion by an in-line associative mechanism. There has been no need to invoke adjacent mechanisms, metaphosphate intermediates, or pseudorotation. Results are summarized in Table 8.1.

Phosphokinases proceed with inversion at phosphorus, and it is thus usually assumed that these reactions involve direct transfer between the two substrates. This has been challenged, however, for acetate kinase: it has been argued that the inversion results from a triple-displacement mechanism with two phosphorylenzyme intermediates and hence three transfers.[48]

Mutases are, in effect, "internal kinases," in that they transfer a phosphoryl group from one hydroxyl to another in the molecule. Unlike kinases, however, their reactions proceed with retention. Since there is generally good corroborative evidence for a phosphorylenzyme in mutase reaction pathways, double-displacement reactions are most likely.

Most phosphodiesterases catalyze transfer of the phosphoryl group to water with retention of configuration. Again, in keeping with other evidence for phosphorylenzyme intermediates, these probably represent double-displacement reactions. One exception occurs with the $3' \rightarrow 5'$ exonuclease of T4 DNA polymerase (Chapter 14); its reactions proceed with inversion.[49]

F. Stereoelectronic control of enzymatic reactions

One of the developments in organic chemistry in recent years is the realization that many reactions are under stereoelectronic control. That is, there is a relationship between the energetics of the electronic changes that occur in bond making and breaking and the conformation or configuration of the reactants. Molecules may thus have an optimal conformation for a particular reaction. This could be of importance in enzymatic reactions for controlling product formation and minimizing side reactions. Stereoelectronic control helps explain the specificity of the reactions of pyridoxal phosphate derivatives. These reactions nicely illustrate the principles involved in the addition or elimination of groups in molecules containing conjugated double bonds.

1. Pyridoxal phosphate reactivity

As discussed in Chapter 2, section C2, pyridoxal phosphate condenses with amino acids to form a Schiff base (structure 8.44). Each of the three groups around the chiral carbon at the top of structure 8.44 may be cleaved to give an anion that is stabilized by delocalization of the electrons over the π orbitals.

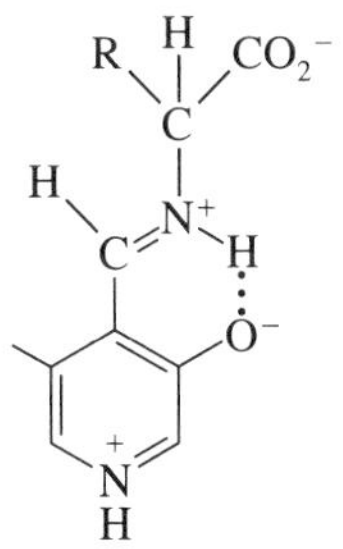

$$\text{(8.44)}$$

An essential feature of such stabilization is that the atoms in the π system are planar. The extended molecular orbital is constructed from atomic orbitals that are perpendicular to the plane. Thus, for the electrons involved in any bond making or breaking processes to be stabilized by delocalization, the bonds that are being made or broken must also be perpendicular to the plane. This criterion may be used by pyridoxal phosphate–utilizing enzymes in choosing which bond to cleave, as may be seen when the intermediate 8.44 is redrawn so that it is perpendicular to the plane of the paper (structures 8.45; the pyridine ring is represented as a solid bar). In each case, the bond that is broken is the one at the top, so that the electrons may be fed into the π system.

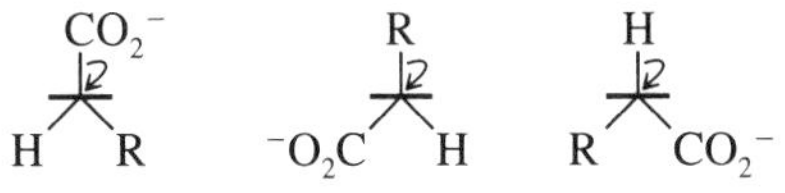

$$\text{(8.45)}$$

The same principle could well be involved in stabilizing other intermediates against unwanted side reactions. For example, any tendency to eliminate phosphate from an enolate anion intermediate of dihydroxyacetone phosphate (structure 8.46) in the reactions of triosephosphate isomerase will be minimized if the P—O bond is coplanar with the carbon skeleton.

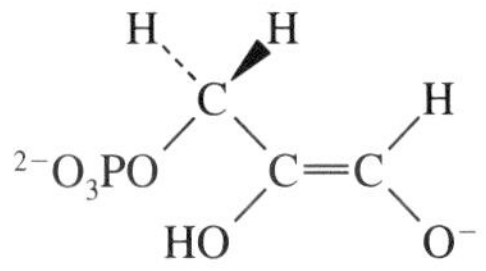

$$\text{(8.46)}$$

Table 8.1 *Stereochemistry of enzyme-catalyzed phosphoryl transfer reactions[a]*

	Retention	Inversion	Refs.
Kinases			
Nucleoside diphosphate kinase	+		1
Acetate kinase		+	2
Adenosine kinase		+	3
Adenylate kinase		+	4
Creatine kinase		+	5
Glucokinase		+	6
Glycerol kinase		+	7, 8
Hexokinase		+	2, 7, 9
Phosphoenolpyruvate carboxykinase		+	10, 11
Phosphofructokinase		+	12
Phosphoglycerate kinase		+	13
Polynucleotide kinase		+	14–17
Pyruvate kinase		+	2, 7, 18
Ribulose-phosphate kinase		+	19
Thymidine kinase		+	20

[a] Adapted from A. C. Hengge, in *Comprehensive biological catalysis,* Vol. I, M. Sinnott (Ed), Academic Press, 517 (1998).

1 K.-E. R. Sheu, J. P. Richard, and P. A. Frey, *Biochemistry* **18**, 5548 (1979).

2 W. A. Blättler and J. R. Knowles, *Biochemistry* **101**, 510 (1979).

3 J. P. Richard, M. C. Carr, D. H. Ives, and P. A. Frey, *Biochem. Biophys. Res. Commun.* **94**, 1052 (1980).

4 J. P. Richard and P. A. Frey, *J. Am. Chem. Soc.* **100**, 7757 (1978).

5 D. E. Hansen and J. R. Knowles, *J. Biol. Chem.* **256**, 5967 (1981).

6 D. Pollard-Knight, B. V. L. Potter, P. M. Cullis, G. Lowe, and A. Cornish-Bowden, *Biochem. J.* **201**, 421 (1982).

7 G. A. Orr, J. Simmons, S. R. Jones, G. J. Chin, and J. R. Knowles, *Proc. Natl. Acad. Sci. USA* **75**, 2230 (1978).

8 D. H. Pliura, D. Schomburg, J. P. Richard, P. A. Frey, and J. R. Knowles, *Biochemistry* **19**, 325 (1980).

9 G. Lowe and B. V. L. Potter, *Biochem. J.* **199**, 227 (1981).

10 K.-F. Sheu, H.-T. Ho, L. D. Nolan, P. Markovitz, J. P. Richard, M. F. Utter, and P. A. Frey, *Biochemistry* **23**, 1779 (1984).

11 J. M. Konopka, H. A. Lardy, and P. A. Frey, *Biochemistry* **25**, 5571 (1986).

12 R. L. Jarvest, G. Lowe, and B. V. L. Potter, *Biochem. J.* **199**, 427 (1981).

13 M.-D. Tsai, and T. T. Chang, *J. Am. Chem. Soc.* **102**, 5416 (1980).

14 M. R. Webb and D. R. Trentham, *J. Biol. Chem.* **255**, 1775 (1980).

15 D. H. Pliura, D. Schomburg, J. P. Richard, P. A. Frey, and J. R. Knowles, *Biochemistry* **19**, 325 (1980).

	Retention	Inversion	Refs.
Phosphatases			
Acid phosphatase	+		21
Alkaline phosphatase	+		22
ATPase (sarcoplasmic reticulum)	+		23
ATPase (mitochondrial)		+	24
ATPase (thermophilic bacterium PS3)		+	25
Glucose-6-phosphatase	+		26
GTPase (elongation factor G)		+	27
Myosin ATPase		+	28
Nucleoside phosphotransferase	+		29
p21 Ras		+	30
Purple acid phosphatase		+	31
Pyrophosphatase		+	32
Mutases			
Phosphoenolpyruvate phosphomutase	+		33
Phosphoglucomutase	+		34
Phosphoglycerate mutase (co-factor independent)	+		35
Phosphoglycerate mutase (co-factor dependent)	+		35

16 F. R. Brant, S. J. Benkovic, D. Sammons, and P. A. Frey, *J. Biol. Chem.* **256**, 5965 (1981).

17 R. L. Jarvest and G. Lowe, *Biochem. J.* **199**, 273 (1981).

18 G. Lowe, P. M. Cullis, R. L. Jarvest, B. V. L. Potter, and B. S. Sproat, *Phil. Trans. R. Soc London Ser. B* **293**, 75 (1981).

19 H. M. Miziorko and F. Eckstein, *J. Biol. Chem. Soc.* **259**, 13037 (1984).

20 J. R. Arnold, M.-S. Cheng, P. M. Cullis, and G. Lowe, *J. Biol. Chem.* **261**, 1985 (1986).

21 M. S. Saini, S. Buchward, R. L. Van Etten, and J. R. Knowles, *J. Biol. Chem.* **256**, 10453 (1981).

22 S. R. Jones, L. A. Kindman, and J. R. Knowles, *Nature, Lond.* **275**, 564 (1978).

23 M. R. Webb and D. R. Trentham, *J. Biol. Chem.* **256**, 4884 (1981).

24 M. R. Webb, C. Grubmeyer, H. S. Penefsky, and D. R. Trentham, *J. Biol. Chem.* **255**, 11637 (1980).

25 P. Senter, F. Eckstein, and Y. Kagawa, *Biochemistry* **22**, 5514 (1983).

26 G. Lowe and B. V. L. Potter, *Biochem. J.* **201**, 665 (1982).

27 M. R. Webb and J. F. Eccleston, *J. Biol. Chem.* **256**, 7734 (1981).

28 M. R. Webb and D. R. Trentham, *J. Biol. Chem.* **255**, 8629 (1980).

29 J. P. Richard, D. C. Prasher, D. H. Ives, and P. A. Frey, *J. Biol. Chem.* **254**, 4339 (1979).

30 J. Feuerstein, R. S. Goody, and M. R. Webb, *J. Biol. Chem.* **264**, 6188 (1989).

31 E. G. Mueller, M. W. Crowder, B. A. Averill, and J. R. Knowles, *J. Am. Chem. Soc.* **115**, 2974 (1993).

32 M. A. Gonzalez, M. R. Webb, K. M. Welsh, and B. S. Cooperman, *Biochemistry* **23**, 797 (1984).

33 S. Freeman, H. M. Seidel, C. H. Schwalbe, and J. R. Knowles, *J. Am. Chem. Soc.* **111**, 9233 (1989).

34 G. Lowe and B. V. L. Potter, *Biochem. J.* **199**, 693 (1981).

35 W. A. Blättler and J. R. Knowles, *Biochemistry* **19**, 738 (1980).

2. Stereoelectronic effects in reactions of proteases

A different type of stereoelectronic control has been found in the breakdown in solution of tetrahedral addition intermediates that arise in ester and amide hydrolysis and other reactions of carboxyl and carbonyl groups. In the case of an intermediate such as structure 8.47, in which there are two atoms with nonbonded electrons (generally O or N), the lowest-energy transition state for breakdown is a conformation in which nonbonded electrons of each are anti to the group being expelled (structures 8.48).[50]

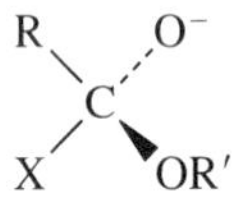

$$(8.47)$$

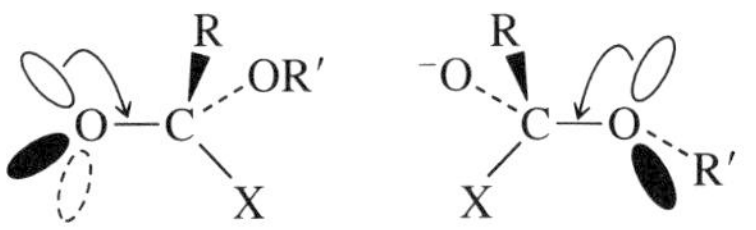

$$(8.48)$$

Two views of lone-pair electron orbitals
on $-O^-$ and $-OR'$ anti to X

It is expected from the principle of microscopic reversibility that during the *formation* of the tetrahedral intermediate, the electrons developing on the two heteroatoms must be anti to the bond being made. This has implications in the mechanism of the serine proteases (Chapters 1 and 15).[51–53] A tetrahedral intermediate is formed during proteolysis by the attack of the hydroxyl of Ser-195 on the amide bond. According to the stereoelectronic theory, the lone pairs produced in the tetrahedral intermediate are anti to the incoming hydroxyl group, so the intermediate is set up to expel Ser-195. Expulsion of the amino moiety of the peptide requires proton transfer from His-57 to the N atom, but its lone pair is seen on examination of the structure of the enzyme–substrate complex to be pointing away from the important general-acid-base catalyst. However, if an inversion of the nitrogen switches around its lone pair and its bonded hydrogen (equation 8.49),

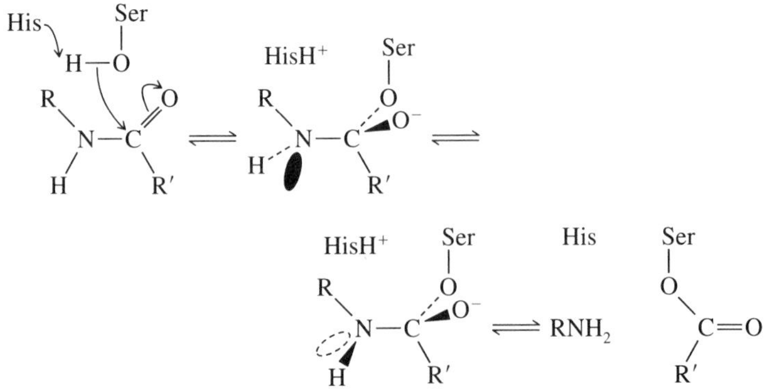

$$(8.49)$$

not only is the N able to accept the necessary proton, but the lone pair of the N is no longer anti to Ser-195. Perhaps inversion is an essential step.

References

1. G. Popjak, *The Enzymes* **2**, 115 (1970).
2. K. R. Hanson, *J. Am. Chem. Soc.* **88**, 2731 (1966).
3. A. Ogston, *Nature, Lond.* **162**, 963 (1948).
4. M. Bergmann and J. S. Fruton, *J. Biol. Chem.* **117**, 189 (1937).
5. F. H. Westheimer, H. Fisher, E. E. Conn, and B. Vennesland, *J. Am. Chem. Soc.* **73**, 2043 (1951).
6. H. F. Fisher, E. E. Conn, B. Vennesland, and F. H. Westheimer, *J. Biol. Chem.* **202**, 687 (1953).
7. H. R. Levy, F. A. Loewus, and B. Vennesland, *J. Am. Chem. Soc.* **79**, 2949 (1957).
8. O. Gawron, A. J. Glaid III, and T. P. Fondy, *J. Am. Chem. Soc.* **83**, 3634 (1981).
9. I. A. Rose and E. L. O'Connell, *Biochim. Biophys. Acta* **42**, 159 (1960).
10. R. Fishbein, P. A. Benkovic, K. J. Schray, I. J. Siewers, J. J. Steffens, and S. J. Benkovic, *J. Biol. Chem.* **249**, 6047 (1974).
11. B. Wurster and B. Hess, *FEBS Lett.* **38**, 257 (1974).
12. T. A. W. Koerner, Jr., E. S. Younathan, A.-L. E. Ashour, and R. J. Voll, *J. Biol. Chem.* **249**, 5749 (1974).
13. D. E. Koshland, Jr., in *Mechanisms of enzyme action*, W. D. McElroy and B. Glass, eds., Johns Hopkins University Press, 608 (1954).
14. K. Wallenfels and G. Kurz, *Biochem. Z.* **335**, 559 (1962).
15. K. Wallenfels and O. P. Malhotra, *Adv. Carbohyd. Chem.* **16**, 239 (1961).
16. J. A. Rupley and V. Gates, *Proc. Natl. Acad. Sci. USA* **57**, 496 (1967).
17. J. J. Pollock, D. M. Chipman, and N. Sharon, *Archs. Biochem. Biophys.* **120**, 235 (1967).
18. J. W. Cornforth, J. W. Redmond, H. Eggerer, W. Buckel, and C. Gutschow, *Nature, Lond.* **221**, 1212 (1969).
19. J. Lüthey, J. Rétey, and D. Arigoni, *Nature, Lond.* **221**, 1213 (1969).
20. J. W. Cornforth, *Chem. Brit.* **6**, 431 (1970).
21. H. Lenz and H. Eggerer, *Eur. J. Biochem.* **65**, 237 (1976).
22. H. H. Floss and M. D. Tsai, *Adv. Enzymol.* **50**, 243 (1979).
23. J. W. Cornforth, in *Structural and functional aspects of enzyme catalysis*, H. Eggerer and R. Huber, eds., Springer-Verlag, 3 (1981).
24. J. R. Knowles, *Ann. Rev. Biochem.* **49**, 877 (1980).
25. G. Lowe, P. M. Cullis, R. L. Jarvest, B. V. Potter, and B. S. Sproat, *Phil. Trans. R. Soc.* **B293**, 75 (1981).
26. P. A. Frey, in *New comprehensive biochemistry*, Vol. 3 (Ch. Tamm, Ed.), Elsevier, p. 201 (1982).
27. A. C. Hengge, in *Comprehensive biological catalysis*, Vol. I, M. Sinnott, ed., Academic Press, 517 (1998).
28. M. Cohn and A. Hu, *Proc. Natl. Acad. Sci. USA* **75**, 200 (1978).
29. G. Lowe and B. S. Sproat, *J. Chem. Soc. Chem. Commun.* **1978**, 565 (1978).
30. G. Lowe, B. V. L. Potter, B. S. Sproat, and W. E. Hull, *J. Chem. Soc. Chem. Commun.* **1979**, 733 (1979).
31. M. D. Tsai, *Biochemistry* **18**, 1468 (1979).
32. J. J. Villafranca and F. M. Raushel, *Ann. Rev. Biophys. Eng.* **9**, 363 (1980).

33. D. A. Usher, D. I. Richardson, and F. Eckstein, *Nature, Lond.* **288**, 663 (1970).
34. D. A. Usher, E. S. Erenrich, and F. Eckstein, *Proc. Natl. Acad. Sci. USA* **69**, 115 (1972).
35. P. M. J. Burgers, F. Eckstein, D. H. Hunneman, J. Baraniak, R. W. Kinas, K. Lesiak, and W. J. Stec, *J. Biol. Chem.* **254**, 9959 (1979).
36. G. A. Orr, J. Simon, S. R. Jones, G. J. Chin, and J. R. Knowles, *Proc. Natl. Acad. Sci. USA* **75**, 2230 (1978).
37. J. P. Richard, H.-T. Ho, and P. A. Frey, *J. Am. Chem. Soc.* **100**, 7756 (1978).
38. S. J. Abbott, S. R. Jonas, S. A. Weinman, and J. R. Knowles, *J. Am. Chem. Soc.* **100**, 2558 (1978).
39. P. M. Cullis and G. Lowe, *J. Chem. Soc. Chem. Commun.* **1978**, 512 (1978).
40. W. A. Blättler and J. R. Knowles, *Biochemistry* **18**, 3927 (1979).
41. M. R. Webb and D. R. Trentham, *J. Biol. Chem.* **255**, 1775 (1980).
42. M. D. Tsai, *Biochemistry* **19**, 5310 (1980).
43. I. A. Rose, *Adv. Enzymol.* **50**, 361 (1979).
44. C. F. Midelfort and I. A. Rose, *J. Biol. Chem.* **251**, 5881 (1976).
45. G. Lowe and B. S. Sproat, *J. Biol. Chem.* **255**, 3944 (1980).
46. I. A. Rose, *Biochem. Biophys. Res. Commun.* **94**, 573 (1980).
47. A. Hassett, W. Blättler, and J. R. Knowles, *Biochemistry* **21**, 6335 (1982).
48. L. B. Spector, *Proc. Natl. Acad. Sci. USA* **77**, 2625 (1980).
49. A. Gupta, C. DeBrosse, and S. J. Benkovic, *J. Biol. Chem.* **257**, 7689 (1982).
50. P. Deslongchamps, *Tetrahedron* **31**, 2463 (1975).
51. S. A. Bizzozero and B. O. Zweifel, *FEBS Lett.* **59**, 105 (1975).
52. S. A. Bizzozero and H. Dutler, *Bioorg. Chem.* **10**, 46 (1982).
53. M. Fujinaga, R. J. Read, A. Sieleck, W. Ardelt, M. Laskowski, Jr., and M. N. G. James, *Proc. Natl. Acad. Sci. USA* **79**, 4846 (1982).

Active-Site-Directed and Enzyme-Activated Irreversible Inhibitors: "Affinity Labels" and "Suicide Inhibitors"

9

Amino acid side chains of proteins react with a variety of chemical reagents to form covalent bonds. In general, the reagents tend to be nonspecific and to react with any accessible amino acid residue that has the appropriate chemical nature, with variable consequences. Covalent modification of an enzyme may lead to an irreversible loss of activity if a catalytically essential residue is blocked, if substrate binding is sterically impeded, or if the protein is distorted or its mobility impaired. Alternatively, modification may not affect activity if an unimportant residue is modified, or the inhibition may be only transitory if the chemical reaction is reversible. In this chapter, we are concerned mainly with specially designed irreversible inhibitors that bind specifically to the active site of an enzyme and utilize components of its catalytic apparatus for their effect. Their mode of inhibition is thus controlled and efficient. Before discussing the design of such highly specific irreversible inhibitors, we first list briefly the types and uses of the chemical modifications most commonly employed, and survey the chemical reactivities expected of amino acid chains.

A.   Chemical modification of proteins

Protein chemistry is an extensive and highly developed area of organic chemistry that deals with the chemical reactions of proteins. Much of this chemistry concerns reactions that occur in aqueous solution at ambient temperatures and neutral pH, that is, under conditions where proteins are stable. The objective is to modify residues in proteins chemically, either to provide mechanistic information or to produce useful alterations of activity. Some of the more frequent modification reactions are listed in Table 9.1. Spectroscopic probes may be covalently

Table 9.1 *Chemical modification reactions referred to in this text*

Residue (reactive ionic form)	Reagents	Products	Comments
$-CO_2H$	$R'-N=C=NR'$ soluble diimide) $+ RNH_2$	$-CONHR$	Surface carboxyls readily blocked; cf. chymotrypsin, pp. 179, 186.
$-CO_2H$	$N_2CHCONHR$ (diazoamide or -ester)	$-CO_2CH_2CONHR$	Reactive in affinity labels; cf. pepsin, pp. 280, 487.
$-CO_2^-$	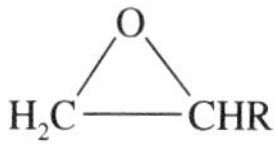 $\overset{O}{\underset{H_2C\overline{\quad\quad}CHR}{\triangle}}$ (epoxide)[a]	$-CO_2CH_2CH(OH)R$	Reactive in affinity labels; cf. glucose 6-phosphate isomerase, p. 279.
$-CO_2^-$	XCH_2COR[b] (X = I, Br, Cl; haloacetates, etc.)	$-CO_2CH_2COR$	Reactive in affinity labels; cf. triosephosphate isomerase, p. 281.
$-NH_2$	Acetic anhydride	$-NHCOCH_3$	Surface groups readily blocked; see pp. 179, 186.
$-NH_2$	Succinic anhydride	$-NHCOCH_2CH_2CO_2^-$	As above, but gives more soluble protein.
$-NH_2$	$RR'CO$, $NaBH_4$ (sodium borohydride)	$-NHCHRR'$	Trapping Schiff base adducts; cf. acetoacetate decarboxylase, p. 78. H_2CO and $NaBH_4$ give reductive methylation of surface amino groups.
$-NH_2$	$CH_3OC(=\overset{+}{N}H_2)CH_3$ (methyl acetimidate)	$-\overset{+}{N}H=C(NH_2)CH_3$	—
$-NH_2$, $-NH_2$	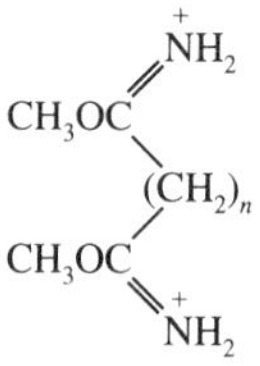 $CH_3OC\overset{\overset{+}{N}H_2}{\underset{(CH_2)_n}{\diagdown}}$ $CH_3OC\underset{\overset{+}{N}H_2}{\diagup}$	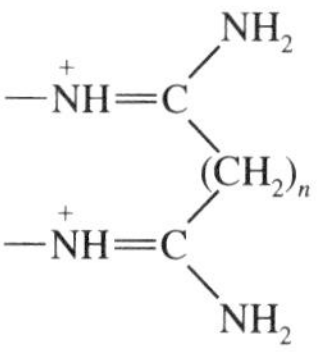 $-\overset{+}{N}H=C\overset{NH_2}{\underset{(CH_2)_n}{\diagdown}}$ $-\overset{+}{N}H=C\underset{NH_2}{\diagup}$	Cross-linking[c] of proteins by linking $-NH_2$ groups, e.g., dimethylsuberimidate; see p. 276.

[a] The $-NH_2$ group also reacts with epoxides.
[b] The α carbon of haloacetates is particularly susceptible to nucleophilic attack. The order of reactivity for different halides is $I > Br > Cl$.

Residue (reactive ionic form)	Reagents	Products	Comments
—NH$_2$	Dansyl chloride	—NH-dansyl	Fluorescent label; cf. immunoglobin; see p. 48.
—S$^-$	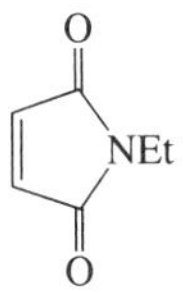 (N-ethylmaleimide)	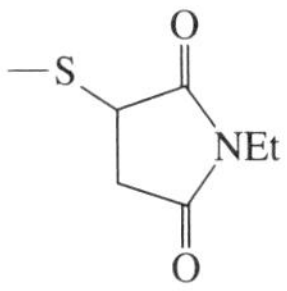	Selective for — SH. Et may be replaced by spin label; cf. pyruvate dehydrogenase, p. 37.
—S$^-$	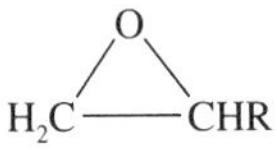	—SCH$_2$—CH(OH)R	Reactive in affinity label; cf. glyceraldehyde 3-phosphate dehydrogenase, p. 281.
—S$^-$	XCH$_2$COR (haloacetate, etc.)	—SCH$_2$COR	S$^-$ sufficiently nucleophilic to react rapidly in solution; see p. 277.
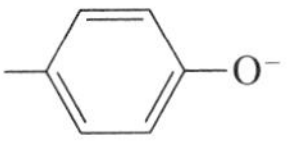	(NO$_2$)$_4$C (tetranitromethane)	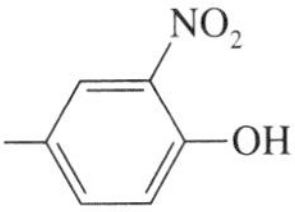	Chromophore, low pK_a (~7); cf. carboxypeptidase; see p. 420.
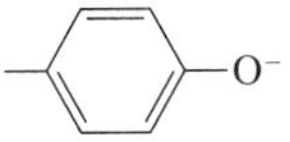	I$_3^-$	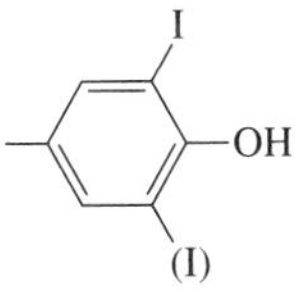 (I)	Radioactive tag; see p. 276.
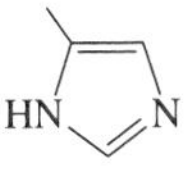	 (EtOC)$_2$O (diethylpyrocarbonate)	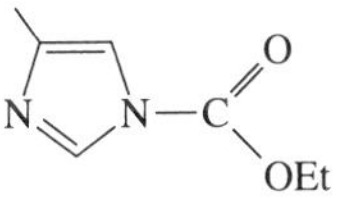	Relatively stable at neutral pH; cf. lactate dehydrogenase, p. 468.
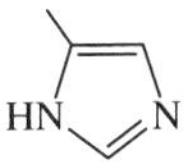	XCH$_2$COR (haloacetate, etc.)	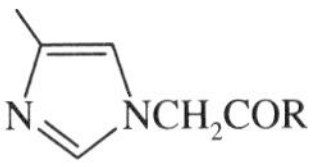	100× less reactive than —S$^-$. Useful in affinity labels; cf. chymotrypsin, p. 278.

[c] Cross-linkers have been synthesized with an —SS— bond that may be cleaved by a reducing agent to reverse the cross-linking.

linked to specific regions of proteins; e.g., fluorescent derivatives such as the dansyl group may be linked to amino groups, spin labels such as nitroxide derivatives may be attached to cysteine residues, and tyrosine residues may be modified by nitration. These probes are sometimes termed *reporter groups* since they examine local structure or overall conformation. Other common examples are: the cross-linking of proteins by bifunctional reagents such as dimethylsuberimidate, in order to measure the degree of association or subunit composition of oligomers; the insertion of radioactive tags into molecules, by such procedures as the iodination of tyrosine side chains by radioactive iodine and the reductive methylation of amino groups with formaldehyde and radioactively labeled (^{3}H) borohydride; searching for unusually reactive groups by their high reaction rates; measuring the degree of exposure of groups to solvent from their activity with reagents; measuring the pK_a values and ionic states of reactive groups from the pH dependence of their reactivities (as described in Chapter 5); assessing whether groups are important in catalysis by noting the effect of modification on reaction rates; irreversibly inhibiting activity by chemical modification. Protein engineering (Chapter 14) has largely replaced chemical modification for assessing the roles of side chains in catalysis and binding.

The subject of protein chemistry is too large to be surveyed systematically in one chapter. Various examples of chemical modifications are covered nonsystematically throughout this volume in discussions of individual enzymes and methods. To aid in locating these examples, some are listed in Table 9.1.

1. The chemical reactivity of amino acid side chains

The principal chemically reactive groups in proteins are *nucleophiles*. These are generally the same groups that are found at the active sites of enzymes and that are responsible for catalysis, as discussed in Chapter 2. The nucleophiles that are potent toward "hard" electrophilic centers, such as the carbonyl, phosphoryl, and sulfuryl groups (Chapter 2, section D2a), include: the —OH of serine, threonine, and tyrosine (the alcohols being activated by general-base catalysis, with the phenols reacting via their ionized forms at neutral and higher pH's); the ϵ-NH$_2$ group of lysine and the α-amino groups of the N-termini; the imidazole ring of histidine; the —S$^-$ of cysteine; and the —CO$_2^-$ of aspartate, glutamate, and the C-termini. These nucleophiles are also reactive in varying degrees toward "soft" electrophiles such as saturated carbon, according to the principles discussed in Chapter 2, section D2b. In addition, the S atom in the side chain of methionine is nucleophilic toward soft electrophiles, as is the aromatic ring of tyrosine. Consequently, the majority of reagents that are used to modify proteins are *electrophiles*.

Two important classes of reagents that are used to modify the nucleophilic side chains are the acylating and similarly reacting agents, and the alkylating agents. For example, acetic anhydride and other activated acyl compounds acylate amino groups to form stable amide derivatives. The phenolic hydroxyl group of tyrosine is also acylated, but the resultant ester is unstable in mild alkali. The acyl derivatives of cysteine (thioester) and histidine (acylimidazole) are also rapidly hydrolyzed. Thus, acylating agents are not generally useful for inhibiting these important catalytic groups irreversibly. The products of alkylation, however, are usually quite stable, and so irreversible inhibitors are often based on alkylating agents.

Alkylating agents are soft electrophiles. As was explained in Chapter 2, section D2b, one of the most reactive common nucleophiles toward soft electrophiles is the $-S^-$ ion. Accordingly, it is found that cysteine residues in proteins are readily alkylated by haloacetates and similar reagents. On the other hand, carboxylate ions are usually so unreactive that they are not modified by simple haloacetates or epoxides. But, as will be seen in the following section, cleverly designed reagents may react specifically and rapidly with otherwise poorly reactive groups.

B. Active-site-directed irreversible inhibitors

An affinity label, or active-site-directed irreversible inhibitor, is a chemically reactive compound that is designed to resemble a substrate of an enzyme, so that it binds specifically to the active site and forms covalent bonds with the protein residues.[1-3] Affinity labels are very useful for identifying catalytically important residues and determining their pK_a values from the pH dependence of the rate of modification.

The reaction of an affinity label with an enzyme involves the initial formation of a reversibly bound enzyme–inhibitor complex followed by covalent modification and hence irreversible inhibition:

$$E + I \underset{}{\overset{K_I}{\rightleftharpoons}} E{\cdot}I \overset{k_I}{\longrightarrow} E{-}I \tag{9.1}$$

This scheme is analogous to that of the Michaelis-Menten mechanism, and the reaction should thus show saturation kinetics with increasing inhibitor concentration. The kinetics were solved in Chapter 4, equation 4.46. For the simple case of pre-equilibrium binding followed by a slow chemical step, the solution reduces to

$$-\frac{d[E]}{dt} = \frac{k_I[E][I]}{K_I + [I]} \tag{9.2}$$

An important consequence of the chemical reaction taking place in the confines of an enzyme–"substrate" complex is that not only is the binding specific, but the rate of the chemical step may be unusually rapid because it is favored entropically over a simple bimolecular reaction in solution, in the same way as is a normal enzymatic reaction. Thus, reagents that are normally only weakly reactive may become very reactive affinity labels.

The principles of the method are very nicely illustrated by one of the first affinity labeling experiments, the reaction of *tos*-L-phenylalanine chloromethyl ketone (TPCK) with chymotrypsin.[1] TPCK resembles substrates like *tos*-L-phenylalanine methyl ester, but the chloromethyl ketone group of TPCK is an alkylating reagent.

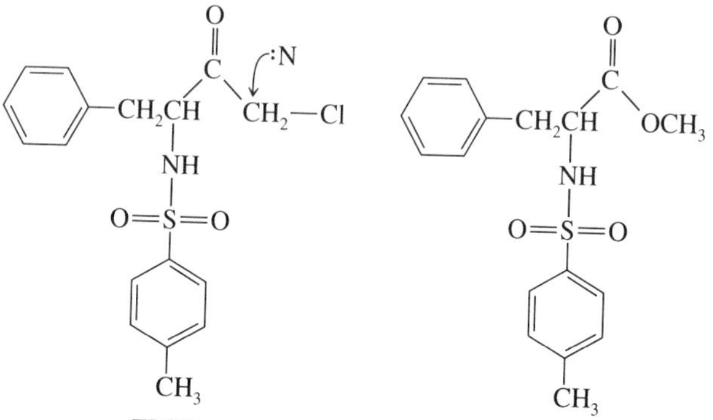

TPCK

tos-L-Phenylalanine
methyl ester

Halomethyl ketones and acids are known to react with thiols and imidazoles. TPCK reacts far more rapidly with chymotrypsin than it does with normal histidine-containing peptides because of its high reactivity as an affinity label. This can be seen in Table 9.2 for an analogous chloromethyl ketone. In addition to this important diagnostic feature, the irreversible inhibition of chymotrypsin by TPCK has four other characteristic features:[1,4]

1. The rate of inactivation is inhibited by reversible inhibitors or substrates of the enzyme.

2. The relative rates of inhibition as a function of pH at low inhibitor concentrations ($[I] \ll K_I$) follow a bell-shaped curve, with the same pK_a values as found for the pH dependence of k_{cat}/K_M for the hydrolysis of substrates.

Table 9.2 *High reactivity in affinity labeling*[a]

Reaction	Second-order rate constant ($s^{-1} M^{-1}$)
Cbz-L-phenylalanine chloromethyl ketone + chymotrypsin (alkylation of His-57)	69
Cbz-L-phenylalanine chloromethyl ketone + acetylhistidine	4.5×10^{-5}

[a] From E. N. Shaw and J. Ruscica, *Archs. Biochem. Biophys.* **145**, 484 (1971).

3. The inactivated enzyme has 1 mol of inhibitor covalently bound per mol of active sites. (This was found by using ^{14}C-labeled inhibitor.)
4. The inhibition follows saturation kinetics (with a "K_M" of about 0.3 mM).[5]

(It was later shown that the site of modification is at His-57.)

Characteristics 1, 2, 3, and 4 are diagnostic tests for an affinity label that is modifying the active-site group whose ionization controls activity. The saturation kinetics (feature 4) show that the label binds to the enzyme—although these may not be observed if the K_I is higher than the solubility of the label. Competitive inhibition by substrates, etc. (feature 1) suggests that the binding is at the active site. The 1:1 stoichiometry (feature 3) shows that the modification is selective. The pH dependence (feature 2) gives important evidence. It was shown in Chapter 5 that the pH dependence of V_{max}/K_M or k_{cat}/K_M gives the pK_a's of the free-enzyme groups that are involved in catalysis and binding of the substrate. Similarly, the pH dependence of k_I/K_I (or the relative rates of reaction at $[I] \ll K_I$, since the rate is proportional to k_I/K_I under this condition) gives the pK_a's of the free-enzyme groups that are involved in binding and reacting with the inhibitor. Thus, identical sets of pK_a's should be obtained from the pH dependence of k_{cat}/K_M and k_I/K_I if the groups that are involved in catalysis are also those that react with the affinity label. The pH dependence of k_I gives the pK_a's of the enzyme–inhibitor complex just as k_{cat} gives the pK_a of the enzyme–substrate complex (subject to the provisos of Chapter 5).

Enzymes do have some tricks up their sleeves (or rather, in their pockets) to spring on enzymologists. For example, the affinity label 1,2-anhydro-D-mannitol 6-phosphate labels a glutamate residue in glucose 6-phosphate isomerase.[6] It shows saturation kinetics, competitive inhibition by substrates, and 1:1 stoichiometry, and the pH dependence of k_I gives pK_a values for the enzyme–inhibitor complex similar to those for enzyme–substrate complexes. However, a crystallographic study has provided some speculative evidence that the group modified was not the one aimed at, but another base that is catalytically important.[7] If this is correct, the unexpected result is more interesting than the expected one, since an important but previously unknown catalytic group has been located.

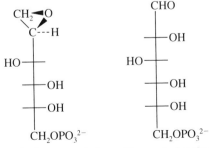

1,2-Anhydro-D-mannitol   Glucose 6-phosphate
6-phosphate

Another good example of the use of affinity labels involves pepsin, and is illustrated in Chapter 16, equations 16.28 and 16.29. The enzyme has two catalytically important aspartic acid residues, one ionized and the other un-ionized. The ionized carboxylate is trapped with an epoxide, which, of course, requires the reaction of a nucleophilic group. The un-ionized carboxyl is trapped with a diazoacetyl derivative of an amino acid ester:

$$\overset{+}{N_2}CHCONHCH(R)CO_2CH_3 \xrightarrow[N_2]{Cu(II)} Cu(II)\overset{..}{C}HCONHCH(R)CO_2CH_3 \qquad (9.3)$$

The reaction is catalyzed by cupric ions and presumably results from a copper-complexed carbene.[8] The electron-deficient carbene with only six electrons in its outer valence shell is known to add across the O—H bonds of un-ionized carboxyl groups to form the methyl ester.

Another general approach is the use of *photoaffinity labels*.[9-11] A compound that is stable in the absence of light but that is activated by photolysis is reversibly bound to an enzyme and photolyzed. The usual reagents are diazo compounds that when photolyzed give highly reactive carbenes, or azides that give highly reactive nitrenes:

$$RC\overset{\displaystyle O}{\underset{\displaystyle CHN_2}{\diagup}} \xrightarrow{h\nu} RC\overset{\displaystyle O}{\underset{\displaystyle CH}{\diagup}} + N_2 \qquad (9.4)$$

$$R-N_3 \xrightarrow{h\nu} R-N: + N_2 \qquad . \qquad (9.5)$$

The photoaffinity labels have been useful in mapping out residues at the active sites of enzymes and the binding sites of proteins such as antibodies. The normal affinity labels are more useful for kinetic work, since they are selective for the basic and nucleophilic groups that are so prevalent in catalysis, and since they also give information on the pK_a's of the groups that are modified.

Many hundreds of affinity labels have been synthesized, the majority based on halomethyl ketones or epoxides (Table 9.3). They are listed each year in the *Specialist Periodical Reports: Amino-Acids, Peptides, and Proteins*, published by the Royal Society of Chemistry (U.K.).

C. Enzyme-activated irreversible inhibitors[12-15]

There is considerable interest in the design of highly specific irreversible enzyme inhibitors because of their potential use as therapeutic agents. Part of the research program of most pharmaceutical companies is the rational design of drugs based on mechanistic ideas from enzymology, biochemistry, and chem-

Table 9.3 *Some affinity labels[a]*

Enzyme	Affinity label	Substrate
Triosephosphate isomerase	$ICH_2CCH_2OPO_3^{2-}$ with $\parallel$ O	$HOCH_2CCH_2OPO_3^{2-}$ with $\parallel$ O
	 $CH_2CHCH_2OPO_3^{2-}$ (epoxide) [b]	$HCCH(OH)CH_2OPO_3^{2-}$ with $\parallel$ O
Lysozyme	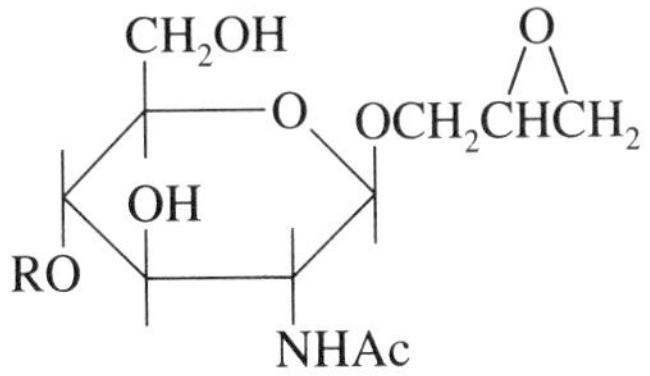 (sugar ring) CH_2OH, OCH_2CHCH_2 (epoxide), OH, RO, $NHAc$	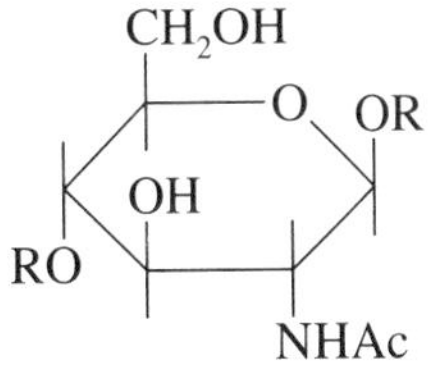 (sugar ring) CH_2OH, O OR, OH, RO, $NHAc$
Isoleucyl-tRNA synthetase	$BrCH_2CO$-Ile-tRNA^{Ile}	Ile-tRNA^{Ile}

[a] From F. C. Hartmann, *Biochem. Biophys. Res. Commun.* **33**, 888 (1968); S. G. Waley, J. C. Miller, I. A. Rose, and E. L. O'Connell, *Nature, Lond.* **227**, 181(1970); E. W. Thomas, J. F. McElvy, and N. Sharon, *Nature, Lond.* **222**, 485 (1969); D. V. Santi and W. Marchant, *Biochem. Biophys. Res. Commun.* **51**, 370 (1973).
[b] Also reacts with glyceraldehyde 3-phosphate dehydrogenase. S. McCaul and L. D. Byers, *Biochem. Biophys. Res. Commun.* **72**, 1028 (1976).

istry. There are serious drawbacks to the use of the affinity labels in this context, since they are, in general, reactive chemical compounds that owe their selectivity to productive binding at the active site of their target enzyme. Although this selectivity is perfectly adequate for specifically modifying the active sites of purified proteins, there are too many possible side reactions for the use of such inhibitors *in vivo*. In particular, the high reactivity of —SH groups in small biological molecules and in other proteins could either destroy an affinity label or cause side effects. There is, however, another class of irreversible inhibitors which, in addition to utilizing the binding specificity of their target enzyme, specifically utilize its catalytic apparatus for chemical activation, with the result that a normally innocuous reversible inhibitor is converted into a powerful irreversible inhibitor. These inhibitors, which are chemically unreactive in the absence of the target enzyme, are commonly called suicide inhibitors[14] (since the enzyme appears to "commit suicide") or k_{cat} inhibitors[13] (since they use the active-site residues for activation). Other terminologies applied to them are mechanism-based, trojan-horse,[16] and enzyme-activated substrate (EASI)[17] inhibitors. They react with the enzyme according to the following scheme:

$$E + I \underset{K_I}{\rightleftharpoons} E \cdot I \xrightarrow{k_{cat}} E \cdot I^* \begin{array}{c} \xrightarrow{k_1} E-I^* \\ \xrightarrow{k_{diss}} E + I^* \end{array} \tag{9.6}$$

An effective suicide inhibitor must have these characteristics:

1. The inhibitor must be chemically unreactive in the absence of enzyme.
2. It must be activated specifically by its target enzyme.
3. It must in its activated form react more rapidly with its target enzyme than it dissociates (i.e., $k_1 \gg k_{diss}$). (This criterion may be relaxed if I* is destroyed by water much faster than it reacts with other enzymes.)

In the preceding section, four diagnostic tests of affinity labeling were listed (inactivation inhibited by substrates, pH dependence of inactivation similar to that of catalysis, labeled inhibitor covalently bound in 1:1 stoichiometry, and saturation kinetics obeyed). The same criteria may be used to diagnose suicide inhibition. In addition, tests must be made to detect any diffusion of the activated intermediate I* into solution. For example, the addition of —SH reagents that rapidly react with electrophiles and hence scavenge them should not slow down the rate of reaction. The suicide inhibitor should not, in any case, react with the thiol at an appreciable rate in the absence of enzyme.

Most suicide inhibitors are based on the generation of an intermediate that has conjugated double bonds and that is susceptible to a Michael addition reaction. A nucleophilic group on the enzyme may then be alkylated by the intermediate (equation 9.7). The conjugated intermediate is usually generated by proton-abstraction by a basic group on the enzyme.

$$E: \overset{R}{\underset{X}{\diagdown}} \longrightarrow E- \overset{BH^+}{\underset{X^-}{\diagdown}} R \longrightarrow E \overset{R}{\underset{X}{\diagdown}} \tag{9.7}$$

(X = O, N, or S)

Often, the basic group that is responsible for the proton abstraction is also the nucleophilic group in the Michael addition. Thus, most of the suicide inhibitors made so far have been aimed at enzymes that catalyze the formation of carbanions or carbanion-like intermediates. Suicide inhibitors are typically based on acetylenic compounds (as in equation 9.8), β, γ-unsaturated compounds (as in equation 9.9), or β-halo compounds (as in equation 9.10). (The α protons in such compounds are acidic because the negative charge in the carbanion is delocalized by the conjugation with X.)

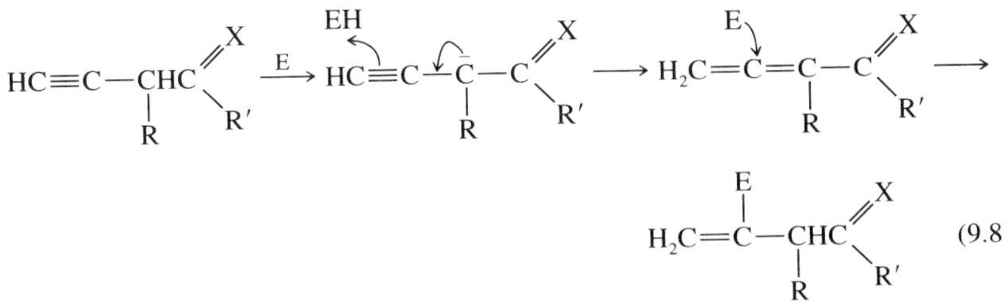
(9.8)

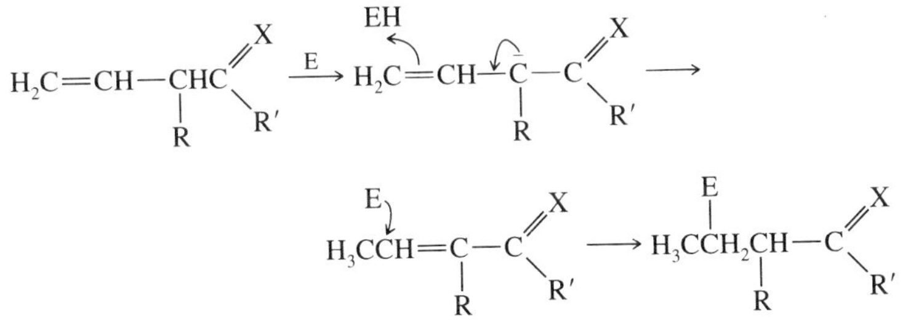
(9.9)

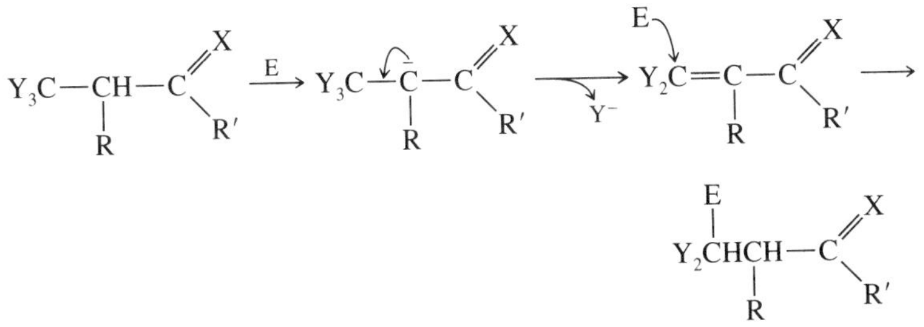
(9.10)

(X = O, N, or S; Y = F or Cl)

A classic example of suicide inhibition is that of β-hydroxyl-decanoyl-dehydrase by 3-decenoyl-*N*-acetylcysteamine.[12] The enzyme catalyzes the reaction

$$CH_3(CH_2)_6CHOHCH_2CO{-}NAC \rightleftharpoons \underset{\text{(cis and trans)}}{CH_3(CH_2)_6CH{=}CHCO{-}NAC} \qquad (9.11)$$

where NAC = $SCH_2CH_2NHCOCH_3$. The suicide inhibitor is activated by the enzyme, which is then alkylated on the active-site histidine:

$$CH_3(CH_2)_5C{\equiv}CCH_2CO{-}NAC \xrightarrow{E} CH_3(CH_2)_5CH{=}C{=}CHCO{-}NAC \longrightarrow CH_3(CH_2)_5CH{=}\underset{\underset{E}{|}}{C}CH_2CO{-}NAC \qquad (9.12)$$

It should be noted that there is a kinetic isotope effect on the normal reaction (9.11) when the α-deuterated compound is used as the substrate. A similar effect is found when the deuterated suicide inhibitor is used. Thus, both reactions involve a proton transfer in the rate-determining step of the reaction. It has also been shown that a sample of the allenic intermediate that is prepared chemically does in fact irreversibly inhibit the enzyme.[18]

1. Pyridoxal phosphate–linked enzymes

Enzymes containing pyridoxal phosphate are prime targets for suicide inhibition because the chemistry is so naturally suitable. As discussed in Chapter 2, section C2, the pyridoxal ring acts as an electron sink that facilitates the formation of carbanions and also forms part of an extended system of conjugated double bonds. For example, vinyl glycine, $CH_2{=}CHCH(NH_3{^+})CO_2{^-}$, condenses with the pyridoxal phosphate of aspartate aminotransferase to form a Schiff base, as described in Chapter 2, equation 2.42.[19] The α proton may be abstracted (as in equation 2.43) so that the isomerization shown in equation 9.13 readily occurs.

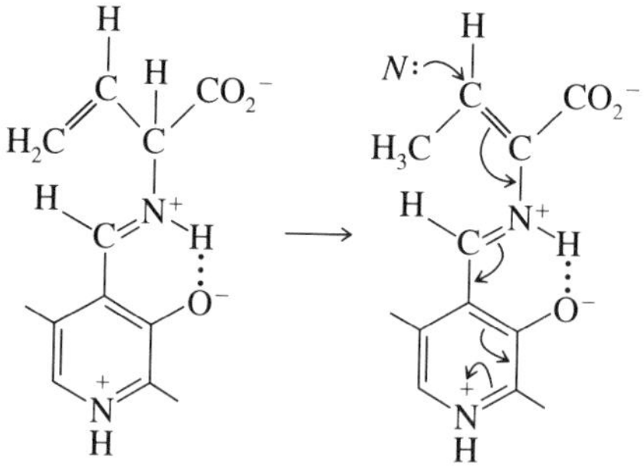

$$(9.13)$$

Michael addition may then take place at the terminus of the conjugated system. Lysine-258 at the active site is alkylated by the suicide inhibitor. The reaction is similar to that of equation 9.7, except that in the Michael addition the pyridoxal ring acts as the electron sink, rather than the N atom originating from the vinyl glycine.

The product of certain elimination reactions, e.g., that of serine dehydratase (Chapter 2, equation 2.47), is structure 9.14.

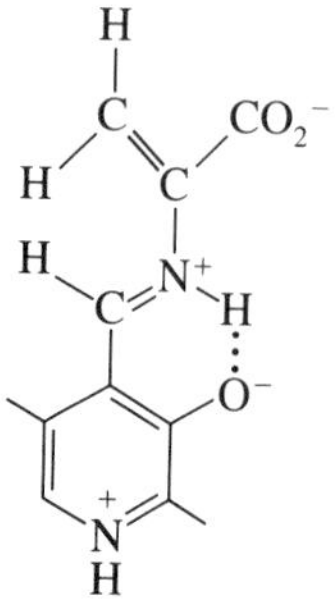

$$(9.14)$$

This is similar to the reactive intermediate of equation 9.13; here, though, the enzyme has evolved to deal with the intermediate and is not inactivated. However, a similar but far more reactive intermediate can be generated from the reaction of the enzyme with β-trifluoroalanine (equation 9.15), which will alkylate the enzyme and inactivate it.[20]

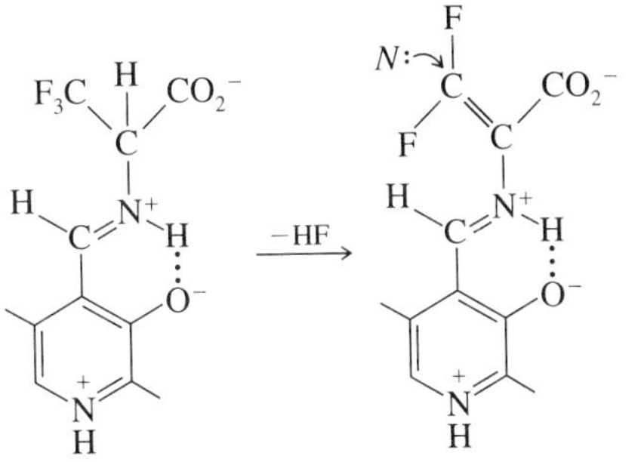

(9.15)

The intermediate 9.14 is probably generated during the suicide inhibition of β-aspartate decarboxylase,[21] aspartate aminotransferase,[22] and alanine racemase[23,24] by β-chloroalanine. These enzymes are inactivated by the intermediate, since they have not evolved to cope with it during the normal course of reaction.

2.   Monoamine oxidases and flavoproteins

Another means of producing Michael acceptors is by the *oxidation* of unsaturated amines of alcohols, such as allyl amine (equation 9.16).

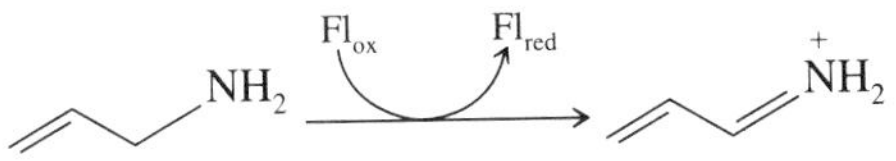

(9.16)

(Fl = flavin)

These reactions, which have provided a means of inhibiting the flavin-linked monoamine oxidases, enable us to end on a clinical note. The monoamine oxidases are responsible for the deamination of monoamines such as adrenaline, noradrenaline, dopamine, and serotonin, which act as neurotransmitters. Imbalances in the levels of monoamines cause various psychiatric and neurological disorders: Parkinson's disease is associated with lowered levels of dopamine, and low levels of other monoamines are associated with depression. Inhibitors of monoamine oxidases may consequently be used to treat Parkinson's disease and depression. The flavin moiety is covalently bound to the enzyme by the thiol group of a cysteine residue (equation 9.17). The acetylenic suicide inhibitor *N,N*-dimethyl-propargylamine inactivates monoamine oxidases by alkylating the flavin on N-5.[25] A likely mechanism for the reaction is the Michael addition of the N-5 of the reduced flavin to the acetylenic carbon:[2]

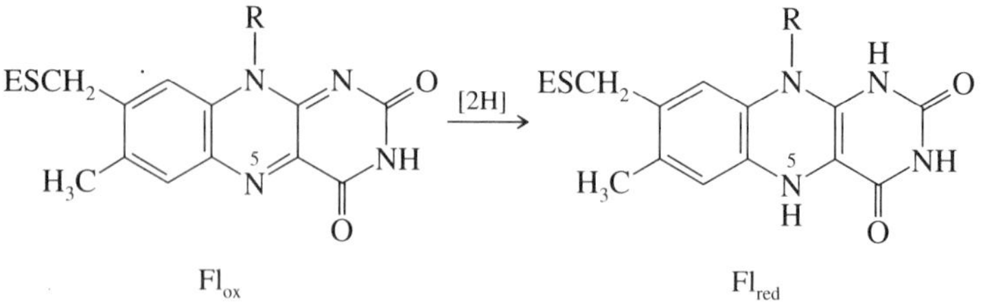
ESCH2
H3C
R
N
N
O
NH
5
N
O
[2H]
ESCH2
H3C
R
N
H
N
O
NH
5
N
H
O
Flox
Flred

(9.17)

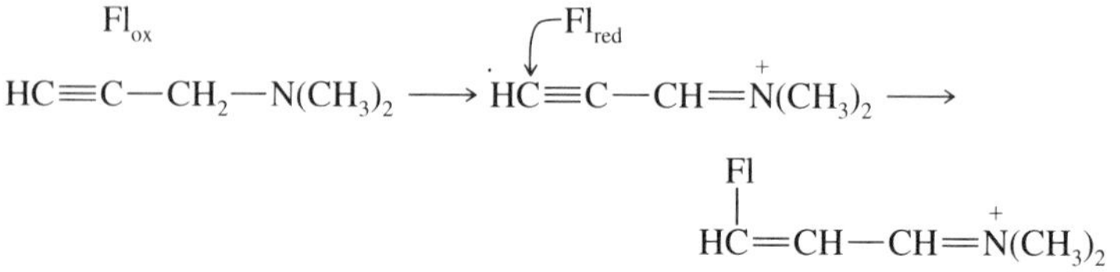
Flox
HC≡C—CH2—N(CH3)2
Flred
HC≡C—CH=N(CH3)2
Fl
HC=CH—CH=N(CH3)2

(9.18)

One drug that is used to treat both Parkinson's disease and depression is (−)deprenyl. It, too, is an acetylenic suicide inhibitor that inhibits the enzyme by binding covalently to the flavin.[27]

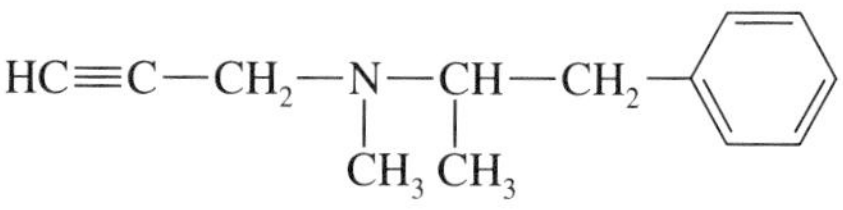
HC≡C—CH2—N—CH—CH2
CH3 CH3

(−) Deprenyl

D. Slow, tight-binding inhibition

Many inhibitors with very low dissociation constants appear to have a slow onset of inhibition when they are added to a reaction mixture of enzyme and substrate. This was once interpreted as the inhibitors having to induce a slow conformational change in the enzyme from a weak binding to a tight binding state. But in most cases, the slow binding is an inevitable consequence of the low concentrations of inhibitor used to determine its K_I. For example, consider the inhibition of trypsin by the basic pancreatic trypsin inhibitor. K_I is 6×10^{-14} M and the association rate constant is $1.1 \times 10^6\ s^{-1}\ M^{-1}$ (Table 4.1). To determine the value of K_I, inhibitor concentrations should be in the range of K_I, where the observed first-order rate constant for association is $(6 \times 10^{-14}\ M) \times (1.1 \times 10^6\ s^{-1}\ M^{-1})$; that is, $6.6 \times 10^{-8}\ s^{-1}$. The half-life is $(0.6931/6.6) \times 10^8$ s, which is more than 17 weeks.

1. Kinetics of slow, tight-binding inhibition[28]

There are two basic mechanisms: mechanism A, competitive inhibition (9.19);

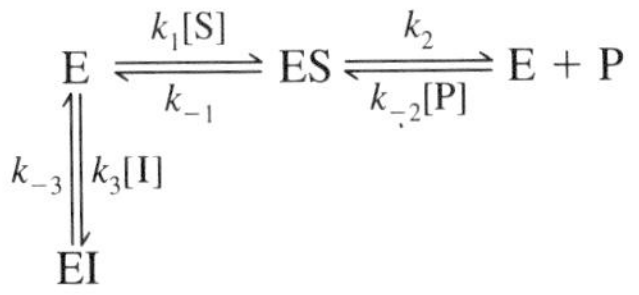

$$\begin{array}{c} E \underset{k_{-1}}{\overset{k_1[S]}{\rightleftharpoons}} ES \underset{k_{-2}[P]}{\overset{k_2}{\rightleftharpoons}} E + P \\[2mm] k_{-3} \big\Vert k_3[I] \\[2mm] EI \end{array}$$

$$(9.19)$$

and mechanism B, competitive inhibition with conformational change of EI complex (9.20).

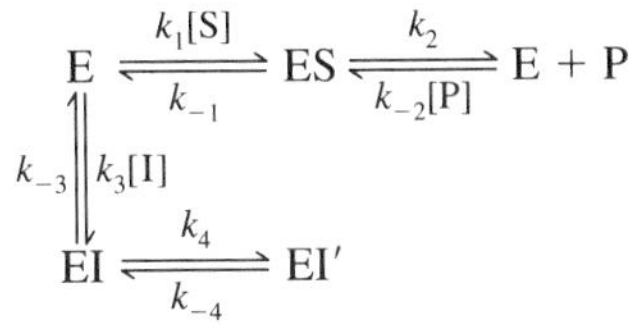

$$\begin{array}{c} E \underset{k_{-1}}{\overset{k_1[S]}{\rightleftharpoons}} ES \underset{k_{-2}[P]}{\overset{k_2}{\rightleftharpoons}} E + P \\[2mm] k_{-3} \big\Vert k_3[I] \\[2mm] EI \underset{k_{-4}}{\overset{k_4}{\rightleftharpoons}} EI' \end{array}$$

$$(9.20)$$

The kinetics of mechanism A are easy to solve when the ES complex is in rapid equilibrium with E and S and the concentration of I is not appreciably changed during the reaction. Phenomenologically, the reaction proceeds at a relatively fast rate in the absence of I; on the addition of I, the rate slows down in an exponential manner to the final steady state rate when the inhibitor is bound at equilibrium. For mechanism A, $[E] + [ES] + [EI] = [E]_0$, $[S][E] - K_M[ES] = 0$, and

$$-k_3[I][E] + k_{-3}[EI] = \frac{d[E]}{dt} = \frac{K_M}{[S]} \frac{d[ES]}{dt} \tag{9.21}$$

Solution of these simultaneous equations yields

$$v = v_s + (v_0 - v_s) \exp(-k_{obs}t) \tag{9.22}$$

where v is the observed rate of reaction at any time, v_0 is the initial rate (before the addition of I), and v_s is the rate in the steady state. The rate constant for the approach to equilibrium, $k_{obs}, = k_{-3}(1 + [I]/K_I + [S]/K_M)$, $K_I = k_{-3}/k_3$, and $K_M = (k_{-1} + k_2)/k_1$. The same equation may also be derived for mechanism B, but in this case $k_{obs} = k_{-4}(1 + [I]/K_I + [S]/K_M)/(k_3[I]/k_{-3})$, and $K_I = k_{-3}k_{-4}/k_2(k_4 + k_{-4})$. The standard equation 3.32 can be applied to the steady state rates to calculate K_I.

In practice, great care must be taken that the reaction is followed for at least 10 half-lives of the exponential phase to ensure that the steady state rate is reached. During that time, the substrate should not be so depleted that the rate falls off as a consequence or there is onset of product inhibition. Even so, it may still be difficult to distinguish between the kinetics of mechanism A and variants of mechanism B if the free enzyme and the two forms of enzyme inhibitor complex are in not in steady state equilibrium.[28]

References

1. G. Schoellmann and E. Shaw, *Biochem. Biophys. Res. Commun.* **7**, 36 (1962); *Biochemistry* **2**, 252 (1963).
2. B. R. Baker, W. W. Lee, E. Tong, and L. O. Ross, *J. Am. Chem. Soc.* **83**, 3713 (1961).
3. L. Wofsy, H. Metzger, and S. J. Singer, *Biochemistry* **1**, 1031 (1961).
4. E. Shaw, *The Enzymes* **1**, 91 (1970).
5. D. Glick, *Biochemistry* **7**, 3391 (1968).
6. E. L. O'Connell and I. A. Rose, *J. Biol. Chem.* **248**, 2225 (1973).
7. P. J. Shaw and H. Muirhead, *FEBS Lett.* **65**, 50 (1976).
8. R. L. Lundblad and W. H. Stein, *J. Biol. Chem.* **244**, 154 (1969).
9. A. Singh, E. R. Thornton, and F. H. Westheimer, *J. Biol. Chem.* **237**, 3006 (1962).
10. V. Chowdhry and F. M. Westheimer, *Ann. Rev. Biochem.* **48**, 293 (1979).
11. J. R. Knowles, *Accts. Chem. Res.* **5**, 155 (1972).
12. K. Bloch, *Accts. Chem. Res.* **2**, 193 (1969).
13. R. R. Rando, *Science* **185**, 320 (1974).
14. R. H. Abeles and A. L. Maycock, *Accts. Chem. Res.* **9**, 313 (1976).
15. C. Walsh, *Tetrahedron* **38**, 871 (1982).
16. F. M. Miesowicz and K. Bloch, *J. Biol. Chem.* **254**, 5868 (1979).
17. E. H. White, L. W. Jelinski, I. R. Politzer, B. R. Branchini, and D. F. Roswell, *J. Am. Chem. Soc.* **103**, 4231 (1981).
18. M. Morisaki and K. Bloch, *Bioorg. Chem.* **1**, 188 (1971).
19. H. Gehring, R. R. Rando, and P. Christen, *Biochemistry* **16**, 4832 (1977).
20. R. B. Silverman and R. H. Abeles, *Biochemistry* **15**, 4718 (1976).
21. E. W. Miles and A. Meister, *Biochemistry* **6**, 1735 (1967).
22. Y. Marin and M. Okamoto, *Biochem. Biophys. Res. Commun.* **50**, 1061 (1973).
23. J. M. Manning, N. E. Merrifield, W. M. Jones, and E. C. Gotschlich, *Proc. Natl. Acad. Sci. USA* **71**, 417 (1974).
24. E. Wang and C. T. Walsh, *Biochemistry* **17**, 1313 (1978).
25. A. L. Maycock, R. H. Abeles, J. L. Salach, and T. P. Singer, *Biochemistry* **15**, 114 (1976).
26. R. H. Abeles, in *Enzyme-activated irreversible inhibitors* (N. Seiler, M. J. Jung, and J. Koch-Weser, Eds.), Elsevier North-Holland, p. 4 (1978).
27. M. B. H. Youdim and J. I. Salach, in *Enzyme-activated irreversible inhibitors* (N. Seiler, M. J. Jung, and J. Koch-Weser, Eds.), Elsevier North-Holland, p. 235 (1978).
28. M. J. Sculley, J. F. Morrison, and W. W. Cleland, *Biochim. Biophys. Acta Protein Structure Molecular Enzymology* **1298**, 78 (1996).

Further reading

J. Eyzaguirre, ed., *Chemical modification of enzymes, active site studies, Ellis Horwood series in biochemistry*, Ellis Horwood Ltd. (1987).

B. I. Kurganov, ed., *Chemical modification of enzymes*, Nova Science Publishers (1996).

J. M. Walker, *The protein protocols handbook*, Humana Press (1996).

Conformational Change, Allosteric Regulation, Motors, and Work

10

Proteins can change their shape on binding ligands or being modified by enzymatic reactions, particularly by phosphorylation. Nature uses the induced conformational changes to control the activity of regulatory proteins and to interconvert mechanical work and physicochemical free energy. By such means, the activities of enzymes may be programmed to optimize the flux through metabolic pathways and to control such complex phenomena as regulation of the cell cycle. The coupling of conformational change and chemical reaction powers movement in the cell; the pumping of ligands against chemical gradients; and, in reverse, the synthesis of ATP from chemical potentials. This chapter focuses on enzymes as physicochemical machines and as sensors. We begin by discussing how the affinity of ligands may be modulated, using the binding of oxygen to hemoglobin as the archetypal example.

A. Positive cooperativity

Many proteins are composed of subunits and have multiple ligand-binding sites. In some cases the ligand-binding curves do not follow the Michaelis-Menten equation (Chapter 6, section D), but instead are *sigmoid*. This is illustrated in Figure 10.1, where the degree of saturation of hemoglobin with oxygen is plotted against the pressure of oxygen (curve *b*). The sigmoid curve may be compared with the hyperbolic curve that is expected from the Michaelis-Menten equation and that is found for the binding to myoglobin (curve *a*). Sigmoid curves are characteristic of the *cooperative* binding of ligands to proteins that have multiple binding sites. Hemoglobin, for example, is composed of four

289

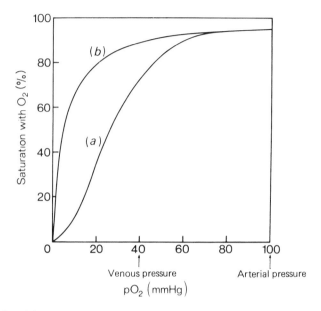

Figure 10.1 The oxygen-binding curves of (a) myoglobin and (b) hemoglobin.

polypeptide chains, each of which is similar to the single polypeptide chain of myoglobin. The hemoglobin binding curve may be fitted to four successive binding constants, as per the Adair equation (Table 10.1).[1] The surprising feature of this is that the affinity for the fourth oxygen that binds is between a hundred and a thousand times higher than for the first oxygen. The increase in affinity with increasing saturation cannot be explained by four noninteracting sites of differing affinities. If this were the case, the high-affinity sites would fill first, so that the partially ligated molecules would be of lower affinity than the free deoxyhemoglobin. The increase in affinity with increasing saturation is due instead to the sites *interacting*, so that binding at one site causes an increase in affinity at another.

A similar cooperativity of substrate binding is found to occur with some enzymes leading to sigmoid plots of v against [S]. These enzymes are usually the metabolic regulatory of control enzymes whose activities are subject to feedback inhibition or activation. The terminology of cooperative interactions stems from the studies on control.[2] The enzymes are termed *allosteric* (Greek, *allos* = other, *stereos* = solid or space), since the allosteric effector (the inhibitor or activator) is generally structurally different from the substrates and binds at its own separate site away from the active site. The term homotropic is sometimes used to denote the interactions among the identical substrate molecules, whereas the term heterotropic is used for the interactions of the allosteric effectors with the substrates.

Table 10.1 *Adair constants for the binding of O_2 to hemoglobin*[a]				
2,3-Diphosphoglycerate (mM)	K_1	K_2	K_3	K_4
		(mmHg)		
0	0.024	~0.074	~0.086	7.4
2.0	0.01	~0.023	~0.008	11.2

[a] At 25°C, pH 7.4, and 0.1-M NaCl. The Adair equation describes the following:

$$\text{Hb} \xrightleftharpoons{K_1,O_2} \text{HbO}_2 \xrightleftharpoons{K_2,O_2} \text{Hb(O}_2)_2 \xrightleftharpoons{K_3,O_2} \text{Hb(O}_2)_3 \xrightleftharpoons{K_4,O_2} \text{Hb(O}_2)_4$$

where

$$K_1 = \frac{[\text{HbO}_2]}{[\text{Hb}][\text{O}_2]}, K_2 = \frac{[\text{Hb(O}_2)_2]}{[\text{HbO}_2][\text{O}_2]}, \text{etc.}$$

[From I. Tyuma, K. Imai, and K. Shimizu, *Biochemistry* **12**, 1491 (1973).]

B. Mechanisms of allosteric interactions and cooperativity

The original attempts to explain the mechanism of cooperativity were based on hemoglobin. The best way to understand them is to consider the structures of deoxyhemoglobin and oxyhemoglobin. Hemoglobin is composed of two pairs of very similar chains, α and β, arranged in a symmetrical tetrahedral manner. The oxygen-binding sites, the hemes, are too far apart to interact directly. When deoxyhemoglobin is oxygenated, the overall tetrahedral symmetry is maintained but there are changes in quaternary structure (Figure 10.2).[3] The two α sub-

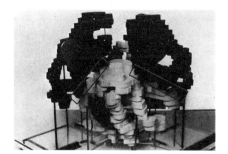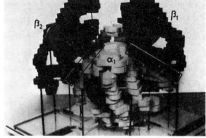

Figure 10.2 Balsa wood models of oxy- and deoxyhemoglobin. The hemes are represented by disks. Note the increased separation of the β subunits on deoxygenation. This is a historic photograph from the heroic early days of x-ray crystallography. [Courtesy of Dr. M. F. Perutz.]

units stay together but rotate relative to each other through 15°. There are no gross changes in tertiary structure that can be seen at low resolution. (Later, high-resolution studies revealed the small but important tertiary structural changes caused by ligand binding.) It is known from the binding measurements that deoxyhemoglobin has a low affinity for oxygen, whereas oxyhemoglobin has a high affinity.

1.   The Monod-Wyman-Changeux (MWC) concerted mechanism[4]

J. Monod, J. Wyman, and J.-P. Changeux showed that cooperativity can be accounted for in a very simple and elegant manner by assuming that a small fraction of deoxyhemoglobin exists in the quaternary oxy structure that binds oxygen more strongly. On the binding of 1 mol of oxygen, the concentration of the oxy structure is increased, since oxygen binds preferentially to it. When enough oxygen molecules are bound, the oxy form is sufficiently stabilized to be the major structure in solution, so that subsequent binding is strong. In order to simplify the mathematical equations and the physical concepts, the quaternary structure of the molecule is assumed to be always symmetrical; a particular partly ligated molecule is assumed to be either in the oxy state or in the deoxy state, so that mixed states do not occur (Figure 10.3). For this reason, the MWC model is often described as "concerted," "all or none," or "two-state."

The model may be generalized to cover other allosteric proteins if the following assumptions are made:

1.   The protein is an oligomer.

2.   The protein exists in either of two conformational states, T (= tense), the predominant form when the protein is unligated, and R (= relaxed); the two states are in equilibrium. They differ in the energies and numbers of bonds

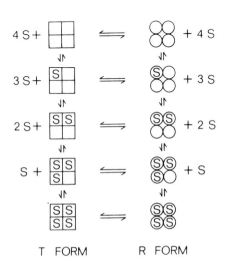

Figure 10.3 The MWC model for the binding of ligands to a tetrameric protein. S = substrate; T = tense conformation; and R = relaxed conformation.

T FORM   R FORM

between the subunits, with the T state being constrained compared with the R state.

3. The T state has a lower affinity for ligands.

4. All binding sites in each state are equivalent and have identical binding constants for ligands (the symmetry assumption).

The sigmoid binding curve of any allosteric protein can be calculated by using just three parameters: L, the allosteric constant, which is equal to the ratio $[T]/[R]$ for the unligated states; and K_T and K_R, the dissociation constants for each site in the T and R states, respectively.

a.   An explanation of control through allosteric interactions

The first achievement of the MWC theory was to provide a theoretical curve (based only on L, K_R, and K_T) that fitted the oxygen-binding curve of hemoglobin with high precision. But, even more impressively, the theory provided a very simple explanation for control. Monod, Wyman, and Changeux noted that it is a common feature of allosteric enzymes to exhibit cooperativity in v-vs.-[S] plots in the absence of their allosteric activators or inhibitors. They reasoned that if this cooperativity is due to binding and an R–T equilibrium, then an explanation of control could be that the effector alters the R–T ratio by preferentially stabilizing one of the forms. An activator functions by binding to the R state and increasing its concentration. An inhibitor binds preferentially to the T state and so causes the transition to the R state to be more difficult.

In extreme cases, an activator will displace the R–T equilibrium to such an extent that the R state predominates. Cooperativity is then abolished so that Michaelis-Menten kinetics will hold. This has been verified experimentally for several enzymes (Figure 10.4).

It is also predicted that an allosteric inhibitor should bind noncooperatively to an enzyme that binds its substrates cooperatively, since the inhibitor binds to the predominant T state. The converse should be true for activators binding to multiple binding sites in the R state.

The feature of the MWC model that is most open to criticism is the symmetry assumption. Clearly, changes in the constraints at the subunit interfaces on ligand binding must be mediated by changes in the tertiary structure of the subunit doing the binding. Because the model minimizes the number of intermediate states, it is only an approximation of reality. But this is also its virtue. The model provides a simple framework in which to rationalize experiments and explain phenomena. The predictions, such as the switch from sigmoid to Michaelis-Menten kinetics in control enzymes at sufficiently high activator concentrations, do not, in any case, depend on the intermediate states. Also, despite its simplicity, the theory accounts for the binding curves of oxygen to a wide series of mutant hemoglobins (see section D2).

The MWC is basically a structural theory. The hypothesis that there are constraints between the subunits in the T state has provided the basis of much of the

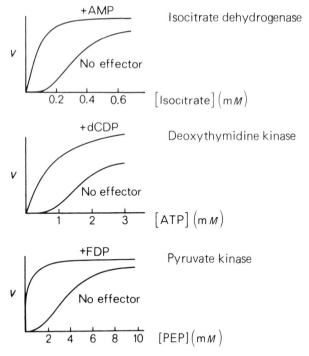

Figure 10.4 The abolition of positive cooperativity on the binding of allosteric effectors to some enzymes. Note the dramatic increases in activity at low substrate concentrations on the addition of adenosine monophosphate to isocitrate dehydrogenase, of deoxycytosine diphosphate to deoxythymidine kinase, and of fructose 1,6-diphosphate to pyruvate kinase; this shows how the activity may be "switched on" by an allosteric effector (PEP = phosphoenolpyruvate). [From J. A. Hathaway and D. E. Atkinson, *J. Biol. Chem.* **238**, 2875 (1963); R. Okazaki and A. Kornberg, *J. Biol. Chem.* **239**, 275 (1964); R. Haeckel, B. Hess, W. Lauterhorn, and K.-H. Würster, *Hoppe-Seyler's Z. Physiol. Chem.* **349**, 699 (1968).]

structural work by M. F. Perutz and others in elucidating the nature and the energies of these constraints in hemoglobin.

b. K systems and V systems

Control in allosteric enzymes may take two extreme forms.[4] In K (= binding) systems, the ones discussed so far, the substrate and the effector molecules have different affinities for the R and T states. The binding of an effector alters the affinity of the enzyme for the substrate, and *vice versa*. The R and the T states can have the same intrinsic value of k_{cat}, and activity is modulated by changes in affinity for the substrate. In V (= rate) systems, the substrate has the same affinity for both states, but one state has a much higher value of k_{cat}. The effector molecule binds preferentially to one of the two states, and so modulates activity by changing the equilibrium position between the two. As the substrate binds equally well to both states, there is no cooperative binding of substrate, and the binding of the allosteric effector does not alter the affinity of the enzyme toward the substrate.

2.   The Koshland-Némethy-Filmer (KNF) sequential model[5]

The KNF model avoids the assumption of symmetry but uses another simplifying feature. It assumes that the progress from T to the ligand-bound R state is a sequential process. The conformation of each subunit changes in turn as it binds the ligand, and there is no dramatic switch from one state to another (Figure 10.5). Whereas the MWC model uses a quaternary structural change, the KNF uses a series of tertiary structural changes.

The two assumptions in the KNF model are:

1. In the absence of ligands, the protein exists in one conformation.
2. Upon binding, the ligand induces a conformational change in the subunit to which it is bound. This change may be transmitted to neighboring vacant subunits via the subunit interfaces.

This model embodies Koshland's earlier idea of *induced fit*, according to which the binding of a substrate to an enzyme may cause conformational changes that align the catalytic groups in their correct orientations.

Using these assumptions, it is possible to describe the binding of oxygen by four successive binding constants. This is formally equivalent to the Adair equation; the KNF model may be considered as a molecular interpretation of that equation. In general, the number of constants required is equal to the number of binding sites, unlike the situation in the MWC model, which always uses three.

In sacrificing simplicity, the KNF model is more general and is probably a better description of some proteins than is the MWC model. In turn, the explanation of phenomena is often somewhat more complicated.

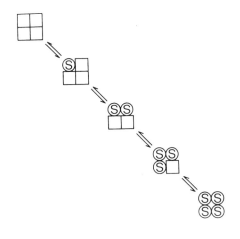

Figure 10.5 The KNF model for the binding of ligands to a tetrameric protein.

3. The general model[6]

M. Eigen has pointed out that the MWC and KNF models are limiting cases of a general scheme involving all possible combinations (Figure 10.6). The scheme is more complicated than the "chessboard" illustrated in the figure, since, for reasons of symmetry, there are 44 possible states for the hemoglobin case. The KNF model moves across the chessboard like a bishop confined to the long diagonal, and the MWC like a rook confined to the perimeters. The general model, combining both the MWC and KNF extremes, along with dissociation of the subunits, etc., has been analyzed.[7] But the results are too complex for general use, and it is far more convenient to interpret experiments in terms of the MWC and KNF simplifications.

4. Nested cooperativity

The molecular chaperone GroEL (Chapter 19, section I) is a 14-mer with two stacked 7-membered rings. There are dramatic changes in structure on binding ATP (see Figure 19.20). The changes exhibit nested cooperativity: interactions within a ring have positive cooperativity and follow the MWC model, but there is negative cooperativity between the two rings, according to the KNF model.[8]

C. Negative cooperativity and half-of-the-sites reactivity[9,10]

Some enzymes bind successive ligand molecules with decreasing affinity. As pointed out before, this can be explained by the presence of binding sites of differing affinities so that the stronger are occupied first. But in many cases, decreasing affinity is found with oligomeric enzymes composed of identical subunits. In two examples, the tyrosyl-tRNA synthetase[11] and the glyceraldehyde

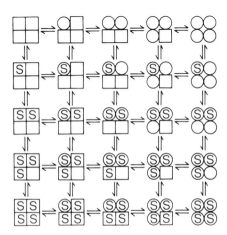

Figure 10.6 Eigen's general scheme for the binding of ligands to a tetrameric protein. The columns on the extreme left and right represent the MWC simplification. The diagonal from top left to bottom right represents the KNF simplification.

3-phosphate dehydrogenase[12] from *Bacillus stearothermophilus*, x-ray diffraction studies on the crystalline enzymes show that the subunits are arranged symmetrically so that all sites are initially equivalent. Yet, the dimeric tyrosyl-tRNA synthetase binds only 1 mol of tyrosine tightly; the binding of the second remains undetected even at millimolar concentrations of tyrosine.[13,14] The 4 mol of NAD^+ that bind to the rabbit muscle glyceraldehyde 3-phosphate dehydrogenase do so with increasing dissociation constants of $< 10^{-11}$, $< 10^{-9}$, 3×10^{-7}, and $3 \times 10^{-5} M$.[15,16] Similar changes in affinity are found for the bacterial enzyme (see Chapter 16, section A4b). This antagonistic binding of molecules is known as negative cooperativity or anticooperativity.

Negative cooperativity cannot be accounted for by the MWC theory; the binding of the first ligand molecule can only stabilize the high-affinity state and cannot increase the proportion of the T state. The KNF model accounts for negative cooperativity by the binding of the ligand to one site, causing a conformational change that is transmitted to a vacant subunit (assumption 2 of the KNF model). Negative cooperativity is thus a diagnostic test of the KNF model.

A related phenomenon is half-of-the-sites or half-site reactivity, by which an enzyme containing $2n$ sites reacts (rapidly) at only n of them (Table 10.2). This can be detected only by pre–steady state kinetics. The tyrosyl-tRNA synthetase provides a good example, in that it forms 1 mol of enzyme-bound tyrosyl adenylate with a rate constant of 18 s^{-1}, but the second site reacts 10^4 times more slowly.[13] However, as will be seen in Chapter 15, section J2b, protein engineering studies on the tyrosyl-tRNA synthetase unmasked a pre-existing asymmetry of the enzyme in solution.

Half-of-the-sites reactivity is inconsistent with the simple MWC theory, since the sites lose their equivalence and symmetry is lost.

A cautionary note must be injected at this point. The diagnosis of half-of-the-sites reactivity depends on knowledge of the exact concentration of binding sites on the enzyme. This depends on determination of the precise concentration of the protein and its purity. Similarly, a heterogeneous preparation of enzyme containing molecules of differing affinities for the ligand will give a binding curve similar to that of negative cooperativity. Failure to check these points has led to spurious reports about subunit interactions. A further artifact that may be misinterpreted as half-of-the-sites reactivity occurs in the reactions of the lactate and alcohol dehydrogenases, and is discussed in Chapter 15: an enzyme-bound intermediate does not appear to accumulate fully because of an unfavorable equilibrium constant.

D. Quantitative analysis of cooperativity

1. The Hill equation: A measure of cooperativity[17]

Consider a case of completely cooperative binding: An enzyme contains n binding sites and all n are occupied simultaneously with a dissociation constant K.

Enzyme	Reaction	Number of subunits	Ref.
Acetoacetate decarboxylase	Inactivation of active-site lysine	12	1
Aldolase	Partial reaction with fructose 6-phosphate	2	2
Aminoacyl-tRNA synthetases (some)	Biphasic formation of aminoacyl adenylate	2	3
Cytidine triphosphate synthetase	Stoichiometry of irreversible inhibition	4	4
Glyceraldehyde 3-phosphate dehydrogenase	Reaction with nonphysiological substrates; *but* full site reactivity with physiological	4	5
Aspartate transcarbamoylase	CTP binding Carbamoyl phosphate binding	6 (regulatory) 6 (catalytic)	5
Glutamine synthetase	Irreversible inhibition	8	6

Table 10.2 *Some enzymes showing half-of-the-sites reactivity*

1 D. E. Schmidt, Jr., and F. H. Westheimer, *Biochemistry* **10,** 1249 (1971), and references therein.
2 O. Tsolas and B. L. Horecker, *Archs. Biochem. Biophys.* **173,** 577 (1976).
3 R. S. Mulvey and A. R. Fersht, *Biochemistry* **15,** 243 (1976).
4 A. Levitzki, W. B. Stallcup, and D.. Koshland, *Biochemistry* **10,** 3371 (1971).
5 F. Seydoux, O. P. Malhotra, and S. Bernhard, *C. R. C. Crit Rev. Biochem.* 227 (1974).
6 S. S. Tate and A. Meister, *Proc. Natl. Acad. Sci. USA* **68,** 781 (1971).

We have

$$E + nS \rightleftharpoons ES_n \tag{10.1}$$

and

$$K = \frac{[E][S]^n}{[ES_n]} \tag{10.2}$$

The degree of saturation Y is given by

$$Y = \frac{[ES_n]}{[E]_0} \tag{10.3}$$

and

$$1 - Y = \frac{[E]}{[E]_0} \qquad (10.4)$$

Equations 10.2 and 10.4 may be manipulated to give

$$\log \frac{Y}{1 - Y} = n \log[S] - \log K \qquad (10.5)$$

A similar equation called the Hill plot (equation 10.6) is found to describe satisfactorily the binding of ligands to allosteric proteins in the region of 50% saturation (10 to 90%) (Figure 10.7).

$$\log \frac{Y}{1 - Y} = h \log[S] - \log K \qquad (10.6)$$

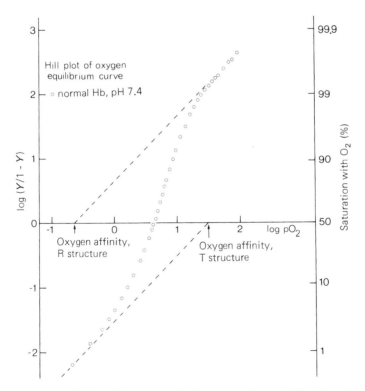

Figure 10.7 A Hill plot of the oxygen-binding curve of hemoglobin. [From J. V. Kilmartin, K. Imai, and R. T. Jones, in *Erythrocyte structure and function*, Alan R. Liss, 21 (1975).]

Outside this region, the experimental curve deviates from the straight line. The value of h obtained from the slope of equation 10.6 in the region of 50% saturation is known as the *Hill constant*. It is a measure of cooperativity. The higher h is, the higher the cooperativity. At the upper limit, h is equal to the number of binding sites. If $h = 1$, there is no cooperativity; if $h > 1$, there is positive cooperativity; if $h < 1$, there is negative cooperativity.

The Hill equation may be extended to kinetic measurements by replacing Y by the rate v, as in

$$\log \frac{v}{V_{max} - v} = h \log[S] - \log K \tag{10.7}$$

2. The MWC binding curve[4]

Let the dissociation constant of the ligand from the R state be K_R and that from the T state be K_T. Then c is defined by

$$c = \frac{K_R}{K_T} \tag{10.8}$$

The R state with x ligand molecules bound is termed R_X, and the equivalent T state, T_X. The allosteric constant L is then defined by

$$L = \frac{[T_0]}{[R_0]} \tag{10.9}$$

The fraction of the protein that is bound with ligand, Y, is calculated from the mass balance equations. Recall that it is necessary to "statistically correct" the binding constants. For example, the dissociation constant for the first O_2 binding to hemoglobin is $K_T/4$, since there are four sites to which the ligand may bind, but there is only one site for dissociation when it is bound. Similarly, the dissociation constant for the second molecule that binds is $2K_T/3$, since there are three sites to which it can bind, but once it is bound there are two bound sites that can dissociate.

Y for a protein containing n sites is given by

$$Y = \frac{([R_1] + 2[R_2] + \cdots + n[R_n]) + ([T_1] + 2[T_2] + \cdots + n[T_n])}{n([R_0] + [R_1] + \cdots + [R_n] + [T_0] + [T_1] + \cdots + [T_n])} \tag{10.10}$$

Solving equation 10.10 in terms of L and c, and using for convenience

$$\alpha = \frac{[S]}{K_R} \tag{10.11}$$

gives

$$Y = \frac{Lc\alpha(1 + c\alpha)^{n-1} + \alpha(1 + \alpha)^{n-1}}{L(1 + c\alpha)^n + (1 + \alpha)^n} \qquad (10.12)$$

The fraction of protein in the R state, R, may be similarly derived:

$$R = \frac{(1 + \alpha)^n}{L(1 + c\alpha)^n + (1 + \alpha)^n} \qquad (10.13)$$

According to the MWC theory, the saturation curve for any oligomeric protein composed of n protomers is defined by only three unknown parameters and the concentration of ligand (i.e., L, K_R, K_T, and [S], with the latter three disguised as c and α in equation 10.12).

Some values of L and c obtained from the computer fitting of equation 10.12 to the binding curves of some proteins are listed in Table 10.3. (Note: This does not imply that the structural changes follow the MWC model, since the KNF model also predicts the same binding curve).

a. The dependence of the Hill constant on L and c[18]

The value of h may be calculated from a computer analysis of equation 10.12. It is found that a plot of h against L at a constant value of c gives a bell-shaped

---

Table 10.3 *Allosteric constants for some proteins*

Protein	Ligand	Number of binding sites(n)	Hill constant (h)	L	c	Ref.
Hemoglobin	O_2	4	2.8	3×10^5	0.01	1
Pyruvate kinase (yeast)	Phosphoenol-pyruvate	4	2.8	9×10^{3a}	0.01^a	2
Glyceraldehyde 3-phosphate dehydrogenase (yeast)	NAD$^+$	4	2.3	60	0.04	3

^{a}Estimated by the author.
1 S. J. Edelstein, *Nature, Lond.* **230,** 224 (1971).
2 R. Haeckel, B. Hess, W. Lauterhorn, and K.-H. Würster, *Hoppe-Seyler's Z. Physiol. Chem.* **349,** 699 (1968); H. Bischofberger, B. Hess, and P. Röschlau, *Hoppe-Seyler's Z. Physiol. Chem.* **352,** 1139 (1971).
3 K. Kirschner, E. Gallego, I. Schuster, and D. Goodall, *J. Molec. Biol.* **58,** 29 (1971).

Figure 10.8 Variation of the Hill constant with L for a tetrameric protein. [From M. M. Rubin and J.-P. Changeux, *J. Molec. Biol.* **21**, 265 (1966).]

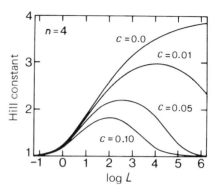

curve (Figure 10.8). The value of h is equal to 1 for L much greater or lower than c, and is a maximum when

$$L = c^{-n/2} \tag{10.14}$$

(where n is the number of binding sites). The reason for this behavior is that when L is low there is initially sufficient protein in the R state to give good binding, and when L is very high, the concentration of the R state is too small to contribute significantly to binding.

The bell-shaped curve is of particular interest in the analysis of how structural changes in a protein affect L. The Hill constants of a wide series of modified and mutant hemoglobins fit such a curve (Figure 10.9).[19]

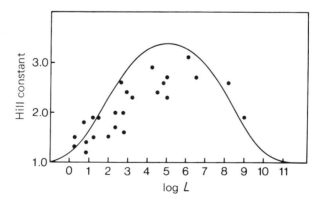

Figure 10.9 Variation of the Hill constant with L for the binding of oxygen to various mutant hemoglobins. [From J. M. Baldwin, *Progr. Biophys. Molec. Biol.* **29**, 3 (1975).]

3. The KNF binding curve

The MWC model gives a simple equation for Y because of the assumption of only two dissociation constants. The KNF model requires a different dissociation constant for every intermediate state, and there is no simple general equation for Y. As many variables are required as there are binding sites.

4. Diagnostic tests for cooperativity, and MWC versus KNF mechanisms

Cooperativity is determined from the value of h in the Hill plot or from the characteristic deviations in the straightforward saturation curves or Scatchard plots (Figure 10.10). The finding of negative cooperativity excludes the simple MWC theory, but positive cooperativity is consistent with both models. Analysis of the shape of the binding curve is generally also ambiguous, since the KNF and MWC models often predict similar shapes. In theory, measurement of the *rates* of ligand binding can distinguish between the two models.[6] The MWC model, for example, predicts fewer relaxation times since fewer states are involved. This approach has been applied with success to the glyceraldehyde 3-phosphate dehydrogenase from yeast, for which it has been shown that the kinetics of NAD^+ binding are consistent with the positive cooperativity due to an MWC model.[20] On the other hand, ligand-binding studies on hemoglobin suggest the importance of intermediate states. For example, the KNF model describes the overall binding curve of oxygen and carp hemoglobin better than does the MWC, since careful analysis shows that there are stages of negative cooperativity.[21] But sophisticated spectroscopic techniques have shown that the β subunits have a slightly

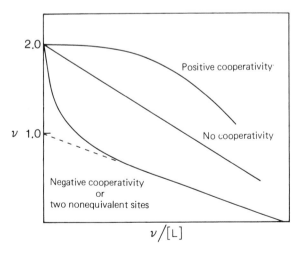

Figure 10.10 Plots of stoichiometry v against $v/[L]$ for the binding of ligand (L) to a dimeric protein.

higher oxygen affinity than do the α.[22] Such pre-existing asymmetry can, of course, give rise to apparent negative cooperativity without invoking the KNF scheme, as was discussed in section C.

The choice of model often depends on the experiments involved. Workers in the area of, say, the effects of structural changes on the oxygen affinity and Hill constant for hemoglobin prefer the MWC model because it is essentially a *structural* theory. It provides a simple framework for the prediction and interpretation of experiments. Application of the theory to the Hill constant and other *equilibrium* measurements gives very acceptable results. Kineticists prefer the KNF model, since the kinetic measurements are more sensitive to the presence of intermediates. There are more variables in the KNF theory and there is more flexibility in fitting data.

E.   Molecular mechanism of cooperative binding to hemoglobin

1.   The physiological importance of the cooperative binding of oxygen

Positive cooperativity is not a device for increasing the affinity of hemoglobin for oxygen; the association constant of free hemoglobin chains for oxygen is far higher than that for the binding of the first mole of oxygen to deoxyhemoglobin, and is about the same as for the binding to oxyhemoglobin. It is instead a means of lowering the oxygen affinity over a very narrow range of pressures, so as to allow the hemoglobin to be saturated with oxygen in the lungs and then to unload about 60% of the oxygen to the tissues. This can be done over a relatively narrow range of oxygen pressures, because of the steepness of the curve of saturation with oxygen against oxygen pressure in the region of 50% saturation. A simple Michaelis-Menten curve would require a far greater range of pressures. (The Hill constant is a measure of the steepness of the saturation curve at 50% saturation; h is close to 3 for the sigmoid curve of hemoglobin, but would be 1 for a hyperbolic curve.)

2.   Atomic events in the oxygenation of hemoglobin[3,23]

The two pairs of α and β subunits are arranged tetrahedrally around a twofold axis of symmetry (see Figure 10.2). The two α subunits make few contacts with each other, as is also the case with the two β subunits. The main interactions are across the $\alpha_1\beta_2$ and the $\alpha_1\beta_1$ interfaces. There is a cavity at the center of the molecule through which the axis passes. Organic phosphates such as 2,3-diphosphoglycerate, which is an allosteric effector, bind in this cavity in deoxyhemoglobin in a 1:1 stoichiometry (Figure 10.11).[24–26] The negatively charged organic phosphate sits between the two β subunits, forming four salt linkages with each (with Val-1, His-2, Lys-82, and His-143). In addition, deoxyhemoglobin is stabilized by four pairs of salt bridges formed by the C-terminal residues

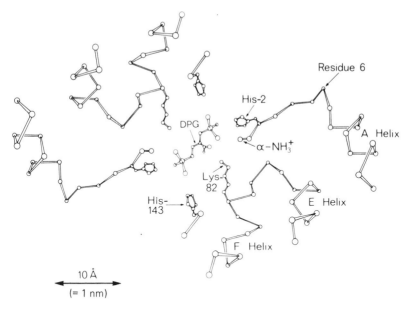

Figure 10.11 Binding of 2,3-diphosphoglycerate (DPG) between the β chains in the central cavity of human hemoglobin. [From A. Arnone and M. F. Perutz, *Nature, Lond.* **249**, 34 (1974).]

of the α and β chains: two pairs between the interfaces of the two α chains and two pairs between the α and β chains.

On oxygenation, the subunits rotate relative to each other by about 15°. The hemes change their angles of tilt by a few degrees and the helical regions move 2 to 3 Å (0.2 to 0.3 nm) relative to each other. The salt bridges between the subunits are broken. The α-amino groups of Val-1 of each β subunit in the central cavity move apart by 4 Å as the cavity narrows, expelling any organic phosphate that is bound. There is also a decrease in the hydrophobic area that is in contact at the interfaces of the subunits.[27] The "constraints" that were predicted between the subunits in the T state by Monod, Wyman, and Changeux are thus the salt bridges and additional hydrophobic interactions.

This explains why deoxyhemoglobin has a low affinity for oxygen compared with myoglobin or artificially prepared single chains.[3] On the binding of oxygen, energy must be expended to disrupt favorable interactions in the quaternary T structure. Salt bridges are broken and hydrophobic surfaces are exposed. The mode of action of the allosteric effectors is also clear. Organic phosphates bind strongly to a well-defined site in the T state and stabilize it relative to the R state.

The breaking of the salt bridges on oxygenation also explains the Bohr effect.[28] On oxygenation at physiological pH, there is a release of about 0.7 proton per heme. This is caused by the raised pK_a values of weak bases such as imidazoles and α-amino groups when they are bound to carboxylate ions

(Chapter 4). On disruption of the bridges, the pK_a's drop to the normal values, releasing protons.

It seems remarkable that the binding of O_2 causes such extensive structural changes, but nature has provided an ingenious trigger mechanism, based mainly on a small movement of the iron relative to the plane of the porphyrin.[29-32] The movement is a result of the change from 5 to 6 coordination and the accompanying change in the spin of the iron atom (Figure 10.12). Oxyhemoglobin is diamagnetic, with a short $Fe-N_P$ (P = porphyrin) bond distance of 1.99 Å; van der Waals repulsion between the porphyrin nitrogens and the proximal imidazole (His-F8 in Figure 10.12) is balanced by the repulsion between those nitrogens and the bound oxygen. As a result, the iron atom is kept close to or in the porphyrin plane. Deoxyhemoglobin is paramagnetic with a spin of S = 2. Repulsion between the occupied $d_{x^2y^2}$ orbitals of the iron and the π orbitals of the porphyrin lengthens $Fe-N_P$ to 2.07 Å, longer by 0.3 to 0.4 Å than the distance of the porphyrin nitrogens from the center of the ring. Moreover, van der Waals repulsion between the porphyrin nitrogens and the proximal imidazole is no longer balanced by a sixth ligand. Both factors conspire to pull the iron atoms out of the porphyrin plane by 0.4 Å and to dome the porphyrins toward the proximal histidines. The movements of the iron atoms cause movements of the helices to which they are linked, and these in turn are transmitted to the nearby subunit boundaries, thus triggering the change in quaternary structure. A second trigger is provided by a valine that obstructs the oxygen-binding site in the β

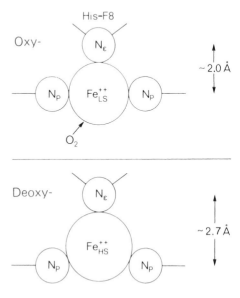

Figure 10.12 Movement of the imidazole ring of His-F8 on the binding of oxygen to deoxyhemoglobin. Steric repulsion between N_ε and the porphyrin nitrogens (N_P), and the slightly too large radius of Fe^{2+} (high-spin), force it out of the porphyrin plane. On the binding of O_2, the steric repulsion between it and the porphyrin in the opposite direction to that of N_ε, and the decrease in the radius of Fe^{2+}, force the iron atom to move close to the plane of the porphyrin ring, dragging the His-F8 with it.

subunits of the T structure. When this valine gives way, it sets in motion changes in tertiary structure that are concerted with those caused by the movement of the iron relative to the porphyrin plane.

The relationship between the structural and the energetic changes has been summarized.[23] It is in remarkable agreement with the original MWC hypothesis.[4] The oligomer has two different modes of close packing of subunit interfaces that can lead to two different quaternary structures. Deoxyhemoglobin is more stable in the T state because there are more hydrophobic, ionic, and van der Waals interactions. On the binding of ligands, strains are set up in the T state that eventually tip the energy balance in favor of the R state.

3. Chemical models of hemes

Chemical model systems, in particular the "picket fence"[33] (Figure 10.13), have been particularly useful in the studies on hemoglobin. The small chemical models may be crystallized and their structures determined with far higher precision than the structures of the actual proteins can be. Binding measurements may be made on the models without the complications that arise from the protein structure. The precise displacement of the iron atom from the heme and the geometry of iron–ligand bonds were first measured in the models.[29,33,34] The Fe—O_2 bond is bent, whereas the Fe—CO bond is linear (structures 10.15). The Fe—O_2 bond is bent at 156° in hemoglobin.[35]

$$
\begin{array}{cc}
\overset{\displaystyle O}{\underset{\displaystyle \diagup}{}} & \overset{\displaystyle O}{\underset{\displaystyle |}{}} \\
\overset{\displaystyle O}{\underset{\displaystyle |}{}} & \overset{\displaystyle C}{\underset{\displaystyle |}{}} \\
{>}\!Fe\!{<} & {>}\!Fe\!{<} \\
{|} & {|}
\end{array}
\qquad (10.15)
$$

Binding studies on models show a far higher affinity for CO than is found for myoglobin or hemoglobin.[36] The proteins have evolved to discriminate against

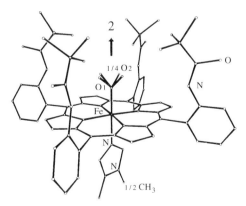

Figure 10.13 Perspective view of a "picket fence" model of the porphyrin ring system of hemoglobin. Note that the oxygen molecule is bound at an angle (there is a four-fold statistical distribution of the terminal O because of rotation about the Fe—O bond). [From J. P. Collman, R. R. Gagne, C. A. Reed, W. T. Robinson, and G. A. Rodley, *Proc. Natl. Acad. Sci. USA* **71**, 1326 (1974).]

the binding of CO by a combination of two factors. (1) There is steric hindrance against the binding of linearly bound ligands such as CO (more so in myoglobin than in hemoglobin, where the binding pocket is more open).[35] (2) Neutron diffraction studies on oxymyoglobin have located a hydrogen bond between the terminal oxygen of the $Fe—O_2$ (which is formally the superoxide $Fe^{3+}O_2^-$) and a histidine.[37] Hemoglobin also has the correct geometry for forming this bond.[35]

In the 30 years since the initial proposal, almost every aspect of Perutz's stereochemical mechanism for the cooperativity of binding oxygen has been challenged. But, as summarized by Perutz et al.,[3] it has emerged virtually unscathed. Even the complex binding kinetics can be fitted to the simple two-state Monod-Wyman-Changeux mechanism.[38]

F. Regulation of metabolic pathways

The rate of chemical flux through a metabolic pathway is often regulated by controlling the activity of a key enzyme in the pathway. The natural point of control is the enzyme whose catalytic activity is the *rate-limiting step* in the pathway. Such an enzyme may be identified by examination in the steady state *in vivo* of the ratio of the concentrations of its products to those of its substrates—the *mass action ratio*. Since the earlier enzymes in the pathway are producing the substrates of the rate-limiting enzyme faster than it can cope with them, and since its products are being rapidly consumed by the following enzymes in the pathway, the rate-limiting reaction is not in equilibrium. The mass action ratio is therefore less than that expected from the equilibrium concentrations, often by 2 to 4 orders of magnitude. In contrast, the mass action ratios of metabolic enzymes whose activities are not rate-limiting tend to be close to the equilibrium values: enzymes that have very high activities catalyze the forward and reverse reactions of their steps at rates far higher than that of the rate-limiting enzyme, but the *net* rate through each step—i.e., the rate of the forward reaction minus the rate of the reverse reaction—is the same for each step. The flux through metabolic pathways is now studied by control analysis.[39–44]

The activity of an enzyme can be controlled by regulation of its *concentration*: for example, by induction and repression of its synthesis in prokaryotes, by compartmentation of enzymes in organs or organelles, and by activation of a zymogen by covalent modification (as in the chymotrypsinogen-to-chymotrypsin conversion described in Chapter 16, section B). In this chapter, we are concerned with the regulation of rate by controlling the *activity* of an enzyme. The two principle means of doing this are:

1. Binding of allosteric effectors.
2. Covalent modification by phosphorylation and dephosphorylation of hydroxyl groups on amino acid side chains.

We shall now discuss one example of each process.

G. Phosphofructokinase and control by allosteric feedback

Glycolysis, the anaerobic degradation of glucose to pyruvate, generates ATP (equation 10.16). The glycolytic pathway is regulated to meet the cellular requirements for this important energy source.

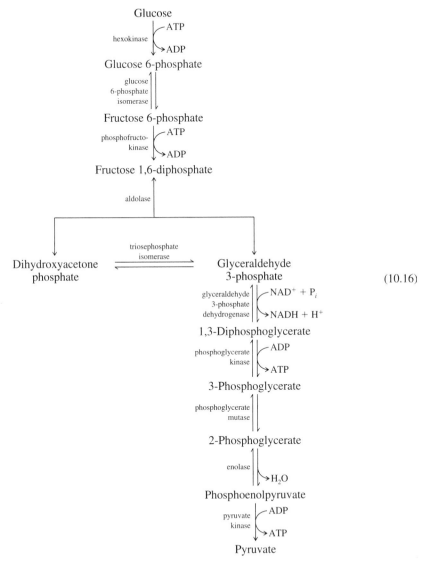

(10.16)

The most important control element is the allosteric enzyme phosphofructokinase. It catalyzes the phosphorylation of fructose 6-phosphate to fructose 1,6-diphosphate,

$$\text{Fructose 6-phosphate} + \text{ATP} \rightleftharpoons \text{fructose 1,6-diphosphate} + \text{ADP} \qquad (10.17)$$

(where both ATP and ADP are bound as their Mg^{2+} salts).

The reaction proceeds with inversion of structure,[45] and, as shown by site-directed mutagenesis,[46] involves the direct attack of the fructose —OH_1 group on the γ-phosphoryl group of ATP, using Asp-127 as a general base catalyst.

Phosphofructokinase from various organisms is subject to feedback control so that the activity of the enzyme is exquisitely sensitive to the energy level of the cell. Fructose 6-phosphate binds to the enzyme in a positive cooperative manner. The enzymes from eukaryotes are activated by AMP, ADP, and 3′,5′-cyclic AMP, and are inhibited by high concentrations of ATP and citrate, a subsequent product of glycolysis. The rate of glycolysis is thus high when the cell requires energy and low when it does not. The enzymes from the prokaryotes *Escherichia coli* and *Bacillus stearothermophilus* are regulated by the same principle but in a slightly less complex manner: they are activated by ADP and GDP, and inhibited by phosphoenolpyruvate only (Figure 10.14). Because protein crystallographic studies of the enzyme from *B. stearothermophilus* are well advanced, and because its kinetics are very similar to those of the enzyme from *E. coli*, which has been studied in even greater depth,[46–51] the remainder of the discussion is devoted to this pair. (The structural details refer to the thermophilic enzyme, whereas the kinetic data are from the *E. coli* enzyme.)

The enzyme from *B. stearothermophilus* is an α_4 tetramer of subunit M_r 33 900. Early kinetic studies indicated that the enzyme acts in a manner that is qualitatively consistent with an MWC two-state model. The enzyme acts as a K system; i.e., both states have the same value of k_{cat} but different affinities for the principle substrate. In the absence of ligands, the enzyme exists in the T state that binds fructose 6-phosphate more poorly than does the R state. In the absence of ADP, the binding of fructose 6-phosphate is highly cooperative, and $h = 3.8$. The positive homotropic interactions are lowered on the addition of the allosteric effector ADP, with h dropping to 1.4 at 0.8-mM ADP.[52] ADP thus binds preferentially to the R state. The allosteric inhibitor phosphoenolpyruvate binds preferentially to the T

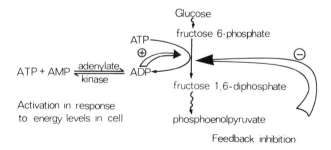

Figure 10.14 Phosphofructokinase and the control of glycolysis in *E. coli*. [From P. R. Evans, G. W. Farrants, and P. J. Hudson, *Phil. Trans. R. Soc.* **B293**, 53 (1981).]

state and stabilizes it. The binding of the effector GDP (which binds only to the effector site and not also to the active site, as does ADP) has been studied directly by equilibrium dialysis.[53] GDP binds with positive cooperativity to the free enzyme. The cooperativity is increased in the presence of the inhibitor phosphoenolpyruvate. The binding behavior is thus entirely in accord with the predictions of the MWC model (section B1).

The structures of the activated R states and inhibited T states have been solved, allowing the allosteric transition to be directly analyzed.[48-51] The R and T are best viewed as families of two different conformations, as each individual structure differs somewhat according to the nature of the bound ligand, but there are large differences between the two families.

1. The structure of the R state

The enzyme has been crystallized in the presence of fructose 6-phosphate. Since the crystals also bind ADP without major structural changes, they are presumably in the R state. The subunits are clearly divided into two domains. Each subunit makes close contacts with only two of the others. The binding of fructose 6-phosphate and of ADP was determined directly. Because ATP was unstable under the conditions of the crystallographic studies, the structure of its complex with the enzyme was determined by model building from extrapolation of ADP binding and by using the analogue AMP-PNP (i.e., ATP in which —NH— replaces the —O— between the β- and γ-phosphoryl groups). There are three binding sites per subunit: Sites A and B form the active site, binding the sugar and the nucleotide, respectively; site C is the effector site, binding ADP or phosphoenolpyruvate.

The active site with its catalytic groups is in an extended cleft between the two domains of a subunit. Interestingly, the 6-phosphate lies between two subunits, bound by His-249 and Arg-252 from its own subunit and by Arg-162 and Arg-243 from the neighbor. (The 1-hydroxyl is in a suitable position for nucleophilic attack on the γ-phosphoryl of ATP. The carboxylate of Asp-127 is in a likely position to act as a general-base catalyst.)

The effector site lies in a deep cleft between two subunits. Again, the phosphate residues are bound by positively charged side chains from both subunits.

These structural features imply a mechanism for the allosteric changes in a K system, given that a change of quaternary structure means that subunits move relative to each other (Chapter 1). The positioning of effector groups and substrate-binding groups *between* subunits means that binding is poised for regulation by changes of quaternary structure. On the other hand, the location of the catalytic residues of the active site *within* a subunit means that k_{cat} need not be affected by a change in quaternary structure.

2. The structure of the T state

There is a major change in quaternary structure caused by a rotation of about 7° of one dimer with respect to the other. Arg-162 in a loop (the "6-F" loop, comprising residues 156–162) makes an important interaction with the γ-phosphoryl group of ATP in the R state. The transition to the T state drastically changes the conformation of the F-loop, causing Glu-161 to occupy the space vacated by Arg-162 and hence electrostatically repelling the phosphate of ATP. The change in conformation of the F-loop is intimately linked to the change in quaternary structure, so that the transition is effectively all-or-none, obeying the MWC mechanism.[4] The allosteric effectors also bind close to the 6-F loop.

H. Glycogen phosphorylase and control by phosphorylation

Regulation by protein phosphorylation is a very effective means of controlling activity: protein kinases catalyze the phosphorylation, and phosphatases remove the phosphoryl group (Figure 10.15). The stereochemical importance of the phosphoryl group is that it is large and so can cause steric repulsion, it is a dianion and so is sensitive to electrostatic effects, and it can make a network of hydrogen bonds. The specific kinases and phosphatases can themselves be exquisitely regulated. All these properties conspire to make phosphorylation a very effective allosteric regulator. For example, the transitions of the eukaryotic cell cycle are triggered by the cyclin-dependent protein kinases, CDKs.[54] The CDK activity is itself controlled by its state of phosphorylation and the binding of an activator, cyclin. The activity of the active complex can be switched off by further phosphorylation or by CDK inhibitory subunits. By having additional regulatory subunits that can bind different ligands, many different signals in the cell can be integrated.

1. Glycogen phosphorylase and regulation of glycogenolysis[55–61]

Glycogen, a major source of energy for muscle contraction, is the principle storage form of glucose in mammalian cells. Glycogen consists mainly of glucose

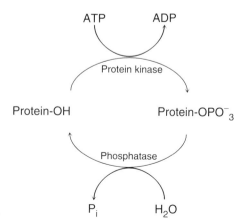

Figure 10.15 Phosphoryl kinase/phosphatase regulation.

units linked by $\alpha(1 \rightarrow 4)$ glycosidic bonds, with branches created by residues linked by $\alpha(1 \rightarrow 6)$ glycosidic bonds. The rate-limiting enzyme in glycogenolysis is glycogen phosphorylase, which is frequently called phosphorylase. It catalyzes the sequential phosphorolysis of the $\alpha(1 \rightarrow 4)$-linked units from the nonreducing end to form glucose 1-phosphate (equation 10.18). Phosphorylase exists in two interconvertible forms, a and b. Phosphorylase b, the form in resting muscle, is an α_2 dimer of $M_r\ 2 \times 97\ 333$, each polypeptide chain containing 841 amino acid residues (rabbit muscle enzyme). This form is inactive but it is activated by the allosteric effectors AMP and IMP. ATP and ADP act as allosteric inhibitors. In response to neural or hormonal signals, phosphorylase b is converted to phosphorylase a by the phosphorylation of Ser-14, catalyzed by phosphorylase kinase (equation 10.19). This is accompanied by a further dimerization in the absence of glycogen to give a tetramer of $M_r 4 \times 97\ 412$, which has access to the active site obstructed. Phosphorylase a is fully active at saturating substrate concentrations, but at low concentrations of P_i its activity is stimulated by AMP. Glucose is an allosteric inhibitor.

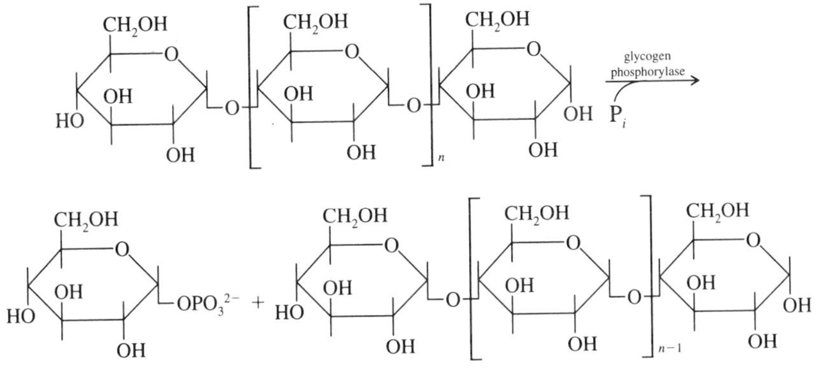

$$(10.18)$$

Crucial events in the regulation of the activity of phosphorylase are thus its phosphorylation and dephosphorylation (equation 10.19), but superimposed on

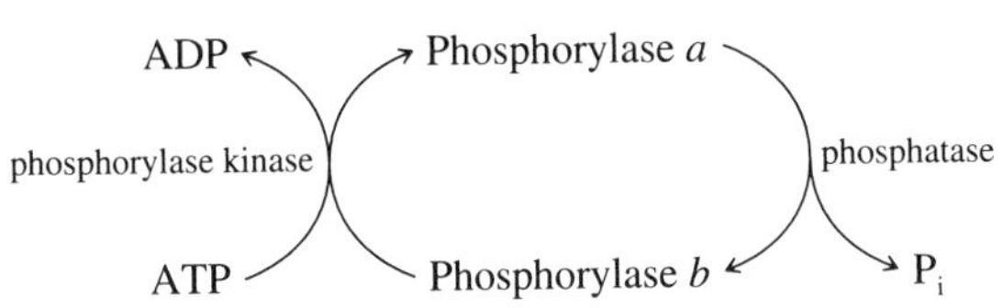

$$(10.19)$$

these are the allosteric effects. The key factor in the phosphorylation state of phosphorylase is the activity of the phosphorylase kinase.[56] This enzyme, of complex structure $(\alpha\beta\gamma\delta)_4$, may be activated by two types of stimuli:

1. *Neural control of the activity of phosphorylase kinase.* The electrical stimulation of muscle is mediated by the release of Ca^{2+} ions. These ions also

activate phosphorylase kinase, which in turn activates phosphorylase and thus accelerates glycogenolysis to provide the necessary ATP for muscle contraction.

2. *Hormonal control of the activity of phosphorylase kinase.* Just as the activity of phosphorylase is increased by phosphorylation, so is the activity of its phosphorylase kinase (which may be phosphorylated on two serine residues, one in an α subunit and one in a β subunit). Hormonal stimulation (β-adrenergic) leads to the production of 3′,5′-cyclic AMP ("second messenger"), which stimulates the activity of the cyclic-AMP-dependent protein kinase that catalyzes the phosphorylation of phosphorylase kinase.

Four different phosphatases in skeletal muscle catalyze the dephosphorylation of the various enzymes involved in glycogenolysis. Their activities are controlled by various inhibitors.

Phosphorylase and phosphorylase kinase are but two enzymes in a complex phosphorylase/glycogen synthase cascade system that regulates glycogen metabolism.[62] In such a cascade, the interconversion of active and inactive forms of enzymes is not an all-or-none phenomenon, but a dynamic process that leads to steady state levels of the active form. The activity of "converter" enzymes such as the kinases and phosphatases along with multisite interactions involving allosteric effectors provide a system that can respond to fluctuations in the concentrations of a multitude of cellular metabolites. This allows for a continuous gradation of enzymatic activity over a wide range of levels to balance the needs of the cell or system.[62] Protein phosphorylation is a major mechanism by which external physiological stimuli control intracellular processes in mammalian tissues.

The yeast enzyme is a homodimer of $M_r 2 \times 102\,500$ and has 49% sequence identity to the muscle enzyme. The yeast enzyme is more simply regulated: feedback inhibition by the allosteric inhibitor glucose-6-phosphate and activation by a 3′-5′-cyclic AMP-dependent protein kinase or a yeast phosphorylase that phosphorylates Thr-10.[55]

2.   The allosteric activation of phosphorylases

Several forms of mammalian and yeast phosphorylases *a* and *b* have been solved at high resolution, allowing the activation mechanism to be solved.[55–61] The effector molecules ATP and AMP bind at sites close to the subunit interfaces of the muscle enzyme, making interactions across the interface, as found for phosphofructokinase and hemoglobin. The effector site is some 30 Å from the catalytic site. The crucial factor in the activation of both yeast and muscle enzymes is the position of the N-terminal tail, containing Ser-14 in the rabbit muscle enzyme and Thr-10 in the yeast enzyme, which are in highly negatively charged regions in the inactive phosphorylase *b*. The tail wraps around from the regulatory face of one subunit to the catalytic face of the other, inhibiting substrate binding.[61] Phosphorylation of Ser-14 or Thr-10 forces the tail of the negatively charged site into a positively charged one. The phosphorylated Thr-10 moves 36 Å, forcing the subunits to move apart, becoming buried in the effector site, and displacing the inhibitory glucose-6-phosphate.[60] Despite the difference in allosteric activation signals between the yeast and muscle enzymes, the downstream "response element" is conserved. It consists of 10 hydrophobic residues over the subunits of the dimer. The phosphorylation leads to hydrophobic condensation, triggering the conformational change.[60]

I. G proteins: Molecular switches[63]

G proteins are GTPases that regulate a wide variety of different processes in a similar manner.[64,65] The G protein·GTP complex binds to a target that is downstream in a pathway and changes its activity. For example: EF-Tu·GTP binds aminoacyl-tRNA and delivers it to the A site on the ribosome during protein synthesis; Ras-GTP activates the Raf kinase in mitogenic signaling pathways; and transducin-α-GTP activates cGMP phosphodiesterase in light-mediated signal transduction during vision. The "active state" of the G protein·GTP-target complex is switched off by the hydrolysis of GTP, which, after the release of P_i, causes the release of the G protein from its target. The G protein then recycles (Figure 10.16).

The half-life of the active complex is determined by the rate constant for the chemical cleavage step, which acts as a "molecular clock." Most G proteins have a low GTPase activity (a long half-life), which is stimulated by a GTPase activating protein (GAP) that binds to it and causes structural changes. Sometimes, as for EF-Tu·GTP, the GAP is the target itself. Recycling to the active conformation is limited by the rate constant for the dissociation of GDP, which is generally very slow and is enhanced by a specific nucleotide exchange protein (NEP); for example, EF-Ts for EF-Tu·GDP.

G proteins have a common core structure, the "G domain" of M_r about 21 000, that is virtually superimposable in all crystal structures so far solved.[66–74]

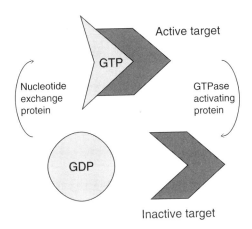

Figure 10.16 G-protein cycle.
[Adapted from R. D. Vale, *J. Cell Biol.* **135**, 291 (1996).]

This domain has grafted onto it the different structures for the different biological activities. There is a loop, which is absolutely conserved, that binds the α,β-phosphates (the phosphate binding loop, G1). Two other conserved regions in loops are called G2 and G3. A threonine in G2 and a glycine residue in G3 form hydrogen bonds with the γ-phosphate of bound GTP. These residues sense whether GDP or GTP is bound and lead to the conformational change between the GTP- and GDP-bound states. Intriguingly, there are structural and mechanistic similarities between the G proteins and the motor proteins[63] discussed in the next section. The G1 loop is an example of the Walker A consensus sequence, discovered first in the F_1–ATPase[75] (section K) and otherwise known as the P-loop,[76] which is a common feature of many GTP or ATP binding proteins (Table 10.4). It typically forms a flexible loop between a β strand and an α helix. Movement of the G3 glycine toward the γ-phosphate of GTP changes the angle of the following helix in a G protein. For Ras protein, this leads to loop movements that shift C^α atoms up to 10 Å.[67,77] For EF-Tu, this leads to internal loop movements similar to those in Ras protein, which cause large domain rearrangements (90° rotation, 40 Å shifts).[72,78] Similar movements are found for transducin-α.[71]

Table 10.4 *Some proteins having the Walker A consensus sequence^a or P-loop^b for binding the α,β-phosphate groups of ATP or GTP*

ATP synthase α and β subunits
Myosin heavy chains
Kinesin heavy chains and kinesin-like proteins
Dynamins and dynamin-like proteins
GTP-binding elongation factors (EF-Tu, EF-1α, EF-G, EF-2, etc.).
Ras family of GTP-binding proteins (Ras, Rho, Rab, Ral, Ypt1, SEC4, etc.)
Adenylate cyclase
Guanylate kinase
Thymidine kinase
Thymidylate kinase
ATP-binding proteins involved in "active transport" (ABC transporters)
DNA and RNA helicases

^{a}J. E. Walker, M. Saraste, M. J. Runswick, and N. J. Gay *EMBO J.* **1,** 945 (1982).
^{b}M. Saraste, P. R. Sibbald, and A. Wittinghofer, *Trends Biochem. Sci.* **15,** 430 (1990). Other data from H. R. Bourne, D. A. Sanders, and F. Mccormick, *Nature, Lond.* **349,** 117 (1991) and E. V. Koonin, *J. Molec. Biol.* **229,** 1165 (1993). The general consensus sequence is G/AxxxxGKS/T. The G proteins have GxxGxGKS/T; myosins, GESxxGKS/T; and kinesins, GQTxxGKS/T, where / = either/or and x = variable residue. Gly is found instead of Ser or Thr in the last position of the P-loop adenylate kinase. The A sequence typically forms a flexible loop between a β strand and an α helix.

A mechanism for the GTPase activity has been proposed from the crystal structure of aluminum fluoride (AlF_4^-) complex with transducin-α.[69] Aluminum fluoride activates transducin-$\alpha\cdot$GDP by binding with a geometry resembling a pentavalent intermediate for GTP hydrolysis.

J. Motor proteins[63,79,80]

There are many biological functions that involve movement of a protein complex along cytoskeletal filaments. Muscle contraction, organelle movement, beating of cilia, cytokinesis, and mitosis, for example, are driven by motor proteins that use the free energy of hydrolysis of ATP to move unidirectionally along the protein scaffolds. There are three superfamilies of motor proteins: the myosin family, which moves along actin filaments, and the kinesin and dynein families, which move along microtubules. Members of each family have similar motor domains of about 30–50% identical residues that can function as autonomous units. The proteins are differentiated by their nonmotor or "tail" domains. The motor domain of myosin consists of about 850 amino acid residues, that of dynein about 1000, and that of kinesin about 340. The best characterized motors are those of kinesin and myosin. They have no significant identity in amino acid sequence, but they have a common fold for binding the nucleotide, with a phosphate binding loop that is topologically nearly identical.[81,82]

The binding affinity of the motor to the cytoskeleton depends on the state of the nucleotide that is bound. For example, myosin$\cdot$ATP and myosin$\cdot$ADP$\cdot P_i$ bind weakly to actin, while myosin$\cdot$ADP binds 10,000 times more tightly.[83] A structural change on going from the weak-binding to the tight-binding state of actomyosin is thought to be responsible for the production of force by the motor and hence the movement that causes muscle contraction.[84,85] The motors have a reaction cycle that is analogous to that of the G proteins (Figure 10.17), but with the important difference that it is the NDP state that binds tightly to its target. The understanding of how these motors work is essentially a structural problem.[80]

Parallels with the G proteins, which have been better characterized, have shed light on the motor proteins.[63] The phosphate binding loop of the motor proteins is very similar to that in the G proteins and can be used as a reference point for superimposing the G domain on the motor domain.[63] There are similarities between the G proteins and the motor proteins in the spatial distribution of residues that bind the nucleotide.[86] In particular, four loops in the motor proteins, N1–N4, have sequence similarities with G1–G4, suggesting that they have the same functions. A serine in N2 and a glycine in N3 form hydrogen bonds in putative transition state analog complexes (myosin$\cdot$ADP$\cdot AlF_4$ and myosin$\cdot$ADP$\cdot VO_4$,[87,88] implying similar movements as in the G proteins, which could lead to the changes in structure responsible for movement.[63] The possible structural changes have been reviewed by Amos and Cross.[80]

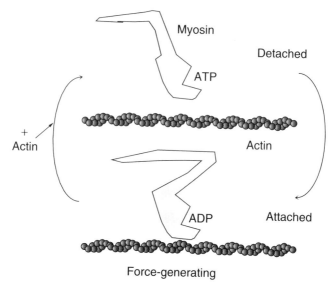

Figure 10.17 Motor-protein cycle. [Adapted from R. D. Vale, *J. Cell Biol.* **135**, 291 (1996).]

The crystal structures of the complexes of myosin-Mg^{2+} with ADP and the two alleged substrate analogues AMP-PNP and ATPγS are all very similar, suggesting that the latter do not induce the ATP-bound state.[89] AMP-PNP and ATPγS are not good substrate mimics when there are conformational transitions that depend crucially on the differences between the different nucleotides that are bound.

The modular design of kinesin motors is nicely illustrated by switching domains between the kinesin and Ncd motors.[90] Kinesin moves unidirectionally to one end of microtubules (the "plus" end), whereas Ncd moves in the opposite direction. Replacing the catalytic domain of kinesin with that of Ncd gives a motor that still moves in the kinesin direction. Kinesin is highly processive, remaining attached to a single filament (see Chapter 14, section A7). The Ncd-kinesin chimera has the lower processivity of Ncd.

K. ATP synthesis by rotary catalysis: ATP synthase and F_1-ATPase

A resting adult turns over about half its body weight in ATP daily, rising 100-fold during hard work. The energy for synthesizing ATP comes from respiration. In the 1960s and early 1970s, many bioenergeticists thought that the energy from respiration was coupled to the formation of high-energy covalent chemical intermediates. In 1961, Peter Mitchell put forward a radically novel theory—the chemiosmotic hypothesis—which, after a turbulent history, is now accepted. He proposed that rather than the energy from respiration being stored as covalent chemical bonds, it is used to set up a difference in hydrogen ion concentration — i.e., a ΔpH — between the inside and outside of the mitochondrial membrane (or chloroplast membrane in photorespiration). This produces a chemical potential across the membrane, proportional to $RT \ln (\Delta pH)$. The gradient was called the proton motive force (pmf; designated as $\Delta\mu_{H+}$).

The enzyme ATP synthase, formerly called F_0F_1-ATPase, makes most of the ATP, using ADP and inorganic phosphate, in mitochondria, eubacteria, and chloroplasts.[91] The F_0 component sits in the membrane and is connected to the F_1 component, which is in the cytosol, by a 45-Å-long stalk (the γ subunit). The coupling of ATP synthesis to the flow of hydrogen ions across the membrane is done by the F_0 part. The bovine F_1 complex is an assembly of five polypeptides, α_3, β_3, γ, δ, and ϵ, of $M_r\,371\,000$. The three α and three β subunits are packed in a cylinder around the γ, alternating α-β-α-β-α-β.[92]

The basic mechanism was worked out by P. Boyer[93] from kinetic studies prior to the crystallographic studies, which spectacularly confirmed and extended it.[92] The essence of the mechanism is that the three β subunits are asymmetric, each having a different conformation and a different bound nucleotide (Figure 10.18). The three conformations in Figure 10.18 are: O ($=$ open) site, with a very low affinity for substrates and catalytically inactive; L ($=$ loose) site, loosely binding substrates and catalytically inactive; T ($=$ tight) site, tightly binding substrates and catalytically active. The pmf powers the asymmetric γ subunit to rotate relative to the $\alpha_3\beta_3$ and forces them to interconvert; γ acts as a crankshaft. The T site with bound ATP is converted into an O site, releasing the bound nucleotide. Concomitantly, an L site with loosely bound ADP and P_i is converted to a T site, where the change in binding energy causes ATP to form. Fresh substrates bind to the new O site converting it to an L site, and so on. The β subunits have the P-loop (Walker A) motif, and so the P-loop is presumably involved in linking the conformational changes with the substrate affinity, as described for the G proteins and the motor proteins.

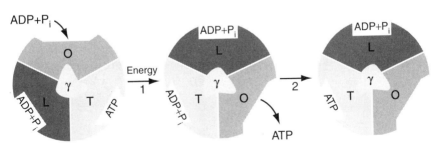

Figure 10.18 Boyer's "Binding Change Mechanism." The asymmetrical γ subunit that causes changes in the structure of the β subunits can be seen in the center. [Courtesy of Dr. J. E. Walker.]

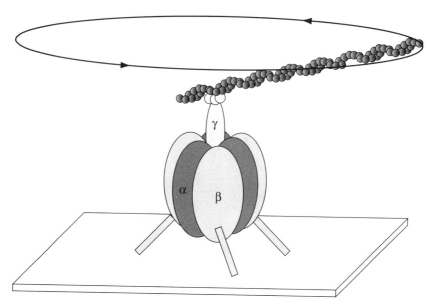

Figure 10.19 ATP-fueled rotation of fluorescently labeled actin that is attached to the γ subunit of the F_1-ATPase by a streptavidin-biotin linker. The β subunits are immobilized on a microscope slide. [Modified from H. Noji, R. Yasuda, M. Yoshida, and K. Kinosita, *Nature* **386**, 299 (1997).]

The thermodynamic basis for the formation of ATP exemplifies how the equilibrium constant for enzyme-bound reagents may be radically different from that for the reagents in solution (Chapter 3, section H2). The free energy of formation of ATP is about 31 kJ/mol (7.5 kcal/mol) in the cell for the process ADP + P_i = ATP. In the active site of an enzyme, formation is inherently more favorable because ADP + P_i are bound to the enzyme, and so less overall translational and rotational entropy is lost on bond formation (Chapter 2, section B4). The free energy of formation of ATP in the active site of myosin is only 6 kJ/mol (1.4 kcal/mol).[83] Consequently, a subunit of F_1 has to bind ATP by 6 kJ/mol more tightly than does myosin for ATP synthesis to be favorable. The role of enzyme–ligand binding energy has such a profound influence on catalysis and recognition that the following two chapters are devoted to this subject. It will be seen in Chapter 15, sections F2 and K, that the tyrosyl-tRNA synthetase changes a free energy of reaction by 31 kJ/mol (7.5 kcal/mol) by tighter binding of the products than the reagents.

In the absence of F_0, the F_1 is an ATPase, known as the F_1-ATPase, and not a synthase. The rotation of the γ subunit has been demonstrated in the ATPase reaction. The most graphic example is in Figure 10.19.[94] The $\alpha_3\beta_3\gamma_1$ components were attached to a Ni^{2+}-coated glass surface at the opposite side from the normal attachment to a membrane by adding a Ni^{2+}-binding sequence (His_{10}) at the

N-terminals of the β subunits. At the other end, a cysteine residue was introduced into the exposed tip of the γ subunit, which was coupled to biotin, and then attached to a fluorescently labeled actin filament via a streptavidin linker. The ATP-dependent anticlockwise rotation of the 1- to 3-μm-long actin filaments was seen in a fluorescence microscope.[94] Smaller probes show that the rotation is consistent with the turnover of ATP by the F_1-ATPase, which is consistent with a three-step motor.[95,96] This implies that ATP synthesis requires that the γ subunit be cranked in a clockwise direction by a rotary motor in F_1.

References

1. G. S. Adair, *J. Biol. Chem.* **63**, 529 (1925).
2. J. Monod, J.-P. Changeux, and F. Jacob, *J. Molec. Biol.* **6**, 306 (1963).
3. M. F. Perutz, A. J. Wilkinson, M. Paoli, and G. G. Dodson, *Ann. Rev. Biophys. Biomol. Structure* **27**, 1 (1998).
4. J. Monod, J. Wyman, and J.-P. Changeux, *J. Molec. Biol.* **12**, 88 (1965).
5. D. E. Koshland, Jr., G. Neméthy, and D. Filmer, *Biochemistry* **5**, 365 (1966).
6. M. Eigen, *Nobel Symp.* **5**, 333 (1967).
7. J. Herzfield and H. E. Stanley, *J. Molec. Biol.* **82**, 231 (1974).
8. O. Yifrach and A. Horovitz, *Biochemistry* **34**, 5303 (1995).
9. A. Levitzki, W. B. Stallcup, and D. E. Koshland, Jr., *Biochemistry* **10**, 3371 (1971).
10. R. A. MacQuarrie and S. A. Bernhard, *Biochemistry* **10**, 2456 (1971).
11. M. J. Irwin, J. Nyborg, B. R. Reid, and D. M. Blow, *J. Molec. Biol.* **105**, 577 (1976).
12. G. Biesecker, J. I. Harris, J. C. Thierry, J. E. Walker, and A. J. Wonacott, *Nature, Lond.* **266**, 328 (1977).
13. A. R. Fersht, R. S. Mulvey, and G. L. E. Koch, *Biochemistry* **14**, 13 (1975).
14. H. R. Bosshard, G. L. E. Koch, and B. S. Hartley, *Eur. J. Biochem.* **53**, 493 (1975).
15. A. Conway and D. E. Koshland, Jr., *Biochemistry* **7**, 4011 (1968).
16. J. Schlessinger and A. Levitzki, *J. Molec. Biol.* **82**, 547 (1974).
17. R. Hill, *Proc. R. Soc.* **B100**, 419 (1925).
18. M. M. Rubin and J.-P. Changeux, *J. Molec. Biol.* **21**, 265 (1966).
19. S. J. Edelstein, *Nature, Lond.* **230**, 224 (1971).
20. K. Kirschner, E. Gallego, I. Schuster, and D. Goodall, *J. Molec. Biol.* **58**, 29 (1971).
21. K. H. Mayo, *J. Molec. Biol.* **146**, 589 (1981).
22. A. Nasuda-Kouyama, H. Tachibana, and A. Wada, *J. Molec. Biol.* **146**, 451 (1983).
23. J. M. Baldwin and C. Chothia, *J. Molec. Biol.* **129**, 175 (1979).
24. M. F. Perutz, *Nature, Lond.* **228**, 734 (1970).
25. A. Arnone, *Nature, Lond.* **237**, 146 (1972).
26. A. Arnone and M. F. Perutz, *Nature, Lond.* **249**, 34 (1974).
27. C. Chothia, S. Wodak, and J. Janin, *Proc. Natl. Acad. Sci. USA* **73**, 3793 (1976).
28. C. Bohr, K. A. Hasselbach, and A. Krogh, *Skand. Arch. Physiol.* **16**, 402 (1904).
29. J. L. Hoard, in *Hemes and hemoproteins* (B. Chance, R. W. Estabrook, and T. Yonetani, Eds.), Academic Press, p. 9 (1966).
30. M. F. Perutz, S. S. Hasnain, P. J. Duke, J. L. Sessler, and J. E. Hahn, *Nature, Lond.* **295**, 535 (1982).
31. A. Warshel, *Proc. Natl. Acad. Sci. USA* **74**, 1789 (1977).
32. B. R. Gelin and M. Karplus, *Proc. Natl. Acad. Sci. USA* **74**, 801 (1977).

33. J. P. Collman, R. R. Gagne, C. A. Reed, W. T. Robinson, and G. A. Rodley, *Proc. Natl. Acad. Sci. USA* **71**, 1326 (1974).
34. G. B. Jameson, F. S. Molinaro, J. A. Ibers, J. P. Collman, J. I. Brauman, B. Rose, and K. S. Suslick, *J. Am. Chem. Soc.* **102**, 3225 (1980).
35. B. Shaanan, *Nature, Lond.* **296**, 5858 (1982).
36. J. P. Collman, J. I. Brauman, I. J. Collins, B. Iverson, and J. L. Sessler, *J. Am. Chem. Soc.* **103**, 2450 (1981).
37. S. E. V. Phillips and B. P. Schoenborn, *Nature, Lond.* **292**, 81 (1981).
38. E. R. Henry, C. M. Jones, J. Hofrichter, and W. A. Eaton, *Biochemistry* **36**, 6511 (1997).
39. J. Puigjaner, B. Rais, M. Burgos, B. Comin, J. Ovadi, and M. Cascante, *FEBS Lett.* **418**, 1 (1997).
40. J. H. Woods and H. M. Sauro, *Computer Applications in the Biosciences* **13**, 123 (1997).
41. J. Sudi, *Biochim. Biophys. Acta Protein Structure and Molecular Enzymology* **1341**, 108 (1997).
42. J. Nielsen, *Biochemical J.* **321**, 133 (1997).
43. C. Giersch, *J. Theoret. Biol.* **189**, 1 (1997).
44. M. Ehlde and G. Zacchi, *Chemical Engineering Science* **52**, 2599 (1997).
45. R. L. Jarvest, G. Lowe, and B. Potter, *Biochemical Journal* **199**, 427 (1981).
46. S. A. Berger and P. R. Evans, *Biochemistry* **31**, 9237 (1992).
47. P. R. Evans, G. W. Farrants, and P. J. Hudson, *Phil. Trans. R. Soc.* **B293**, 53 (1981)
48. P. R. Evans, *Activity and allosteric regulation in bacterial phosphofructukinase*, in *Conference on Chemical Research XXXVI, Regulation of Proteins by Ligands*, Robert A. Welch Foundation, Houston, 39 (1992).
49. T. Schirmer and P. R. Evans, *Nature, Lond.* **343**, 140 (1990).
50. Y. Shirakihara and P. R. Evans, *J. Molec. Biol.* **204**, 973 (1988).
51. W. R. Rypniewski and P. R. Evans, *J. Molec. Biol.* **207**, 805 (1989).
52. D. Blangy, H. Buc, and J. Monod, *J. Molec. Biol.* **31**, 13 (1968).
53. D. Blangy, *Biochimie* **53**, 135 (1971).
54. D. O. Morgan, *Ann. Rev. Cell and Developmental Biol.* **13**, 261 (1997).
55. K. Lin, P. K. Hwang, and R. J. Fletterick, *J. Biol. Chem.* **270**, 26833 (1995).
56. L. N. Johnson, E. D. Lowe, M. E. M. Noble, and D. J. Owen, *FEBS Lett.* **430**, 1 (1998).
57. E. P. Mitchell, S. G. Withers, P. Ermert, A. T. Vasella, E. F. Garman, N. G. Oikonomakos, and L. N. Johnson, *Biochemistry* **35**, 7341 (1996).
58. K. Lin, V. L. Rath, S. C. Dai, R. J. Fletterick, and P. K. Hwang, *Science* **273**, 1539 (1996).
59. J. L. Buchbinder and R. J. Fletterick, *J. Biol. Chem.* **271**, 22305 (1996).
60. K. Lin, P. K. Hwang, and R. J. Fletterick, *Structure* **5**, 1511 (1997).
61. V. L. Rath and R. J. Fletterick, *Nature Structural Biology* **1**, 681 (1994).
62. P. B. Chock, S. G. Rhee, and E. R. Stadtman, *Ann. Rev. Biochem.* **49**, 813 (1980).
63. R. D. Vale, *J. Cell Biol.* **135**, 291 (1996).
64. H. R. Bourne, D. A. Sanders, and F. Mccormick, *Nature, Lond.* **348**, 125 (1990).
65. H. R. Bourne, D. A. Sanders, and F. Mccormick, *Nature, Lond.* **349**, 117 (1991).
66. M. B. Mixon, E. Lee, D. E. Coleman, A. M. Berghuis, A. G. Gilman, and S. R. Sprang, *Science* **270**, 954 (1995).

67. M. V. Milburn, L. Tong, A. M. Devos, A. Brunger, Z. Yamaizumi, S. Nishimura, and S. H. Kim, *Science* **247**, 939 (1990).

68. L. Tong, A. M. Devos, M. V. Milburn, and S. H. Kim, *J. Mol. Biol.* **217**, 503 (1991).

69. J. Sondek, D. G. Lambright, J. P. Noel, H. E. Hamm, and P. B. Sigler, *Nature, Lond.* **372**, 276 (1994).

70. D. G. Lambright, J. P. Noel, H. E. Hamm, and P. B. Sigler, *Nature, Lond.* **369**, 621 (1994).

71. J. P. Noel, H. E. Hamm, and P. B. Sigler, *Nature, Lond.* **366**, 654 (1993).

72. H. Berchtold, L. Reshetnikova, C. Reiser, N. K. Schirmer, M. Sprinzl, and R. Hilgenfield, *Nature, Lond* **365**, 368 (1993).

73. P. Nissen, M. Kjeldgaard, S. Thirup, G. Polekhina, L. Reshetnikova, B. Clark, and J. Nyborg, *Science* **270**, 1464 (1995).

74. Y. Wang, Y. X. Jiang, M. MeyeringVoss, M. Sprinzl, and P. B. Sigler, *Nature Structural Biology* **4**, 650 (1997).

75. J. E. Walker, M. Saraste, M. J. Runswick, and N. J. Gay, *EMBO J.* **1**, 945 (1982).

76. M. Saraste, P. R. Sibbald, and A. Wittinghofer, *Trends Biochem. Sci* **15**, 430 (1990).

77. I. Schlichting, S. C. Almo, G. Rapp, K. Wilson, K. Petratos, A. Lentfer, K. Wittinghofer, W. Kabsch, E. F. Pai, G. A. Petsko, and R. S. Goody, *Nature, Lond.* **345**, 309 (1990).

78. M. Kjeldgaard, P. Nissan, S. Thirup, and J. Nyborg, *Structure* **1**, 35 (1993).

79. R. D. Vale and R. J. Fletterick, *Ann. Rev. Cell and Developmental Biol.* **13**, 745 (1997).

80. L. A. Amos and R. A. Cross, *Curr. Opin. Struct. Biol.* **7**, 239 (1997).

81. F. J. Kull, E. P. Sablin, R. Lau, R. J. Fletterick, and R. D Vale, *Nature, Lond.* **380**, 550 (1996).

82. E. P. Sablin, F. J. Kull, R. Cooke, R. D. Vale, and R. J. Fletterick, *Nature, Lond.* **380**, 555 (1996).

83. H. D. White and E. W. Taylor, *Biochemistry* **15**, 5818 (1976).

84. E. Eisenberg and T. L. Hill, *Science* **227**, 999 (1985).

85. E. W. Taylor, *Mechanism and energetics of actomyosin ATPase*, in *The heart and cardiovascular system*, Vol. 2., H. Fozzard, ed., Raven Press, 1281 (1992).

86. C. A. Smith, and I. Rayment, *Biophysical J.* **70**, 1590 (1996).

87. A. J. Fisher, C. A. Smith, J. B. Thoden, R. Smith, K. Sutoh, H. M. Holden, and I. Rayment, *Biochemistry* **34**, 8960 (1995).

88. C. A. Smith and I. Rayment, *Biochemistry* **35**, 5404 (1996).

89. A. M. Gulick, C. B. Bauer, J. B. Thoden, and I. Rayment, *Biochemistry* **36**, 11619 (1997).

90. R. B. Case, D. W. Pierce, N. Hom-Booher, C. L. Hart, and R. D. Vale, *Cell* **90**, 959 (1997).

91. D. G. Nicholls and S. J. Ferguson, in *Bioenergetics 2*, Academic Press (1992).

92. J. P. Abrahams, A. G. W. Leslie, R. Lutter, and J. E. Walker, *Nature, Lond.* **370**, 621 (1994).

93. P. D. Boyer, *Ann. Rev. Biochem.* **66**, 717 (1997).

94. H. Noji, R. Yasuda, M. Yoshida, and K. Kinosita, *Nature, Lond.* **386**, 299 (1997).

95. D. Sabbert, S. Engelbrecht, and W. Junge, *Nature, Lond.* **381**, 623 (1996).

96. D. Sabbert and W. Junge, *Proc. Natl. Acad. Sci. USA* **94**, 2312 (1997).

Forces Between Molecules, and Binding Energies

11

The distinguishing feature of enzyme catalysis is that the enzyme binds the substrate so that the reactions proceed in the confines of the enzyme–substrate complex. In order to gain insight into the strength and specificity of the binding, we shall discuss in a somewhat empirical and phenomenological manner the interactions between nonbonded atoms. In particular, we shall concentrate on the magnitudes of the energies involved. Besides being responsible for binding, the noncovalent interactions are important in further ways. One important consideration to be discussed in Chapter 12 is that these interactions can be used for lowering the activation energy of a chemical step instead of directly contributing to the binding energy. In general, noncovalent interactions are responsible for all molecular recognition and for determining and maintaining the three-dimensional structure of proteins.

There is a crucial distinction between the binding energies that are observed in solution and the fundamental forces between enzymes and substrates. Water binds to all reagents in solution and has to be displaced when the reagents bind to one another. The observed binding energy is the net change in energy, which measures the affinity of the reagents for one another in competition with water. In addition, there are significant changes in entropy on binding because of the various displacements of solvent molecules. These two phenomena are of the utmost importance in understanding binding and are a recurring theme of the discussion. We discuss first the forces and then the binding energies.

A. Interactions between nonbonded atoms

1. Electrostatic interactions

All forces between atoms and molecules are electrostatic in origin, even those between nonpolar molecules. But we shall reserve the term electrostatic for interactions that occur between charged or dipolar atoms and molecules.

Electrostatic interaction energies vary according to the nature of the charges and the dielectric constant D of the medium. Between ions with net charges, the energy falls off with distance r as $1/Dr$. This is a long-range effect because energy decreases slowly with r. Between randomly oriented permanent dipoles, the energy falls off as $1/Dr^6$. The same is true for the energy between a permanent dipole and a dipole induced by it. The r^{-6} dependence marks these out as short-range forces that fall off rapidly with distance. The energy between an ion and a dipole induced by it falls off as $1/Dr^4$.

The value of D is also very important, being about 80 in water and 2–4 in a protein. A simple model for the effects of a dielectric is very instructive. The charge of $-q$ in Figure 11.1 is surrounded by a very large volume of dielectric. The charge polarizes the dielectric so that dipoles are formed in it that point their positive charges toward $-q$. If the dipoles are small, then they can be considered to be just a series of positive charges with their neutralizing negative charges at the boundaries of the dielectric, which are so far away that they can be ignored. The $\delta+$ charges partly neutralize $-q$ so that its effective charge is

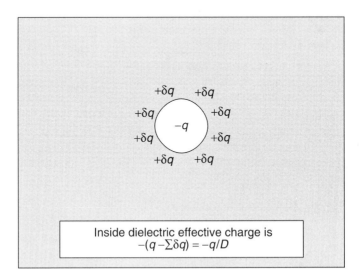

Inside dielectric effective charge is
$$-(q - \Sigma \delta q) = -q/D$$

Figure 11.1 Partial neutralization of charge by polarization of dielectric. The "$-\delta q$" components are effectively at infinity.

$-(q - \Sigma\delta q) = -q/D$. Suppose, as in Figure 11.2, that the charge is at the interface of two different dielectric media. Then the effective charge is $-q(1/2D + 1/2D')$.

a. Long-range charge-charge interactions

The calculation of electrostatic interactions in proteins has been difficult because of the local heterogeneity in their structure and the presence of the protein–water interface, which causes the dielectric constant to vary wildly. But there has been a major breakthrough in the past 10 years in calculating long-range electrostatic interactions. Computer methods have been introduced that have been benchmarked on experimental data for proteins in which electrostatic interactions have been altered by protein engineering (see Chapter 5, section D). The Warwicker-Watson algorithm, implemented in the commercial package *DelPhi*, divides the protein and solvent into cubes of 1 Å^3 or less and assigns to each the appropriate dielectric constant. The appropriate fractional charge is put at the corner of each cube, and the classical electrostatic equations are applied iteratively to the system. This procedure calculates in a fraction of a second long-range interaction energies in barnase and subtilisin to $\pm 10\%$ (Table 11.1). The procedure becomes more difficult over short distances, and so short-range electrostatic interactions are more difficult to calculate. Note that D can be greater than the value of 80 found in water, because parts of the protein can act as lenses to focus or defocus charges or to align water molecules.

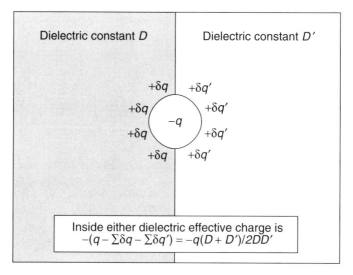

Figure 11.2 Partial neutralization of charge by polarization of dielectric for a charge at the interface of two different dielectrics.

Table 11.1 *Measured and calculated interactions of charged residues in subtilisin with His-64* [a]

Charged residue	Measured interaction energy		Calculated interaction energy	
	kJ/mol	kcal/mol	kJ/mol	kcal/mol
Asp-36	1.1	0.25	1.6	0.38
Asp-99	2.3	0.55	1.9	0.45
Lys-136	0.63	0.15	0.42	0.10
Glu-156	2.2	0.53	2.3	0.52
Lys-170	1.1	0.27	1.0	0.23
Lys-213	0.79	0.19	0.75	0.18

[a] The interaction energy between the residue and His-64 was measured from the shift in its pK_a on mutating that residue to a neutral one. The calculated energies are form *DelPhi*. [From R. Loewenthal, J. Sancho, T. Reinikainen and A. R. Fersht *J. Mol. Biol.* **232**, 574–583 (1993).]

2. Nonpolar interactions (van der Waals or dispersion forces)

Typical potential energy curves for the interaction of two atoms are illustrated in Figure 11.3. There is characteristically a very steeply rising repulsive potential at short interatomic distances as the two atoms approach so closely that there is interpenetration of their electron clouds. This potential approximates to an inverse twelfth-power law. Superimposed upon this is an attractive potential due mainly to the London *dispersion forces*. This follows an inverse sixth-power law. The total potential energy is given by

$$U = \frac{A}{r^{12}} - \frac{B}{r^6} \tag{11.1}$$

The distance dependence of $1/r^6$ for the attractive potential is characteristic of the interaction between dipoles. This is because the attractive dispersion forces result from the mutual induction of electrostatic dipoles. Although a nonpolar molecule has no net dipole averaged over a period of time, at any one instant there will be dipoles due to the local fluctuations of electron density. Because the energies depend on the induction of a dipole, polarizability is an important factor in the strength of the interaction between any two atoms.

The value of B in equation 11.1 for the interaction of two atoms i and j may be calculated from the Slater-Kirkwood equation,

$$B = \frac{\frac{3}{2}e(\hbar/m^{1/2})\alpha_i\alpha_j}{(\alpha_i/N_i)^{1/2} + (\alpha_j/N_j)^{1/2}} \tag{11.2}$$

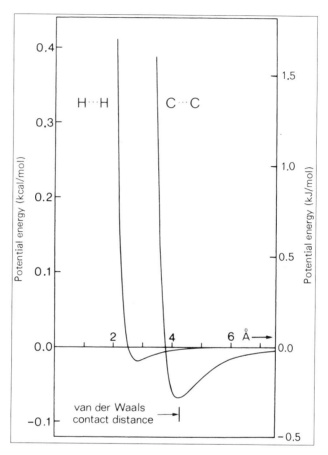

Figure 11.3 The interaction energies of two hydrogen atoms and two (tetrahedral) carbon atoms in a protein. (Calculated from the data in Table 11.3.)

where α is the polarizability, e is the charge on the electron, m is its mass, $\hbar$ is Dirac's constant, and N is the effective number of outer-sphere electrons. Some values of α and N are listed in Table 11.2. Note the low polarizability of O and the high value for S, indicating the relative importance of dispersion energies in the interactions of these atoms.

The interatomic distance at the bottom of the potential well, the most favorable distance of separation, is known as the *van der Waals contact distance*. A particular atom has a characteristic van der Waals radius (Table 11.3). These radii are additive, so that the optimal distance of contact between two atoms may be found by the addition of their two van der Waals radii. The van der Waals radii are not as sharply defined as covalent bond radii. This is because the potential energy wells are so shallow that contact distances may vary by 0.1 Å (0.01 nm) or so

Table 11.2 *Polarizabilities and van der Waals attractions of some atoms and groups*

Atom/group	Polarizability[a] (mL × 10^{24})	N^b	B (for self-interaction)[c] (kcal · Å^6/mol)
—H	0.40	1	46
—O—	0.63	6	220
—OH	0.99	7	470
—CH_2—	1.80	7	1160
—S—	3.00	16	3760
—SH	3.34	17	4560

[a]From A. R. Fersht and C. Dingwall, *Biochemistry* **18**, 1245 (1979), plus additional calculated values.
[b]The effect number of outer-sphere electrons. [From J. A. McCammon, P. G. Wolynes, and M. Karplus, *Biochemistry* **18**, 927 (1979).]
[c]The value of *B*, the attractive potential, in the "6–12" equation (11.1), as calculated from the Slater-Kirkwood equation for two of the indicated atoms or groups interacting.

with little change in energy. Thus, in a crystal structure, it is found that there is a range of values, depending upon the local steric constraints.

Although the attractive forces are weak and the van der Waals energies low, they are additive and make significant contributions to binding when they are summed over a molecule. For example, it is found experimentally from heat-of-sublimation data that each methylene group in a crystalline hydrocarbon has 8.4 kJ/mol (2 kcal/mol) of van der Waals energy, and that each CH group in benzene crystals has 6.7 kJ/mol (1.6 kcal/mol). The van der Waals energy between the D subsite of lysozyme and the occupying glucopyranose ring is about − 58 kJ/mol (− 14 kcal/mol).[1]

3. The hydrogen bond

A particularly important bond in biological systems is the hydrogen bond. This consists of two electronegative atoms, one of which is usually oxygen, bound to the same proton. The bonds in the systems of interest are asymmetric, the proton being at its normal covalent bond distance from the atom to which it is formally bonded, and at a distance from the other somewhat shorter than the usual van der Waals contact distance. The optimal configuration is linear, but bending causes only small energy losses. In the hydrogen bond between the =NH group and the carbonyl oxygen in amide crystals, the O···N distance is typically 2.85 to 3.00 Å, and the O···H distance is 1.85 to 2.00 Å.[2,3] The variation of the potential

Table 11.3 *Van der Waals radii and energies[a]*

Atom	Radius (Å)	Energy[b]	
		kcal/mol	kJ/mol
H	1.44	0.016	0.067
O (hydroxyl)	1.72	0.21	0.88
O (carbonyl)	1.66	0.21	0.88
O^- (carboxyl)	1.66	0.21	0.88
N (amide)	1.82	0.17	0.71
NH_3^+	1.88	0.17	0.71
CH (tetrahedral)	1.91	0.11	0.46
C (trigonal)	1.91	0.086	0.36
CH (aromatic)	1.91	0.086	0.36
S	2.00	0.25	1.0

[a]From W. D. Cornell, P. Cieplak, C. I. Bayly, I. R. Gould, K. M. Merz, D. M. Ferguson, D. C. Spellmeyer, T. Fox, J. W. Caldwell, and P. A. Kollman *J. Am. Chem. Soc.* **117,** 5170 (1995).
[b]van der Waals well depth (ϵ) for the self-interaction of the atom. The energy for two different atoms i and j is given by $\epsilon_{ij} = (\epsilon_i \epsilon_j)^{1/2}$. These data can be inserted into the van der Waals term in equation 11.3 by using R_{ij} = sum of the radii of i and j, $A_{ij} = \epsilon_{ij} (R_{ij})^{12}$, and $B_j = 2\epsilon_{ij} (R_{ij})^6$.

energy with the N···O distance is similar to that in Figure 11.3, the minimum distance of approach being about 2.4 to 2.5 Å. In the H_2O···H—OH hydrogen bond, the distance between the two oxygens is 2.76 Å and the O···H distance is 1.77 Å.

The energies of hydrogen bonds have been variously estimated to be between 12 and 38 kJ/mol (3 and 9 kcal/mol), including 21 kJ/mol (5 kcal/mol) for the amide-amide NH···O bond.[2-4] Bonds of this strength are of particular importance, since they are stable enough to provide significant binding energy, but sufficiently weak to allow rapid dissociation. If the activation energy for the breaking of the bond is the whole of the bond strength, then transition state theory may be used to calculate that bonds of energies 12.5, 25.0, and 37.6 kJ/mol (3, 6, and 9 kcal/mol) dissociate with rate constants of 4×10^{10}, 3×10^8, and 2×10^6 s^{-1}, respectively.

The proton in most hydrogen bonds sits in a double potential energy well, one associated with the donor and one with the acceptor. The barrier separating the two wells becomes lower as the donor and acceptor atoms come closer, eventually disappearing to leave a single-well potential. It has been proposed that the hydrogen bond becomes unusually strong when the barrier becomes very low. There is a lively debate on whether these *low barrier hydrogen bonds* occur in enzyme–substrate interactions and, if they do, whether they are important in catalysis.[5–14]

The backbone NH groups of the enzyme are not only used for binding the substrate, but they may also act as the solvation shell for negative charges developing in the transition state. In particular, as was pointed out in Chapter 2, a binding site for the carbonyl oxygen of the substrates of the serine proteases consists of two backbone amido NH groups. When the oxygen becomes negatively charged in the transition state of the reaction, it is stabilized by the dipole moments of the amide groups.

4. Force fields for simulating energies in proteins and complexes

Computer simulations are regularly performed of the energetics of protein structure and their interactions with ligands using potential energy functions for the forces described so far. The energy functions are of necessity simplifications, but they are calibrated on experimental data. A "minimalist" model uses equation 11.3.[15]

$$
\begin{aligned}
E_{total} = &\sum_{bonds} K_r(r - r_{eq})^2 + \sum_{angles} K_\theta(\theta - \theta_{eq})^2 + \\
&\sum_{dihedrals} \frac{V_n}{2}[1 + \cos(n\phi - \gamma)] + \sum_{i<j}\left[\frac{A_{ij}}{R_{ij}^{12}} - \frac{B_{ij}}{R_{ij}^6} + \frac{q_i q_j}{DR_{ij}}\right]
\end{aligned} \quad (11.3)
$$

The term $\sum_{bonds}K_r(r - r_{eq})^2$ represents the stretching of covalent bonds from the equilibrium bond length r_{eq}, where K_r is the constant of proportionality. (The bottom region of all potential energy wells approximates to a parabola. That is, the energy varies as δr^2, and so there is simple harmonic motion about very small displacements, δr, from equilibrium. At higher amplitudes, the approximation breaks down and the motion becomes anharmonic.) The term $\sum_{angles}K_\theta(\theta - \theta_{eq})^2$ similarly represents the bending of bonds from the equilibrium angle θ. The term $\sum_{dihedrals}(V_n/2)[1 + \cos(n\phi - \gamma)]$ is for the twisting of bonds from the equilibrium dihedral angle ϕ ($V_n = 0$ for a single bond). The term $\sum_{i<j}[A_{ij}/R_{ij}^{12} - B_{ij}/R_{ij}^6]$ calculates the van der Waals energies and $\sum_{i<j}[q_i q_j/DR_{ij}]$ the electrostatic interactions. Full or partial electrostatic charges are assigned to each atom from quantum mechanical calculations. The hydrogen bond energies are included in the electrostatic interactions.

The next chapter describes how utilization of the changes in the binding energy between an enzyme and a substrate as the reaction proceeds is the hallmark of enzyme catalysis. The forces and the energies of interaction described so far are the directly relevant quantities for calculating the binding energies *within* enzyme–substrate complexes and the consequent changes during the reaction. But, as described in the next section, the calculation of binding energies relevant to the association and dissociation of substrates and ligands is far more complex because we have to consider the transfer of the substrate from water to the active site.

B.   The binding energies of proteins and ligands

The essence of binding between enzymes and substrates is the matching of complementary surfaces; polar residues match up with neutralizing partners, nonpolar residues pair with polar residues. In doing so, the water of solvation is sloughed off. Thus, the extrapolation of the energetics discussed in the last section to the energetics of binding is difficult. We have to calculate not only the energies between enzymes and substrates or other ligands but also their interaction energies with water and the changes in entropy during the reaction. The net binding energies are the small differences between two larger numbers. Although it is difficult to calculate the values de novo, there is now much experimental data available on these energies.

We begin with the hydrophobic bond. Purists prefer to call this an "effect" rather than a bond because it is not a direct interaction but one that involves the displacement of water. But, as will be seen, all bonding in solution follows the same pattern!

1.   The hydrophobic bond[16–19]

The hydrophobic bond is a way of describing the tendency of nonpolar compounds such as hydrocarbons to transfer from an aqueous solution to an organic phase.[16] The classic theory of the hydrophobic bond is that it results not so much from the direct interaction of solvent and solute molecules as from the reorganization of the normal hydrogen-bonding network in water by the presence of a hydrophobic compound. Water consists of a dynamic, loose network of hydrogen bonds. The presence of a nonpolar compound causes a local rearrangement in this network. In order to preserve the number of hydrogen bonds, each one having an energy of about 25 kJ/mol (6 kcal/mol), the water molecules line up around the nonpolar molecule. The hydrophobic solute therefore does not cause large enthalpy changes in the polar solvent but instead decreases the entropy of the system due to the increase in local order. A hydrophobic molecule is driven into the hydrophobic region of a protein by the regaining of entropy by water.

There is, however, a second component that results from dispersion energies. Dispersion forces are weak in water because of the low polarizability of oxygen (Table 11.2) and because of the low atom density (the dispersion energies are additive). This is an additional factor favoring the self-association of hydrocarbons, as they have a higher atom density and the polarizability of $-CH_2-$ is greater than that of O.

One convenient way of measuring the hydrophobicity of a molecule is to measure its partition between the organic and aqueous phases when it is shaken with an immiscible mixture of an organic solvent, often n-octanol, and water.

The distribution of the solute between the two phases depends on the competing tendencies of the hydrophobic regions to be squeezed into the organic phase by the hydrophobic bond, and the polar regions to be solvated and drawn into the aqueous phase.

a. Simplified macroscopic model for comparing the free energies of transfer of side chains from water to hydrocarbons with transfer to proteins

It is instructive to analyze the hydrophobic bond by a simple classical model of surface tension. The model is correct in principle but not in detail. It does, however, illustrate the basic physical principles and is useful in comparing hydrophobic binding in model systems using the transfer of solutes from water to hydrocarbons with proteins and model experiments. The transfer of a hydrocarbon solute from water to a hydrocarbon solvent consists of the following notional steps (Figure 11.4): (1) removal of the hydrocarbon to vacuum; (2) creation of a cavity in the hydrocarbon solvent; and (3) transfer of the solute to the

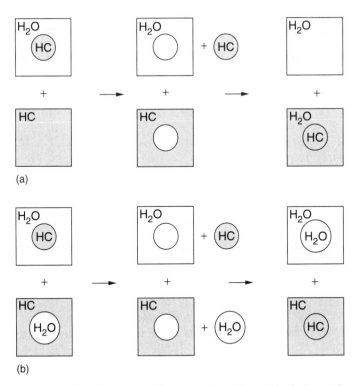

Figure 11.4 Steps in a simple model for comparison of transfer of a ligand from water to a preformed pocket in a protein with the transfer from water to a hydrocarbon solvent. [Modified from A. R. Fersht, S. E. Jackson, and L. Serrano, *Phil. Trans. R. Soc. London Series A* **345**, 141 (1993).]

hydrocarbon solvent and the collapse of the cavity in water. According to the Dupré equation, the "work of adhesion" per unit area, w, between two immiscible liquids is given by

$$w = \gamma_1 + \gamma_2 - \gamma_{1/2} \tag{11.4}$$

where γ_1 is the surface tension of liquid 1, γ_2 is that of liquid 2, and $\gamma_{1/2}$ is the interfacial surface tension of the two liquids. Equation 11.4 may be applied to the notional steps in Figure 11.4. For a typical hydrocarbon, $\gamma_w = 435$, $\gamma_{HC} = 108$, and $\gamma_{HC/w} = 304$ J/mol/$Å^2$ (1.82, 0.45, and 1.27 kcal/mol/$Å^2$).

Case A.   Transfer of hydrocarbon from water to hydrocarbon. The transfer of a hydrocarbon solute of surface area A from water to hydrocarbon removes the hydrocarbon–water interface ($w = -\gamma_{HC/w}$) and creates a hydrocarbon–vacuum and a water–vacuum interface ($w = \gamma_{HC} + \gamma_w$); formation of a cavity in the hydrocarbon solvent creates a hydrocarbon–vacuum interface ($w = \gamma_{HC}$); transfer of the hydrocarbon solute to the hydrocarbon, and collapse of the water–vacuum cavity, loses two hydrocarbon–vacuum surfaces and a water–vacuum surface ($w = -2\gamma_{HC} - \gamma_w$). The free energy of transfer, ΔG_{trans}, is given by $A \times \Sigma w$; i.e.,

$$\Delta G_{trans} = A(-\gamma_{HC/w} + \gamma_{HC} + \gamma_w + \gamma_{HC} - 2\gamma_{HC} - \gamma_w) \tag{11.5}$$

$$\Delta G_{trans} = -A\gamma_{HC/w} \tag{11.6}$$

This may be seen directly by comparing the right-hand side of Figure 11.4 with the left-hand side, which shows that the net process is the loss of one water–hydrocarbon interface.

Case B.   Transfer of hydrocarbons from water to hydrocarbon (protein) with preformed water-filled cavity. Applying the same procedure to a protein in which there is a preformed hydrophobic cavity that is filled with water gives

$$\Delta G_{trans} = A(-2\gamma_{HC/w} + 2\gamma_{HC} + 2\gamma_w - 2\gamma_w - 2\gamma_{HC}) \tag{11.7}$$

$$\Delta G_{trans} = -2A\gamma_{HC/w} \tag{11.8}$$

That is, the free energy of transfer is *twice* that calculated for case A. This is expected from comparing the right-hand side of Figure 11.4 with the left-hand side, which shows that the net process in transfer is the loss of two water–hydrocarbon interfaces.

Two important principles arise from this analysis:

1. The energy of the hydrophobic effect should be proportional to the surface area that is buried.

2. The effect should be twice as strong for the binding of hydrocarbons in preformed cavities in proteins.

It has been found experimentally that the hydrophobic effect is proportional to buried surface area for the transfer of small molecules to hydrophobic solvents. The energy of transfer is $80-100$ kJ/mol per Å^2 of solute surface area that becomes buried ($20-25$ cal/mol/Å^2).[20–23]

The structuring of water molecules around hydrophobic groups leads to large specific heat effects, with interesting consequences for protein folding. This, and the thermodynamics of the hydrophobic effect, is discussed in Chapter 17, section A1. Another example of hydrophobic interactions being proportional to buried surface area is given in Chapter 17, section D1a, where it will be seen that this is an important factor in rationalizing helix stability in proteins. The hydrophobic effect is a dominant interaction in protein folding and its kinetics.

b. Hydrophobicity of small groups: The Hansch equation[24,25]

From determinations of the partitioning of several series of substituted compounds between *n*-octanol and water, C. Hansch and coworkers found that many substituents make a constant, and additive, contribution to the hydrophobicity of the parent compound. If the ratio of the solubility of the parent compound (H—S) in the organic phase to that in the aqueous phase is P_0, and that of the substituted compound (R—S) is P, then the hydrophobicity constant for the substituent R, π, is defined by

$$\pi = \log \frac{P}{P_0} \tag{11.9}$$

Note that $RT \ln(P/P_0)$, i.e., $2.303RT\pi$, is the incremental Gibbs free energy of transfer of the group R from *n*-octanol to water (relative to the hydrogen atom H). Some values of π are listed in Table 11.4. Points to be noted are:

1. The values of π for groups that are not strongly electron-donating or -withdrawing are virtually constant and independent of the group to which they are attached. Further, their effects are additive. For example, the methylene group has a π of 0.5, and the addition of each additional methylene adds a further increment of 0.5 to π. The 0.5 log unit is equal to a change of 2.84 kJ/mol (0.68 kcal/mol) in the Gibbs free energy. (In this context, the substitution of a methyl group for a hydrogen is the same as the addition of a methylene group, since it is equivalent to the interposing of a methylene group between the hydrogen and the rest of the molecule.)

2. The behavior of groups that can conjugate with the benzene ring, such as the nitro and amino groups, is variable and depends on the other groups attached to the ring.

We use the term "incremental" Gibbs free energy because we measure the change in free energy of transfer of a parent compound on the addition of an extra group. The absolute free energy of transfer of the parent molecule is unimportant; it is the change in energy that matters. The change in energy, of which π is a measure, is reproducible from one parent molecule to the next for the same

Table 11.4 *Some values of π^a*

Group[b]	π	Incremental Gibbs free energy of transfer from n-octanol to water	
		kJ/mol	kcal/mol
—CH$_3$	0.5	2.85	0.68
—CH$_2$CH$_3$	1.0	5.71	1.36
—(CH$_2$)$_2$CH$_3$	1.5	8.56	2.05
—(CH$_2$)$_3$CH$_3$	2.0	11.41	2.73
—(CH$_2$)$_4$CH$_3$	2.5	14.26	3.41
—CH(CH$_3$)CH$_3$	1.3	7.42	1.77
—CH(CH$_2$CH$_3$)CH$_3$	1.8	10.27	2.45
—CH$_2$Ph	2.63	15.00	3.59
—OH[c]	−1.16	−6.62	−1.58
—NHCOCH$_3$[c]	−1.21	−6.90	−1.65
—OCOCH$_3$[c]	−0.27	−1.54	−0.37

[a] From C. Hansch and E. Coats, *J. Pharm. Sci.* **59**, 731 (1970).
[b] Relative to the hydrogen atom.
[c] Bound to aliphatic compounds.

solvent. The values also may be scaled for different solvents. For example, the values are uniformly larger in octane, which is more hydrophobic than octanol. π is a very useful indicator of the hydrophobicity of a side chain, and values of π derived from the model systems are used as references for plotting or calculating the effects of changes in hydrophobic interactions in other reactions.

c.   Aqueous solvation energies

The values of π are derived from the transfer of a side chain from water to a hydrophobic solvent and so measure the relative energies of solvation of water relative to the organic solvent. R. Wolfenden and coworkers have measured the incremental free energies of transfer of amino acid side chains from water to near vacuum (low gas pressures) by measuring the vapor pressure of a series of substituted compounds (Table 11.5). These numbers are useful in understanding, for example, how mutations in proteins affect the energies of the denatured state (Chapter 17).

2. Hydrogen bonds, salt bridges, and the hydrogen bond inventory

A typical hydrogen bonding reaction is equation 11.10. On the left, a hydrogen bond donor,

$$E-XH \cdots OH_2 + OH_2 \cdots B-S \rightleftharpoons E-XH \cdots B-S + H_2O \cdots H_2O \quad (11.10)$$

—XH, on an enzyme makes a hydrogen bond with the O of water, and an acceptor B on the substrate makes a hydrogen bond with the H of another water

Table 11.5 *Free energies of transfer of amino acid side chains to water from cyclohexane or the gas phase*

Side chain of	Cyclohexane $\rightarrow H_2O$		Vapor $\rightarrow H_2O$	
	kcal/mol	kJ/mol	kcal/mol	kJ/mol
Leu	4.92	20.59	2.28	9.54
Ile	4.92	20.59	2.15	9.00
Val	4.04	16.91	1.99	8.33
Pro	3.58	14.98	1.50	6.28
Phe	2.98	12.47	−0.76	−3.18
Met	2.35	9.83	−1.48	−6.19
Trp	2.33	9.75	−5.88	−24.61
Ala	1.81	7.57	1.94	8.12
Cys	1.28	5.36	−1.24	−5.19
Gly	0.94	3.93	2.39	10.00
Tyr	−0.14	−0.59	−6.11	−25.57
Thr	−2.57	−10.75	−4.88	−20.42
Ser	−3.40	−14.23	−5.06	−21.18
His	−4.66	−19.50	−10.27	−42.98
Gln	−5.54	−23.18	−9.38	−39.25
Lys	−5.55	−23.23	−9.52	−39.84
Asn	−6.64	−27.79	−9.68	−40.51
Glu	−6.81	−28.50	−10.24	−42.85
Asp	−8.72	−36.49	−10.95	−45.82
Arg	−14.92	−62.44	−19.92	−83.36

Free energies of transfer at 25°C and pH 7.0, adjusted for the effects of ionization, assuming that ionized forms move entirely into the aqueous phase. The difference between these two sets (i.e., the value for vapor→cyclohexane transfer) is presumably a measure of susceptibility to van der Waals attractions and seems to be simply related to surface area. A. Radzicka and R. Wolfenden, *Biochemistry* **27**, 1664–1670 (1988); R. Wolfenden, L. Andersson, P. M. Cullis, and C. C. B. Southgate, *Biochemistry* **20**, 849–855 (1981); P. R. Gibbs, A. Radzicka, and R. Wolfenden, *J. Am Chem. Soc.* **118**, 6105 (1996); **113**, 4714–4715 (1991); and A. Radzicka, G. B. Young, and R. Wolfenden, *Biochemistry* **32**, 6807–6809 (1993).

molecule. On the right, —XH bonds with B, and the released water molecules each make a hydrogen bond with water, which is formally equivalent to binding to each other.

A crude, but very effective, way of understanding the energetics of hydrogen bonding is to perform a hydrogen bond inventory; i.e., count the number and nature of hydrogen bonds on each side of the chemical equation.[26,27] Suppose that —XH is a good donor and B a good acceptor. Then, on each side of the equation, there are two good hydrogen bonds. There is then little change in enthalpy in the reaction. Suppose that —XH is a poor hydrogen bond donor, but B is a good acceptor. Then on the left there is one strong bond and one weak one. The same is true on the right. The same inventory can be performed if —XH is a good donor and B is a poor acceptor. Again, there should be little change in enthalpy during the reaction. Thus, hydrogen bonding should be relatively isoenthalpic.

Although the hydrogen bond inventory is zero, hydrogen bonding is energetically favorable in the formation of enzyme–substrate complexes because of the increase in entropy on the release of bound water molecules. This is analogous to the discussion in Chapter 2, section B4, of the merits of intramolecular reactions versus intermolecular reactions (also see section D later in this chapter). When two molecules associate, they lose overall translational and rotational entropy, and this has to be deducted from the energy balance on binding. But if they are already held together in a complex, there is no further loss of translational and rotational entropy if an extra hydrogen bond is built in.

It is further instructive to take the hydrogen bond inventory to extreme situations. Suppose we make—XH such a weak acceptor that it becomes a nonpolar group (equation 11.11). The hydrogen bond inventory is 1 for each side of the equation, and so the reaction

$$E/OH_2 + OH_2 \cdots B{-}S \rightleftharpoons E/B{-}S + H_2O \cdots H_2O \qquad (11.11)$$

should still be isoenthalpic, despite the presence of an unpaired hydrogen bond acceptor. This situation is mimicked by protein engineering experiments (Chapter 15). These show directly that the removal of a hydrogen bond donor or acceptor weakens binding by only 2–6 kJ/mol (0.5–1.5 kcal/mol), providing there are no unpaired ions formed, despite the enthalpy of an individual bond being some 20 kJ/mol (5 kcal/mol). The proton inventory model is basically sound and stands up to critical analysis.[28]

Another extreme is the formation of a salt bridge between, say, an ammonium ion and a carboxylate (equation 11.12). There are two good hydrogen bonds on each side,

$$E{-}NH_3^+ \cdots OH_2 + OH_2 \cdots {}^-O_2C{-}S \rightleftharpoons E{-}NH_3^+ \cdots {}^-O_2C{-}S + H_2O \cdots H_2O \qquad (11.12)$$

but there is a favorable electrostatic interaction on the right. We know that electrostatic energies are weak in water. In practice, the energies are indeed approximately balanced (see Chapter 7).

It is not easy to calculate binding energies between enzymes and substrates because the energies are delicately balanced for the formation of hydrogen bonds and salt bridges, and so much hinges on the entropy terms. But there is good experimental evidence on the binding energies of different types of groups.

C. Experimental measurements of incremental energies

1. Binding versus specificity

Protein chemists measure the incremental binding energy of a group R— on a substrate by measuring its affinity for an enzyme relative to that of a parent substrate that has H— at that position. There is a crucial difference between doing this and measuring the partitioning of the two compounds between an organic solvent and water. The solvent can always adjust its structure to fit around the substrate, but the enzyme has a predetermined geometry. The ideal scenario for measurements with an enzyme is Figure 11.5; there is a series of subsites that can be successively filled by the addition of increasingly larger groups to a

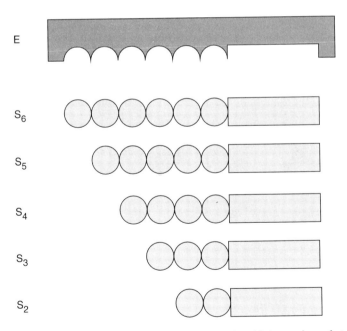

Figure 11.5 Filling up the binding subsites on a protein with increasing substrate size. This is achieved in practice by using substrates smaller than the natural one. [Modified from A. R. Fersht, *Trends in Biochemical Sciences* **12**, 301 (1987).]

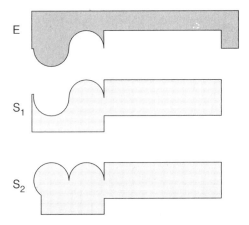

Figure 11.6 Comparison of the binding of a substrate S_2 with that of the natural substrate S_1 can be dominated by unfavorable interactions between S_2 and the protein. [Modified from A. R. Fersht, *Trends in Biochemical Sciences* **12**, 301 (1987).]

parent compound. It is easy to do this in reverse; take a large substrate and progressively remove groups from it. Unfortunately, many experiments are not done this way because it is often easier to substitute one group for another, as in Figure 11.6, and there can be steric clashes. For example, suppose there is an $>NH \cdots O<$ between a substrate and an enzyme. The substitution of an $>O$ for the $>NH$ leads to a repulsive $>O \cdots O<$ interaction because the sum of the van der Waals radii of two oxygen atoms is less than that of the $>NH \cdots O<$ bond length. Thus, a measure of the relative affinities of the two substrates will have a term for the loss of the $>NH \cdots O<$ hydrogen bond plus a large contribution from the $>O \cdots O<$ repulsion. Yet results from experiments such as this one have been claimed as evidence for large hydrogen bond strengths.

A measure of the relative affinities of two substrates for an enzyme, or a parent and modified enzyme for a single substrate, is always a measure of the *specificity* of an interaction and is not a true measure of the binding energy. But in the ideal circumstances of Figure 11.5, we can approximate the energy to an incremental binding energy.[29] Always bear in mind that the energetics could result from an instability in one state rather than a stability in another. The golden rule is that it is far safer to remove groups from a larger substrate than to add to or substitute groups. The same rule applies to protein engineering experiments, discussed later in Chapters 15 and 17.

2. Estimation of increments in binding energy from kinetics

One way of measuring the contribution of a substituent R in a substrate R—S to binding, $\Delta\Delta G_b$, is to compare the dissociation constants of R—S and H—S from the enzyme:

$$\Delta G_{b(\text{R—S})} = -RT \ln K_{S(\text{R—S})} \tag{11.13}$$

$$\Delta G_{b(\text{H—S})} = -RT \ln K_{S(\text{H—S})} \tag{11.14}$$

so that

$$\Delta\Delta G_b = -RT \ln \frac{K_{S(\text{H—S})}}{K_{S(\text{R—S})}} \tag{11.15}$$

But, as will be seen in the next chapter, this often underestimates the binding energy of the larger substrate, since enzymes frequently use binding energy to lower the activation energies of reactions rather than using it to give tighter K_M's. A much better method is to compare the values of k_{cat}/K_M for the two substrates. This quantity includes both the activation energies and the binding energies, and avoids the underestimates that result from using dissociation constants alone. It is shown in Chapter 13, equation 13.3, that

$$\Delta\Delta G_b = -RT \ln \frac{(k_{\text{cat}}/K_M)_{(\text{R—S})}}{(k_{\text{cat}}/K_M)_{(\text{H—S})}} \tag{11.16}$$

$(-\Delta\Delta G_b$ is the incremental Gibbs free energy of transfer of R from the enzyme to water, relative to H. This is related to π of section A4b.)

Note that the ratio of k_{cat}/K_M for two substrates is the correct measure for specificity (Chapter 3, equation 3.44), emphasizing that we are really measuring the energetics of specificity.

a. Intrinsic binding energies

The intrinsic binding energy of a group R is the maximum binding energy possible when there is perfect complementarity between it and its binding cavity in a protein. In practice, enzymes do not necessarily utilize all of the intrinsic binding energy of a group, for various reasons. Consider the utilization of the binding energy of the hydroxyl group of tyrosine with various enzymes. At one extreme, there is the binding to the phenylalanyl-tRNA synthetase. Here, as will be discussed in Chapter 13, the enzyme has evolved to bind tyrosine as weakly as possible; there is no binding site for the hydroxyl group, and there will be steric repulsion. In the middle of the range is chymotrypsin. This enzyme has approximately equal specificities for phenylalanine and tyrosine, so it uses only a fraction of the intrinsic binding energy. At the other extreme, the tyrosyl-tRNA synthetase has evolved to bind the —OH of tyrosine as tightly as possible to maximize the specificity against phenylalanine.[30]

b. Estimation of the upper limits of binding energies by measurements on the aminoacyl-tRNA synthetases

The evolutionary pressure on the aminoacyl-tRNA synthetases to bind the distinctive features of their correct substrates as tightly as possible (Chapter 13)

Table 11.6 *Binding energies of aminoacyl-tRNA synthetases and various groups[a]*

Group[b]	Incremental Gibbs energy of transfer from enzyme to water[b]	
	kJ/mol	kcal/mol
—CH$_3$	14	3.2
—CH$_2$CH$_3$	27	6.5
—CH(CH$_3$)$_2$	40	9.6
—S—	23	5.4

[a]From A. R. Fersht, J. S. Shindler, and W.-C. Tsui, *Biochemistry* **19**, 5520 (1980): W.-C. Tsui and A. R. Fersht, *Nucl. Acids Res.* **9**, 4627 (1981).
[b]Relative to the hydrogen atom.

affords an experimental method of measuring the maximum possible binding energies. This is done by measuring the values of k_{cat}/K_M for the pyrophosphate exchange reaction (Chapter 7, section D1) and using equation 11.16. Fortunately, one of the assumptions in the derivation of that equation—that the group R has a negligible inductive effect on the reaction—holds well here. The data are listed in Table 11.6. Note the following:

A —CH$_3$ group contributes about 13 kJ/mol (3.2 kcal/mol) of binding energy, as determined by experiments on many different enzymes. This is some 5 times higher than the values for hydrophobic binding listed in Table 11.4 for transfer from *n*-octanol to water. (The values in Table 11.6 are for experiments equivalent to those in 11.4, except that the transfer is from the enzyme to water.) The Gibbs energy of transfer for a larger side chain, e.g., that of valine, is also 5 times higher from an enzyme to water than from *n*-octanol to water.

c. Typical binding energies from measurements on chymotrypsin

Chymotrypsin, an enzyme of broad specificity, has a hydrophobic binding pocket (Chapter 1). Kinetic measurements on this enzyme give a general idea of the strength of hydrophobic binding with enzymes.

The values of k_{cat}/K_M for the chymotrypsin-catalyzed hydrolysis of a series of esters of the form R—CH(NHAc)CO$_2$CH$_3$, where R is an unbranched alkyl chain, increase with increasing hydrophobicity of R.[8] The decrease in the activation energy is 2.2 times greater than the free energy of transfer of the alkyl groups from water to *n*-octanol (Figure 11.7).[31] The hydrophobic binding pocket appears to be 2.2 times more hydrophobic than *n*-octanol.

The inhibition constants of a series of substituted formanilides increase with increasing hydrophobicity. A plot of the logarithms of the constants against π

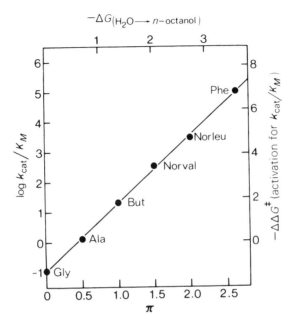

Figure 11.7 The relationship between the hydrophobicity of the side chain of the amino acid and k_{cat}/K_M for the hydrolysis of N-acetyl-L-amino acid methyl esters by chymotrypsin. Energies are in kcal/mol. [From V. N. Dorovskaya, S. D. Varfolomeyev, N. F. Kazanskaya, A. A. Klyosov, and K. Martinek, *FEBS Lett.* **23**, 122 (1972).]

yields a straight line of slope -1.5.[32] This again shows that the active site of chymotrypsin is more hydrophobic than n-octanol (Figure 11.8).

d. Why are enzymes more hydrophobic than organic solvents?

Chymotrypsin fits the simple model of section B1a for the hydrophobic effect very well: it has a preformed binding pocket that is filled with 16 water molecules. It is expected that the hydrophobic bond should be twice as effective, because two unfavorable hydrophobic–water interfaces are removed on formation of the enzyme–substrate complex, one around the substrate and one inside the enzyme (see equations 11.6 and 11.8).

The tight binding of methylene groups by the aminoacyl-tRNA synthetases has an additional cause. When valine occupies the binding site of the isoleucyl-tRNA synthetase, for example, there is a "hole" in the complex that would be occupied by the additional methylene group in isoleucine if it were bound. It was pointed out earlier that the dispersion energy of a methylene group in a crystalline hydrocarbon is 8.4 kJ/mol (2 kcal/mol). Since proteins are as closely packed as solids,[33] an empty "hole" adds a further 8.4 kJ/mol to the hydrophobic bond energy of the methylene group. The hydrophobic binding in these examples is clearly dominated by the contribution of dispersion energies, as opposed

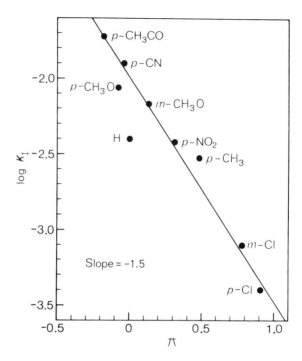

Figure 11.8 Variation of the dissociation constants of substituted formanilides with π for the partition between n-octanol and water.

to the classic theory of the hydrophobic bond. Since dispersion energies are additive and so are dependent on the density of atoms, dispersion contributions will be important in enzymatic binding where specific, close-packed interactions are involved. This is shown experimentally in Chapter 17, Figure 17.8.

e. Buried charged groups and high apparent energies.

Many interactions between proteins and ligands involve salt linkages between positively and negatively charged groups. Removal of one of the partners to give a buried charge often gives very large changes in binding energy, typically 12–20 kJ/mol (3–5 kcal/mol); for example, from comparing the dissociation constants for equations 11.17 and 11.18.

$$\text{E—CO}_2^- \cdots \text{OH}_2 + \text{OH}_2 \cdots {}^+\text{H}_3\text{N—S} \rightleftharpoons \text{E—CO}_2^- \cdots {}^+\text{H}_3\text{N—S} + \text{H}_2\text{O} \cdots \text{H}_2\text{O} \tag{11.17}$$

$$\text{E—CO}_2^- \cdots \text{OH}_2 + \text{H—S} \rightleftharpoons \text{E—CO}_2^-/\text{H—S} + \text{H}_2\text{O} \cdots \text{H}_2\text{O} \tag{11.18}$$

This is because the energy of desolvation of a charged group is huge. It is so large that the structure will rearrange so that water can penetrate the protein to solvate the strength of the salt bridge. The figure is not so much a measure of the stabilization of the salt bridge in equation 11.17 as the instability of a buried charge in equation 11.18.

The same argument is true for the removal of a hydroxyl group on a substrate that solvates a buried negative charge in the protein. For example, the tyrosyl-tRNA synthetase has a specific binding site for the —OH of tyrosine to maximize the discrimination against phenylalanine (see Figure 15.1), consisting of the carboxylate of Asp-76 and the —OH of Tyr-34. In the absence of substrate, a water molecule occupies the site. Tyrosine binds to the tyrosyl-tRNA synthetase 10^5 times more tightly than does phenylalanine, equivalent to 29 kJ/mol (7 kcal/mol).[30] This does not result so much from a very strong hydrogen bond to tyrosine as from a very unfavorable interaction in the enzyme–phenylalanine complex. The high changes in energy in these examples are measures of specificity; that is, relative rather than incremental binding energies.

D. Entropy and binding[34]

The pairing of monomeric complementary bases, such as A (adenosine) and U (uracil), is not detectable in aqueous solution due to the competition of hydrogen bonding with water molecules. However, a triplet of bases binds strongly to a complementary anticodon of a tRNA. The triplet UUC (uracil-uracil-cytosine) binds to tRNA^{Phe} with an association constant of 2×10^3 M^{-1}.[35,36] Two tRNA molecules with complementary anticodons associate even more strongly; the association constant between tRNA^{Phe} and tRNA$_2{}^{Glu}$, is 2×10^7 M^{-1}.[37] The reason for this increase in the strength of hydrogen bonding is entropy. When a single A and U associate, they gain the energy of the complementary base pairing but lose the energy of hydrogen bonding with water. They also lose their independent entropies of translation and overall rotation, but in turn there is a gain of entropy as the hydrogen-bonded water molecules are released. When a complementary pair of triplets bind, three times as many molecules of water are freed from hydrogen bonding, but there is still the loss of only one set of entropies of translation and overall rotation. The situation is very reminiscent of the advantages of an intramolecular reaction over its intermolecular counterpart (Chapter 2, section B4d). The lesson is that although single hydrogen bonds are weak in solution, multiple hydrogen bonds may be very stable.

A related phenomenon is that the binding of a dimer $X—Y$ to an enzyme may be far greater than expected from the binding of X and Y separately, because the dimer loses only one set of translational and rotational entropies overall.

This phenomenon has been known for many years to inorganic chemists as the *chelate effect*. The magnitude may be illustrated by one of their examples, the replacement by ammonia and polyamines of some or all of the six water molecules that are coordinated to the Ni^{2+} ion. It is seen in Table 11.7 that there are enormous increases in the association constants of the ligands as the number of amino groups increases.

Table 11.7 *The chelate effect on the binding of amines to Ni^{2+a}*

Ligand	Association constants (M^{-1})					
	K_1	K_2	K_3	K_4	K_5	K_6
NH_3	468	132	41	12	4	0.8
$H_2N(CH_2)_2NH_2$	2×10^7	1×10^6	2×10^4	—	—	—
$H_2N(CH_2)_2NH(CH_2)_2NH_2$	6×10^{10}	1×10^8	—	—	—	—
$H_2N(CH_2)_2NH(CH_2)_2NH(CH_2)_2NH_2$	2×10^{14}	—	—	—	—	—
$(H_2N(CH_2)_2NH(CH_2)_2)_2NH$	3×10^{17}	—	—	—	—	—

[a]From the Chemical Society (London) Special Publications 17 (1964) and 25 (1971).

E. Enthalpy–entropy compensation

We have seen throughout this chapter that entropy changes dominate many of the interactions between proteins and ligands. It would appear important to dissect out measured values of ΔG into contributions from entropy and enthalpy. Unfortunately, we cannot do this because of enthalpy–entropy compensation in water; a phenomenon first described in the 1930s but invariably forgotten.[38–40] It is illustrated from the thermodynamics of ionization of acetic and formic acids. Formic acid is the stronger acid by one pK_a unit (5.7 kJ/mol, 1.4 kcal/mol). Elementary chemistry tells us that this is due to the inductive (electron donating) effect of the α-methyl group of acetic acid, which destabilizes the carboxylate ion. This is an enthalpic contribution. But the enthalpies of ionization of acetic and formic acids are identical; the differences are in the entropy. This is because water changes its local structure when it surrounds solutes. These changes minimize its free energy so that $\Delta G \sim 0$, but there are larger and compensating changes in ΔH_{water} and ΔS_{water}. Thus, measured values of ΔH and ΔS for reactions in solution contain contributions from ΔH_{water} and ΔS_{water} that may obscure them but that cancel out in ΔG because ΔG_{water} ($= \Delta H_{water} - T\Delta S_{water}$) is zero.

Dunitz has analyzed entropy–enthalpy compensation semiquantitatively and has shown that it extends to all weak interactions.[40] A weak interaction is a bond with a low vibrational frequency, which gives it an appreciable entropy (Chapter 2, section B4, Table 2.3). In general, a particle in a shallow, wide, energy well has more entropy than one in a deep, narrow well, and so there is a trade-off between enthalpy and entropy. For interactions of about 20 kJ/mol (5 kcal/mol) at 300 K, the changes in entropy and enthalpy by this process balance, and so there is nearly complete enthalpy–entropy compensation when the bond is broken.

ΔG is the thermodynamic quantity that is the most accurately measured from binding constants in solution, is subject to the least uncertainty in interpretation, and the most relevant in calculations of equilibria.

F. Summary

The dissociation constant of an enzyme and substrate reflects the *relative* stabilities of the substrate when it is bound to the enzyme and when it is free in solution. The constant depends on: the strength of the hydrogen bonding of the enzyme and the substrate to each other, compared with the combined strengths of their separate hydrogen bonding to water molecules; the stability of the salt linkages between the enzyme and the substrate, compared with the tendency of the individual ions to be solvated by water; the dispersion energies, compared with those in water; and also the hydrophobic bonding. These differences in energy, when they are summed over a whole molecule, can be quite considerable. It is of interest that the hydrogen bond and the salt linkage with the enzyme as well as the hydrophobic bond are favored entropically by the release of constrained water molecules. These principles apply to all molecular recognition phenomena in solution: protein–protein; protein–ligand; protein–nucleic acid; and nucleic acid–nucleic acid interactions.

The *absolute* energy of interaction between the enzyme and the substrate depends on the dispersion forces and the absolute energies of the hydrogen bonds and salt linkages. The hydrogen bonds and the salt linkages provide the strongest attractive forces. Since these are also able to stabilize unfavorable charge formation in the transition state, they are especially important in catalysis.

Although opinions differ on the overall contribution of hydrogen bonding and salt linkage to the total stabilization energy of noncovalent complexes, and it has been suggested that many association constants are entirely accounted for by hydrophobic bonding, there is no doubt that hydrogen bonds and electrostatic interactions are important for specificity.[38,41,42] Whatever the positive contribution of correctly formed hydrogen bonds and salt bridges in the "correct" complexes, the presence of *unpaired* hydrogen-bond donors/acceptors and ions in "incorrect" complexes provides considerable driving energy for their dissociation.

References

1. A. Warshel and M. Levitt, *J. Molec. Biol.* **103**, 227 (1976).
2. A. T. Hagler, S. Lifson, and E. Huler in *Peptides, polypeptides, and proteins* (E. R. Blout, F. A. Bovey, M. Goodman, and N. Lotan, Eds.), Wiley, p. 35 (1974).
3. P. Dauber and A. T. Hagler, *Accts. Chem. Res.* **13**, 105 (1980).
4. A. T. Hagler, P. Dauber, and S. Lifson, *J. Am. Chem. Soc.* **101**, 5131 (1979).
5. W. W. Cleland and M. M. Kreevoy, *Science* **264**, 1887 (1994).
6. P. A. Frey, *Science* **268**, 189 (1995).
7. B. Schwartz and D. G. Drueckhammer, *J. Am. Chem. Soc.* **117**, 11902 (1995).
8. O. Hur, C. Leja, and M. F. Dunn, *Biochemistry* **35**, 7378 (1996).
9. J. A. Gerlt, M. M. Kreevoy, W. W. Cleland, and P. A. Frey, *Chemistry & Biology* **4**, 259 (1997).
10. A. Warshel, A. Papazyan, and P. A. Kollman, *Science* **269**, 102 (1995).
11. A. Warshel and A. Papazyan, *Proc. Natl. Acad. Sci. USA* **93**, 13665 (1996).

12. S. O. Shan, S. Loh, and D. Herschlag, *Science* **272**, 97 (1996).
13. J. P. Guthrie, *Chemistry & Biology* **3**, 163 (1996).
14. E. L. Ash, J. L. Sudmeier, E. C. DeFabo, and W. W. Bachovchin, *Science* **278**, 1128 (1997).
15. W. D. Cornell, P. Cieplak, C. I. Bayly, I. R. Gould, K. M. Merz, D. M. Ferguson, D. C. Spellmeyer, T. Fox, J. W. Caldwell, and P. A. Kollman, *J. Am. Chem. Soc.* **117**, 5170 (1995).
16. W. Kauzmann, *Adv. Protein Chem.* **14**, 1 (1959).
17. A. Ben-Naim, *ACS Symposium Series* **568**, 371 (1994).
18. A. Ben-Naim, *Pure and Applied Chemistry* **69**, 2239 (1997).
19. A. Ben-Naim and R. Mazo, *J. Phys. Chem. B* **101**, 11221 (1997).
20. R. B. Hermann, *J. Phys. Chem.* **76**, 2754 (1972).
21. M. J. Harris, T. Higuchi, and J. H. Rytting, *J. Phys. Chem.* **77**, 2694 (1973).
22. J. A. Reynolds, D. B. Gilbert, and C. Tanford, *Proc. Natl. Acad. Sci. USA* **71**, 2925 (1974).
23. C. Chothia, *Nature, Lond.* **248**, 338 (1974).
24. T. Fujita, J. Iwasa, and C. Hansch, *J. Am. Chem. Soc.* **86**, 5175 (1964).
25. A. Leo, C. Hansch, and D. Elkins, *Chem. Rev.* **71**, 525 (1971).
26. A. R. Fersht, J. P. Shi, J. Knill Jones, D. M. Lowe, A. J. Wilkinson, D. M. Blow, P. Brick, P. Carter, M. Waye, and G. Winter, *Nature* **314**, 235 (1985).
27. A. R. Fersht, *Trends in Biochemical Sciences* **12**, 301 (1987).
28. F. Lau and M. Karplus, *J. Molec. Biol.* **236**, 1049 (1994).
29. A. R. Fersht, *Biochemistry* **27**, 1577 (1988).
30. A. R. Fersht, J. S. Shindler, and W.-C. Tsui, *Biochemistry* **19**, 5520 (1980).
31. V. N. Dorovskaya, S. D. Varfolomeyev, N. F. Kazanskaya, A. A. Klyosov, and K. Martinek, *FEBS Lett.* **23**, 122 (1972).
32. J. Fastrez and A. R. Fersht, *Biochemistry* **12**, 1067 (1973).
33. Y. Harpaz, M. Gerstein, and C. Chothia, *Structure* **2**, 641 (1994).
34. W. P. Jencks, *Adv. Enzymol.* **43**, 219 (1975).
35. J. Eisinger, B. Feuer, and T. Yamane, *Nature New Biology, Lond.* **231**, 126 (1971).
36. O. Pongs, R. Bald, and E. Reinwald, *Eur. J. Biochem.* **32**, 117 (1973).
37. J. Eisinger and N. Gross, *Biochemistry* **14**, 4031 (1975).
38. W. P. Jencks, *Catalysis in chemistry and enzymology*, McGraw-Hill, pp. 351, 399 (1969).
39. R. Lumry and S. Rajender, *Biopolymers* **9**, 1125 (1970).
40. J. D. Dunitz, *Chemistry & Biology* **2**, 709 (1995).
41. C. Chothia and J. Janin, *Nature, Lond.* **256**, 705 (1975).
42. J. Janin and C. Chothia, *J. Molec. Biol.* **100**, 197 (1976).

Enzyme – Substrate Complementarity and the Use of Binding Energy in Catalysis

12

We saw in Chapter 2 how chemists have mimicked some of the features of enzyme catalysis in model reactions. Significant rate enhancements can be achieved by simple acid-base catalysis, avoidance of unfavorable entropic effects, and the use of nucleophilic groups on the enzyme to change mechanism. But catalysis in model systems is still many orders of magnitude short of that found in enzymes. We saw in the last chapter that there is considerable energy of binding between enzymes and substrates. Enzymes have evolved to use the binding energy between enzymes and substrates to provide the additional catalysis. It was suggested by J. B. S. Haldane as long ago as 1930, before it was universally accepted that enzymes are proteins, that enzymes may use binding energy to distort substrates to resemble products.[1] Ever since then, theoreticians have explored the ways in which the binding energy can be used to lower chemical activation energies.[2-6] Many of the early ideas were based on speculation about what would happen if hypothetical changes were made to an enzyme. But the advent of protein engineering (Chapter 14) allowed the speculation to be put into practice. This chapter should be read in conjunction with Chapter 15, in which the ideas have been put to the test.

We begin by applying transition state theory to catalysis in the simplest Michaelis-Menten mechanism (where $K_M = K_S$; Chapter 3). The theory is used to calculate changes in activation energies rather than absolute values, so it is quite rigorous. It will be seen that the use of binding energy automatically lowers the activation energy of k_{cat}/K_M. But this alone is not sufficient for high rates, and some of the binding energy must be used to lower the activation energy of k_{cat}.

A. Utilization of enzyme–substrate binding energy in catalysis[4]

1. Binding energy lowers the activation energy of k_{cat}/K_M

$$E + S \underset{\Delta G_S}{\overset{K_M}{\rightleftharpoons}} ES \xrightarrow[\Delta G^\ddagger]{k_{cat}} products \qquad (12.1)$$

$$E + S \underset{\Delta G_T^\ddagger}{\overset{k_{cat}/K_M}{\rightleftharpoons}} ES^\ddagger \qquad (12.2)$$

Recall from Chapter 3 that the rate constant for free enzyme reacting with free substrate to give products is k_{cat}/K_M (equation 12.1). Expressed in terms of transition state theory (Chapter 2), the equilibrium constant between E + S and the transition state $ES^\ddagger$ is proportional to the activation energy $\Delta G_T^\ddagger$ of k_{cat}/K_M (equation 12.2). This activation energy is composed of two terms, an energetically unfavorable term $\Delta G^\ddagger$, due to the activation energy of the chemical steps of bond making and breaking, and a compensating energetically favorable term ΔG_S, due to the realization of the binding energy. That is,

$$\Delta G_T^\ddagger = \Delta G^\ddagger + \Delta G_S \qquad (12.3)$$

(where $\Delta G_T^\ddagger$ and $\Delta G^\ddagger$ are algebraically positive and ΔG_S is negative). This is illustrated in Figure 12.1 for the simple Michaelis-Menten mechanism.

Substituting equation 12.3 into equation 2.5 to express k_{cat}/K_M in terms of transition state theory gives

$$RT \ln \frac{k_{cat}}{K_M} = RT \ln \frac{kT}{h} - \Delta G^\ddagger - \Delta G_S \qquad (12.4)$$

2. Interconversion of binding and chemical activation energies

The maximum binding energy between an enzyme and a substrate occurs when each binding group on the substrate is matched by a binding site on the enzyme. In this case the enzyme is said to be complementary in structure to the substrate. Since the structure of the substrate changes throughout the reaction, becoming first the transition state and then the products, the structure of the *undistorted* enzyme can be complementary to only one form of the substrate. We shall show (section A3b) that it is catalytically advantageous for the enzyme to be complementary to the structure of the transition state of the substrate rather than to the original structure. If this happens, the increase in binding energy as the structure changes to that of the transition state lowers the activation energy of k_{cat}. On the other hand, if the enzyme is complementary to the structure of the unaltered sub-

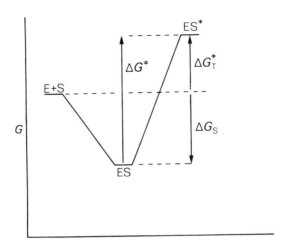

Figure 12.1 Gibbs free energy changes for the scheme

$$E + S \underset{K_S}{\rightleftharpoons} ES \xrightarrow{k_{cat}} \text{products}$$

(ΔG_S is algebraically negative, and $\Delta G^{\ddagger}$ and $\Delta G_T^{\ddagger}$ are positive.)

strate, the decrease in binding energy on the formation of the transition state will increase the activation energy of k_{cat}.

Because these ideas are crucial to the understanding of the role of binding energy in enzyme catalysis, they are amplified in the remainder of this section. Suppose that we are able to increase the binding energy of a particular enzyme with its substrate by adding an extra group to one of the amino acid side chains. For example, we could add a hydroxyl group to an alanine residue, to convert it to a serine that could hydrogen-bond with the substrate and contribute a binding energy of ΔG_R. Now we examine the consequences of controlling at which stage in the reaction the binding actually takes place.

First we consider Figure 12.2, which represents the condition of [S] greater than K_M, so that the enzyme is saturated with substrate and the reaction rate is $v = k_{cat}[E]_0$. There are three extreme cases. The simplest is when the full binding energy of the group is realized equally well in the enzyme–substrate (ES) and the transition state (ES‡) complexes, so that the Gibbs free energies of both are lowered equally. In other words, there is both enzyme–substrate and enzyme–transition state complementarity. This is sometimes termed a *uniform* binding energy effect.[5] As shown in Figure 12.2, this decreases K_M but it does not affect the value of k_{cat}. Thus, when the enzyme is saturated with substrate and the binding energy is realized equally in ES and ES‡, the rate of reaction is not increased and

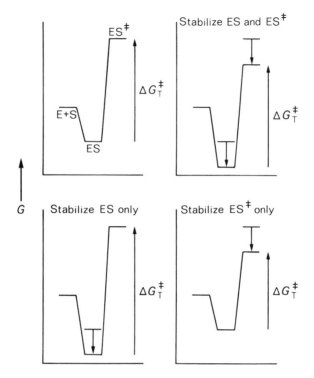

Figure 12.2 Where to utilize some extra binding energy ΔG_R when $[S] > K_M$? The Gibbs energy changes are for the reaction under the experimental condition of saturating $[S]$, so that $v = k_{cat}[E]_0$. On stabilization of only $ES^{\ddagger}$, the activation energy is lowered by ΔG_R, whereas stabilization of only ES leads to an increase of activation energy by that amount. Stabilization of ES and $ES^{\ddagger}$ equally has a neutral effect.

there is no catalytic advantage to adding the group. In the second case, we suppose that the binding energy is realized only in the enzyme–substrate complex. Its energy is lowered whereas that of the transition state remains the same. The activation energy of k_{cat} increases, so the rate of reaction decreases. This example corresponds to enzyme–substrate complementarity. In the third case, which corresponds to enzyme–transition state complementarity, the binding energy is realized only in the transition state complex. Its energy is lowered while that of the enzyme–substrate complex remains the same. This lowers the activation energy of k_{cat} and so increases the rate of reaction. The latter two examples represent *differential* binding effects.[5]

We consider now the analogous experiments under the condition of $[S]$ less than K_M, as in Figure 12.3. Here the Gibbs free energy of ES is greater than that of $E + S$, and the reaction rate is $v = (k_{cat}/K_M)[E]_0[S]$. Under this condition, both the

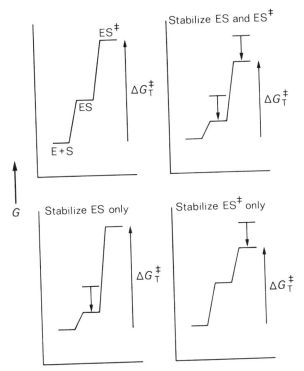

Figure 12.3 Where to utilize some extra binding energy ΔG_R when $[S] < K_M$? The Gibbs free energy changes are for the reaction condition of subsaturating $[S]$, so that $v = (k_{cat}/K_M)[E]_0[S]$. The activation energy is lowered by ΔG_R on the stabilization of only $ES^{\ddagger}$, or of ES and $ES^{\ddagger}$ (in the latter case, as long as $[S]$ remains below K_M; otherwise, there is a transition to Figure 12.2). Stabilization of ES only does not affect $\Delta G_T^{\ddagger}$.

cases of enzyme–transition state and of combined enzyme–substrate and enzyme–transition state complementarity lower the activation energy by ΔG_R, and hence increase rate. Where the additional binding energy of the extra group stabilizes only the enzyme–substrate complex—i.e., where there is enzyme–substrate complementarity only—there is no rate advantage from the additional binding energy.

The situation in which the enzyme has a group that can bind to the substrate only in the transition state is an example of *transition state stabilization*. It is important to note that the presence on the enzyme of a group that can bind only to the transition state of the substrate decreases the activation energy for a chemical step, and, further, does not have to distort the substrate to do so.

Examples and consequences of uniform and differential binding effects are discussed in Chapter 15, section H. Uniform binding energy changes are effective in increasing k_{cat}/K_M when $[S] < K_M$, and this will increase the reaction rate. A series of uniform binding energy changes will lower K_M until $[S] > K_M$. Then, differential binding changes that cause transition state stabilization are required to increase the rate further.

In the following section, we shall analyze quantitatively by transition state theory the consequences of enzyme–substrate complementarity. To do this we shall divide the components of activation energies into two groups: those arising from chemical terms and those arising from changes in the binding energy as the reaction proceeds.

The idea of complementarity in enzyme–substrate interactions was introduced by E. Fischer with his famous "lock and key" analogy.[7] In modern terminology this would represent enzyme–substrate complementarity. The currently favored concept of enzyme–transition state complementarity was introduced by Haldane[1] and elaborated by L. Pauling.[2]

3. Enzyme complementarity to transition state implies that k_{cat}/K_M is at a maximum

a. Enzyme complementarity to the initial substrate

Suppose that the maximum amount of intrinsic binding energy available is ΔG_b. In this case the full ΔG_b is realized in the initial enzyme–substrate complex, so binding will be good; that is, K_M, the dissociation constant of the enzyme–substrate complex, will be low. But the formation of the transition state will lead to a reduction in binding energy as the substrate geometry changes to give a poorer fit, and so will lower k_{cat} as described above. If the adverse energy change caused by the poorer fit is ΔG_R, and the free energy of activation due to the chemical bond making and breaking involved in k_{cat} is $\Delta G_0^{\ddagger}$ (Figure 12.4), then the observed free energy of activation for k_{cat} is given by

$$\Delta G^{\ddagger} = \Delta G_0^{\ddagger} + \Delta G_R \tag{12.5}$$

and

$$\Delta G_S = \Delta G_b \tag{12.6}$$

The Gibbs energy of activation for k_{cat}/K_M is given by $\Delta G^{\ddagger} + \Delta G_S$, i.e.,

$$\Delta G_T^{\ddagger} = \Delta G_0^{\ddagger} + \Delta G_R + \Delta G_b \tag{12.7}$$

b. Enzyme complementarity to the transition state

Here the full binding energy ΔG_b is realized only in the transition state. The adverse energy term ΔG_R in the initial enzyme–substrate complex will increase K_M, but the gain in binding energy as the reaction reaches the transition state will increase k_{cat}. Thus,

$$\Delta G^{\ddagger} = \Delta G_0^{\ddagger} - \Delta G_R \qquad (12.8)$$

and

$$\Delta G_S = \Delta G_b + \Delta G_R \qquad (12.9)$$

Again, the Gibbs free energy of activation for k_{cat}/K_M is given by

$$\Delta G_T^{\ddagger} = \Delta G^{\ddagger} + \Delta G_S$$

i.e.,

$$\Delta G_T^{\ddagger} = \Delta G_0^{\ddagger} + \Delta G_b \qquad (12.10)$$

because ΔG_R cancels out.

Comparison of equations 12.7 and 12.10 shows that k_{cat}/K_M is higher by a factor of exp $(\Delta G_R/RT)$ when the enzyme is complementary to the transition state rather than to the initial substrate. It should be noted that in enzyme–transition state complementarity, k_{cat}/K_M is independent of the interactions in the initial enzyme–substrate complex, as shown by the term ΔG_R dropping out of the equations.

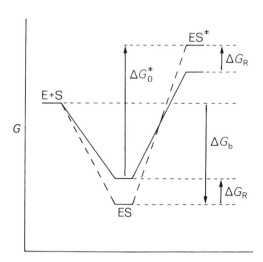

Figure 12.4 Gibbs free energy changes for the schemes in Figures 12.1 and 12.2, when the enzyme is complementary in structure to (1) the substrate (dashed curve), and (2) the transition state (solid curve). (ΔG_b is algebraically negative, and $\Delta G_0^{\ddagger}$ and ΔG_R are positive.)

B. Experimental evidence for the utilization of binding energy in catalysis and enzyme – transition state complementarity

1. Classic experiments: Structure – activity relationships of modified substrates

Some of the most instructive evidence for the utilization of binding energy comes from kinetic experiments on the serine proteases. We recall from Chapter 1 that these enzymes have a series of subsites for binding the amino acid residues of their polypeptide substrates. Table 12.1 shows that as larger groups occupy the leaving group site in chymotrypsin, their binding energy is used to increase k_{cat}/K_M. Similarly, increasing the length of the polypeptide chain of substrates of elastase increases k_{cat}/K_M. Interestingly, in the examples given, the binding energy of the additional groups does not lower K_M; i.e., the binding energy is not used to bind the substrate, but rather to increase k_{cat}.

Similar behavior is observed with pepsin. As shown in Table 12.1, the hydrolyses of a wide range of substrates are catalyzed with K_M's of about 0.1 mM. The additional binding energy of the groups on the larger substrates is again used to increase the k_{cat} rather than to decrease the K_M. The k_{cat}/K_M is accordingly higher for the larger substrates.

A striking series of examples occurs with the chymotrypsin-catalyzed hydrolysis of synthetic ester substrates (Table 12.2). As the size of the hydrophobic side chain that fits into the hydrophobic primary binding site of the enzyme is increased, k_{cat}/K_M increases over a range of 10^6. The increase in the binding energy of the larger substrates is distributed between lowering the dissociation constant of the enzyme – substrate complex and increasing the acylation and deacylation rate constants.

Section C explains how this use of binding energy to increase k_{cat} rather than to lower K_M gives higher reaction rates.

2. Transition state analogues: Probes of complementarity[8,9]

Direct evidence for enzyme – transition state complementarity has come from x-ray diffraction experiments on the serine proteases and lysozyme (section D5c of this chapter, and also Chapter 1), and from studies on the binding of transition state analogues. This approach was suggested by Pauling in the 1940s.[2] Chemists, with their knowledge of organic reaction mechanisms, can guess at the structure of the transition state of an enzymatic reaction. Compounds that mimic the transition state may then be synthesized and their binding to the enzyme compared with that of the substrate.

Table 12.1 *Interconversion of activation and binding energies*

Enzyme and substrate	k_{cat} (s^{-1})	K_M (mM)	k_{cat}/K_M $(s^{-1} M^{-1})$
Chymotrypsin[a]			
Ac-Tyr-NH$_2$	0.17	32	5
Ac-Tyr-Gly-NH$_2$	0.64	23	28
Ac-Tyr-Ala-NH$_2$	7.5	17	440
Ac-Phe-NH$_2$	0.06	31	2
Ac-Phe-Gly-NH$_2$	0.14	15	10
Ac-Phe-Ala-NH$_2$	2.8	25	114
Elastase[b] (cleavage at —NH$_2$)			
Ac-Ala-Pro-Ala-NH$_2$	0.09	4.2	21
Ac-Pro-Ala-Pro-Ala-NH$_2$	8.5	3.9	2.2×10^3
Ac-Gly-Pro-Ala-NH$_2$	0.02	33	0.5
Ac-Pro-Gly-Pro-Ala-NH$_2$	2.8	43	64
Pepsin[c] (cleavage of Phe-Phe bond in A-Phe-Phe-OP4P)			
Phe-Gly	0.5	0.3	1.7×10^3
Z-Phe-Gly	25	0.11	2.2×10^5
Z-Ala-Gly	145	0.25	5.8×10^5
Z-Ala-Ala	282	0.04	7×10^6
Z-Gly-Ala	409	0.11	3.7×10^6
Z-Gly-Ile	13	0.07	1.8×10^5
Z-Gly-Leu	134	0.03	4.2×10^6
Phe-Gly-Gly	6	0.6	1×10^4
Z-Phe-Gly-Gly	127	0.13	9.8×10^5
Mns[d]	0.002	0.1	20
Mns-Gly[d]	0.13	0.03	3.7×10^3
Mns-Gly-Gly[d]	16	0.07	2.3×10^5
Mns-Ala-Ala[d]	112	0.06	2×10^6

[a] At 25°C, pH 7.9. [From W. K. Baumann, S. A. Bizzozero, and H. Dutler, *FEBS Lett.* **8**, 257 (1970); *Eur. J. Biochem.* **39**, 381 (1973).]
[b] At 37°C, pH 9. [From R. C. Thompson and E. R. Blout, *Biochemistry* **12**, 51 (1973).]
[c] At 37°C, pH 3.5; OP4P = 3-(4-pyridyl)propyl-1-oxy. The N-terminal portions "A" are listed. [From G. P. Sachdev and J. S. Fruton, *Biochemistry* **9**, 4465 (1970).]
[d] At 25°C, pH 2.4; Mns = mansyl. [From G. P. Sachdev and J. S. Fruton, *Proc. Natn. Acad. Sci. USA* **72**, 3424 (1975).]
Further compilations for chymotrypsin, elastase, and α-lytic protease are given by: C. A. Bauer, *Biochemistry* **17**, 375 (1978); C. A. Bauer, G. D. Brayer, A. R. Sielecki, and M. N. G. James, *Eur. J. Biochem.* **120**, 289 (1981); S. A. Bizzozero, W. K. Baumann, and H. Dutler, *Eur. J. Biochem.* **122**, 251 (1982).

Table 12.2 *Kinetic parameters for the hydrolysis of N-acetyl amino acid methyl esters by chymotrypsin[a]*

$$(E + RCO_2CH_3 \xrightleftharpoons{K_S} E \cdot RCO_2CH_3 \xrightarrow{k_2} RCO-E \xrightarrow{k_3} E + RCO_2H)$$

Amino acid	k_2 (s^{-1})	k_3 (s^{-1})	K_s (mM)	k_{cat}/K_M $(s^{-1} M^{-1})$
Gly	0.49	0.14	3.38×10^3	0.13
But	8.8	1.7	417	21
Norval	35.6	5.93	100	360
Norleu	103	19	34	3×10^3
Phe	796	111	7.6	1×10^5
Tyr[b]	5×10^3	200	17	3×10^5

[a] α-Chymotrypsin at 25°C, pH 7.8. (Data from Table 7.3.)
[b] Ethyl ester.

a. Lysozyme and glucosidase

Lysozyme catalyzes the hydrolysis of the polysaccharide component of plant cell walls and synthetic polymers of $\beta(1 \rightarrow 4)$-linked units of N-acetylglucosamine (NAG) (Chapter 1). It is expected from studies on nonenzymatic reactions that one of the intermediates in the hydrolytic reaction is a oxocarbenium ion in which the conformation of the glucopyranose ring changes from a full-chair to a sofa (half-chair) conformation (Chapter 1). The transition state analogue I, in which the lactone ring mimics the carbonium ion–like transition state II, binds tightly to lysozyme: $K_{diss} = 8.3 \times 10^{-8} M$.[10]

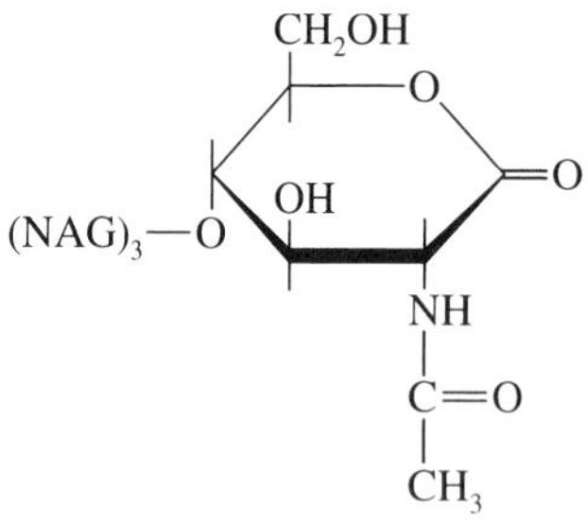

I 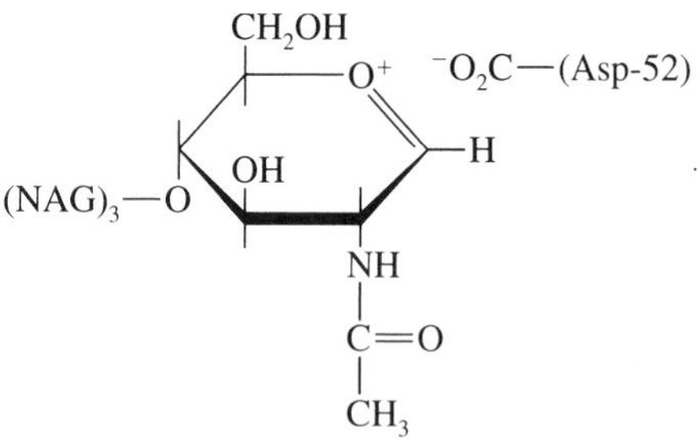   II

This may be compared with the dissociation constants of $10^{-5} M$ and $5 \times 10^{-6} M$ for $(NAG)_4$ and NAG-NAM-NAG-NAG binding in the A, B, C, and D subsites.[10,11] The 100-fold tighter binding of the transition state analogue may be due in part to the electrostatic interaction of the negatively charged Asp-52 with the partial positive charge on the carbonyl carbon of the lactone.

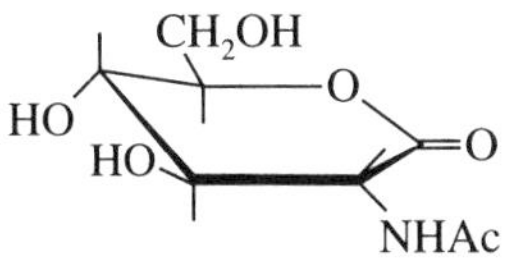

III
Lactone
(sofa conformation)

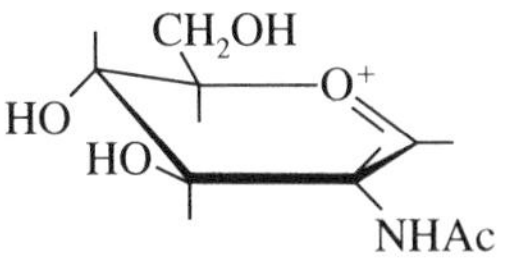

IV
Oxocarbenium ion
(sofa conformation)

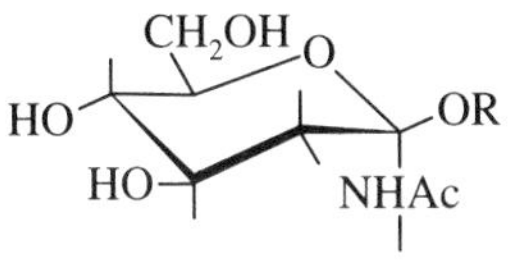

V
Pyranoside
(full-chair conformation)

More striking is the binding of the lactone III to β-N-acetyl-D-glucosaminidase. The dissociation constant of $5 \times 10^{-7} M$ is 4000 times smaller than the K_M of $2 \times 10^{-3} M$ for the pyranoside substrate V.[12] However, it is possible in this example that the enzyme forms a covalent bond with the analogue so that the tight binding does not result solely from noncovalent binding.[13]

b. Proline racemase

During the racemization of proline (structure VI), the chiral carbon must at some stage become trigonal. In accordance with a trigonal transition state, both structures VII and VIII bind 160 times more tightly than proline.[14,15]

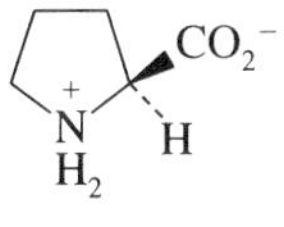

VI

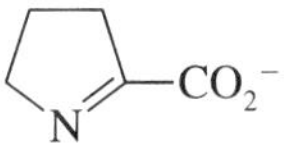

VII

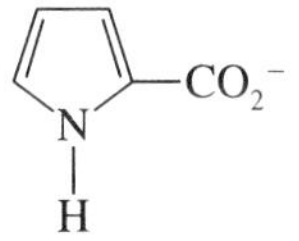

VIII

c. Cytidine deaminase

The dissociation constant of tetrahydrouridine (structure IX) is about 10 000 times smaller than the combined constants for the reaction products, uridine (structure X) and ammonia. Tetrahydrouridine presumably resembles the transition state, which is similar to the tetrahedral intermediate XI.[16,17]

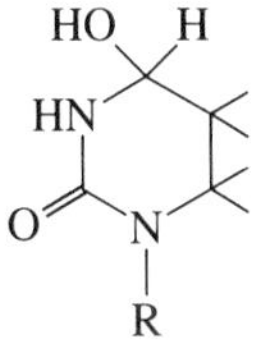

IX

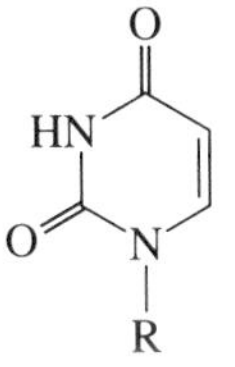

X

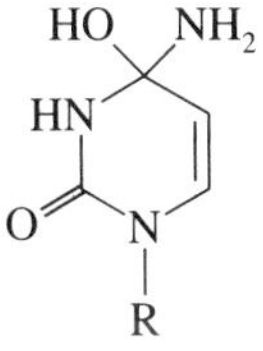

XI

d. Assessment of the results of binding transition state analogues

The transition state analogues that have been designed so far give a measure of the part of the catalysis that is due to the difference in complementarity of the enzyme for the transition state and the substrate. In the four examples given above, the transition state analogues bind between 10^2 and 10^4 times more tightly than the original substrates. This is good evidence that enzymes have evolved to be complementary in structure to the transition state. Furthermore, it shows that k_{cat} may be increased by a factor of 10^2 to 10^4 at the expense of increasing K_M. Considering that these synthetic analogues might be extremely crude in mimicking the transition state structure, it seems likely that the increase in complementarity between the substrate and the transition state is worth at least 20 kJ/mol (5 kcal/mol). As will be discussed in Chapter 16, the binding site for the carbonyl oxygen of a substrate of a serine protease is deficient by one hydrogen bond. When the transition state for the reaction is formed, the additional hydrogen bond is made. This increase in complementarity must also be worth a factor of about 20 to 25 kJ/mol (5 to 6 kcal/mol).

e. Some possible errors in interpreting transition state analogue data

The effects of enzyme–transition state complementarity on the binding of transition state analogues may be masked by extraneous binding artifacts. Chapter 11 showed that small groups can involve large binding energies when a specific binding site is involved. For example, a methylene group may contribute 12 kJ/mol (3 kcal/mol). Also, where specific binding sites are not involved, fairly large energies may come from general hydrophobic effects: the substitution of a chloro group for an acetyl on a phenyl ring on an anilide substrate causes it to bind 50 times more tightly to the hydrophobic pocket of chymotrysin.[18]

Difficulties also arise with analogues for multisubstrate reactions due to the chelate effect. It was pointed out in Chapter 11 that multidentate ligands, such as EDTA, bind tightly to metal ions whereas unidentate ligands do not. The difference is due to entropy; the binding of six unidentate ligands leads to the loss of six sets of translational and rotational entropies. The same applies to the binding of a "multisubstrate" or a "multiproduct" analogue to an enzyme. Assume, for instance, that A and B bind separately and adjacently to an enzyme with Gibbs free energies of association x and y kJ, respectively. Then if A–B binds in an identical manner, its free energy of association is given by

$$\Delta G_{ass} = x + y + S \qquad (12.11)$$

where S is an energetically favorable term because only one set of entropies is lost on the binding of A–B, compared with two on the binding of A and B separately. The binding of multisubstrate analogues should be very tight, without any effects from enzyme–transition state complementarity. For example, the binding of structure XII to aspartate transcarbamoylase is very tight; $K_{diss} = 2.7 \times 10^{-8}\ M$. However, this value is only equal to the product of the dissocia-

$$
\begin{array}{ccc}
 & \text{NH} & \text{CH}_2\text{PO}_3^{2-} \\
 & \diagdown & \diagup \\
\text{CH}_2\text{—CH} & \text{C} & \\
| \qquad | & \| & \\
\text{CO}_2^- \ \text{CO}_2^- & \text{O} &
\end{array}
$$

XII

tion constants of succinate and carbamoyl phosphate $(9 \times 10^{-4} \times 2.7 \times 10^{-5} \, M^{-2})$, which approximate to the fragments of XII.[19] The transition state structure is derived from XIII.

$$
\begin{array}{c}
\text{PO}_3^{2-} \\
\diagup \\
\text{O} \\
\diagdown \\
\text{C}{=}\text{O} \\
\diagup \\
\text{NH}_2 \\
\text{NH}_2 \\
\text{CH}_2\text{—CH} \\
| \qquad | \\
\text{CO}_2^- \ \text{CO}_2^-
\end{array}
$$

XIII

3. Catalytic antibodies (abzymes)

Antibodies are proteins that are produced by the immune system to recognize and bind to antigens. Very early studies on antibodies by Ehrlich laid down the basic principle of biological recognition: the importance of complementary interactions. Pauling wrote in the 1940s that an enzyme could be considered an antibody that specifically recognizes the transition state of a reaction.[2] The advent of monoclonal antibodies spurred molecular biologists and chemists to raise antibodies against transition analogues (section 2). These are called catalytic antibodies or abzymes.[20–22] Weak catalytic activity can be found for a number of reactions, limited to about a factor of 10^2 to 10^4, and occasionally 10^6, in rate enhancement over the rate constant for the uncatalyzed reaction, compared to the 10^{16}- to 10^{17}-fold rate enhancement by enzymes (Chapter 2, Table 2.1). The analysis of section 2 for transition state analogues applies equally well to catalysis by abzymes.[23] For example, just as multisubstrate analogues bind well because of entropy considerations, so the binding of two reagents can help catalyze a ligation reaction without necessarily stabilizing the transition state. Similarly, just as transition state analogues do not bind as tightly as transition states because they are not perfect analogues, so they are not perfect analogues for raising antibodies. Enzymes are very sophisticated catalysts, often using the binding energy of subsites far removed from the seat of reaction, often using residues spread widely around the active site (Chapter 15), and sometimes using the induced-fit process (section D2), and so it is difficult to mimic the combined effects of all their tricks.

4. Structure–activity experiments on engineered enzymes

The side chains of amino acid residues in a protein may be changed at will by protein engineering (Chapter 14) and the consequent effects on binding and catalysis studied directly. This is the subject of Chapter 15, where it will be seen how the equations derived so far actually hold in practice and are used to analyze the data. There is direct evidence for enzyme–transition state complementarity.

C. Evolution of the maximum rate: Strong binding of the transition state and weak binding of the substrate[4]

In the last section we showed that enzymes have evolved to bind the transition states of substrates more strongly than the substrates themselves. It will now be seen that it is catalytically advantageous to bind substrates *weakly.*

Although enzyme–transition state complementarity maximizes k_{cat}/K_M, this is not a sufficient criterion for the maximization of the overall reaction rate. The reason is that the maximum reaction rate for a particular concentration of substrate depends on the individual values of k_{cat} and K_M. It can be seen in Table 12.3, where some rates are calculated for various values of k_{cat} and K_M (subject to k_{cat}/K_M being kept constant), that maximum rates are obtained for K_M greater

Table 12.3 *Illustration of the importance of the evolution of k_{cat} and K_M at constant values of k_{cat}/K_M and $[S]^a$*

$(k_{cat}/K_M = 10^6 \, M^{-1} \, s^{-1}, [S] = 10^{-3} \, M)$

K_M (M)	k_{cat} (s^{-1})	Rate[b] (s^{-1})
10^{-6}	1	1
10^{-5}	10	9
10^{-4}	10^2	90
10^{-3}	10^3	500
10^{-2}	10^4	909
10^{-1}	10^5	990
1	10^6	999

[a] The hypothetical processes are: (1) The enzyme has evolved to be complementary to the transition state of the substrate, so that k_{cat}/K_M is maximized. (2) While maintaining k_{cat}/K_M, the enzyme evolves to *increase* K_M. The values assigned to k_{cat}/K_M and [S] are arbitrary.

[b] Moles of product produced per mole of enzyme per second.

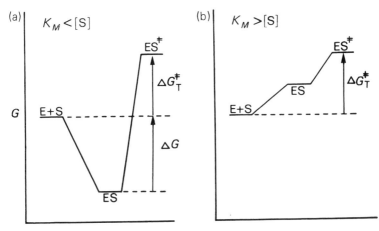

Figure 12.5 Two cases of enzyme evolution. In both cases the enzymes bind the transition states equally well, but in (a) the substrate is bound strongly, and in (b) the enzyme has evolved to bind the substrate weakly ([S] is the same in both graphs). The activation energy in (a) is for ES → ES‡, i.e., $\Delta G_T^{\ddagger} + \Delta G$, whereas in (b) it is for E + S → ES‡, i.e., $\Delta G_T^{\ddagger}$. (The changes in Gibbs free energies are for the concentration of substrate used in the experiment, and not for standard states of 1 M.)

than [S]. The maximization of rate requires *high* values of K_M. That is, enzymes should have evolved to bind substrates weakly.

1. The principle of maximization of K_M at constant k_{cat}/K_M[4]

This principle contradicts the widely held belief that strong binding, or a low K_M, is an important component of enzymatic catalysis. The two additional proofs that follow emphasize the importance of high K_M's, and the figures indicates the physical reason.

a. Graphical illustration of the importance of high K_M's

Suppose that, as in Figure 12.5a, the substrate is at a higher concentration than the K_M for the reaction. The enzyme–substrate complex is at a lower energy than the free enzyme and substrate, and the activation energy is $\Delta G_T^{\ddagger} + \Delta G$. However, if, as in Figure 12.5b, everything is the same except that K_M is now higher than [S], so that the ES complex is at a higher energy than E + S, the activation energy is at the lower value of $\Delta G_T^{\ddagger}$.

The low K_M leads to a thermodynamic "pit" into which the reaction falls and has to climb out. The high K_M leads to the enzyme–substrate complex being "a step up the thermodynamic ladder."

b. An algebraic illustration

In Chapter 3 it was shown that the Michaelis-Menten equation may be cast into the useful form

$$v = [E][S] \frac{k_{cat}}{K_M} \qquad (12.12)$$

to relate the reaction rate to the concentration of free enzyme, $[E]$.

The evolution of an enzyme to give maximum rate may be divided into two hypothetical steps based on equation 12.12:

1. The value of k_{cat}/K_M is maximized by having the enzyme be complementary to the transition state of the substrate.
2. The $[E]$ is maximized by having the K_M be high, so that as much of the enzyme as possible is in the unbound form.

There is an evolutionary pressure on K_M to increase with a consequent increase in k_{cat}. This evolutionary pressure rapidly decreases as K_M becomes greater than $[S]$. At $K_M = [S]$, half of the enzyme is unbound, so according to equation 12.12 the rate is 50% of the maximum possible. At $K_M = 5[S]$, five-sixths of the enzyme is unbound, so the rate is 83% of the maximum. Any further increase in K_M gives only a marginal increase in the rate.

How far a K_M evolves relative to the substrate concentration depends on the change in structure when the substrate becomes the transition state. A limit must eventually be reached; at this point, any increase in K_M must be matched by a weakening of transition state binding. The problem will be most severe for large metabolites present at high concentrations.

c.   Exceptions to the principle of high K_M's: Control enzymes

Implicit in the preceding arguments for high K_M's is the assumption of the maximization of rate. Although this is valid for most enzymes, there are cases in which rate is subordinate to *control*. Metabolic pathways are characterized by their regulation, which is usually maintained by the activity of certain key enzymes on the pathway (Chapter 10, section F). The activities of these control enzymes are themselves often controlled by variations in the K_M's of critical substrates via allosteric effects. The K_M's for control enzymes have evolved for the purposes of regulation, and are not necessarily subject to the rate arguments of the previous section.

A low K_M could sometimes be advantageous for the first enzyme on a metabolic pathway. This would then control the rate of entry to the pathway and prevent it from being overloaded and accumulating reactive intermediates. For example, hexokinase, the first enzyme in glycolysis, has a K_M for glucose of 0.1 mM, whereas the concentration of glucose in the human erythrocyte is about 5 mM. A tenfold increase or decrease in the glucose concentration will hardly alter the rate of glycolysis.

2.   Experimental observations on K_M's

As was shown in Chapter 11, the binding energy of an enzyme and substrate is potentially very high. However, K_M's are usually found to be relatively high. An extreme example of this is NAD^+. This large substrate has two ribose moieties, one adenine ring, a nicotinamide residue, and a pyrophosphate linkage. If all the intrinsic binding energy of these groups were realized, a dissociation constant of less than 10^{-20} M could be attained. Indeed, the dissociation constant for the binding of the first NAD^+ from the tetrameric glyceraldehyde 3-phosphate dehydrogenase has been found to be immeasurably strong at less than 10^{-11} M.[24] Yet the K_M's and dissociation constants of NAD^+ with dehydrogenases are often found to be in the range of 0.1 to 1 mM. Even more striking is the dissociation constant of 10^{-13} M for ATP and myosin,[25] which may be compared with the K_M's of 0.1 to 10 mM that are often found for ATP.

The comparison of K_M's with physiological substrate concentrations is difficult in many cases due to a lack of knowledge of the concentrations, but there are some well-characterized examples. One particular case is carbonic anhydrase, because the concentrations of carbon dioxide and bicarbonate in the blood are easily measured. Under physiological conditions, the enzyme is only about 6% saturated with each substrate, and the K_M of carbon dioxide is too high to be measured.[26]

a. Substrate concentrations and K_M's in glycolysis

Good data are available for glycolysis. The glycolytic enzymes are particularly well studied and understood, and metabolite concentrations have been determined for three diverse types of cells (brain, erythrocyte, and muscle). Data for the nonregulatory glycolytic enzymes are listed in Table 12.4 and illustrated in Figure 12.6. The histogram in the figure shows that the K_M's tend to be in the range of 1 to 10 and 10 to 100 times the substrate concentrations. Notable

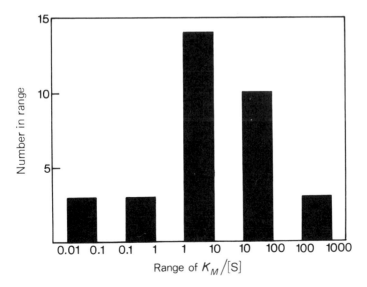

Figure 12.6 Distribution of the values of $K_M/[S]$ found in glycolysis.

Table 12.4 *Metabolite concentrations and K_M's for some glycolytic enzymes[a]*

Enzyme	Source	Substrate	Concentration (μM)	K_M (μM)	K_M/[S]
Glucose	Brain	G6P	130	210	1.6
6-phosphate	Muscle[b]	G6P	450	700	1.6
isomerase		F6P	110	120	1.1
Aldolase	Brain	FDP	200	12	0.06
	Muscle[c]	FDP	32	100	3.1
		G3P	3	1000	333
		DHAP	50	2000	40
Triosephosphate	Erythrocyte[d]	G3P	18	350	19
isomerase	Muscle[e]	G3P	3	460	153
		DHAP	50	870	17
Glyceraldehyde	Brain	G3P	3	44	15
3-phosphate	Muscle[f]	G3P	3	70	23
dehydrogenase		NAD	600	46	0.08
		P_i	2000		$>10^g$

[a] Abbreviations: G6P = glucose 6-phosphate, F6P = fructose 6-phosphate, FDP = fructose 1,6-diphosphate, G3P = glyceraldehyde 3-phosphate, DHAP = dihydroxyacetone phosphate, P_i = orthophosphate, 1,3DPG = 1,3-diphosphoglycerate, 3PG = 3-phosphoglycerate, 2PG = 2-phosphoglycerate, PEP = phosphoenolpyruvate, Pyr = pyruvate, Lac = lactate (all D-sugars); Gly-P = L-glycerol phosphate. Mouse brain enzymes and mouse brain metabolites from O. H. Lowry and J. V. Passonneau, *J. Biol. Chem.* **239**, 31 (1964). Human erythrocyte metabolites from S. Minakami, T. Saito, C. Suzuki, and H. Yoshikawa, *Biochem. Biophys. Res. Commun.* **17**, 748 (1964). Human erythrocyte enzymes: see below. Rat diaphram metabolites from E. A. Newsholme and P. J. Randle, *Biochem. J.* **80**, 655 (1961); H. J. Hohorst, M. Reim, and H. Bartels, *Biochem. Biophys. Res. Commun.* **7**, 137 (1962). Rabbit skeletal muscle enzymes: see below. Metabolite concentrations were calculated on an intramolecular water content of 60% for brain and muscle cells, and 70% for erythrocytes. No allowance has been made for compartmentation in the muscle and brain cells, but gross metabolite concentrations are usually close to those in the cytosol [A. L. Greenbaum, K. A. Gumaa, and P. McLean, *Archs. Biochem. Biophys.* **143**, 617 (1971)]. The values for mouse brain are those immediately on decapitation. The use of peak levels does not cause significant differences.

[b] From J. Zalitis and I. T. Oliver, *Biochem. J.* **102**, 753 (1967).

[c] From W. J. Rutter, *Fedn. Proc.* **23**, 1248 (1964); P. D. Spolter, R. C. Adelman, and S. Weinhouse, *J. Biol. Chem.* **240**, 1327 (1965).

[d] From A. S. Schneider, W. N. Valentine, M. Hattori, and H. L. Heins. *New Engl. J. Med.* **272**, 229 (1965).

[e] From P. M. Burton and S. G. Waley, *Biochem. Biophys. Acta* **151**, 714 (1968).

[f] From M. Oguchi, E. Gerth, B. Fitzgerald, and J. H. Park. *J. Biol. Chem.* **248**, 5571 (1973).

[g] The K_M of ~6 mM for P_i refers to high G3P concentrations where the acylenzyme accumulates. At low concentrations of G3P, the K_M is immeasurably high [P. J. Harrigan and D. R. Trentham,

Enzyme	Source	Substrate	Concentration (μM)	K_M (μM)	$K_M/[S]$
Phosphoglycerate kinase	Brain	1,3DPG	<1	9	>9
		ADP	1500	70	0.05
	Erythrocyte[h]	3PG	118	1100	9.3
	Muscle[i]	3PG	60	1200	200
		ADP	600	350	0.6
Phosphoglycerate mutase	Brain	3PG	40	240	6
	Muscle[j]	3PG	60	5000	83
Enolase	Brain	2PG	4.5	33	7
	Muscle[k]	2PG	7	70	10
Pyruvate kinase[l]	Erythrocyte[m]	PEP	23	200	9
		ADP	138	600	4.4
Lactate dehydrogenase	Brain	Pyr	116	140	1.2
	Erythrocyte[n]	Pyr	51	59	1.2
		Lac	2900	8400	2.9
		NADH	0.01[o]	10[p]	100
		NAD	33	150	4.6
Glycerol phosphate dehydrogenase	Mouse	Gly-P	170	37	0.22
	Muscle[q]	Gly-P[r]	220	190	0.9
		DHAP	50	190	3.8

Biochem. J. **143**, 353 (1974)]. Note: The *unhydrated* forms of G3P and DHAP are probably the substrates of the reactions. The concentrations tabulated are for both the hydrated and the unhydrated forms, but the values of K_M for the unhydrated forms and their concentrations are overestimated in the same ratio [D. R. Trentham, C. H. McMurray, and C. I. Pogson, *Biochem. J.* **114**, 19 (1969); S. J. Reynolds, D. W. Yates, and C. I. Pogson, *Biochem. J.* **122**, 285 (1971)].

[h] From A. Yoshida and S. Watanabe, *J. Biol. Chem.* **247**, 440 (1972).

[i] From D. R. Rao and P. Oesper, *Biochem. J.* **81**, 405 (1961).

[j] From R. W. Cowgill and L. I. Pizer, *J. Biol. Chem.* **223**, 885 (1956); S. Grisolia and W. W. Cleland, *Biochemistry* **7**, 1115 (1968).

[k] From F. Wold and R. Barker, *Biochim. Biophys. Acta* **85**, 475 (1964).

[l] It is debatable whether or not this is a control enzyme; PEP is certainly well below the K_M in any case. The data quoted are for the presence of 500-u*M* FDP, in which case Michaelis-Menten kinetics hold. In the absence of FDP, sigmoid kinetics holds with a $K_{0.5}$ of 650 μM.[m]

[m] From S. E. J. Staal, J. F. Koster, H. Kamp, L. van Milligan-Boersma, and C. Veeger, *Biochim. Biophys. Acta* **227**, 86 (1971).

[n] From J. S. Nisselbaum and O. Bodansky, *J. Biol. Chem.* **238**, 969 (1963).

[o] Calculated from the lactate/pyruvate ratio, assuming NAD and NADH at equilibrium, and using an equilibrium constant of 1.11×10^{-4}. [From R. L. Veech, L. V. Eggleston, and H. A. Krebs, *Biochem. J.* **115**, 609 (1969).]

[p] From S. Rapoport, *Essays in Biochemistry* **4**, 69 (1969).

[q] From T. P. Fondy, L. Levin, S. J. Sollohub, and C. R. Ross, *J. Biol. Chem.* **243**, 3148 (1968).

[r] From R. M. Denton, R. E. Yorke, and P. J. Randle, *Biochem. J.* **100**, 407 (1996).

among these enzymes is triosephosphate isomerase. This well-studied enzyme, which has been described as "evolutionarily perfect"[5] because of its catalytic efficiency, has very high K_M's for both substrates.

It is not known at present whether the examples in which the K_M's are below the substrate concentrations are due to the enzymes' inability to evolve further or for metabolic reasons.

3. The perfectly evolved enzyme for maximum rate

We can use the two hypothetical steps of section C1b; i.e., that k_{cat}/K_M be maximized and that K_M be greater than [S], to set up criteria for judging the state of evolution of an enzyme whose function is to maximize rate. We recall from Chapter 3 that the maximum value of k_{cat}/K_M is the rate constant for the diffusion-controlled encounter of the enzyme and substrate, and from Chapter 4 that this is about 10^8 to 10^9 s^{-1} M^{-1}. A perfectly evolved enzyme should have a k_{cat}/K_M in the range of 10^8 to 10^9 s^{-1} M^{-1}, and a K_M greater than [S]. Using the data for k_{cat}/K_M listed in Table 4.4 and the substrate concentrations and K_M values mentioned in this chapter, it appears that carbonic anhydrase and triosephosphate isomerase are perfectly evolved for the maximization of rate, which agrees with the conclusions of W. J. Albery and J. R. Knowles on triosephosphate isomerase.[5]

There is one further consideration. The value of k_{cat}/K_M cannot be at the diffusion-controlled limit for a reaction that is thermodynamically unfavorable. This point stems from the Haldane equation (Chapter 3, section H), which states that the equilibrium constant for a reaction in solution is given by the ratio of the values of k_{cat}/K_M for the forward and reverse reactions. Clearly, k_{cat}/K_M for an unfavorable reaction cannot be at the diffusion-controlled limit, since k_{cat}/K_M for the favorable reverse reaction would have to be greater than the diffusion-controlled limit to balance the Haldane equation. The value of k_{cat}/K_M for an unfavorable reaction is limited by the diffusion-controlled limit multiplied by the unfavorable equilibrium constant for the reaction.

It was pointed out in Chapter 3, section A3, that when k_{cat}/K_M is at the diffusion-controlled limit, Briggs-Haldane rather than Michaelis-Menten kinetics are obeyed. Thus, the more advanced an enzyme is toward the evolution of maximum rate, the more important are Briggs-Haldane kinetics.

There have been more sophisticated attempts to derive optimal values for rate constants for the "perfectly evolved enzyme." Without any structural constraints, the perfect enzyme is bound to be one that has all the rate constants in the favorable direction equal to the maximum possible (6×10^{12} s^{-1} at 25°C, equation 2.2) and all the association rate constants at the diffusion-controlled limit.

D. Molecular mechanisms for the utilization of binding energy

We have discussed in general terms the catalytic advantages of enzyme–transition state complementarity combined with high K_M's, and have seen that this combination tends to occur in practice. We shall now deal with the specific mechanisms that are used to achieve it.

1. Strain

Strain is the classic concept of Haldane and Pauling.[1,2] The enzyme has an active site whose structure is complementary to the structure of the transition state of the substrate rather than to the substrate itself. On binding, the substrate is strained or distorted. In Haldane's words: "Using Fischer's lock and key simile, the key does not fit the lock perfectly but exercises a certain strain on it" (Figure 12.7). Nowadays, the modified concept of *transition state stabilization* has gained favor. According to this idea, it is not that the substrate is distorted but rather that the transition state makes better contacts with the enzyme than the substrate does, so that the full binding energy is not realized until the transition state is reached. Nevertheless, we shall use the term strain to cover the general situation of enzyme–transition state complementarity.

2. Induced fit

Chapter 10 showed how the induced-fit theory nicely describes some of the phenomena associated with allosteric enzymes. This theory had been introduced earlier to account for specificity in simple enzymes. It includes the notion that in the absence of substrate, the enzyme does not have a structure complementary to that of the transition state. However, the enzyme is floppy and the substrate rigid, so that when the enzyme–substrate complex is formed, the catalytic groups on the enzyme are aligned in their optimal orientations for catalysis: that is, the structure of the enzyme is complementary to the transition state only after binding has occurred. In the classic strain mechanism, the K_M is increased by binding energy being used to distort the substrate; in induced fit, it is increased by binding energy being used to distort the enzyme.

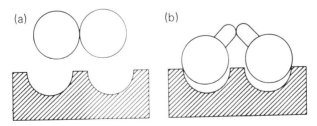

Figure 12.7 Haldane's picture of strain. The binding site on the enzyme (a) stretches the substrate toward the structures of the products, and (b) compresses the products toward the structure of the substrate.

a. Disadvantages of the induced-fit mechanism in nonallosteric enzymes

In the strain mechanism, the value of k_{cat}/K_M is at a maximum, since the undistorted enzyme is complementary to the undistorted transition state. Induced fit lowers the k_{cat}/K_M because it is only the *distorted* enzyme that is complementary to the undistorted transition state: the k_{cat}/K_M is decreased by the energy required to distort the enzyme.

An alternative way of viewing the induced-fit process is to divide it into hypothetical steps. We can suppose that there is an equilibrium between the inactive form of the enzyme, E_{in}, which is the major species ($[E_{in}] \approx [E]_0$), and a small fraction of active enzyme, E_{act}, in which the catalytic groups are correctly aligned:

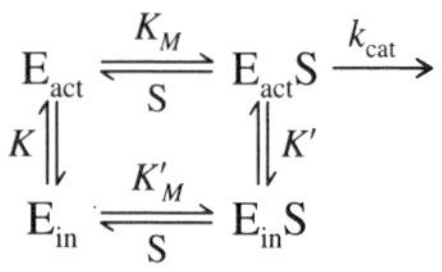

$$
\begin{array}{ccc}
E_{act} & \xrightleftharpoons[S]{K_M} & E_{act}S \xrightarrow{k_{cat}} \\
K \updownarrow & & \updownarrow K' \\
E_{in} & \xrightleftharpoons[S]{K'_M} & E_{in}S
\end{array}
$$

Scheme 1

By treating the K_M's in scheme 1 as simple dissociation constants and defining $K = [E_{act}]/[E_{in}]$, it may be shown that

$$
v = \frac{[E]_0[S]k_{cat}K'/(1 + K')}{[S] + (K_M K'/K)(1 + K)/(1 + K')} \tag{12.13}
$$

and

$$
\left(\frac{k_{cat}}{K_M}\right)_{obs} = \frac{k_{cat}}{K_M}\left(\frac{K}{1 + K}\right) \tag{12.14}
$$

Since $K \ll 1$,

$$
\left(\frac{k_{cat}}{K_M}\right)_{obs} = K\frac{k_{cat}}{K_M} \tag{12.15}
$$

The observed value of k_{cat}/K_M is much less than if all the enzyme is in the active conformation.

Equation 12.13 may be further simplified if $K' \gg 1$, i.e., if virtually all of the enzyme is in the active form when it is bound with substrate. Under these conditions,

$$
(k_{cat})_{obs} = k_{cat} \tag{12.16}
$$

and

$$(K_M)_{obs} = \frac{K_M}{K} \qquad (12.17)$$

The value of k_{cat} is the same as if all the enzyme were in the active conformation, but the value of K_M is far higher.

Induced fit thus increases K_M without increasing k_{cat}, and so decreases k_{cat}/K_M. It mediates against catalysis.

Equation 12.15 shows that as far as k_{cat}/K_M is concerned, it is as if a small fraction of the enzyme is permanently in the active conformation. This is the same for all substrates since K is substrate-independent.

b. Importance of induced fit in catalysis

The transition state of a substrate often needs to be completely surrounded by groups on the enzyme to be stabilized; for example, as in the tyrosyl-tRNA synthetase, which is discussed in Chapter 15. This then poses the problem of how the substrate gets into such an active site. Induced fit provides the means of access. The active site is open to solvent to allow the substrate to enter. Then, when the substrate binds, it uses some of its binding energy to pay for the active site to close around it.

3. Nonproductive binding

Although nonproductive binding is not a mechanism for increasing K_M, it is appropriately discussed here since it gives rise to effects that are qualitatively similar to those of strain and induced fit. This theory was originally invoked to account for specificity in the relative reactivities of larger, specific substrates compared with smaller, nonspecific substrates. It is assumed that as well as the productive binding mode at the active site, there are alternative, nonproductive modes in which the smaller substrates may bind and not react.

An example is the binding of polysaccharide substrates to lysozyme. In order for the reaction to occur, the substrate must bind across sites D and E of the six subsites A, B, C, D, E, and F. There is some strain associated with binding in subsite D, and occupying it does not increase the overall binding energy. Trimers and tetramers bind nonproductively in A, B, and C (Figure 12.8) and in A, B, C,

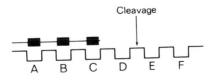

Figure 12.8 Nonproductive binding with substrates and lysozyme. Small substrates may bind at alternative sites along the extended active site of lysozyme, avoiding the cleavage site, which has a lower affinity.

and D. However, the favorable binding energy of occupying sites E and F causes hexamers to bind productively in subsites A through F.

4. The unimportance of strain, induced fit, and nonproductive binding in specificity

Specificity, in the sense of discrimination between competing substrates, is independent of the above three effects. The reasons are discussed in detail in Chapter 13. The basic reason is that specificity depends on k_{cat}/K_M, and strain and nonproductive binding do not affect the value of k_{cat}/K_M because it is independent of interactions in the *ES* complex (equations 12.10 and 3.36). Equation 12.16 shows that induced fit does alter k_{cat}/K_M for the active conformation, but equally for all substrates (i.e., by a factor of K).

5. Strain, induced fit, nonproductive binding, and steady state kinetics

Although the results of steady state kinetic measurements are often interpreted as supporting one of the three mechanisms, with the exact one depending on the whim of the experimentalist, the evidence is usually ambiguous. The approach generally used is to compare the k_{cat} and K_M values for a series of substrates, as in Tables 12.1 and 12.2, to see if the specificity is manifested in increasing k_{cat} rather than decreasing K_M. If this is found, it is good evidence that one of the processes is occurring, but it does not indicate which one. The problem is that all three mechanisms predict the same result: that binding energy is converted into chemical activation energy. The following arguments may be made:

Strain. The additional groups on the larger, specific substrates are used to strain the substrate rather than to provide binding. (In the transition state stabilization model this is modified to: "The additional binding energy is not realized until the transition state is reached.")

Induced fit. The additional groups on the larger, specific substrates are used to provide energy for the distortion of the enzyme. The smaller, nonspecific substrates bind predominantly to the inactive form of the enzyme. The k_{cat} is lower for these since less is productively bound, but the K_M is correspondingly lower since the binding energy is not used to convert the inactive conformation to the active one.

Nonproductive binding. The larger substrate binds in the productive mode only, but the smaller one, in addition to binding more weakly in the productive mode, binds in nonproductive modes, lowering the K_M. The k_{cat} is correspondingly lower.

6. Conclusions about the nature of strain: Strain or stress?

Although strain may be manifested in some cases by a genuine distortion of the substrate, it is likely that the strain will generally be distortionless. This could be due either to the substrate and the enzyme having unfavorable interactions that are relieved in the transition state, or to the transition state having additional binding interactions that are not realized in the enzyme–substrate complex. In both cases there would be forces that *tend* to distort the substrate toward the transition state. Since nonbonded interactions have weak force constants (apart from van der Waals repulsion) and enzymes and substrates are flexible, it is improbable that the substrate would be distorted by its interactions with the enzyme. Rotation about single bonds, as in conformational changes in lysozyme substrates, is possible, but the stretching of single bonds or the twisting of double bonds appears to be less likely, as these require strong forces. There is ample binding energy available, but for it to be used to distort a substrate the energy would have to change greatly in magnitude over a short distance, i.e., provide a strong *force*. On the basis of energy calculations, Levitt has suggested the following tentative rule: "Small distortions of a substrate conformation that cause large increases in strain energy cannot be caused by binding to the enzyme."[11,27] He suggests also that the largest forces that can be exerted are less than 12 kJ/mol/Å (3 kcal/mol/Å or 120 kJ/mol/nm), so that to strain a substrate by about 12 kJ/mol (3 kcal/mol), atoms must be moved by 1 Å (0.1 nm). In extreme cases genuine distortion might occur; but in general, strain will involve the subtle interplay of favorable and unfavorable interactions. It is more likely that the enzyme rather than the substrate is distorted, because the enzyme is less rigid. Indeed, as we have seen, there are many examples of the distortion of an enzyme on the binding of a substrate. Perhaps some of these distortions are an unavoidable consequence of the flexibility of proteins, and of the impossibility of constructing an active site that is precisely complementary to the substrate. Also, a low-energy conformational transition in which part of the active site closes over the substrate (as with peptides and carboxypeptidase, and with NAD^+ and dehydrogenases—Chapter 16) may be a small price for the enzyme to pay to provide easy access to its active site.

The term strain has a specific meaning in physics and engineering: it implies that an object is physically distorted. Its companion term, stress, means that an object being subjected to forces is not distorted by them. *Using these precise physical terms, it is probably apt to say that in the enzyme–substrate complex, the enzyme is often strained, whereas the substrate is often stressed.*

Strain and stress in enzymes arise from several different causes. We have seen in this chapter, and we shall see further in Chapters 15 and 16, that stress and strain may be divided into two processes, substrate destabilization and transition state stabilization. Substrate destabilization may consist of: steric strain, where there are unfavorable interactions between the enzyme and the substrate (e.g., with proline racemase, lysozyme); desolvation of the enzyme (e.g., by displacement of two bound water molecules from the carboxylate of Asp-52 of lysozyme); and desolvation of the substrate (e.g., by displacement of any bound water molecules from a peptide[28]). Transition state stabilization may consist of: the presence of transition state binding modes that are not available for the substrate (e.g., with tyrosyl-tRNA synthetase, serine proteases, cytidine deaminase); relief of steric strain; and reestablishment of the solvation of the enzyme or substrate (or the formation of electrostatic bonds) in the transition state.

E. Effects of rate optimization on accumulation of intermediates and internal equilibria in enzymes

1. Accumulation of intermediates

We saw in Chapter 7 that much effort has been put into the detection of chemical intermediates in enzymatic reactions. It has been found, though, that these do not accumulate in the reactions of many of the most common hydrolytic enzymes with their natural substrates under physiological conditions (Table 12.5).

Table 12.5 *Enzymes and intermediates*

Enzyme (class)	Substrate	Intermediate	Accumulation[a]
Chymotrypsin (serine proteases)	Peptides	Acylenzyme	−
Pepsin (acid proteases)	Peptides	Acylenzyme(?)[b] Aminoenzyme(?)[b]	− −
Carboxypeptidase (Zn^{2+} proteases)	Peptides	?[b]	−
Papain (thiol proteases)	Amides	Acylenzyme	−
Pig liver esterase (liver esterases)	Aliphatic esters	Acylenzyme	−
Acetylcholine esterase (choline esterases)	Acetylcholine	Acylenzyme	+
Acid phosphatase	Phosphate monoesters	Phosphorylenzyme	+
Lysozyme (glycosidases)	Polysaccharides	Carbonium ion (ester)	−[c]

[a] Whether or not the physiologically relevant substrates involve the accumulation of an enzyme-bound intermediate at saturating substrate concentrations. It should be noted that if the physiological concentration of the substrate is below its K_M value, an intermediate does not accumulate, even if it would at saturating concentrations.
[b] Evidence is against covalent intermediates in these reactions (Chapter 16).
[c] An intermediate accumulates for an artificial leaving group (*p*-nitrophenol) with glycosidase.

Where intermediates do accumulate, it is often through the use of synthetic, highly reactive substrates, such as esters with chymotrypsin, or through the effects of pH with alkaline phosphatase. There is a good theoretical explanation. It is a corollary of the principle of maximization of rate by the mutual increasing of k_{cat} and K_M that the accumulation of intermediates lowers the reaction rate.[4] Any intermediate that does accumulate lowers the K_M for the reaction, causing saturation at lower substrate concentrations. In other words, the accumulation of an intermediate means that the reaction has fallen into a "thermodynamic pit."

The "accumulation problem" is most severe for enzymes such as the digestive enzymes, which have to cope with pulses of high substrate concentrations. If the concentration of the substrate is below the K_M for the reaction under physiological concentrations, no intermediate accumulates *in vivo* in any case, since the enzyme is unbound. But in a test tube experiment in which the experimenter can use artificially high concentrations of substrate, an intermediate can sometimes be made to accumulate. An example occurs with glyceraldehyde 3-phosphate dehydrogenase. As Table 12.4 shows, the concentration of the aldehyde is below the K_M *in vivo*. But in the laboratory, the acylenzyme accumulates at saturating substrate concentrations.

2. Balanced internal equilibria

Although an enzyme cannot alter the equilibrium constant for a reaction in solution, there are no similar constraints on the equilibrium constant for the reagents when they are bound to the enzyme; i.e., there are no constraints on the internal equilibrium constant (Chapter 3, section H). There is, in fact, an inherent tendency for the internal equilibrium constant for the formation of an unstable intermediate or product to be more favorable than that in solution.[5] This stems from the Hammond postulate (Chapter 2, section A2) and the principle of enzyme–transition state complementarity: the transition state resembles the unstable intermediate or product, so the enzyme is closer in complementarity to the unstable species than to the substrate. The enzyme thus binds the unstable species more tightly than the substrate, and consequently increases the internal equilibrium constant in favor of the intermediate or product. However, the unstable species must not be bound too tightly or the reaction may fall into a "thermodynamic pit." For reversible reactions, the optimal situation is when all the intermediates have the same Gibbs free energy, so that the internal equilibria are balanced with each equilibrium constant being 1.[29] A compilation of internal equilibrium constants for a range of phosphotransferases reveals that most are indeed close to 1.[29]

Most of the speculations in this chapter have been shown to occur in practice in Chapter 15, where it will be seen in section F how the tyrosyl-tRNA synthetase changes the equilibrium constant for the formation of tyrosyl-adenylate from 3.5×10^{-7} in solution to 2.5 in the internal equilibrium.

References

1. J. B. S. Haldane, *Enzymes*, Longmans, Green and Co., p. 182 (1930). M.I.T. Press (1965).
2. L. Pauling, *Chem. Engng. News* **24**, 1375 (1946); *Am. Scient.* **36**, 51 (1948).
3. W. P. Jencks, *Adv. Enzymol.* **43**, 219–410 (1975).
4. A. R. Fersht, *Proc. R. Soc.* **B187**, 397 (1974).
5. W. J. Albery and J. R. Knowles, *Biochemistry* **15**, 5627, 5631 (1976); *Angewandte Chemie* **16**, 285 (1977).
6. S. A. Benner, *Chemical Reviews* **89**, 789 (1989).
7. E. Fischer, *Ber. Dt. Chem. Ges.* **27**, 2985 (1894).
8. G. E. Lienhard, *Science, N. Y.* **180**, 149 (1973).
9. R. Wolfenden, *Accts. Chem. Res.* **5**, 10 (1972).
10. I. I. Secemski and G. E. Lienhard, *J. Biol. Chem.* **249**, 2932 (1974).
11. M. Levitt, in *Peptides, polypeptides and proteins* (E. R. Blout, F. A. Bovey, M. Goodman, and N. Lotan, Eds.), Wiley, p. 99 (1974).
12. D. H. Leaback, *Biochem. Biophys. Res. Comm.* **32**, 1025 (1968).
13. G. Legler, M. L. Sinnott, and S. G. Withers, *J. Chem. Soc. Perk. II,* 1374 (1980).
14. G. J. Cardinale and R. H. Abeles, *Biochemistry* **7**, 3970 (1968).
15. M. V. Keenan and W. L. Alworth, *Biochem. Biophys. Res. Comm.* **57**, 500 (1974).
16. R. M. Cohen and R. Wolfenden, *J. Biol. Chem.* **246**, 7561 (1971).
17. S. B. Xiang, S. A. Short, R. Wolfenden, and C. W. Carter, *Biochemistry* **34**, 4516 (1995).
18. J. Fastrez and A. R. Fersht, *Biochemistry* **12**, 1067 (1973).
19. K. D. Collins and G. R. Stark, *J. Biol. Chem.* **246**, 6599 (1971).
20. F. Kohen, J. B. Kim, H. R. Lindner, Z. Eshhar, and B. Green *FEBS Lett.* **111**, 427 (1980).
21. P. G. Schultz and R. A. Lerner, *Science* **269**, 1835 (1995).
22. E. M. Driggers and P. G. Schultz, *Adv. Prot. Chem.* **49**, 261 (1996).
23. J. D. Stewart and S. J. Benkovic, *Nature, Lond.* **375**, 388 (1995).
24. J. Schlessinger and A. Levitzki, *J. Molec. Biol.* **82**, 547 (1974).
25. H. G. Mannherz, H. Schenck, and R. S. Goody, *Eur. J. Biochem.* **48**, 287 (1974).
26. J. C. Kernohan, W. W. Forrest, and F. J. W. Roughton, *Biochim. Biophys. Acta* **67**, 31 (1963).
27. M. Levitt, Ph.D. Thesis, University of Cambridge (England), p. 270 (1972).
28. R. Wolfenden, *Biochemistry* **17**, 201 (1978).
29. A. Hassett, W. Blättler, and J. R. Knowles, *Biochemistry* **21**, 6335 (1982).

Specificity and Editing Mechanisms

Specificity is a grossly overworked and often misused word. The most important meaning for the enzymologist refers to an enzyme's *discrimination* between several substrates competing for an active site: for example, the specificity of a particular aminoacyl-tRNA synthetase for a particular amino acid and a particular tRNA in a mixture of all the amino acids and all the tRNAs. This is the definition of specificity that is relevant to biological systems. It concerns the situation in which a desired and an undesired substrate are competing for an enzyme, and deals with the problem of the relative rate of reaction of the undesired substrate and desired substrate in a mixture of the two. Specificity in this sense is a function of both substrate binding and catalytic rate: if the undesired substrate and the enzyme have a k_{cat} that is 1000 times lower than the k_{cat} for the desired substrate, but the undesired substrate binds 1000 times more tightly, the preferential binding will compensate for the lower rate. For this reason, as discussed below, the k_{cat}/K_M is the important kinetic constant in determining specificity, since it combines both the rate and the binding terms.

A meaning of specificity that is really a misuse of the term refers to the activity of an enzyme toward an alternative substrate in the *absence* of a specific substrate, as can happen in an experiment *in vitro*. In such a test tube experiment, a substrate is often described as "poor" because it involves either a high value of K_M or a low value of k_{cat}. In biological systems both k_{cat} *and* K_M are important.

The difference between the two meanings is crucial to the status of strain, induced fit, and nonproductive binding in catalysis. As we discussed in Chapter 12 and as we shall amplify below, these do not affect biological specificity, since they alter k_{cat} and K_M in a mutually compensating manner without altering k_{cat}/K_M.

A. Limits on specificity

The basic problem in specificity is: How does an enzyme discriminate against a substrate that is smaller than, or the same size as (isosteric with), the specific substrate? There is no difficulty in discrimination against a substrate that is larger than the specific substrate, since the binding cavity at the active site may be constructed to be too small to fit the larger competitor. But a smaller competitor must always be able to bind, and it cannot be excluded by steric hindrance.[1] There will just be less binding energy available to be used for catalysis. Crude examples of this have been discussed in regard to the serine proteases. The larger aromatic amino acid derivatives cannot bind in the small binding pocket of elastase, but the smaller amino acid derivatives can bind to and react with chymotrypsin (Chapter 1). However, as discussed at the beginning of Chapter 12, the reactions of the smaller substrates involve much lower values of k_{cat} and k_{cat}/K_M. There is also no difficulty in discrimination against substrates with the wrong stereochemistry. As was pointed out in Chapter 8, the substitution of an L-amino acid by a D-amino acid leads to an interchange of two groups around the chiral carbon, so that the substrate cannot be bound productively because of steric effects.

The areas in which high specificity is most essential are DNA replication and protein biosynthesis, because of the necessity of maintaining the genetic information encoded in DNA and faithfully translating it into protein structure. Because of the evolutionary pressure on the enzymes involved in these processes to be as accurate as possible, measurements on the key polymerases and synthetases delineate the maximum possible practical limits on specificity. We used measurements on these enzymes in Chapter 11, section B1, to tabulate maximum values for the binding energies of small groups to proteins.

Good examples of discrimination occur in the reactions of the aminoacyl-tRNA synthetases (see Chapter 7, section D, for their mechanism). These enzymes are responsible for the selection of amino acids during protein synthesis, and so have to discriminate among a multitude of often very similar substrates with high precision. For example, the isoleucyl-tRNA synthetase has to discriminate between isoleucine and valine, and the valyl-tRNA synthetase between valine and threonine (Figure 13.1). Valine, being shorter by one methylene group than isoleucine, binds to the isoleucyl-tRNA synthetase, but 150 times more weakly.[2] Threonine, although it is isosteric with valine, binds 100 to 200 times more weakly to the valyl-tRNA synthetase, because of the burying of the hydroxyl group in the hydrophobic pocket normally occupied by a methyl group of valine.[3] Similarly, the alanyl-tRNA synthetase discriminates against glycine by a factor of 250.[4] Thus, a precisely tailored active site can recognize the absence of a methylene group on a substrate by a relative rate factor of up to 250, though a factor of 10 or so is more usual. The limits on steric exclusion of a larger substrate from a smaller binding site can be estimated from the relative rates of activation of isoleucine and valine by the valyl-tRNA synthetase. Isoleucine, which

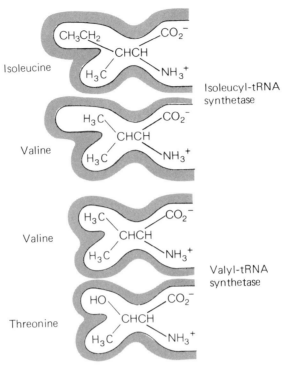

Figure 13.1 The binding cavity at the active site of the isoleucyl-tRNA synthetase must be able to bind valine as it binds the larger isoleucine. The active site of the valyl-tRNA synthetase cannot exclude threonine, because it is isosteric with valine.

again is too large by one methylene group to fit the site tailored for valine, reacts at a rate that is up to 2×10^5 times lower.[5]

The specificity against binding a small substrate to an active site constructed for a larger one may be analyzed by transition state theory in the same way as enzyme–substrate complementarity was in Chapter 12. The activation energy of the reaction is divided into contributions from the chemical activation energy and from the enzyme–substrate binding energy. We shall see that if the smaller or isosteric substrate differs from the specific substrate by lacking an element that has a structure R and a potential binding energy of $\Delta\Delta G_b$, the maximum possible discrimination due to this difference in stereochemistry is $\exp(-\Delta\Delta G_b/RT)$, and this cannot be amplified by strain, by induced fit, by a series of conformational changes, by an additional series of chemical steps, or by two (or more) sites functioning simultaneously. We shall first demonstrate this limit for specific examples in Michaelis-Menten kinetics, and then generalize for any mechanism. It will be seen that specificity is due only to transition state binding.

1. Michaelis-Menten kinetics

It was shown in Chapter 3, section G2, that specificity for competing substrates is controlled by k_{cat}/K_M. If the rate of reaction of the specific substrate A is v_A, and that of the competitor B is v_B, then

$$\frac{v_A}{v_B} = \frac{[A](k_{cat}/K_M)_A}{[B](k_{cat}/K_M)_B} \tag{13.1}$$

This is translated into terms of binding energy by using equations 12.3, 12.4, and 12.10; i.e.,

$$\ln \frac{k_{cat}}{K_M} = \ln \frac{kT}{h} - (\Delta G_0^{\ddagger} + \Delta G_b) \tag{13.2}$$

where $\Delta G_0^{\ddagger}$ is the chemical activation energy and ΔG_b is the binding energy of the enzyme and transition state. If the additional group R on A is not directly chemically involved in the reaction, $\Delta G_0^{\ddagger}$ will be the same for both A and B, apart from inductive effects. Ignoring these for convenience and setting the difference in binding energy at $\Delta\Delta G_b$ gives, from equation 13.2,

$$\frac{(k_{cat}/K_M)_A}{(k_{cat}/K_M)_B} = \exp\left(-\frac{\Delta\Delta G_b}{RT}\right) \tag{13.3}$$

(where $\Delta\Delta G_b$ is algebraically negative).

Equation 13.3 quantifies the *maximum* effect the additional binding energy can have. If the rate of the slow step in the reaction of A is lowered to such an extent in B that another step becomes rate-determining, then the activation energy has not been lowered by the full amount, $\Delta\Delta G_b$. Also, if B binds and reacts in alternative modes, these will be in addition to the mode of A and they will add to the overall rate of reaction of B.

One example in which specificity may be lost is when Briggs-Haldane kinetics are occurring (Chapter 3, section A3a). Under these conditions, k_{cat}/K_M is equal to the rate constant for the association of the enzyme and the substrate. Since it is usually found that the higher dissociation constants for smaller substrates arise from a higher rate of dissociation rather than from a lower rate of association, there will be a partial or complete loss of specificity.

The following mechanisms have been suggested as causes of specificity, but we will see why they cannot be so.

1. *Strain.* As was explained in Chapter 12, strain does not affect k_{cat}/K_M but just causes compensating changes in k_{cat} and K_M without altering their ratio.

2. *Induced fit.* Chapter 12 also pointed out that k_{cat}/K_M for enzymes involving induced fit is just the value of k_{cat}/K_M for the active conformation scaled down by a constant factor for all substrates (equations 12.15 and 12.18). Induced fit does not alter the relative values of k_{cat}/K_M from what they would be if all of the enzyme were in the active conformation, and thus does not affect specificity.

3. *Nonproductive binding.* It was shown in Chapter 3, section E, that nonproductive binding does not alter k_{cat}/K_M but decreases both k_{cat} and K_M while maintaining their ratio. Specificity is unaffected.

4. *A series of sequential reactions.* Specificity cannot be amplified by there being a series of steps in which $\Delta\Delta G_b$ is utilized at each one. A simple way of seeing this is to recall from Chapter 3, section F, that in the following equation k_{cat}/K_M is always equal to k_2/K_S, irrespective of the number of additional intermediates:

$$E + S \underset{K_S}{\rightleftharpoons} ES \xrightarrow{k_2} ES' \xrightarrow{k_3} ES'' \xrightarrow{k_4} ES''' \xrightarrow{k_5} \text{etc.} \tag{13.4}$$

2. The general case

The following formal thermodynamic approach can be used quite generally for analyzing binding energy contributions.

Consider any series of reactions

$$E + S \rightleftharpoons ES \rightleftharpoons E'S' \rightleftharpoons E''S'' \rightleftharpoons \rightleftharpoons \rightleftharpoons E^\ddagger S^\ddagger \xrightarrow{\text{rate-determining}} \tag{13.5}$$

where E, E', E'', etc. are different states of the enzyme, and S, S', S'', etc. are different states of the substrates (covalently altered, etc.). The rate of the reaction may be calculated from transition state theory by ignoring all the intermediate steps and just considering the energetics of the process $E + S \rightarrow E^\ddagger S^\ddagger$. The Gibbs free energy of activation may be considered to be composed of three terms: $\Delta G_E^\ddagger$, a free energy change representing the energy difference between E and $E^\ddagger$; $\Delta G_S^\ddagger$, representing the energy difference between S and $S^\ddagger$; and ΔG_b, the binding energy of $E^\ddagger$ and $S^\ddagger$.

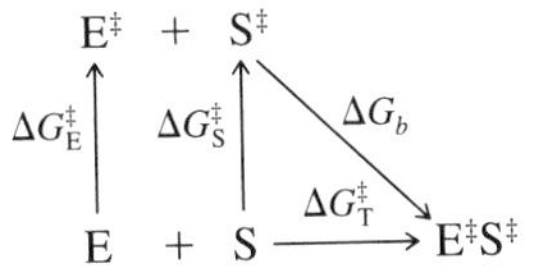

Scheme 1

The thermodynamic cycle in scheme 1 shows that the activation energy $\Delta G_T^\ddagger$ is given by

$$\Delta G_T^\ddagger = \Delta G_E^\ddagger + \Delta G_S^\ddagger + \Delta G_b \qquad (13.6)$$

The rate of reaction is given from transition state theory:

$$v = \frac{kT}{h}[E][S]\exp\left(-\frac{\Delta G_E^\ddagger + \Delta G_S^\ddagger + \Delta G_b}{RT}\right) \qquad (13.7)$$

The relative reaction rates of the two substrates A and B are given by substituting the values of the Gibbs energies for A and B into equation 13.7 and taking the ratio:

$$\frac{v_A}{v_B} = \frac{[A]}{[B]}\exp\left(-\frac{\Delta\Delta G_b + \Delta G_A^\ddagger - \Delta G_B^\ddagger}{RT}\right) \qquad (13.8)$$

The difference in the binding energy of A and B, $\Delta\Delta G_b$, comes into the equations only once. As mentioned in the preceding section, $\Delta\Delta G_b$ cannot be used in a cumulative manner over a series of steps to amplify the differences. A reaction might involve several steps, but the specificity due to $\Delta\Delta G_b$ will just be spread out over them.

For similar substrates, $\Delta G_A^\ddagger$ will be similar to $\Delta G_B^\ddagger$, so that

$$\frac{v_A}{v_B} = \frac{[A]}{[B]}\exp\left(-\frac{\Delta\Delta G_b}{RT}\right) \qquad (13.9)$$

This procedure may be extended to include common cosubstrates, e.g., NAD^+ or ATP. The Gibbs free energy changes involving these cancel out, as does $\Delta G_E^\ddagger$, when the ratios of rates are found from equation 13.7 or its equivalent.

3. Interacting active sites

Can specificity be increased by more than one molecule of substrate binding to an enzyme with multiple binding sites? Such an increase could appear to happen in a test tube experiment with only one substrate present, since absolute rate and not discrimination would be measured. But in a biological experiment with both the specific and the competitive substrates present, it would become clear that specificity cannot be increased in this way, because of the reaction of mixed complexes containing the enzyme and both types of substrates. This may be proved by the formal thermodynamic approach. In the following example we consider the case of "half-of-the-sites reactivity," in which one molecule of substrate S* binds without giving products during the turnover of the enzyme, but the second molecule S goes on to react. By comparing the reaction rates of pairs of complexes, e.g., E·A·A* with E·B·A*, it will be seen that the additional binding energy of the larger substrate, $\Delta\Delta G_b$, can be used only "once":

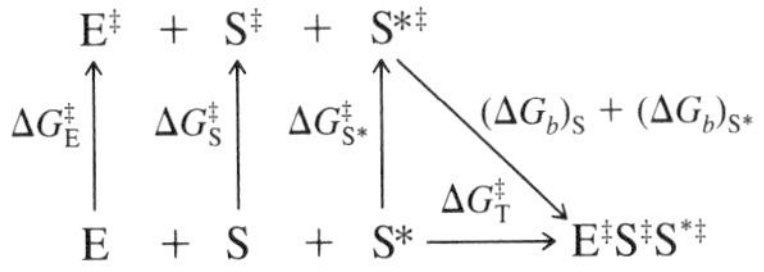

$$E^\ddagger \; + \; S^\ddagger \; + \; S^{*\ddagger}$$

$$\Delta G_E^\ddagger \Big\uparrow \quad \Delta G_S^\ddagger \Big\uparrow \quad \Delta G_{S*}^\ddagger \Big\uparrow \quad \diagdown \; (\Delta G_b)_S + (\Delta G_b)_{S*}$$

$$\Delta G_T^\ddagger$$

$$E \; + \; S \; + \; S^* \longrightarrow E^\ddagger S^\ddagger S^{*\ddagger}$$

Scheme 2

According to transition state theory, the rate is given by

$$v = \frac{kT}{h}\,[E][S][S^*]\exp\left[-\frac{\Delta G_E^\ddagger + \Delta G_S^\ddagger + \Delta G_{S*}^\ddagger + (\Delta G_b)_S + (\Delta G_b)_{S*}}{RT}\right] \qquad (13.10)$$

If A can bind more strongly than B because the difference in structure contributes a binding energy of $\Delta\Delta G_b$, and if v_{AB*} is the reaction rate when A is bound at the chemically reacting site and B is bound at the other, etc., then substituting the Gibbs energies into equation 13.7 and assuming that $\Delta G_A^\ddagger = \Delta G_B^\ddagger$ gives

$$\frac{v_{AA*}}{v_{BA*}} = \frac{[A]}{[B]}\exp\left(-\frac{\Delta\Delta G_b}{RT}\right) \qquad (13.11)$$

and

$$\frac{v_{AB*}}{v_{BB*}} = \frac{[A]}{[B]}\exp\left(-\frac{\Delta\Delta G_b}{RT}\right) \qquad (13.12)$$

so that

$$\frac{v_{AA*} + v_{AB*}}{v_{BA*} + v_{BB*}} = \frac{[A]}{[B]}\exp\left(-\frac{\Delta\Delta G_b}{RT}\right) \qquad (13.13)$$

Owing to the mixed complexes of A and B that bind to the enzyme, the specificity cannot be enhanced by the binding of two molecules of substrate simultaneously. Substrate A can enhance activity more than B does by binding at a non-catalytic site, but it enhances activity with B as well as with itself.

All of the above derivations are for the steady state. Pre–steady state and oscillating reactions can be problematic.

4. The stereochemical origin of specificity

Specificity between competing substrates depends on the relative binding of their transition states to the enzyme. Enzyme–transition state complementarity maximizes specificity because it ensures the optimal binding of the desired transition state. This is also the criterion for the optimal value of k_{cat}/K_M, which is not surprising, since specificity is determined by k_{cat}/K_M. Maximization of rate parallels maximization of specificity, as long as there is no changeover to Briggs-Haldane kinetics.

The reason why hexokinase phosphorylates glucose in preference to water is that glucose binds well in the transition state, whereas water does not. Whatever the mechanism of the reaction, be it strain or induced fit, the competition between glucose and water is the same. If, for the sake of argument, the V_{max} for the phosphorylation of glucose were the same as that for water, and the binding energy of the glucose were used to give a very low K_M, the glucose would be preferentially phosphorylated due to its preferential binding to the active site. If, on the other hand, all the binding energy of the glucose were used in lowering the activation energy of V_{max}, this would also lead to its preferential phosphorylation due to its greater reactivity when bound.

There is one case in which strain or induced fit could be useful in a type of specificity. These mechanisms are unimportant where competition between substrates is concerned. But given a situation in which there is *no* specific substrate present, these mechanisms could be of use in providing a low absolute activity of the enzyme toward, say, water. For example, induced fit could prevent hexokinase from being a rampant ATPase in the *absence* of glucose (although its absence is extremely unlikely).

B. Editing or proofreading mechanisms

There are two fundamental types of interactions that limit the accuracy of DNA replication and protein biosynthesis. The first and more general type is the complementary base pairing that controls specificity in DNA replication and transcription (Chapter 14). Theory suggests[6] and experiment shows[7] that base pairing is accurate to about 1 part in 10^4 to 10^5. The second type of interaction is the intrinsic binding energy of amino acid side chains to proteins, because this limits the accuracy of amino acid selection. As was discussed at length at the beginning of this chapter and in Chapter 11, amino acid selection is accurate only to about 1 part in 10^2 because of the enzyme's difficulties in using steric exclusion to reject substrates that are isosteric with or slightly smaller than the correct substrate. Yet the overall error rate in the replication of DNA in *Escherichia coli* is only 1 mistake per 10^8 to 10^{10} nucleotides polymerized, and the overall error rate in transcription of the DNA and translation of the message into protein is in general only about 1 per 10^3 to 10^4 amino acid residues incorporated. Given the limits on the accuracy of base pairing and amino acid recognition, this specificity is beyond the theoretical thermodynamic limits for simple enzymes. It is possible only because of the evolution of *editing* or *proofreading* mechanisms. Certain key enzymes involved in polymerization have evolved, in addition to their active site for synthesis, a second, hydrolytic active site which is used to destroy incorrect intermediates or products as they are formed. The synthesis is thus double-checked at each step so that errors may be removed before they are permanently incorporated.

The crux of an editing mechanism is the formation of a high-energy intermediate that is unstable with respect to hydrolysis (equation 13.14). This allows an element of *kinetic* control of product formation to be introduced: the intermediate is at a branch point in the reaction pathway so that it may be channeled either to further synthesis or to hydrolytic products.

$$E + S \rightleftharpoons ES \longrightarrow EI \overset{\text{further synthesis}}{\underset{\text{hydrolytic destruction}}{\diagup\diagdown}} \tag{13.14}$$

Without these editing mechanisms, the errors occurring during the replication of the genetic material and the synthesis of proteins would be at an intolerably high level.

1. Editing in protein synthesis

In the absence of editing mechanisms, the least accurate component of protein synthesis would be amino acid selection. For example, the extra methylene group of isoleucine causes the isoleucyl-tRNA synthetase to favor the activation of isoleucine over valine by a factor of only 100 to 200. This, combined with the 5-fold higher concentration of valine over isoleucine *in vivo,* would give an error rate of 1 in 20 to 40. Yet the overall error rate found for the mistaken incorporation of valine in positions normally occupied by isoleucine is only 1 in 3000.[8] The phenomenon of editing was first discovered in relation to this enzyme.[9] The isoleucyl-tRNA synthetase was found to form a stable enzyme–bound valyl adenylate complex in the absence of tRNA. But, whereas the addition of tRNA^{Ile} to the correct complex gives Ile-tRNA^{Ile} (equation 13.15), the addition of tRNA^{Ile} to the incorrect complex of valyl adenylate leads to quantitative hydrolysis to valine and AMP and gives no Val-tRNA^{Ile} (equation 13.16). In the presence of valine and tRNA^{Ile}, the isoleucyl-tRNA synthetase is an ATP pyrophosphatase that wastefully, but necessarily, catalyzes the hydrolysis of ATP to AMP via the activation reaction.

$$E \xrightarrow{\text{Ile, ATP}} E \cdot \text{Ile-AMP} \xrightarrow{\text{tRNA}^{\text{Ile}}} \text{Ile-tRNA}^{\text{Ile}} + \text{AMP} + E \tag{13.15}$$

$$E \xrightarrow{\text{Val, ATP}} E \cdot \text{Val-AMP} \xrightarrow{\text{tRNA}^{\text{Ile}}} \text{Val} + \text{tRNA}^{\text{Ile}} + \text{AMP} + E \tag{13.16}$$

There are two high-energy intermediates on the reaction pathway that could be edited by hydrolysis: the enzyme-bound aminoacyl adenylate and the aminoacyl-tRNA. A mechanistic study must distinguish between the two. A pathway involving the mischarged tRNA involves the formation of a covalent intermediate—the aminoacylated tRNA—so the three rules of proof may be considered (Chapter 7, section A1). These criteria have been rigorously applied to the rejection of threonine and other amino acids by the valyl-tRNA synthetase (Chapter 7, section D2).[3,10] The enzyme has a distinct and separate hydrolytic site for the deacylation of Thr-tRNA^{Val}; the k_{cat}/K_M for formation of a threonyl adenylate complex is some 600 times lower than that for the activation of valine. Thr-tRNA^{Val} is formed but it is rapidly deacylated with a rate constant of 40 s^{-1} before it can dissociate from the enzyme. The deacylation site must presumably have a hydrogen-bond donor and acceptor that binds the hydroxyl group of threonine (Figure 13.2). The cor-

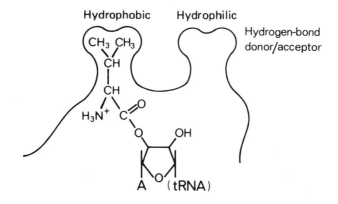

(a) ACYLATION   HYDROLYSIS

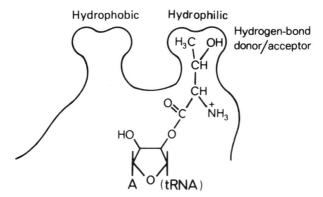

(b) ACYLATION   HYDROLYSIS

Figure 13.2 A possible specificity mechanism for the prevention of the misacylation of tRNA^{Val} with threonine. (a) The hydrophobic acylation site discriminates against threonine. (b) The hydrolytic site specifically uses the binding energy of the hydroxyl of threonine for a binding or catalytic effect. The translocation may occur as illustrated via a 2'- → 3'-hydroxyl acyl transfer. [From A. R. Fersht and M. Kaethner, *Biochemistry* **15**, 3342 (1976).]

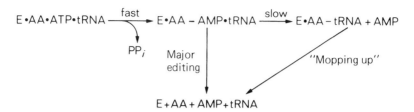

Figure 13.3 A possible double-check mechanism, with the major editing step occurring before the transfer of the amino acid to tRNA. [From A. R. Fersht, *Biochemistry* **16**, 1025 (1977); see also W. Freist and F. Cramer, *Eur. J. Biochem.* **131**, 65 (1983).]

rectly charged Val-tRNA^{Val} is deacylated ~ 3000 times more slowly, presumably because of the unfavorable energetics associated with the wrong amino acid occupying the hydrophilic region tailored for threonine. Nature has thus evolved a specificity mechanism that uses the structural differences twice, but in different ways: threonine binds less well to the activation site but better to the deacylation site than does valine.

The failure to trap mischarged tRNA during editing by other aminocyl-tRNA synthetases leaves open the possibility that editing may also occur at the level of the aminoacyl adenylate. The isoleucyl-tRNA synthetase provides an example. The decomposition of the E· Val-AMP complex (equation 13.16) requires the addition of tRNA^{Ile}; further, Val-tRNA^{Ile} that is prepared synthetically is rapidly deacylated by the isoleucyl-tRNA synthetase. But, in contrast to the mischarged tRNA formed by the valyl-tRNA synthetase, the Val-tRNA^{Ile} could not be detected during the course of the editing reaction.[11] It is possible that there is a double-check mechanism in which most of the valyl adenylate is destroyed prior to the transfer to tRNA, and in which most of the quantity that *is* transferred is mopped up by the esterase activity of the enzyme (Figure 13.3).

The methionyl-tRNA synthetase edits misactivated homocysteine before the transfer to tRNA^{Met}. The homocysteinyl adenylate complex is rapidly decomposed in the absence of tRNA. The reaction is, however, idiosyncratic, in that it invokes the cyclization of the intermediate to give the thiolactone.[12]

An analogy may be made between the amino acid selection process and a "double sieve" (Figure 13.4).[5] This crudely illustrates both the basic principles of selection for editing and how just two active sites can sort the whole range of amino acids by invoking first size and then specific chemical characteristics. Amino acids larger than the correct one are rejected by the first ("coarse") sieve by steric exclusion. All smaller or isosteric amino acids are activated, but at reduced rates because of poorer binding. Finally, the second ("fine") sieve accepts the products of activation or of transfer of the smaller or isosteric amino acids and excludes the products of the correct one, either because it is too large or because it lacks the specific interactions that allow an isostere to bind in the hydrolytic active site.

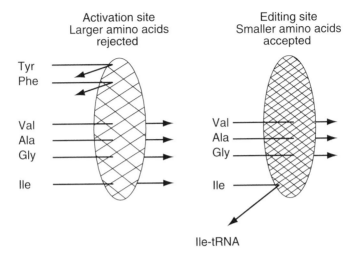

Figure 13.4 The "double sieve" analogy for the editing mechanism of the isoleucyl-tRNA synthetase. The active site for the formation of the aminoacyl adenylate can exclude amino acids that are larger than isoleucine but not those that are smaller. On the other hand, a hydrolytic site that is just large enough to bind valine can exclude isoleucine while accepting valine and all the smaller amino acids. (In some enzymes, the hydrolytic site offers specific chemical interactions that enable it to bind isosteres of the correct amino acid as well as smaller amino acids.)

Not all aminoacyl-tRNA synthetases have editing sites. The cysteinyl- and tyrosyl-tRNA synthetases bind the correct substrates so much more tightly than their competitors that they do not need to edit.[13,14] Similarly, since the accuracy of transcription of DNA by RNA polymerase is better than the overall observed error rate in protein synthesis at about 1 part in 10^4, RNA polymerases do not need to edit.[15] The same should be true for codon–anticodon interactions on the ribosome. However, it is possible that accuracy has been sacrificed to achieve higher rates in this case, which is analogous to a change from Michaelis-Menten to Briggs-Haldane kinetics, and so an editing step is required.[16]

a. Molecular mechanism of the double sieve

The structure of the isoleucyl-tRNA synthetase (IleRS) from *Thermus thermophilus* (1045 residues, M_r 120 000) has been solved, as well as its complexes with Ile and Val.[17] The protein contains a nucleotide binding fold (Chapter 1) that binds ATP. The fold has two characteristic ATP binding motifs: His-54-Val-55-Gly-56-His-57 and Lys-591-Met-592-Ser-593-Lys-594. In the L-Ile·IleRS complex, a single Ile is bound at the bottom of the ATP cleft, with the hydrophobic side chain in a hydrophobic pocket, surrounded by Pro-46, Trp-518, and Trp-558. L-Leucine cannot fit into this pocket because of the steric hindrance of one of its terminal methyl groups. Larger amino acids are similarly excluded from this site. In the L-Val·IleRS complex, Val is bound to the same site, but the contact area with Pro-46 and Trp-558 is lower. This site is the coarse sieve. (Interestingly, the equivalent binding site of the valyl-tRNA synthetase is very similar except that the Gly-45 of the IleRS is a proline residue in ValRS, which would make the cavity smaller and so exclude isoleucine).

There is a long polypeptide sequence that is inserted into the nucleotide binding fold. This forms a four-domain structure, which has been shown by mutagenesis to contain the editing side.[18] In the L-Val·IleRS complex, a second molecule of Val is bound here, whereas no electron density was observed for Ile in the L-Ile·IleRS complex.[17] The binding cavity is surrounded by Trp-32 and Tyr-386 and is too small to accommodate Ile. Deletion of 47 residues that constitute this site in the IleIRS from *E. coli* abolishes the editing activity, and the deletant efficiently catalyzes the formation of Val-tRNA[Ile]. Similarly, mutation of the equivalent to Thr-230 to Ala abolishes editing. It was proposed that Thr-230 is a nucleophile in the hydrolytic activity. The inserted sequence is the fine sieve.

The editing activity is induced by the addition of tRNA[Ile] to the IleRS·Val-AMP complex. Model building suggests that one of the domains in the insert rotates on the addition of the tRNA, so that it and the two active sites form a closed cavity in which the aminoacylation and editing clefts face each other.

2.   Editing in DNA replication

[Chapter 14, sections A1 and A2, may be read at this stage for an introduction to DNA replication and the associated enzymes.] There are two fundamental differences between selection of amino acids during protein synthesis and the matching of base pairs during the replication of DNA.

The first is that whereas each amino acid has its own activating enzyme precisely tailored to it, a single DNA polymerase with just one active site for synthesis has to cope with all four correct base-pair combinations. The specificity of the reaction is largely delegated to the specificity of the base pairing itself, with its inherent error rate of 1 in 10^4 to 10^5. An even higher accuracy would be obtained if there were four separate enzymes (or just four active sites), each precisely tailored for its own base pair of AT, TA, GC, or CG. But it has been found expedient in evolution for errors to be corrected by editing mechanisms and for a common polymerase to be used.

The second difference enables errors in DNA replication to be corrected with relative ease. During protein synthesis, the growing end of the polypeptide chain is activated and transferred to the next amino acid in the sequence (Figure 13.5). There is no means of removing an incorrectly added residue and reactivating the polypeptide. Error correction has to be made before polymerization. But in the synthesis of DNA, the monomeric nucleotide is activated and added to the unactivated growing chain. This has enabled the evolution of a mechanism for the editing of errors after polymerization has occurred.

DNA synthesis proceeds in the $5' \rightarrow 3'$ direction, with the nucleotides being added to the $3'$-hydroxyl of the polynucleotide. At the same time, all prokaryotic DNA polymerases have a $3' \rightarrow 5'$ exonuclease activity that works in the

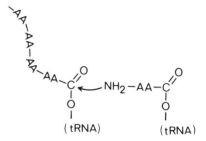

Figure 13.5 Protein synthesis involves the transfer of the activated polypeptide chain to the next amino acid residue.

opposite direction (Figure 13.6). There is strong evidence that this is an editing function for the excision of incorrect, mismatched bases.[19] First, there is the evidence that the exonuclease activity is greatest for mismatched bases or single-stranded DNA.[19,20] Second, the mutation frequency in the T4 bacteriophage correlates with the measured rate of the exonuclease activity catalyzed by the DNA polymerase. "Mutator" mutants have a very high mutation rate and some code for a DNA polymerase with a low exonuclease activity[21,22] (Table 13.1). "Antimutators" have a high resistance to mutation and a high exonuclease activity. As Table 13.1 shows, the antimutators are very wasteful in the high amount of deoxynucleoside triphosphate hydrolyzed to incorporate a single base. Third, whereas the accuracy of DNA replication *in vitro* catalyzed by prokaryotic DNA polymerases is much higher than would be expected from the known frequencies of base mispairing, eukaryotic DNA polymerases that lack the $3' \rightarrow 5'$ exonuclease activity are less accurate and make errors at the expected rate.[23,24] Fourth, as described in the next section, the kinetics of error induction by prokaryotic DNA polymerases are consistent with the active participation of an editing mechanism.[25]

a. Kinetics of polymerase accuracy: "Kinetic genetics"

Prokaryotic DNA polymerases are so accurate that special kinetic assays have had to be introduced to detect errors *in vitro*. These depend on replicating under controlled conditions the circular DNA of a small bacteriophage that contains a

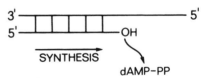

Figure 13.6 DNA synthesis involves the transfer of the activated deoxynucleotide monophosphate from its triphosphate to the 3'-hydroxyl of the growing chain. Editing takes place in the opposite direction.

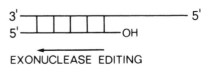

Table 13.1 *Correlation of mutation and exonuclease activity of DNA polymerases of T4 phage[a]*

Strain of phage	Phenotype	dTTP wastefully hydrolyzed ÷ dTMP incorporated
L56	Mutator	0.005
L98	Mutator	0.01
74D	Wild type	0.04
L42	Antimutator	1.6
L141	Antimutator	13

[a] From N. Muzyczka, R. L. Poland, and M. J. Bessman, *J. Biol. Chem.* **247**, 7116 (1972).

single-point mutation.[26,27] The accuracy of copying can be measured by producing viable phage from the synthetic DNA and then scoring by classic plaque-counting assays the proportion of revertant phage produced: each revertant corresponds to an error in replication. This is exactly analogous to the geneticists' method of measuring spontaneous mutation rates *in vivo*. The error rate of replication of bacteriophage ϕX174 *in vitro* was found to be very similar to the rate of spontaneous mutation *in vivo*: about 1 in 10^6 to 10^7. This shows that errors in base substitution during DNA replication are largely rate-determining in spontaneous mutation.[28]

The rate law for incorporation of incorrect nucleotides during DNA replication catalyzed by a polymerase that has an exonuclease activity is more complicated than that for the simple case of two substrates competing for an active site.[25,29,30] There has to be an additional term that allows for the partitioning of the newly added mismatched nucleotide between hydrolytic excision and permanent incorporation via elongation. For example, suppose that an incorrect deoxynucleoside triphosphate, $dNTP_i$, and a correct one, $dNTP_c$, compete for pairing at a site. Then the relative rate of insertion (i.e., of adding to the 3' terminus) is given by

$$R = \frac{[\text{dNTP}]_i(k_{cat}/K_M)_i}{[\text{dNTP}]_c(k_{cat}/K_M)_c} \tag{13.17}$$

But if in addition there is partitioning, as in equation 13.18, then the concentration of the nucleotide that next follows in the sequence to be incorporated, $dNTP_f$, comes into the calculations.

$$\text{E·DNA·dNTP}_i \rightarrow \text{E·DNA-N}_i \underset{}{\overset{\text{dNTP}_f}{\rightleftharpoons}} \text{E·DNA-N}_i\text{·dNTP}_f \rightarrow \text{E·DNA-N}_i\text{-N}_f \rightarrow$$

$$\downarrow \text{editing}$$

$$\text{E·DNA} + \text{dNMP}_i \tag{13.18}$$

In general, the concentration of $dNTP_f$ follows Michaelis-Menten-type kinetics. If the partitioning ratio equals the fraction F of inserted intermediate giving products, then F is of the form

$$F = \frac{[dNTP]_f}{K_M + \alpha[dNTP]_f} \tag{13.19}$$

where K_M has the dimensions of a Michaelis constant.[25,30] The overall misincorporation frequency is given by RF_i/F_c. In general, F_c is close to 1, since excision of the correct nucleotide is slow. Thus, the misincorporation frequency is close to RF_i. The accuracy consequently depends on the concentration of $dNTP_f$ in a predictable manner. At low concentrations, i.e., where $[dNTP]_f$ is less than K_M, partitioning favors editing. Hence the accuracy is high and the error rate is proportional to $[dNTP]_i[dNTP]_f/[dNTP]_c$. At high $[dNTP]_f$, F_i saturates, so the error rate is proportional simply to $[dNTP]_i/[dNTP]_c$. The accuracy is lower under these conditions, and when $\alpha = 1$ there will be no editing at all.

The dependence of overall misincorporation on the concentration of the next nucleotide to be incorporated provides a diagnostic test for the active participation of an editing mechanism.

b. Molecular mechanism of DNA polymerase editing

Detailed structures of several DNA polymerases are available, including those with double- or single-stranded DNA bound.[31–35] Site-directed mutagenesis (Chapter 14) has been used to identify residues that are in the active sites for polymerization and exonuclease activity.[36,37] Mutants that are defective in the exonuclease activity are used for the production of stable complexes with single-stranded DNA or single-stranded $3'$ ends of duplexes bound in the exonuclease site.[31,33] $2',3'$-Dideoxynucleotides (ddNTPs) that lack the nucleophilic $3'$-OH have been used to make templates that are inert so that productive ternary enzyme·DNA·ddNTP complexes may be observed.[34,35,38]

The so-called Klenow fragment of DNA polymerase 1 of *E. coli* (Chapter 14, section A1) contains the $5'$-$3'$-polymerization and the $3'$-$5'$-exonuclease domains. Detailed pre–steady state kinetics have been made of the polymerization and exonuclease activities.[39–43] The editing site is 35 Å away from the polymerization site.[32] The mechanism of the polymerization activity (Figure 13.7) is very similar to that for hydrolysis (Figure 13.8). The key to both is the presence of two metal ions, 3.9 Å apart, that stabilize the developing charges on the transition state and metal-bound HO^- or RO^- ions (see Chapter 2, section B7).[44,45]

There is an equilibrium between DNA binding to the two sites. Time-resolved fluorescence spectroscopy of labeled duplex DNA bound to the Klenow fragment shows that 12% occupies the editing site and 88% the polymerization.[42] The presence of mismatches both slows down elongation and increases the occupancy of the editing site.[43] Comparison of all the steps for correct and incorrect nucleotide incorporation confirms that the fidelity stems from discrimination in

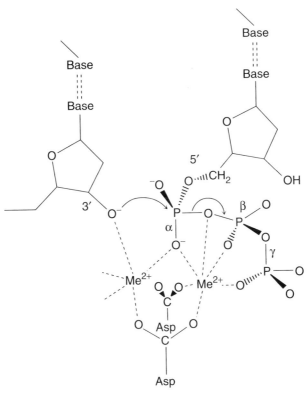

Figure 13.7 The "two-metal-ion" mechanism for polynucleotide polymerases. One metal ion (usually Mg^{2+}) activates the 3′-OH group of the primer terminus and stabilizes one of the partly negatively charged equatorial oxygen atoms of the phosphoryl group, whereas the other binds the phosphoryl oxygen and the oxygen atoms of the pyrophosphate leaving group. The two metal ions are 3.9 Å apart. This mechanism fits both RNA and DNA polymerases. [Modified from T. A. Steitz, *Nature, Lond.* **391**, 231 (1998).]

the chemical step for elongation, combined with increased exonuclease activity toward mismatches because of the longer time spent by the incorrectly elongated product in the reaction sequence,[41] as well as the 3- to 4-fold increased occupancy of mismatched DNA in the sieving site.[43]

c. DNA repair mechanisms

It is possible for base mispairing in duplex DNA to be corrected by repair mechanisms because the information content is duplicated in the two complementary strands. For example, aberrant base pairing arising from environmental damage such as x-rays, ultraviolet radiation, oxidation, and chemical modification may be repaired as follows.

The damaged strand is cut by an endonuclease, bases are removed and then replaced by the polymerase, and the join is sealed by DNA ligase.[46] In these

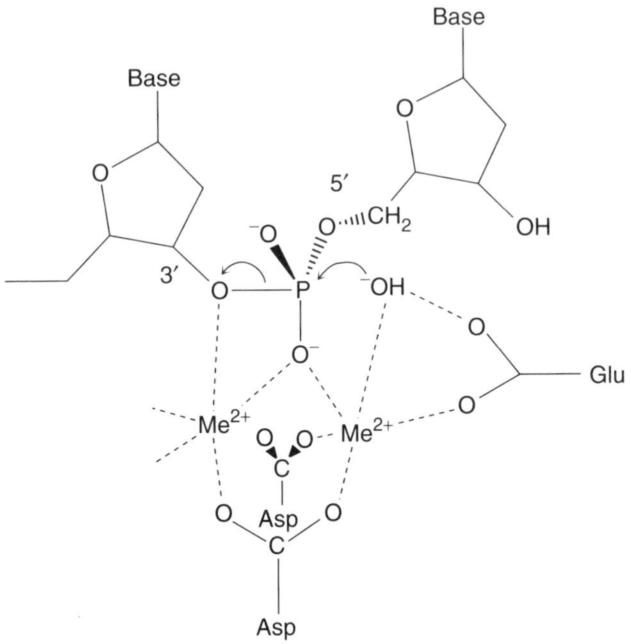

Figure 13.8 The 3'-5' exonuclease activity of DNA polymerases. [Modified from T. A. Steitz, *Current Opinions in Structural Biology* **3**, 31 (1993).]

cases the incorrect base is recognized by its chemical differences from the four naturally occurring ones in DNA. There is also a system that can remove deoxyuridine residues that have been mistakenly incorporated instead of thymidine.[47] This situation is similar to the isoleucine/valine case in protein synthesis, with uracil being smaller than thymine by one methylene group.

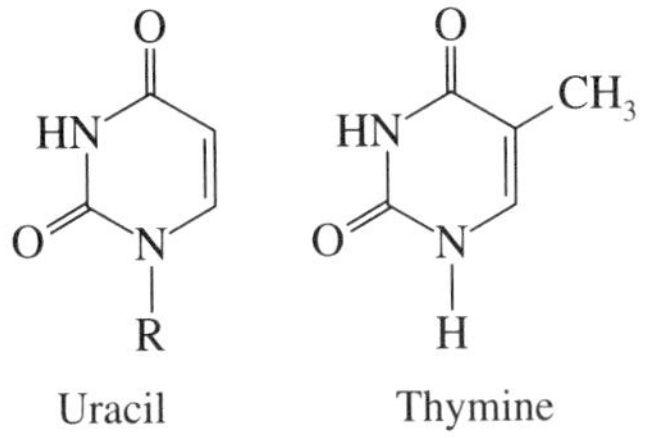

Uracil is removed from the DNA by a uracil glycosidase which excises the base from the sugar ring. This activity is analogous to the hydrolytic activity of the isoleucyl-tRNA synthetase toward Val-tRNA[Ile]. In both cases the hydrolytic site is too small by the size of one methylene group to accommodate the substrate that is to be left intact. In DNA synthesis, the editing is performed by a separate enzyme, since the editing can wait until after polymerization. As this luxury is not permitted in protein synthesis, the hydrolytic function is on the synthetase, so that correction can occur before the misacylated tRNA leaves the enzyme.

Postreplicational mismatch repair has been found to correct errors in base substitution occurring during DNA replication in prokaryotes.[48] This lowers the error rate for the polymerase from 1 in 10^6 to 10^7 to the observed range of values of 1 in 10^8 to 10^{10} in *E. coli*. How does the repair system know in this case which base in a mispair is the incorrect one? The answer appears to be that the parent strand is *tagged* by methylation. A small proportion, some 0.2%, of the cytosine residues are methylated at the 5 position, and a similar proportion of the adenine residues are methylated at the 6 position. As methylation is a postreplicative event, the daughter strand is temporarily undermethylated after replication.

Why is the editing mechanism of the polymerase unable to achieve the necessary accuracy by itself: why does it need the postreplicative mismatch repair mechanism? The answer comes from analyzing the *cost* of editing.

C. The cost of accuracy

It is clear from the data in Table 13.1 on the hydrolysis of substrates accompanying DNA replication that editing costs energy.[22] Not only are the products of incorrect insertions removed, but also some of the correct substrate is wastefully hydrolyzed through insertion followed by exonuclease editing. The fraction of correct substrate wastefully hydrolyzed is defined as the cost. It is further seen that the more efficient the editing, the more the cost. The relationship between cost and accuracy has been analyzed in depth and at various levels of sophistication.[49-53] The following treatment[52,53] gives a very simple equation that provides answers to the questions: *What are the limits of editing? What does it cost? How do these depend on mechanism of selection?*

1. The cost-selectivity equation for editing mechanisms

Most editing mechanisms can be reduced to the sequence outlined in equation 13.20. The substrate first reacts to form the high-energy intermediate EI, which

$$E + S \underset{K_M}{\overset{f}{\rightleftharpoons}} ES \overset{k_{cat}}{\longrightarrow} EI \quad \begin{array}{l} \overset{f'}{\underset{k_s}{\nearrow}} \text{synthesis} \\ \underset{f''}{\overset{k_d}{\searrow}} \text{destruction} \end{array} \qquad (13.20)$$

is then partitioned between further synthesis (with a rate constant k_s) and hydrolytic editing (with a rate constant k_d). (Further synthesis could be elongation in DNA replication, escape of the charged tRNA from the aminoacyl-tRNA synthetase in protein synthesis, etc.) The overall accuracy of the enzyme can be expressed in terms of the cost C and three discrimination factors, f, f', and f'', which are defined in the equations that follow. Kinetic quantities for the correct substrate are labeled with the subscript c, and those for the incorrect substrate are labeled i. The discrimination factor f represents the preferential rate of formation of the correct intermediate EI; f is defined by

$$f = \frac{(k_{cat}/K_M)_c}{(k_{cat}/K_M)_i} \tag{13.21}$$

The f' factor represents the higher rate of further synthesis for the correct substrate: f' is defined by

$$f' = \frac{(k_s)_c}{(k_s)_i} \tag{13.22}$$

The f'' factor represents the higher rate of destruction of the incorrectly formed intermediate EI; f'' is defined by

$$f'' = \frac{(k_d)_i}{(k_d)_c} \tag{13.23}$$

The cost is seen from equation 13.20 to be

$$C = \frac{(k_d)_c}{(k_d + k_s)_c} \tag{13.24}$$

whereas the partitioning to give products is given by

$$F_c = \frac{(k_s)_c}{(k_d + k_s)_c} \tag{13.25}$$

$$F_i = \frac{(k_s)_i}{(k_d + k_s)_i} \tag{13.26}$$

Substituting equations 13.22 and 13.23 into 13.26 gives

$$F_i = \frac{(k_s)_c}{(f'f''k_d + k_s)_c} \tag{13.27}$$

Dividing equation 13.25 by 13.27 and substituting 13.24 gives the relative partitioning ratio for the correct and incorrect substrates:

$$\boxed{\frac{F_c}{F_i} = 1 + (f'f'' - 1)C \tag{13.28}}$$

The relative partitioning ratio is the factor by which specificity is increased by editing. In the absence of editing, v_c/v_i would be equal to $f[S]_c/[S]_i$. In the presence of editing, the overall ratio of the rates of incorporation of correct and incorrect substrates is equal to the ratio of the rates of formation of the intermediate EI multiplied by the relative partitioning ratio; that is,

$$\frac{v_c}{v_i} = \frac{f[S]_c}{[S]_i}\left(\frac{F_c}{F_i}\right) \tag{13.29}$$

We now introduce a new specificity term called the *selectivity, S*, which describes the overall specificity for competing substrates when editing is taking place. *S* is defined by

$$\frac{v_c}{v_i} = S\frac{[S]_c}{[S]_i} \tag{13.30}$$

Comparing equation 13.30 with 13.29 and substituting 13.28 for the partitioning ratio gives the *cost-selectivity* equation.[52]

$$S = f[1 + (f'f'' - 1)C] \tag{13.31}$$

When there is significant editing, the relative partitioning ratio of equation 13.28 is much greater than 1 and reduces to $f'f''C$. The additional specificity caused by editing is thus proportional both to the cost and to the product $f'f''$ of the two discrimination factors. This pair of discrimination factors may be considered as a single value, say f_{ed}, which represents the overall discrimination in the editing step. The efficiency of editing depends crucially on the magnitude of this factor. Mechanisms may be classified according to its value relative to f.

2. Single-feature recognition: $f = f'f''$

Certain editing processes, such as nucleotide selection in DNA replication, appear to use the same structural feature in selection for editing as is used in the initial selection for synthesis. For example, the incoming dNTP is checked in the insertion reaction by testing its base pairing with that of the template. Selection for editing repeats this process, and the same base pairing is checked again. In these cases, $f'f''$ is unlikely to be greater than f, and in many cases the two discrimination factors will be equal. If so, the partitioning ratio F_c/F_i is equal to $1 + (f - 1)C$. This means that the increase in specificity by editing is limited to a further factor of f, since $C \leqslant 1$. In addition, the factor of f is attained only in the ridiculous situation of $C = 1$, when all of the correct substrate is edited! Similarly, S is limited to f^2, which is attained only when $C = 1$. This significantly limits the usefulness of editing.

As an example, the cost-selectivity equation is plotted in Figure 13.9 for $f = f'f'' = 10^4$ or 10^5, the measured discrimination factors for base pairing. Significant increases in specificity are reached only when the cost becomes appreciable. The measured cost for the replicational DNA polymerase of *E. coli in vitro* is 6 to 13%, depending on the dNTP concerned.[53] This must be at the limits of the tolerable. Under these conditions, $S \approx 10^7$ to 10^9; these values are too low

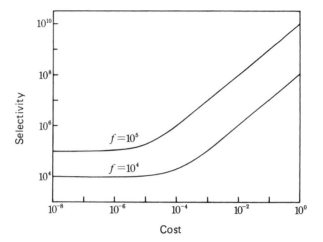

Figure 13.9 Plot of selectivity against cost for single-feature recognition, where $f = f'f'' = 10^4$ or 10^5.

to achieve the error frequency of 1 in 10^8 to 10^{10} observed in *E. coli in vivo*. Hence the necessity of postreplicative mismatch repair mechanisms for attaining the desired error rate. In general, higher selectivity at lower cost can be achieved by multistage editing.[49,51]

J. J. Hopfield has suggested a general mechanism called "kinetic proofreading," in which there is no hydrolytic site on the enzyme; instead, the desired intermediates diffuse into solution, where they hydrolyze nonenzymatically.[54] An example is in the selection of amino acids by the aminoacyl-tRNA synthetases (equation 13.32).

$$
\begin{array}{ccccc}
& \text{ATP} \quad \text{PP}_i & & & \\
\text{E·AA} & \xrightarrow{} & \text{E·AA-AMP} & \xrightarrow{\ k_3\ } & \text{AA-tRNA} \\
\Big\updownarrow k_1 \,\Big\| k_{-1} & & \Big\updownarrow k_4 \,\Big\| k_{-4} & & \\
\text{E + AA} & & \text{E + AA-AMP} & \xrightarrow{\ k_h\ } & \text{AA + AMP}
\end{array}
\tag{13.32}
$$

In this case, $f = f'f''$. For the isoleucyl-tRNA synthetase, $f \approx 150$, so any subsequent increase in specificity is limited to a factor of less than 15 if the cost is to be kept tolerable. In fact, the measured cost is less than 0.05, and it may be predicted from the cost-selectivity equation that if the Hopfield mechanism were operating it would be increasing specificity by a factor of less than 8.5.[55] We show next that "double sieving" is not limited in the same way.

3. Double-feature recognition: $f'f'' > f$

The double-sieve mechanism is an example of a double-feature selection process whereby different structural features of the substrates are used for f and for $f'f''$. To return to the example of the binding of valine and isoleucine to the isoleucyl-tRNA synthetase, $f \approx 150$ because only the weak forces of binding are being invoked. But in the checking step, if the editing site is tailored for the smaller valine, than the strong forces of steric exclusion can be invoked to prevent the binding of isoleucine. The value of $f'f''$ could be as high as 10^5 or so. This very great discrimination at the editing step means that high specificity can be obtained at low cost. Double-feature recognition is thus inherently far more cost-effective than single-feature.

References

1. L. Pauling, *Festschrift Arthur Stoll*, Birkhäuser Verlag AG, Basel, p. 597 (1958).
2. R. B. Loftfield and E. A. Eigner, *Biochim. Biophys. Acta* **130**, 426 (1966).
3. A. R. Fersht and M. Kaethner, *Biochemistry* **15**, 3342 (1976).
4. W.-C. Tsui and A. R. Fersht, *Nucl. Acids Res.* **9**, 4627 (1981).
5. A. R. Fersht and C. Dingwall, *Biochemistry* **18**, 2627 (1979).
6. M. D. Topal and J. R. Fresco, *Nature, Lond.* **263**, 235 (1976).
7. A. R. Fersht, J.-P. Shi, and W.-C. Tsui, *J. Molec. Biol.* **165**, 655 (1983).
8. R. B. Loftfield and M. A. Vanderjagt, *Biochem. J.* **128**, 1353 (1972).
9. A. N. Baldwin and P. Berg, *J. Biol. Chem.* **241**, 831 (1966).
10. A. R. Fersht and C. Dingwall, *Biochemistry* **18**, 1238 (1979).
11. A. R. Fersht, *Biochemistry* **16**, 1025 (1977).
12. H. Jakubowski and A. R. Fersht, *Nucl. Acids Res.* **9**, 3105 (1981).
13. A. R. Fersht and C. Dingwall, *Biochemistry* **18**, 1245 (1979).
14. A. R. Fersht, J. S. Shindler, and W.-C. Tsui, *Biochemistry* **19**, 5520 (1980).
15. C. F. Springgate and L. A. Loeb, *J. Molec. Biol.* **97**, 577 (1975).
16. R. C. Thompson and A. M. Karim, *Proc. Natl. Acad. Sci. USA* **79**, 4922 (1982).
17. O. Nureki, D. G. Vassylvev, M. Tateno, A. Shimada, T. Nakama, S. Fukai, M. Konno, T. L. Hendrickson, P. Schimmel, and S. Yokoyama, *Science* **280**, 578 (1998).
18. L. Lin, S. P. Hale, and P. Schimmel, *Nature, Lond.* **384**, 33 (1996).
19. D. Brutlag and A. Kornberg, *J. Biol. Chem.* **247**, 241 (1972).
20. H. Koessel and R. Roychoudhury, *J. Biol. Chem.* **249**, 4094 (1974).
21. Z. W. Hall and I. R. Lehman, *J. Molec. Biol.* **36**, 321 (1968).
22. N. Muzyczka, R. L. Poland, and M. J. Bessman, *J. Biol. Chem.* **247**, 7116 (1972).
23. G. Seal, C. W. Shearman, and L. A. Loeb, *J. Biol. Chem.* **254**, 5229 (1979).
24. F. Grosse, G. Krauss, J. W. Knill-Jones, and A. R. Fersht, *EMBO J.* **2**, 1515 (1983).
25. A. R. Fersht, *Proc. Natl. Acad. Sci. USA* **76**, 4946 (1979).
26. L. A. Weymouth and L. A. Loeb, *Proc. Natl. Acad. Sci. USA* **75**, 1924 (1978).
27. C. C. Liu, R. L. Burke, U. Hibner, J. Barry, and B. Alberts, *Cold Spring Harbor Symp. Quant. Biol.* **43**, 469 (1979).
28. A. R. Fersht and J. W. Knill-Jones, *Proc. Natl. Acad. Sci. USA* **78**, 4251 (1981).
29. F. Bernardi and J. Ninio, *Biochimie* **60**, 1083 (1978).

30. A. R. Fersht and J. W. Knill-Jones, *J. Molec. Biol.* **165**, 633 (1983).
31. L. S. Beese and T. A. Steitz, *EMBO J.* **10**, 25 (1991).
32. L. S. Beese, V. Derbyshire, and T. A. Steitz, *Science* **260**, 352 (1993).
33. T. A. Steitz, *Current Opinions in Structural Biology* **3**, 31 (1993).
34. J. R. Kiefer, C. Mao, J. C. Braman, and L. S. Beese, *Nature, Lond.* **391**, 304 (1998).
35. S. Doublié, S. Tabor, A. M. Long, C. C. Richardson, and T. Ellenberger, *Nature, Lond.* **391**, 251 (1998).
36. V. Derbyshire, P. S. Freemont, M. R. Sanderson, L. Beese, J. M. Friedman, C. M. Joyce, and T. A. Steitz, *Science* **240**, 199 (1988).
37. A. H. Polesky, T. A. Steitz, N. Grindley, and C. M. Joyce, *J. Biol. Chem.* **265**, 14579 (1990).
38. H. Pelletier, M. R. Sawaya, A. Kumar, S. H. Wilson, and J. Kraut, *Science* **264**, 1891 (1994).
39. B. T. Eger, R. D. Kuchta, S. S. Carroll, P. A. Benkovic, M. E. Dahlberg, C. M. Joyce, and S. J. Benkovic, *Biochemistry* **30**, 1441 (1991).
40. T. L. Capson, J. A. Peliska, B. F. Kaboord, M. W. Frey, C. Lively, M. Dahlberg, and S. J. Benkovic, *Biochemistry* **31**, 10984 (1992).
41. B. T. Eger and S. J. Benkovic, *Biochemistry* **31**, 9227 (1992).
42. C. R. Guest, R. A. Hochstrasser, C. G. Dupuy, D. J. Allen, S. J. Benkovic, and D. P. Millar, *Biochemistry* **30**, 8759 (1991).
43. T. E. Carver, R. A. Hochstrasser, and D. P. Millas, *Proc. Natl. Acad. Sci. USA* **91**, 10670 (1994).
44. T. A. Steitz, *Nature, Lond.* **391**, 231 (1998).
45. T. A. Steitz, S. J. Smerdon, J. Jager, and C. M. Joyce, *Science* **266**, 2022 (1994).
46. A. Kornberg and T. A. Baker, *DNA replication*, W. H. Freeman (1991).
47. T. Lindahl, *Proc. Natl. Acad. Sci. USA* **71**, 3649 (1974).
48. B. W. Glickman and M. Radman, *Proc. Natl. Acad. Sci. USA* **77**, 1063 (1980).
49. M. A. Savageau and R. Freter, *Biochemistry* **18**, 3486 (1979).
50. D. J. Galas and E. W. Branscomb, *J. Molec. Biol.* **124**, 653 (1978).
51. M. Ehrenberg and C. Blomberg, *Biophys. J.* **31**, 333 (1980).
52. A. R. Fersht, *Proc. R. Soc.* **B212**, 351 (1981).
53. A. R. Fersht, J. W. Knill-Jones, and W.-C. Tsui, *J. Molec. Biol.* **156**, 37 (1982).
54. J. J. Hopfield, *Proc. Natl. Acad. Sci. USA* **71**, 4135 (1974).
55. R. S. Mulvey and A. R. Fersht, *Biochemistry* **16**, 4731 (1977).

Recombinant DNA Technology

14

G enetic engineering has revolutionized protein science. First, recombinant DNA technology has enabled the production of large quantities of proteins that were previously either unknown or available only in small quantities. Second, it has allowed rapid sequencing of proteins from their DNA. Third, it has allowed the facile modification of the protein structures by mutation of their genes. This has led to *protein engineering*. Protein engineering is of fundamental importance in analyzing structure–function relationships in proteins and producing novel proteins for biomedical and biotechnological use—the subject of the next chapter. Here, we outline the basic principles of DNA properties and enzymology that enable it to be so readily manipulated for gene cloning, expression, and mutagenesis.

A. The structure and properties of DNA[1]

DNA, the genetic material, is a long, unbranched polymer containing the four deoxynucleoside monophosphates: deoxyadenosine monophosphate (dAMP), deoxythymidine monophosphate (dTMP), deoxyguanosine monophosphate (dGMP), and deoxycytidine monophosphate (dCMP) (Figure 14.1). These are linked in the polymer by an ester bond between the $5'$-phosphate of the nucleotide and the $3'$-hydroxyl of the sugar of the next (Figure 14.2). An essential feature of the bases is that the purine A pairs with the pyrimidine T by hydrogen bonding, as does the purine G with the pyrimidine C (*Watson-Crick pairing rules*, Figure 14.3). Interestingly and importantly, the pairs AT, TA, GC, and CG are almost identical in

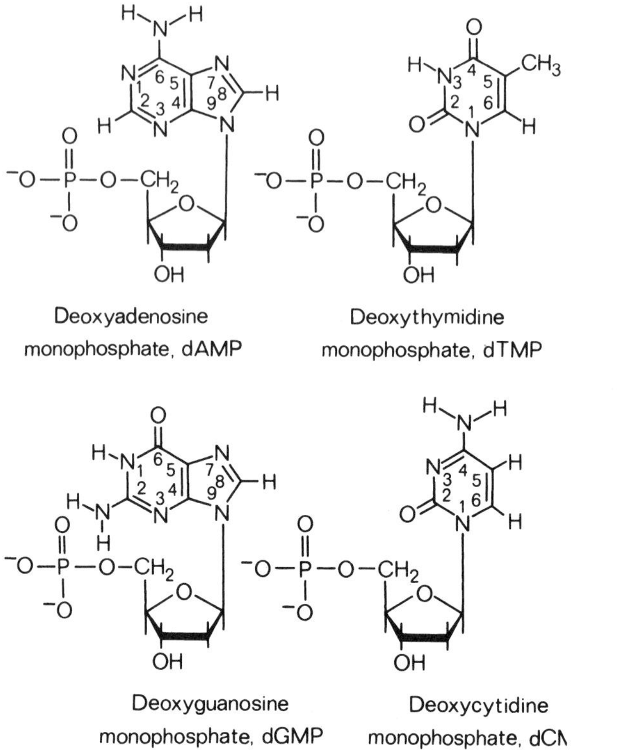

Figure 14.1 The four deoxynucleoside monophosphates that constitute DNA.

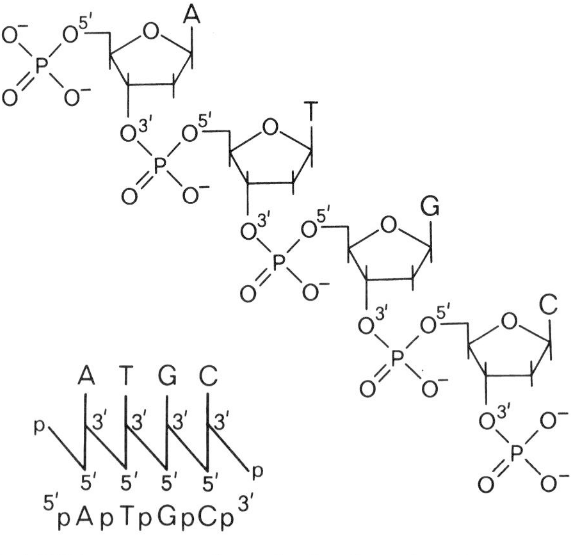

Figure 14.2 A DNA sequence and two shorthand notations for it.

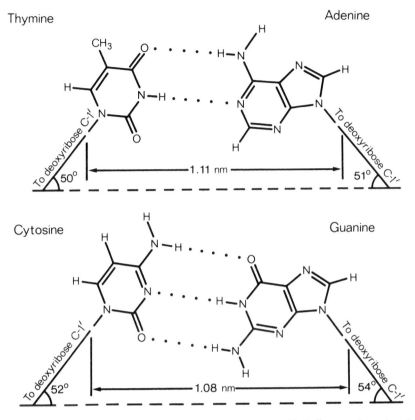

Figure 14.3 Complementary base pairing in the DNA double helix or at the active site of a DNA polymerase.

overall size and shape. This base pairing enables two complementary strands of DNA to form a duplex, as, for example, in the following structure:

$$-T-A-T-G-C-A-C-G-$$
$$| \ | \ | \ | \ | \ | \ | \ |$$
$$-A-T-A-C-G-T-G-C-$$

The duplex is a right-handed double helix with 10 bases per turn. The diameter of the helix is 20 Å (2 nm) and the pitch is 34 Å (3.4 nm). The sugar–phosphate backbone is on the outside of the helix, and the two antiparallel chains are connected by the hydrogen-bonded bases. The DNA in prokaryotes and eukaryotes is generally found in the duplex form, although there are some single-stranded DNA viruses. DNA is a very robust molecule in comparison with many proteins. The simple double-helical secondary structure is readily reassembled after denaturation, unlike the complex tertiary protein structures that can denature irreversibly (Chapters 17 and 19). This feature and the duplex nature of DNA give it the properties that are so important to its manipulation in the laboratory: the information in one strand is duplicated in the other, and the two strands are complementary, so that they spontaneously anneal.

As any elementary textbook on molecular biology will relate, the sequences of proteins are stored in DNA in the form of a triplet code. Each amino acid is encoded by one or more triplet combinations of the four bases A, T, G, and C. For example, tryptophan is coded by the sequence TGG. The sequence of triplets is converted into a protein by a process in which DNA is first transcribed into mRNA. This message is then translated into protein on the ribosomes in conjunction with tRNA and the aminoacyl-tRNA synthetases. In prokaryotes, there is a one-to-one relationship between the sequence of triplets in the DNA and the sequence of amino acids in the protein. In eukaryotes, the DNA often contains stretches of intervening sequences or *introns* which are excised from the mRNA after transcription (Chapter 1).

1. DNA may be replicated: DNA polymerases

DNA is replicated by DNA polymerases. The characteristics of this reaction are:

1. There is an absolute requirement for a template strand of DNA.
2. There is an absolute requirement for a DNA or RNA primer, which is annealed to the template and which possesses a 3'-hydroxyl group on the (deoxy)ribose. (Some very large eukaryotic enzymes have been found to have a "primase" activity by which they synthesize their own RNA primers.)
3. Replication proceeds exclusively in the $5' \rightarrow 3'$ direction, with the four 5'-deoxynucleoside triphosphates serving as the source of monomers.
4. The replication is directed by the template according to the Watson-Crick pairing rules.

DNA polymerases have just one binding site for all four combinations of base pairing — AT, TA, GC, and CG. The specificity of these sites is dictated by the Watson-Crick pairing rules, in that the sites themselves appear to recognize just the overall shape of a correct purine-pyrimidine pair, with the precise specificity resulting from the complementary nature of the base pairing. The polymerase catalyzes the transfer of a complementary deoxynucleoside monophosphate from its triphosphate to the 3'-hydroxyl of the primer terminus (equation 14.1).

$$
\begin{array}{ccc}
& \text{dTMP-PP} & \\
\text{HO} & & \text{HO} \\
5'\text{ -T-A-C-G} & \xrightarrow{\text{polymerase}} & 5'\text{ -T-A-C-G-T} \\
\quad | \ | \ | \ | & & \quad | \ | \ | \ | \ | \\
3'\text{ -A-T-G-C-A-T-G-C-} & & 3'\text{ -A-T-G-C-A-T-G-C-}
\end{array} \quad + \ \text{PP}_i \qquad (14.1)
$$

As is generally true of nucleotide transfer reactions that release pyrophosphate, there is an absolute requirement for Mg^{2+}. The true substrate in the reaction is the complex between the deoxynucleoside triphosphate and Mg^{2+}. The mechanism and fidelity of DNA polymerases[2] are discussed in Chapter 13.

There are many different types of DNA polymerases, and they vary greatly in their activities and in the nature of the reactions they catalyze. Some polymerases are involved mainly in the replication of DNA. Others are used for the repair of damaged DNA. There is also an important difference between the enzymes isolated from eukaryotes and those isolated from prokaryotes. Most of the eukaryotic DNA polymerases that have been isolated so far have just the simple $5' \rightarrow 3'$ polymerization activity shown in equation 14.1. Prokaryotic DNA polymerases, however, are multifunctional. In addition to their $5' \rightarrow 3'$ polymerase activity, they possess a $3' \rightarrow 5'$ exonuclease activity that can excise incorporated deoxynucleoside monophosphates. As was explained in Chapter 13, this is a proofreading or editing activity that enhances specificity by the removal of mismatches. The best known polymerase, DNA polymerase I (Pol I) of *Escherichia coli*, has a further exonuclease activity in the $5' \rightarrow 3'$ direction. This enzyme is a single polypeptide chain of relative molecular mass (M_r) 109 000. The $5' \rightarrow 3'$ polymerase and $3' \rightarrow 5'$ exonuclease activities may be separated by proteolysis. For example, the chain may be cleaved by subtilisin to give a large fragment of M_r 76 000 (the Klenow enzyme) containing the $3' \rightarrow 5'$ exonuclease and polymerase activities, and a small fragment of M_r 36 000 containing the $5' \rightarrow 3'$ exonuclease activity. Viral polymerase, such as that from HIV, lack the proofreading exonuclease activity. They are thus error-prone, which causes HIV to mutate frequently and so become drug resistant.

2. Gaps in DNA may be sealed: DNA ligases

Certain nicks in duplex DNA may be resealed by a DNA ligase. These enzymes will form a phosphodiester bond between a 5'-phosphoryl group and a directly adjacent 3'-hydroxyl, using either ATP or NAD^+ as an external energy source (Figure 14.4). The ligase from *E. coli* is well known. It is a single polypeptide chain of M_r 77 000 and it uses NAD^+ as the energy source. *E. coli* infected with phage T4 provides a useful ligase that has an M_r of $\sim 65\,000$ and that uses ATP. An extended gap in duplex DNA may thus be repaired by first filling in the gap from the 3'-hydroxyl up to the 5'-phosphoryl group with a DNA polymerase, and then resealing the strand with a DNA ligase.

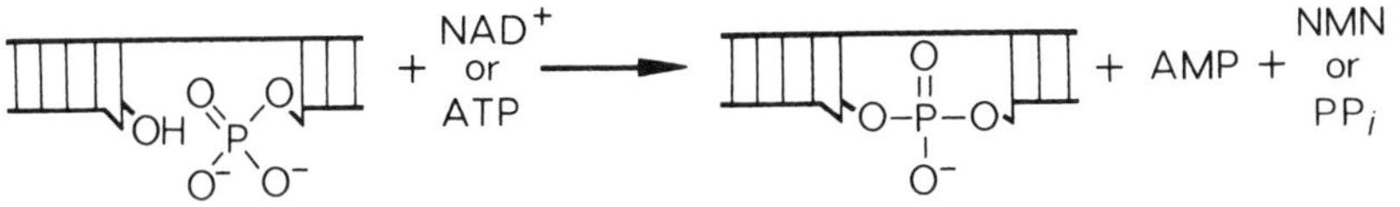

Figure 14.4 The DNA ligase reaction. NMN = nicotinamide mononucleotide.

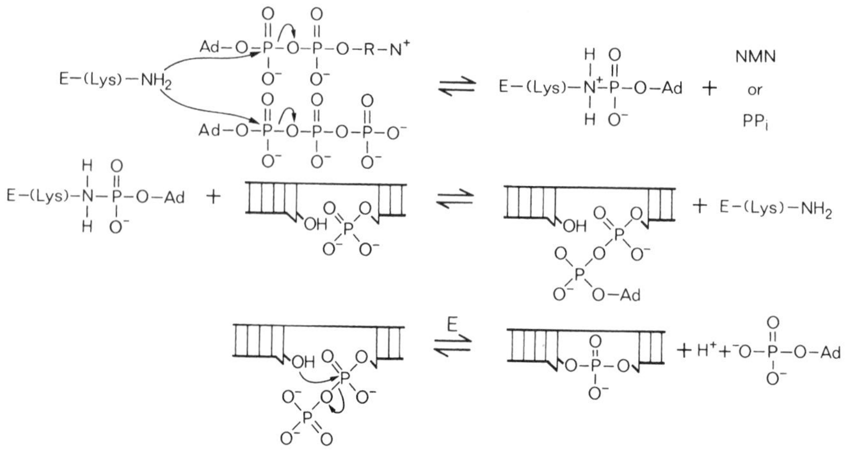

Figure 14.5 The mechanism of DNA ligation.

a. The mechanism of DNA ligase[3]

The ligase reaction proceeds via the two covalent intermediates illustrated in Figure 14.5. A ligase adenylate with a phosphoamide bond is formed by the nucleophilic attack of a lysine side chain on either the ATP or the NAD^+, generating either PP_i or NMN (nicotinamide mononucleotide) with the phage T4 or *E. coli* enzyme, respectively. The adenylate is then transferred to the 5′-phosphoryl terminus of the DNA by nucleophilic attack on the α-phosphate of the AMP moiety to give the second intermediate, in which the 5′-phosphoryl group is activated. The ligase catalyzes the nucleophilic attack of the 3′-hydroxyl group on the activated phosphate to seal the nick, releasing AMP. The ligase adenylate may be readily isolated in 1:1 stoichiometry in the absence of DNA, but the DNA adenylate may be isolated only in small amounts from the complete reaction mixture, where it is formed at a steady state concentration.

3. Duplex DNA may be cleaved at specific sequences: Restriction endonucleases

Just as proteins may be cleaved by specific proteases at defined residues, so duplex DNA may be cleaved in regions of defined sequence by a class of endonucleases known, for historical reasons, as restriction endonucleases. The class of restriction endonucleases we are interested in is the Type II. These enzymes require only Mg^{2+} as a cofactor, and have M_r's of 20 000 to 100 000. They recognize specific DNA sequences that are generally several nucleotides long and that are symmetrical in a rotational sense, as shown in Table 14.1. Both strands of the DNA are cleaved in a symmetrical manner to give nicks with 3′-hydroxyl and 5′-phosphoryl termini. The cuts may be staggered; for example, *Eco*RI (Table 14.1) generates a protruding 5′ terminus (equation 14.2), and *Pst*I generates a

Table 14.1	*Specificity of Type II restriction endonucleases*

Enzyme	DNA sequence cleaved

Axis of symmetry

*Eco*RI
$$5' \quad -G \overset{\downarrow}{-} A-A \mid T-T-C- \quad 3'$$
$$-C-T-T \mid A-A \underset{\uparrow}{-} G-$$

*Bam*I
$$-G \overset{\downarrow}{-} G-A \mid T-C-C-$$
$$-C-C-T \mid A-G \underset{\uparrow}{-} G-$$

*Hae*III
$$-C-C \overset{\downarrow}{\mid} G-G-$$
$$-G-G \underset{\uparrow}{\mid} C-C-$$

*Pst*I
$$-C-T-G \mid C-A \overset{\downarrow}{-} G-$$
$$-G \underset{\uparrow}{-} A-C \mid G-T-C-$$

$$
\begin{array}{ccccc}
5'\text{ -G-A-A-T-T-C-} & \xrightarrow{Eco\text{RI}} & 5'\text{ -G} & & \text{A-A-T-T-C-} \\
\text{ | | | | | | } & & \text{ | } & + & \text{ | } \\
\text{-C-T-T-A-A-G-} & & \text{-C-T-T-A-A-} & & \text{G-}
\end{array}
\qquad (14.2)
$$

protruding 3' terminus (equation 14.3). Other enzymes, such as *Hae*III, give flush (or "blunt") ends because they cut at the center of symmetry.

$$
\begin{array}{ccccc}
5'\text{ -C-T-G-C-A-G-} & \xrightarrow{Pst\text{I}} & 5'\text{ -C-T-G-C-A-} & & \text{G-} \\
\text{ | | | | | | } & & \text{ | } & + & \text{ | } \\
\text{-G-A-C-G-T-C-} & & \text{-G} & & \text{A-C-G-T-C-}
\end{array}
\qquad (14.3)
$$

4. DNA fragments may be joined by using enzymes

An essential process in recombinant DNA technology is the joining of a fragment of DNA from one genome with that from another. For example, the staggered cleavages generated by restriction enzymes in equations 14.2 and 14.3 give ends that are complementary and thus mutually cohesive. This means that restriction fragments (i.e., lengths of DNA produced by cutting with restriction enzymes) generated from different pieces of DNA by the same restriction enzyme may be annealed and sealed with DNA ligase, since they contain the requisite 3'-hydroxyl and 5'-phosphoryl ends.

An important extension of this method is the procedure of *blunt-end ligation*: high concentrations of the T4 ligase will catalyze the ligation of DNA fragments containing flush ends (equation 14.4). Flush ends may be generated by a number

$$
\begin{array}{ccccc}
5' \boxed{||||} \text{-OH} & & \text{P-} \boxed{||||} \text{-OH} & \xrightarrow{\text{T4 ligase}} & 5' \boxed{||||||||} \text{-OH} \\
\text{HO-} \boxed{} \text{-P} & + & \text{HO-} \boxed{} & & \text{OH-} \boxed{} \\
\end{array}
\qquad (14.4)
$$

of enzymatic processes in addition to cleavage by enzymes such as *Hae*III. For example, the protruding 3′ ends generated by *Pst*I in equation 14.3 may be excised by using the 3′ → 5′ exonuclease activity of the T4 DNA polymerase. The protruding 5′ ends generated by *Eco*RI (equation 14.2) may be removed by using a nuclease such as SI, which is specific for the cleavage of single-stranded DNA. Alternatively, the 3′ ends in the *Eco*RI fragments may be extended by "filling in" with the large subfragment of Pol I so that they become flush with the 5′ ends. In combination with these reactions, blunt-end ligation becomes a general means of joining a restriction fragment generated by one restriction enzyme from one piece of DNA with a fragment generated by any other enzyme from another piece.

5. Joining DNA by complementary homopolymeric tails: Terminal transferase

There is an alternative method of generating cohesive tails. The enzyme calf thymus terminal (deoxynucleotidyl) transferase adds deoxynucleoside monophosphate residues from 5′-deoxynucleoside triphosphates to protruding 3′-hydroxyl termini in the absence of a template. For example, as shown in equations 14.5 to 14.7,

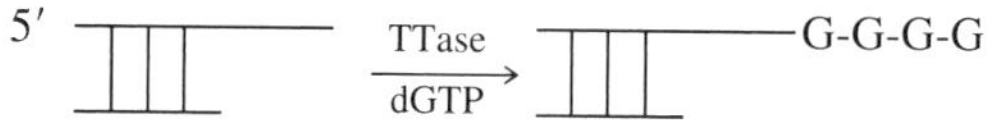

$$ (14.5) $$

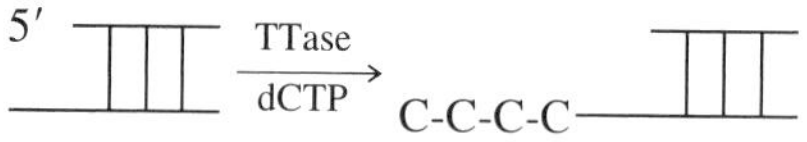

$$ (14.6) $$

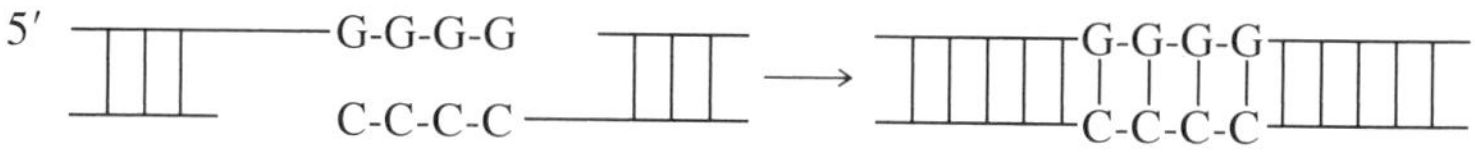

$$ (14.7) $$

complementary ends can be generated and can then be annealed, with the resultant gap filled in by a DNA polymerase and sealed by a ligase.

6. Amplifying DNA by the polymerase chain reaction (PCR)

One of the most important developments in gene technology has been the polymerase chain reaction (PCR),[4] which can amplify regions of DNA, genes, or whole plasmids, as illustrated in Figure 14.6. Double-stranded DNA is dissociated by heating ($\sim 94°C$). It is then cooled to about 55°C in the presence of excess oligonucleotide primers, one of which is complementary to a region of one strand and the other to a region of the other strand. The primers bind to the DNA strands faster than the strands reassociate, because of the higher concentrations of the primers and the second-order kinetics of association. In the presence of dNTPs and, preferably, a thermostable DNA polymerase, such as that from *Thermus aquaticus (Taq)* at $\sim 72°C$, the strands are replicated from the 3′-OH

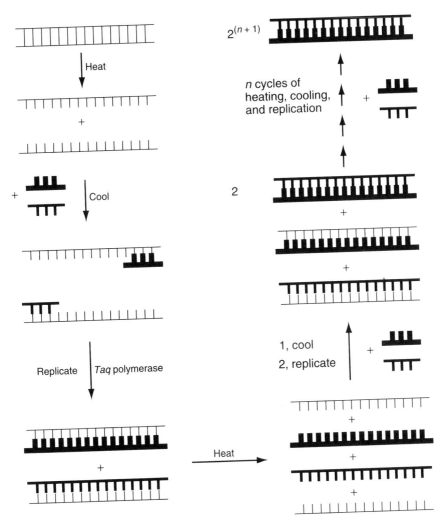

Figure 14.6 The polymerase chain reaction.

ends of the primers. This procedure is then repeated, with the number of strands of DNA doubling at each cycle. There is a proportion of errors introduced at each stage. Typically, 20 cycles are used to give a 10^6-fold amplification, a level at which errors tend not to be intrusive. The amplification is used to detect small amounts of DNA in, for example, DNA fingerprinting for forensic work or for a variety of analytical purposes. As will be seen in the following sections, amplification is invaluable for producing the genes to make proteins. Indeed, the combination of PCR and the knowledge of complete genome sequences has revolutionized recombinant DNA technology. For the protein scientist, this revolution has made many of the old procedures in sections B and C for cloning and mutagenesis obsolete.

7. Processive versus distributive polymerization

Enzyme-catalyzed polymerization reactions have an important characteristic that is not found elsewhere. Once the enzyme has added a monomeric unit to the growing chain, it can either dissociate and recombine at random with other growing termini, or it can remain attached to the same chain and add further residues. Enzymes that dissociate between each addition and distribute themselves among all the termini are termed *distributive*. Those that process along the same chain without dissociating are termed *processive*. These terms apply also to degradative enzymes such as exonucleases.

A typical processive enzyme is terminal transferase. It adds on deoxynucleoside monophosphates randomly to exposed 3'-hydroxyl termini so that the final products are formed in a statistical distribution. The distribution follows *Poisson's law*.[5] Suppose that the enzyme adds on an average of x residues per chain. Then the probability of a particular chain having k residues added [i.e., $p(k)$] is given by

$$p(k) = \frac{x^k}{k!} e^{-x} \tag{14.8}$$

(where $k!$ = factorial $k = 1 \times 2 \times 3 \times \ldots \times k$). For example, if an average of 1 residue has been added per chain, then it may be calculated from equation 14.8 that $p(0) = 0.368$ ($0! = 1$), $p(1) = 0.368$, $p(2) = 0.184$, $p(3) = 0.061$, $p(4) = 0.015$, etc. That is, 36.8% of the chains have no residues added, 36.8% have 1, etc. Similarly, for $x = 2$, $p(0) = 0.135$, $p(1) = 0.271$, $p(2) = 0.271$, $p(3) = 0.180$, $p(4) = 0.090$, $p(5) = 0.036$, etc.

The balance between processivity and distributivity clearly depends on the ratio of rate constants for polymerization (or degradation) and dissociation. DNA polymerases *in vitro* vary from being distributive to having some degree of processivity. *In vivo*, however, the polymerase is part of a multienzyme DNA polymerizing complex (a "replisome") that is almost certainly highly processive.

B. Cloning enzyme genes for overproduction

The biosynthesis of enzymes is under strict control so that they are produced in the correct quantities for the optimal viability of cells. The "housekeeping" enzymes—those involved in high-volume metabolic routes such as glycolysis—are produced in large amounts, whereas others are produced only in small quantities. Consequently, much of enzymology in the past has concentrated upon the easily accessible metabolic enzymes. Now, however, because of the advent of gene cloning, it is possible to obtain large quantities of proteins that were previously rare.

The gene coding for a particular protein from an organism such as *E. coli* may be cloned by the following strategy: the genomic DNA of that organism is cut into small fragments; these are inserted into a *vector*, a double-stranded DNA molecule that can replicate after the foreign DNA has been inserted into it; the vector is allowed to replicate in a host, such as *E. coli*, and is screened for production of the desired protein (Figure 14.7).[6-8]

The basic enzymology for insertion of fragments of DNA into a vector was covered in section A. The genomic DNA is usually fragmented by partly digesting it with a restriction enzyme. This produces a series of restriction fragments, some of which may contain the desired gene. If the vector DNA may be cut at a single site with the same restriction enzyme, then the restriction fragments may be inserted into the cleaved vector by annealing its "sticky ends" and sealing with ligase as described in section A2. This gives a *gene library* or a *gene bank*. An alternative method of producing DNA fragments is to shear the DNA by physical forces and use the terminal transferase procedure of section A5. Nowadays, the sequence of the gene and the regions of DNA that flank it are likely to be known from genome sequencing projects. If so, the gene may be readily amplified directly from genomic DNA, obviating the need for a gene library. Cloning genes or enzymes from eukaryotes is more complicated partly because of problems caused by the introns. These difficulties can be avoided by copying the mRNA (from which the introns have been excised) by using a polymerase to give the so-called cDNA, and then cloning this. However, there are often problems in obtaining a high level of expression of eukaryotic genes in prokaryotes.

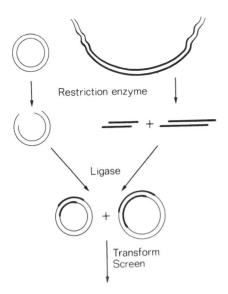

Figure 14.7 A scheme for gene cloning.

1. Vectors

Perhaps the most common vector is the plasmid pBR322 and its derivatives. Plasmids are genetic elements that are not part of the major bacterial chromosome, and that replicate autonomously in the cytoplasm. The plasmid pBR322 is a small, covalently closed circle of double-stranded DNA containing 4362 base pairs. It carries genes for resistance to the antibiotics ampicillin and tetracycline, and replicates autonomously in *E. coli* by using the host replication proteins. Many copies of pBR322 may exist in each cell, and it is thus termed a multicopy plasmid. Partly because of the high copy number, a gene cloned into such a plasmid may produce large amounts of enzyme. The precise degree of overproduction depends also on the promoter for the gene, since the DNA from the cloned protein is transcribed from its own promoter. (The DNA sequence of the promoter, the stretch of DNA to which RNA polymerase binds to initiate transcription, controls the efficiency of transcription: a strong promoter gives high production of mRNA, with the possibility of high levels of protein: a weak promoter gives low production.) Another factor is whether or not the organism can tolerate the overproduction. In extreme cases, *E. coli* is found to tolerate up to 50% of its soluble protein being in the form of the cloned protein. In other cases, overproduction is lethal.

2. Screening

The rapid advances in PCR technology and genome sequence data enable most genes of interest to be directly targeted. Where this is not possible, we have to employ other strategies to obtain the targeted gene. A gene library consists of thousands of different restriction fragments cloned into a vector. Only a few of these will contain the desired gene, so screening methods are necessary to find the needles in the haystack. The classical way of screening the gene library for the gene of the desired protein is by means of a *complementation assay*. This can be performed if there is a mutant of the host that lacks the desired enzyme activity. A vector that contains the missing gene may allow the mutant host to grow under nonpermissive conditions if the necessary protein is expressed from the vector. For example, an available mutant of *E. coli* is temperature-sensitive in its tyrosyl-tRNA synthetase: the bacterium grows at 30°C but not at 42°C. The mutant can be used to screen a gene library of DNA from *E. coli* or *Bacillus stearothermophilus* in pBR322 as follows.[9] The plasmid, which contains the gene for ampicillin resistance, is introduced into the mutant bacterium (i.e., the cells are *transformed*) by mixing the plasmid and bacterium in the presence of $CaCl_2$ under certain conditions. The cells are then grown in the presence of ampicillin at 30°C. This selects the cells that have been transformed and that have the antibiotic resistance coded by the plasmid, and also allows the copy number of the plasmid to increase. The cells are then incubated at 42°C, so that only those that have the plasmids containing the restriction fragments expressing the tyrosyl-tRNA synthetase grow.

If a complementation assay cannot be performed, brute force must be applied to locate the desired gene or its product. One of the oldest procedures is a *DNA hybridization assay*. If the sequence of the protein is available, a short length of DNA corresponding to the sequence may be synthesized. This may then be labeled with ^{32}P by using a phosphokinase and [γ-^{32}P]ATP. The vector DNA from individual colonies of transformed cells is baked onto nitrocellulose sheets, and the denatured DNA is bathed in a solution of the radioactive probe. The probe will hybridize to any DNA containing the complementary sequence. After the surplus probe is washed off, an autoradiograph of the nitrocellulose sheet locates the colonies containing the desired DNA.[10] In practice, because of the degenerate nature of the genetic code, a mixture of probes must be used.

Another important procedure is to screen for production of the desired protein by using immunological methods with antibodies raised against it.

C. Site-specific mutagenesis for rational design

There are a variety of methods for specifically mutating the codon for a specific amino acid in a protein. Most are based on the procedure of *oligodeoxynucleotide-directed mutagenesis*.[11–14] The basic experimental procedure is illustrated in Figure 14.8. The gene is cloned into a double-stranded vector, and one of the constituent circles of single-stranded DNA is isolated. A short oligodeoxynucleotide has already been synthesized to be complementary to the region of the gene to be mutated, except for a single-base (or double-base) mismatch. The mismatch is designed to change the codon for the target amino acid residue into the codon for the desired mutant residue. The oligodeoxynucleotide is annealed to the gene in the single-stranded vector and becomes a primer for Pol I (the Klenow fragment) to use in replicating the rest of the genome. The replicated strand is ligated, and the result is a heteroduplex containing one strand of mutant and one strand of wild-type (unmutated) DNA. The heteroduplex is used to transform a host and produce colonies of cells that each contain either the vector with the mutant or the vector with the wild-type gene. The colonies used to be screened by introducing the original oligodeoxynucleotide primer, now radioactively labeled as a probe in a DNA hybridization assays (section B2). The probe anneals preferentially to the mutant DNA, to which it is fully complementary, rather than to the wild-type DNA, with which it has a mismatch. In the early experiments, the yields of mutants were very low, typically on the order of $1-10\%$, because the mutagenesis was inefficient and the cell selectively repairs the mutation in the double-stranded DNA when it is transfected into *E. coli* (see Chapter 13). The mutants had to be detected laboriously by the DNA hybridization assays. Over the years, the efficiency of oligodeoxynucleotide-directed mutagenesis has been improved, so that the yield of mutants is greater than 90%. One or two colonies are picked and their sequences determined. One of the best methods for mutagenesis uses PCR, a circular plasmid, and two mutant primers that are complementary in sequence at the site of mutation.[15] In the first cycle of mutation, one primer introduces the mutation into

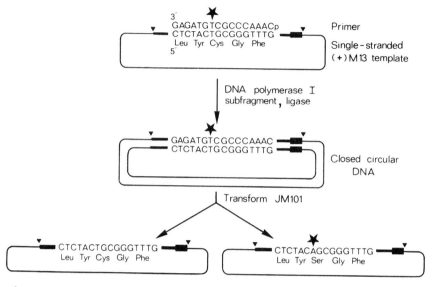

Figure 14.8 The original scheme for oligodeoxynucleotide-directed mutagenesis. The mismatched primer is designed to mutate the codon for cysteine (TGC) to that for serine (AGC).

The gene selected for mutation had to be obtained in the form of single-stranded DNA before the primer could be annealed. This was done by subcloning the gene from the double-stranded pBR322 into a new vector, bacteriophage M13. The genome of this filamentous phage is a covalent circle of single-stranded DNA that goes through a covalent circular duplex termed RF (= replicative form) during replication in its host, *E. coli.* The subcloning is performed on the RF DNA, which then yields single-stranded viral DNA. Note that the viral DNA is termed the (+) strand as it has the same sense as the message. The complementary strand is termed (−). JM101 is a strain of *E. coli* that is a host for M13. JM101 may be transformed by M13 DNA to produce mature phage. [From G. Winter, A. R. Fersht, A. J. Wilkinson, M. Zoller, and M. Smith, *Nature, Lond.* **299**, 756 (1982).] Nowadays, PCR methods have replaced this procedure.

one strand; the other primer causes the complementary mutation in the other strand. In the following rounds of amplification, the mutants are subsequently amplified because all the primers are complementary to the mutants. After 20 cycles of PCR, there should be all mutant progeny.

Mutants of subtilisin were made by cassette mutagenesis, whereby a double-stranded oligodeoxynucleotide was synthesized and inserted into the subtilisin gene using restriction enzymes and ligation.[16] It is not difficult to synthesize the genes for proteins *de novo* using cassettes of double-stranded oligodeoxynucleotides with sticky ends (equation 14.9). Alternatively, longer, single-stranded oligodeoxynucleotides can be linked together by shorter strands that are complementary to the end of one long strand and the beginning of the next. The gaps can then be filled in by a polymerase, followed by ligation.

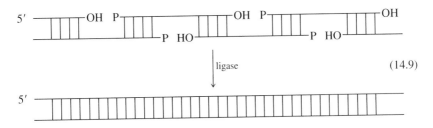

$$(14.9)$$

The production of mutants by site-specific mutagenesis is the basis of the rational and systematic analysis of proteins and their redesign by protein engineering, as described in the next chapter for the tyrosyl-tRNA synthetase and subtilisin.

D. Random mutagenesis and repertoire selection

The theoretical and empirical basis for the rational design of novel binding and catalytic activities is still very limited. Some examples of successful changes in specificity are given in Chapters 15 and 16. But the design of new catalytic activities or binding sites by systematic changes of single amino acids is still impossible. The alternative approach is to mimic nature and use random (or semirandom) mutation followed by selection. The Darwinian approach is the most promising at present for designing novel activities. Methods of selection are necessary because of the huge numbers of mutants that are generated after a few residues have been simultaneously mutated: there are 20^n combinations of n random mutations. There are more than 3×10^6 mutants on the random mutation of just five residues, far too many to assay systematically by conventional procedures.

1. Random mutagenesis

The two most popular procedures for introducing random errors into DNA are error-prone PCR[17] and mutagenic oligodeoxynucleotides. *Taq* polymerase becomes error-prone in the presence of Mn^{2+} ions or at abnormal concentrations of Mg^{2+}. After many rounds of error-prone PCR, mutations build up in a genome. This is a very good procedure for scattering mutations over an extensive region of DNA, although the introduction of mutations is not purely random, because there is some bias.

Mutagenic oligodeoxynucleotides can contain completely scrambled codons at some positions or can be semitargeted. For example, because the codons in the mRNA for Phe, Leu, Ile, Met, and Val all have U in the middle, randomizing positions 1 and 3 while retaining the DNA to code for U at position 2 will restrict the mutants to being large hydrophobics. The mutagenic primer is made by solid phase synthesis. A pure activated monomer of T is used in the synthesis at the position to be conserved, whereas a mixture of activated A, G, C, and T is used at the positions to be scrambled.[18] Different ratios of A:G:C:T can be used at the scrambled positions to minimize the retention of wild-type codon or to bias the pool. This procedure has been used to scramble the hydrophobic core of the ribonuclease barnase.[19] A low level of mutations can be induced by "spiking" or "doping" specific positions.[20] For example, 1% of incorrect bases can be added during the solid-phase synthesis at selected positions so that there will be a low level of mutants. These procedures provide a repertoire of mutants that require screening for selection.

2. Repertoire selection: Phage display

The classical procedure for selecting mutants is the complementation of genetic defects, which was described in section B2 for screening protein expression. When applicable, it is extremely powerful and is the method of choice. Complementation has been used to select for thermostable variants of a protein by cloning of the mesophilic protein followed by random mutagenesis and selection in a thermophile.[21] Unfortunately, most of the products from random mutagenesis experiments are generally not required for bacterial growth, and so alternative methods of selection are required.

The essence of biological selection is that the phenotype comes packaged with its genotype so that when the gene product is selected, so is its DNA. An artificial system for selection has been developed using filamentous phages, initially for displaying antigenic peptides[22] and later for selecting antibodies[23] or other proteins.[24] The principle is simple and elegant. The gene of the target protein is mutated and the products cloned into the gene for a surface protein of a bacteriophage. The gene product is displayed on the surface of the phage and so can be selected by its binding or other activities. Multiple rounds of selection and repropagation of phage allow for the isolation of very rare phage from large mutant repertoires. The desired protein may then be isolated with its genome, which can be amplified by replication and its sequence determined.

Filamentous phages, such as M13 and fd, have three to five copies of the gene III protein at their tips (Figure 14.9). This three-domain protein is responsible for the attachment of the phage to the F pilus of *E. coli*, through which it enters the cell. Gene III protein is a suitable vehicle for displaying a desired target as a fusion protein. For example, human growth hormone was fused to the C-terminal domain of the gene III protein of M13.[24] The gene fusion was cloned into a plasmid containing the origins of replication for *E. coli* and the filamentous phage, packaged into phagemid (phage/plasmid) particles upon infection by an M13 helper phage (which packages its own DNA poorly). The human growth hormone–gene III phagemid particles were enriched more than 5000-fold from the phagemids lacking the hormone by binding to beads coated with the hormone receptor, washing, and then eluting the tightly bound phagemids (Figure 14.10). This procedure was subsequently used to prepare growth hormone of higher affinity.[25]

An artificial immune system was constructed based on phage display to make antibodies without using rodents and hybridoma technology.[23,26,27] The gene III

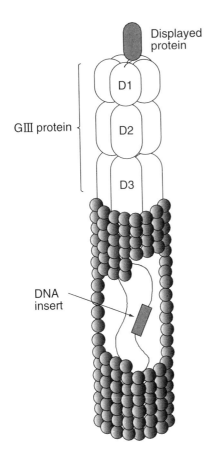

Figure 14.9 Filamentous phage, M13 or fd, displaying a protein fused to a gene III protein.

Displayed protein

GIII protein

D1

D2

D3

DNA insert

protein of fd was used to display antibody fragments. The desired products from random combinatorial libraries were selectively enriched by affinity purification on immobilized antigen. A wide range of useful recombinant antibodies has been made in this way.[27]

A variation on the theme has been to map out protease specificity.[28] A library of fusion proteins was constructed in a modular manner. The synthetic protein had an N-terminal domain that binds very tightly to an affinity column. This domain was connected to the C-terminal domain of M13 gene III by a randomized peptide sequence. The phages were then bound to the affinity support and treated with a protease. Phages that had a protease-susceptible site were cleaved from the support and eluted. This procedure was subsequently used to map out the specificity of furin,[29] which is described in the next chapter.

Phage display has been extremely successful for obtaining novel binding activities. It will be interesting to see if it can be used to make novel catalytic activities.

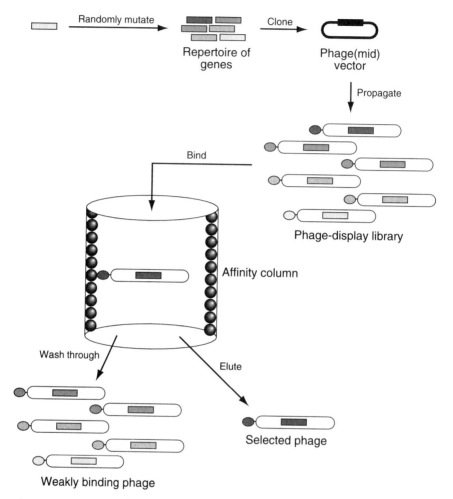

Figure 14.10 Selection of phage displaying a mutant protein using affinity chromatography.

References

1. A. Kornberg and T. H. Baker, *DNA replication*, W. H. Freeman (1991).
2. T. A. Steitz, *Nature, Lond.* **391**, 231 (1998).
3. I. R. Lehman, *Science* **186**, 790 (1974).
4. K. Mullis, F. Faloona, S. Scharf, R. Saiki, G. Horn, and H. Erlich, *Cold Spring Harbor Symp. Quant. Biol.* **51**, 1 (1986).
5. F. N. Hayes, V. E. Mitchell, R. L. Ratliffe, and D. L. Williams, *Biochemistry* **6**, 2488 (1967).
6. D. A. Jackson, R. H. Symons, and P. Berg, *Proc. Natl. Acad. Sci. USA* **67**, 2904 (1972).

7. P. E. Lobban and A. D. Kaiser, *J. Molec. Biol.* **78**, 453 (1973).
8. S. N. Cohen, *Science* **195**, 654 (1977).
9. D. Barker, *Eur. J. Biochem.* **125**, 357 (1982).
10. M. Grunstein and D. S. Hogness, *Proc. Natl. Acad. Sci. USA* **72**, 3961 (1975).
11. H. Schott and H. Kössel, *J. Am. Chem. Soc.* **95**, 3778 (1973).
12. C. A. Hutchinson III, S. Phillips, M. H. Edgell, S. Gillam, P. Jahnke, and M. Smith, *J. Biol. Chem.* **253**, 6551(1978).
13. A. Razin, T. Hirose, K. Itakura, and A. D. Riggs, *Proc. Natl. Acad. Sci. USA* **75**, 4269 (1978).
14. G. Winter, A. R. Fersht, A. J. Wilkinson, M. Zoller, and M. Smith, *Nature, Lond.* **299**, 756 (1982).
15. A. Hemsley, N. Arnheim, M. D. Toney, G. Cortopassi, and D. J. Galas, *Nucleic Acids Research* **17**, 6545 (1989).
16. J. A. Wells, M. Vasser, and D. B. Powers, *Gene* **34**, 2 (1985).
17. D. W. Leung, E. Chen, and E. V. Goeddel, *Technique* **1**, 11 (1989).
18. W. A. Lim and R. T. Sauer, *J. Molec. Biol.* **219**, 359 (1991).
19. D. D. Axe, N. W. Foster, and A. R. Fersht, *Proc. Natl. Acad. Sci. USA* **93**, 5590 (1996).
20. K. M. Derbyshire, J. J. Salvo, and N. Grindley, *Gene* **46**, 2 (1986).
21. H. Liao, T. Mckenzie, and R. Hageman, *Proc. Natl. Acad. Sci. USA* **83**, 576 (1986).
22. G. P. Smith, *Science* **228**, 1315 (1985).
23. J. Mccafferty, A. D. Griffiths, G. Winter, and D. J. Chiswell, *Nature, Lond.* **348**, 552 (1990).
24. S. Bass, R. Greene, and J. A. Wells, *Proteins Structure Function and Genetics* **8**, 309 (1990).
25. H. B. Lowman and J. A. Wells, *J. Molec. Biol.* **234**, 564 (1993).
26. T. Clackson, H. R. Hoogenboom, A. D. Griffiths, and G. Winter, *Nature, Lond.* **352**, 624 (1991).
27. G. Winter, A. D. Griffiths, R. E. Hawkins, and H. R. Hoogenboom, *Ann. Rev. Immunol.* **12**, 433 (1994).
28. D. J. Matthews and J. A. Wells, *Science* **260**, 1113 (1993).
29. D. J. Matthews, L. J. Goodman, C. M. Gorman, and J. A. Wells, *Protein Science* **3**, 1197 (1994).

Protein Engineering | 15

PART 1

Dissection of the structure, activity, and mechanism of an enzyme: The tyrosyl–tRNA synthetase

Enzymologists have always wanted to test mechanisms by altering the groups involved in catalysis. Until the 1980s, however, the only way to do this was by chemical modification (Chapter 9). This very limited technique is restricted to chemically reactive side chains, is often nonspecific, and frequently involves making significant changes that add steric bulk or grossly alter the properties of the side chain. A famous example was testing the role of the phenolic hydroxyl of Tyr-248 in carboxypeptidase by nitration of the aromatic ring. A much more precise experiment would be to remove the hydroxyl by converting the tyrosine into a phenylalanine. And so a pipe dream of the enzymologist was to be able to mutate one amino acid residue into another at will. The last chapter described how this became possible by site-directed mutagenesis. Given the gene for a protein and a means for its expression, one can change any of its amino acid side chains by changing its codon, add amino acid residues by inserting the corresponding sequences of DNA, and delete sections of the polypeptide chain by deleting sections of the gene. Ironically, one of the early mutagenesis experiments was to convert Tyr-248 of carboxypeptidase into Phe-248.[1] This experiment showed that the famous chemical modification experiment gave the wrong results about the role of the tyrosine, an experience that has been repeated elsewhere! Site-directed mutagenesis has enabled us to transcend our role as mere observers of enzymes and become molecular engineers. The rational design and construction of novel proteins using structural information is known as protein engineering.

Protein engineering is being used for two purposes: (1) dissection of the structure and activity of existing proteins by making systematic alterations of their structures and examining the changes in their properties; and (2) production of novel proteins for use in medicine and industry. We do not yet know how to design functional proteins *de novo*, but we can make minor alterations of existing proteins to give useful changes in activity. We can cut and paste segments from one protein to another. Even such minor alterations can have dramatic consequences. A good example is the "humanization" of monoclonal antibodies. Monoclonal antibodies are produced from rodents, which limits their therapeutic use because the human antibody system may recognize a rodent antibody as foreign and neutralize it. It has been shown, however, that the antigen-binding loops from a mouse monoclonal antibody may be grafted onto a human framework with hardly any loss of binding activity.[2] Such chimeric antibodies have been used to achieve long-term remission in cancer patients and to treat systemic illnesses.

In this chapter, we examine in detail the first two enzymes to be systematically analyzed by protein engineering, because they illustrate its two major purposes. In the first part, we shall see how protein engineering experiments on the tyrosyl-tRNA synthetase have illuminated the role of enzyme–substrate binding energy in catalysis: the experiments demonstrate enzyme–transition state complementarity; uncover enzyme–intermediate complementarity; detect an induced-fit process; unravel the catalytic mechanism for the activation of tyrosine; find that catalysis is delocalized over the whole binding site; show how internal equilibrium constants are balanced; and probe the fine-tuning of the activity of the tyrosyl-tRNA synthetase in processes that may mimic the final stage of its evolution. The gross structure and allosteric properties of the enzyme are also dissected by mutagenesis. In the second part, we shall see how a serine protease, subtilisin, has been redesigned for practical purposes and, in the process, has illuminated the factors responsible for catalysis.

A. Mechanistic goals

What can we learn about mechanism from protein engineering that cannot be learned from classical enzymology? Chapter 7 begins with the statement: "The mechanism of an enzymatic reaction is ultimately defined when all the intermediates, complexes, and conformational states of an enzyme are characterized and the rate constants for their interconversion are determined." The classical delineation of a mechanism would have been achieved when the general nature of intermediates on a pathway and the type of catalysis had been determined. But we know from Chapter 12 that the fundamental difference between an enzyme-catalyzed reaction and its uncatalyzed counterpart in solution is the use of the binding energy between enzyme and substrate to provide specificity and rate enhancement. *Thus, to understand enzyme catalysis, we must know the interaction energies between the enzyme and substrate throughout the whole course of the reaction. We then see how binding energy is used to lower activation energies, optimize equilibrium constants, and determine specificity.* In other words, to understand an enzyme reaction, one must characterize the complexes of the bound substrates, intermediates, products, and transition states on the reaction pathway and determine the interaction energies between them and the enzyme as the reaction proceeds. Protein engineering is eminently suitable for this purpose because it can be used to alter systematically the multitude of interactions between an enzyme and its substrates. We shall also see how protein engineering can be used to alter the interactions within a protein so that its allosteric and structural properties can be analyzed.

B. The tyrosyl–tRNA synthetase

In 1982, the tyrosyl-tRNA synthetase became the first enzyme to be studied by protein engineering.[3] Its catalytic mechanism, like that of all the other aminoacyl-tRNA synthetases—a heterogeneous class of enzymes[4]—was then completely unknown. Many years of investigation of the 20 different aminoacyl-tRNA synthetases by classical protein chemistry and kinetics had failed to reveal acid or basic groups involved in catalysis. The crystal structure of the tyrosyl-tRNA synthetase had given no clues to its reaction mechanism or the groups involved in catalysis.[5] But just as the solution of the first x-ray structure of an enzyme, lysozyme, elucidated a previously unknown mechanism, so the first application of protein engineering to an enzyme revealed its mechanism. The studies gave direct experimental evidence about fundamental theories of enzyme catalysis that were previously untested speculation.

The tyrosyl-tRNA synthetase from *Bacillus stearothermophilus* crystallizes as a symmetrical dimer of $M_r 2 \times 47\,316$. It catalyzes the aminoacylation of $tRNA^{Tyr}$ in a two-step reaction. Tyrosine is first activated (equation 15.1) to form a very stable enzyme-bound tyrosyl adenylate complex. Tyrosine is then transferred to tRNA (equation 15.2).[6]

$$E + Tyr + ATP \rightleftharpoons E \cdot Tyr\text{-}AMP + PP_i \tag{15.1}$$

$$E \cdot Tyr\text{-}ATP + tRNA \longrightarrow Tyr\text{-}tRNA + E + AMP \tag{15.2}$$

The chemical mechanism of activation (scheme 1) involves the nucleophilic attack of the carboxylate of Tyr on the α-phosphate of ATP to generate either a 5-coordinate transition state or a high energy intermediate. This then expels PP_i

(as its salt with Mg^{2+}). There is stereochemical inversion about the α-phosphorus (see Chapter 8, section E).[7]

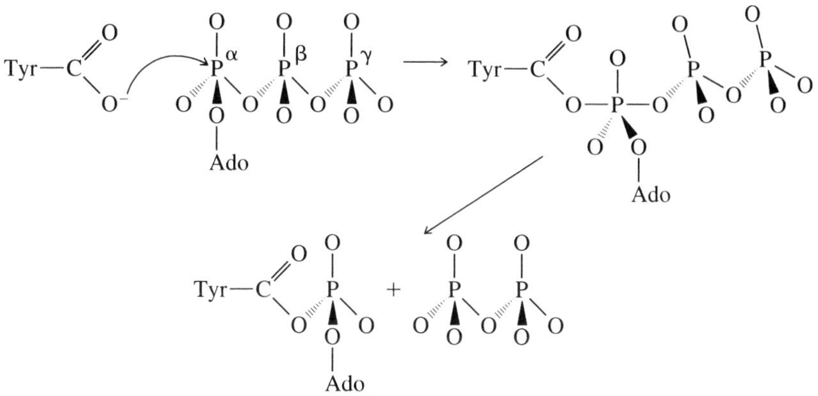

Scheme 1

The crystal structures of the E·Tyr and E·Tyr-AMP complexes have also been solved.[8] Although two moles of tyrosine bind to the crystalline enzyme, only one mole binds to the enzyme in solution. Further, only one mole of tyrosyl adenylate is formed rapidly per mole of dimer, and only one mole of tRNA is bound. The enzyme exhibits half-of-the-sites activity (Chapter 10, section C).[9]

The accumulation of E·Tyr-AMP in the absence of tRNA and the stability of the complex were crucial for the initial success of protein engineering. These factors allow active-site titration and pre–steady state kinetics. Further, the long-term stability of E·Tyr-AMP enables the direct solution of its structure by x-ray crystallography.

C. Requirements for systematic site-directed mutagenesis studies

The prerequisite for protein engineering studies is that the enzyme has been cloned and expressed. Further, unless only relatively crude information is required, it is essential that the structure has been solved at high resolution. Accurate structure–activity studies require even more stringent criteria: absolute values of rate constants. The two following procedures, which were discussed earlier (Chapter 4, section E), must be available. Both depend on the accumulation of an enzyme-bound intermediate or product on the reaction pathway.

1. Active-site titration

The activity of an enzyme often varies from batch to batch because of contamination with the enzyme's denatured products. In systematic studies on a single enzyme, this variation in activity is often compensated for by normalizing the rate constants relative to those measured under a defined, standard set of conditions. This procedure is not suitable for comparing the activity of mutants with wild-type enzyme, because absolute values of rate constants must be known. It is of paramount importance in steady state kinetic studies to have a means of active-site titration to determine how much of the enzyme is active and so generate absolute values of rate constants.

2. Pre–steady state kinetics

An even better way to determine absolute rate constants is to use pre–steady state kinetics to measure the rate constants for the formation or decay of enzyme-bound intermediates (Chapter 4). The rate constants for first-order exponential time courses are independent of enzyme concentration and so are unaffected by the presence of denatured enzyme. The impurity just lowers the amplitude of the trace. Pre–steady state kinetics are also less prone to artifacts, discussed next, that are caused by the presence of small amounts of contaminants that have a much higher activity than the mutant being analyzed. The steady state kinetics of a weakly active mutant could be dominated by a fraction of a percent of wild type. In pre–steady state kinetics, however, that contaminant would contribute only a fraction of a percent of the amplitude of the trace. This would be either lost in the noise or observed as a minor fast phase.

a. Does the very low activity of a mutant result from traces of wild type?

The activity of a very weak active mutant measured by steady state kinetics could result from traces of a wild-type or more active mutant in the preparation either as a contaminant or because of natural errors of misincorporation. The error rate in protein biosynthesis can be as high as one part in 100 or one part in 1000.[10] The presence of a small amount of wild-type enzyme in an inactive mutant would give a low value of k_{cat} (which is directly proportional to the concentration of wild type) but the K_M value for the wild-type enzyme. Thus, the finding of a low value of k_{cat} and the wild-type K_M for a mutant is very suspicious. The presence of an active impurity cannot be ruled out when the K_M for the weak mutant is different from that of wild type, because the contaminant could be a more active mutant. These problems were circumvented in studying the tyrosyl-tRNA synthetase by the use of pre–steady state kinetics.

3. Starting point: The crystal structure of the E·Tyr-AMP complex

A high-resolution structure of a native enzyme is an admirable basis for any mechanistic study relating activity to precise details of structure. It is even better when structures of complexes with substrates and intermediates are available, as is the case with the tyrosyl-tRNA synthetase and tyrosyl adenylate (Figure 15.1). The E·Tyr-AMP complex has two remarkable features. The first is the absence of groups that are candidates for roles in classical catalysis. The second is the

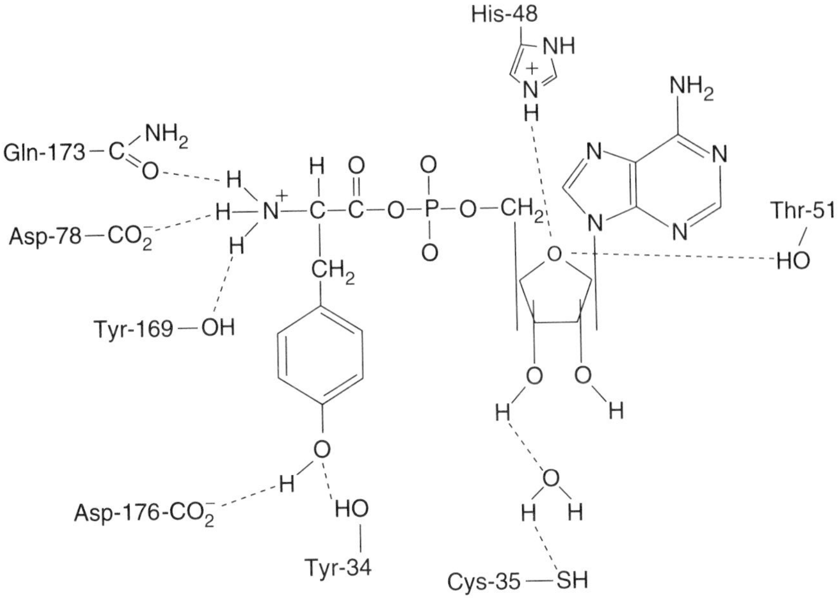

Figure 15.1 Residues of the tyrosyl-tRNA synthetase that form hydrogen bonds with tyrosyl adenylate.

number of hydrogen bonds that are made between Tyr-AMP and the *side chains* of the enzyme. Of course, it is the side chains of the enzyme, and not the backbone, that can be readily modified by protein engineering.

D. Choice of mutation

A bewildering variety of mutations can be made—19 at any one position. There are simple rules, however, to produce mutants that have a chance of being analyzed simply. The goal is to minimize reorganization of the structure of the enzyme, either locally or globally. Reorganization or distortion of structure is accompanied by unknown energy changes that complicate changes arising from the direct interactions of the target side chain. The choice of mutation really follows the same rules that govern specificity, discussed in Chapter 13: an enzyme or enzyme complex can generally tolerate a cavity within it because there is just the loss of the noncovalent interaction energies, but the presence of too large a group sets up high repulsion energies that must be accommodated by a distortion of structure (Figure 15.2).

1. *Choose a mutation that deletes part of a side chain or leads to an isosteric change.* Deletions are preferred to mutations that increase the size of the side chain, especially in the interior of a protein or at an enzyme–substrate

Figure 15.2 Deletion mutations are superior to additions.

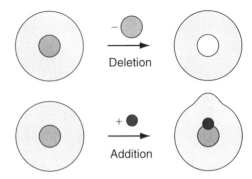

interface. Any increase in volume of the side chain is liable to distort the structure of the enzyme or the enzyme–substrate complex, whereas small cavities in the structure can be tolerated.

2. *Avoid creating buried unpaired charges.* Removal of a group that solvates a buried charge is dangerous. Solvation energies of ions are so high that charged groups must be solvated. Unless there is open access of solvent to buried charge residues, the structure of mutant proteins will rearrange so that there is solvation by water or other groups on the enzyme.

3. *Delete the minimal number of interactions.* It is difficult enough to analyze the change of just one interaction. Avoid the deletion of multiple interactions where possible.

4. *Do not add new functional groups to side chains.* The addition of a new group on a side chain can cause local reorganization of structure if the new group can make novel interactions.

5. *All previous rules may be disobeyed when appropriate.*

The ideal mutation is a *nondisruptive deletion*: that is, one that simply removes an interaction without causing a disruption or reorganization of structure. A formal analysis shows that measurements on nondisruptive deletants provide information on the energetic changes in the wild-type enzyme.[11]

The following mutations are highly suitable for a simple analysis.

Ile → Val, Ala → Gly, Thr → Ser. The loss of a —CH_2— group is a good probe of hydrophobic interactions. There is no change of stereochemistry or increase in branching, and only a tiny cavity is created.

Ile → Ala, Val → Ala, Leu → Ala. There is a larger loss of energy and a greater likelihood of surrounding side chains moving into the cavity or of the ingress of solvent than there is on smaller deletions.

Ser → Ala, Tyr → Phe, Cys → Ala are good for probing hydrogen bonds.

His → Asn or *His → Gln* is sometimes suitable because an NH of the —$CONH_2$ of Asn can sometimes substitute for the $N^\delta H$ of the imidazole ring, or an NH of the —$CONH_2$ of Gln can substitute for the $N^\epsilon H$ as hydrogen bond donors.

Mutation of Asp → Asn, Glu → Gln and *vice versa* appears at first sight to be an isosteric substitution of one polar residue for another. Such a substitution may be acceptable on the surface of a protein, although it can be dangerous even there because the $-CO_2^-$ group is just a hydrogen bond acceptor, whereas the $-CONH_2$ group is both a donor and an acceptor. One warning example is the mutation of Asp-102 in the catalytic triad of chymotrypsin or trypsin. In the native enzyme, one oxygen of the $-CO_2^-$ group of Asp-102 accepts two hydrogen bonds from the protein and the other oxygen accepts one hydrogen bond from the NH of the imidazole of His-57. In the mutant Asp → Asn-102, however, the oxygen of the $-CONH_2$ group binds normally to the protein, and so the $-NH_2$ is directed toward His-57. This requires the unprotonated N of the imidazole to make a hydrogen bond with Asn-102. The NH of His-57, which is no longer a base, faces the nucleophilic Ser-195, and so activity is lost because of the reversal of protonation.

When in doubt, mutate to alanine. This is generally a deletion mutation, and alanine has few quirks. Glycine, because of its wider freedom of conformations, can have odd effects on structure.

E. Strategy: Free energy profiles and difference energy diagrams

The basic procedure of protein engineering is to mutate the side chains that interact with the substrate and measure the changes in activity. The obvious experiment is to locate an acidic or basic residue that appears to be involved in acid-base catalysis and mutate it. A radical drop in activity confirms that the residue is important—although the loss of activity could result from a gross change in structure of the enzyme, and so structural studies must be performed to check this conclusion. The presence of so many mutable side chains in the tyrosyl-tRNA synthetase that interact with Tyr-AMP (Figure 15.1) points to a more subtle strategy: mutate side chains that simply bind to the substrate and are not obviously catalytic, and measure the small changes in activity. The initial studies on the tyrosyl-tRNA synthetase, like those on enzymes in general, involved just the measurement of the effects on simple steady state kinetics. The fully developed strategy is to measure the complete free energy profiles for wild-type and mutant proteins. For the tyrosyl-tRNA synthetase, this entails measuring the rate and equilibrium constants in scheme 2.

$$E \underset{K_T}{\overset{Tyr}{\rightleftharpoons}} E{\cdot}Tyr \underset{K_A'}{\overset{ATP}{\rightleftharpoons}} E{\cdot}Tyr{\cdot}ATP \underset{k_{-3}}{\overset{k_3}{\rightleftharpoons}} E{\cdot}Tyr\text{-}AMP{\cdot}PP_i \underset{K_{PP}}{\rightleftharpoons} E{\cdot}Tyr\text{-}AMP + PP_i$$

Scheme 2

The constants that must be measured in scheme 2 are K_T, the dissociation of the E·Tyr complex (by equilibrium dialysis or kinetics); K_A', the dissociation constant of ATP from the ternary complex, E·Tyr·ATP (from kinetics); k_3, the rate

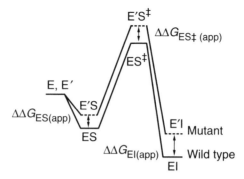

Figure 15.3 Superposition of free energy profiles for wild-type (E) and mutant (E′) enzymes. The reaction involves the formation of an ES complex followed by a transition state for a chemical step, ES‡, and then the formation of an enzyme-bound intermediate EI. Note that the free enzymes are arbitrarily assigned the same free energy. This is valid for comparing the changes in interaction energies at different stages of the reaction (see Chapter 3, section L3, and A. R. Fersht, A. Matouschek, and L.Serrano, *J. Molec. Biol.* **224**, 771 (1992)).

constant for the chemical step (from pre–steady state kinetics, using stopped-flow and the mixing of E·Tyr with ATP); k_{-3}, the rate constant for pyrophospho-rolysis of E·Tyr-AMP to E·Tyr·ATP (from pre–steady state kinetics, using stopped-flow and the mixing of E·Tyr-AMP with PP$_i$); and K_{PP}, the dissociation constant of PP$_i$ from the E·Tyr-AMP·PP$_i$ complex (from the kinetics of the py-rophosphorolysis step). The free energy profile is then calculated for wild-type protein using equilibrium thermodynamics with the equilibrium constants and transition state theory with the rate constants. The same procedure is then re-peated for mutant protein. The two are plotted, assigning, for convenience, the unligated enzymes as the standard states (Figure 15.3).[11,12]

Figure 15.3 shows not the absolute values of the free energy levels but the differences between the levels for wild-type and mutant enzymes. These differ-ences are then replotted as *difference energy diagrams*.[11,12] We sometimes call the difference energy the *apparent binding energy* of a group ($\Delta\Delta G_{app}$) when the mutation deletes an interaction. The importance of difference energy diagrams is now illustrated by a series of examples.

F. Results from difference energy diagrams for the activation of tyrosine

1. Demonstration of enzyme–transition state complementarity[13]

A difference energy diagram constructed for nondisruptive deletion mutations in wild-type enzyme shows at a glance the role of the target side chains in the wild type. This is seen most clearly in the difference energy diagrams plotted in Fig-ure 15.4 for the mutation of Thr-40 to Ala or Gly and for the mutation of His-45 to Ala, Gly, Asn, and Gln. On the left, it is seen that mutation of Thr-40 has essentially no effect on the binding energies of Tyr and ATP to the enzyme. But there is a massive raising of the energy level of the transition state by some 20 kJ/mol (5 kcal/mol). The effect on the binding energy of pyrophosphate could not be determined precisely, but it is weakened by more than 12 kJ/mol (3 kcal/mol). There are negligible effects on the binding of Tyr-AMP.

These results show clearly that Thr-40 binds Tyr and ATP only in the transition state of the reaction. The results represent the first direct demonstration by

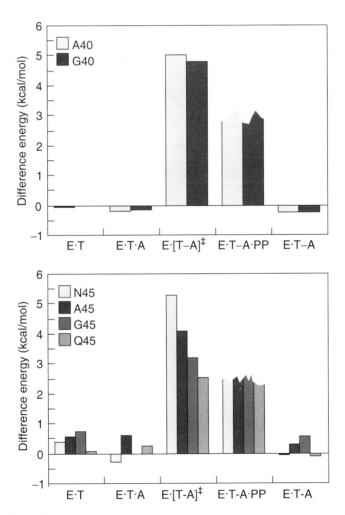

Figure 15.4 Difference energy diagrams (energy of mutant minus energy of wild type) for mutation at residues Thr-40 and His-45. The states are E·T, the E·Tyr complex; E·T·A, E·Tyr·ATP; E·[T-A]‡, the transition state for formation of Tyr-AMP; E·T-A·PP, E·Tyr-AMP·PP$_i$; and E·T-A, E·Tyr-AMP.

experiment of the Haldane-Pauling postulate of enzyme—transition state complementarity.[12] Thr-40 also binds PP_i after the reaction has occurred. But it does not bind Tyr-AMP.

There are similar consequences on mutation of His-45 (right side of Figure 15.4). Although mutation of His-40 to Ala or Gly gives similar values of difference energies, there are significant variations on mutation to Asn or Gln. This result emphasizes that difference energies can depend on the nature of the mutation and that the deviations are greater when new functional groups are introduced.

a. Thr-40 and His-45 form a binding site for the γ-phosphate of ATP in the transition state

What are the chemical roles of Thr-40 and His-45? These residues are not seen in Figure 15.1, which shows the residues implicated by x-ray crystallography as binding to Tyr-AMP. Model-building studies, however, indicate that Thr-40 and His-45 form part of a binding site for the γ-phosphate of ATP in the pentacovalent transition state or intermediate. This site contributes no binding energy with ATP before the reaction takes place. Thus, the enzyme catalyzes the reaction by taking advantage of the subtle change in bond angles at the α-phosphate during the reaction. These subtle changes cause a rotation of the PP_i moiety with a consequent movement of the γ-phosphate (scheme 3).

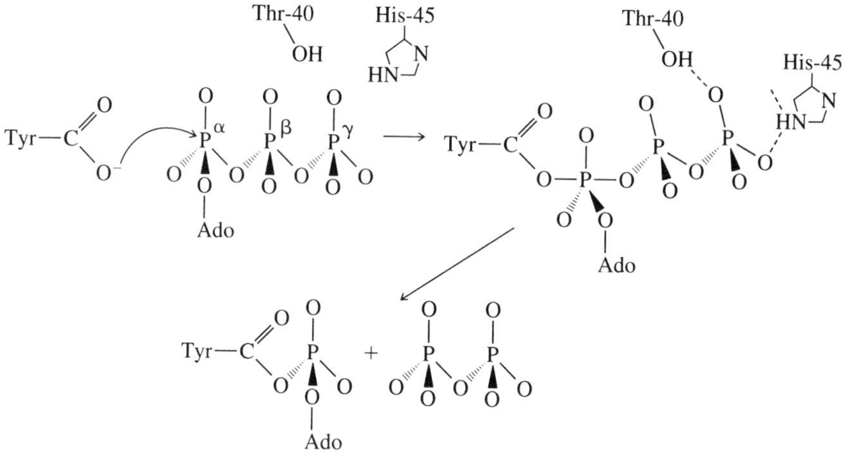

Scheme 3

2. Discovery of enzyme–intermediate complementarity: Balancing internal equilibrium constants; sequestration of unstable intermediates[12]

Mutagenesis of the residues Cys-35 and His-48 that bind to the ribose ring in the E·Tyr-AMP complex (Figure 15.1) revealed that they contribute little or no interaction energy with ATP in the E·Tyr·ATP ground state complex (Figure 15.5).

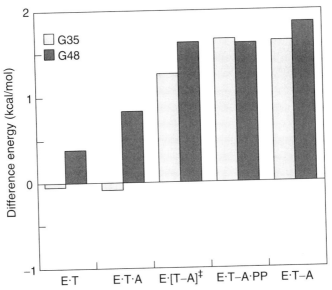

Figure 15.5 Difference energy diagrams (energy of mutant minus energy of wild type) for mutation at residues Cys-35 and His-48.

They do, however, contribute significant stabilization energy with ATP in the transition state and so contribute to catalysis because their binding energy is used to lower the energy difference between the ground state and the transition state. But they contribute even more stabilization energy in the E·Tyr-AMP complex. That is, there is enzyme–intermediate complementarity, in apparent contradiction to Pauling and Haldane! Transition state complementarity is always part of the optimization of rate for reactions that have just a single important transition state and where substrates and products diffuse rapidly into solution. Other criteria, however, can apply for multistep processes or when products or intermediates remain bound to the enzyme. Maximization of the rate of a single step of a reaction may then not necessarily lead to the optimization of the over-all rate of the process. There are two good reasons for enzyme–intermediate complementarity.

1. *Enzyme–product complementarity changes the equilibrium constant for highly unfavorable reactions.* The formation of Tyr-AMP from Tyr and ATP in solution is a very unfavorable reaction. The equilibrium constant for equation 15.4 (= [Tyr-AMP][PP$_i$]/[Tyr][ATP])

$$Tyr + ATP \rightleftharpoons Tyr\text{-}AMP + PP_i \tag{15.4}$$

is 3.5×10^{-7} (see section K). But the internal equilibrium constant for the enzyme-bound reaction (equation 15.5) is 2.3 ($= [E \cdot Tyr\text{-}AMP \cdot PP_i]/$ $[E \cdot Tyr \cdot ATP]$). The enzyme increases the equilibrium constant nearly 10^7-fold.

$$E \cdot Tyr \cdot ATP \rightleftharpoons E \cdot Tyr\text{-}AMP \cdot PP_i \qquad\qquad (15.5)$$

This increase is necessary for catalysis because the rate-determining step *in vivo* is the attack of $tRNA^{Tyr}$ on the $E \cdot Tyr\text{-}AMP$ complex. The enzyme accomplishes the enormous increase in the equilibrium constant by binding Tyr-AMP far more tightly than Tyr + ATP (see Chapter 3, section H, and Chapter 12, section E).

2. *The enzyme increases the yield of reaction by minimizing side reactions and sequestering the highly reactive intermediate.* The optimization of rate in a complex reaction pathway requires the maximization not only of individual rate constants but also of yield by eliminating unwanted side reactions. Aminoacyl adenylates are highly reactive intermediates. If they were to diffuse off the aminoacyl-tRNA synthetases, they would hydrolyze within a few seconds or, more seriously, aminoacylate reactive side chains of proteins. The enzyme minimizes the dissociation rate constant by enzyme–aminoacyl adenylate complementarity. The adenylate is also protected against hydrolysis by groups on the enzyme that interfere with the attack of water or hydroxide ion.

3. Detection of an induced-fit process[14]

All arginine and lysine residues were mutated at random to uncharged side chains to investigate the effects on catalysis. It was discovered that many of these residues contribute to catalysis specifically by binding the transition state. The surprising feature is that some of these residues (Lys-230 and Lys-233) are on a loop (the "KMSK" loop) that is too far away to interact with the model-built transition state. The problem was resolved on examination of the crystallographic temperature factors (B values, Figure 15.6), which are measures of either motional freedom or random disorder (see Chapter 1). The loop containing residues 230 and 233 is very mobile and is able to wrap around the transition state as the reaction occurs. This provides a rationale for the induced-fit process. If the loop in the native enzyme were in the orientation that is optimal for binding the transition state, it would block access of substrates to the active site. Flexibility and induced fit represent a compromise between enzyme–transition state complementarity and open access of substrates to the binding site. (The roles and interactions of the individual and collective effects of the residues in the loop have subsequently been measured by a series of multimutant cycles.[15])

4. The catalytic mechanism for activation of tyrosine

The results of the difference energy diagrams are summarized in Table 15.1. It is obvious that a host of different side chains are involved in catalysis, that cataly-

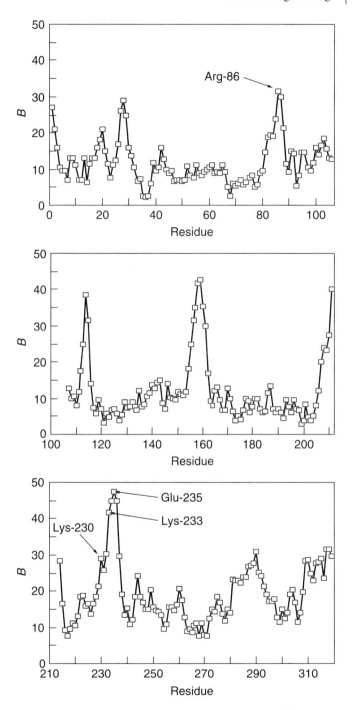

Figure 15.6 *B* values, averaged for the backbone of the tyrosyl-tRNA synthetase.

Table 15.1 *Interaction energies of side chains of tyrosyl-tRNA synthetase with reagents*

	Interaction energy of side chains in complex with:[a]				
Residue	Tyr	ATP	[Tyr-ATP]‡	PP$_i$	Tyr-AMP
Tyrosine binding site					
Tyr-34	+	0	+	0	+
Asp-78	++++	++^{b}	++++	++^{b}	++++
Tyr-169	++++	0	++++	0	++++
Gln-173	++++	++^{b}	++++	+^{b}	++++
Nucleotide and pyrophosphate site					
Cys-35	0	0	++	0	+++
Thr-40	0	0	++++	++++	0
His-45	0	0	++++	++++	0
His-48	0	0	+++	0	+++
Thr-51	0	0	0	0	–
Lys-82	0	++	++++	++++	0
Arg-86	0	0	++++	++++	–
Asp-194	0	0	++++	+	+++
Lys-230	0	0	++++	++++	0
Lys-233	0	++++	++++	++++	0

^{a} Apparent stabilization energy from the side chain in kcal/mol (kJ/mol): $0 = -0.5$ to $+0.5$ (-2 to $+2$); $+ = 0.5$ to 1.0 (2 to 4); $++ = 1.0$ to 1.5 (4 to 6); $+++ = 1.5$ to 2.0 (6 to 8); $++++ = >2.0$ (>8); $- = -0.5$ to -1.0 (-2 to -4).
^{b} Evidence for some disruption of protein structure on mutation.

sis is not attributable to any one residue, and that the whole active site contributes to catalysis. *Catalysis results solely from the use of binding energy without any assistance from the classical mechanisms of acid-base or nucleophilic catalysis.* The minimal mechanism is sketched in Figure 15.7. In the ground state E·Tyr·ATP complex (*top left*), ATP binds to Lys-82 and Lys-233. There is insignificant binding energy with Cys-35, Thr-40, His-45, His-48, Thr-51, Lys-82, Arg-86, and Lys-230. In the transition state (*top right*), the charged groups and other groups interact with the pyrophosphate moiety of ATP and its ribose ring. In the E·Tyr-AMP· PP$_i$ complex (*bottom right*), the groups still interact with the reagents. In the E·Tyr-AMP complex (*bottom left*), the Lys-230/233 loop moves away from the adenylate after dissociation of PP$_i$, and Lys-82 binds to the α-phosphate of Tyr-AMP via a bridging water molecule. It is possible that there is an additional E′·Tyr·ATP complex between the ground state E·Tyr·ATP complex and the transition state, in which the enzyme and

ATP have geometries closer to that in the transition state. ATP and PP_i react as their magnesium salts. The position of the Mg^{2+} in the complexes is not yet known.

5. Mechanism of transfer step

There are no acidic or basic groups that are suitably placed to catalyze the attack of the ribose 2'-OH of the 3' end of tRNA^{Tyr} on the $>C=O$ of Tyr-AMP. But Tyr-AMP is an extremely activated substrate, with a $t_{1/2}$ for hydrolysis in solution of one minute.[16] The intramolecular attack of the 2'-OH in the ternary complex should be very rapid. It has been suggested from examining the crystal structures of the glutaminyl-[17] and aspartyl-tRNA[18] synthetases that a phosphoryl oxygen of the Tyr-AMP accepts the proton from the —OH.

G. Relationship between apparent binding energies from difference energies and incremental binding energies[11,19]

The apparent binding energies of the different groups in Table 15.1 for the transition state ([Tyr-ATP]‡) add up to more than 80 kJ/mol (20 kcal/mol). This raises the question of what these energies mean. The apparent binding energies always measure the relative energies of two different enzymes binding the same species; that is, the relative energies of two mutants (or mutant and wild type) binding a substrate or a transition state. The apparent binding energies are, in effect, a measure of the specificity of binding. The apparent binding energy, in general, is not equal to the true incremental binding energy of a group that is deleted, and it can vary according to the mutation that is made. For example, suppose we mutate a tyrosine to a phenylalanine (e.g., Tyr $\rightarrow$ Phe-34) and remove the hydrogen bond donor to a substrate. If there is no accompanying rearrangement of structure and no access of water to the cavity that is formed at the site of mutation (MUT1 in Figure 15.8), then the binding energy of the substrate and the enzyme is lowered by the particular amount (the energy terms are described in detail in Chapter 11). Suppose, instead, that there is open access of water to the cavity (MUT2). In that case, there will be a different change in energy. If there is unrestricted access of water, the apparent binding energy will represent the difference in energy between the substrate making a hydrogen bond with tyrosine and the substrate making a bond with water. In this case, the apparent binding energy will approach the incremental binding energy described in Chapter 11. If we make a mutation that induces a different interaction (MUT3), then the apparent binding energy will reflect the differences between the interactions in wild type and in mutant. Superimposed upon these energies will be any terms resulting from reorganization of the enzyme structure if it does change or mutation.

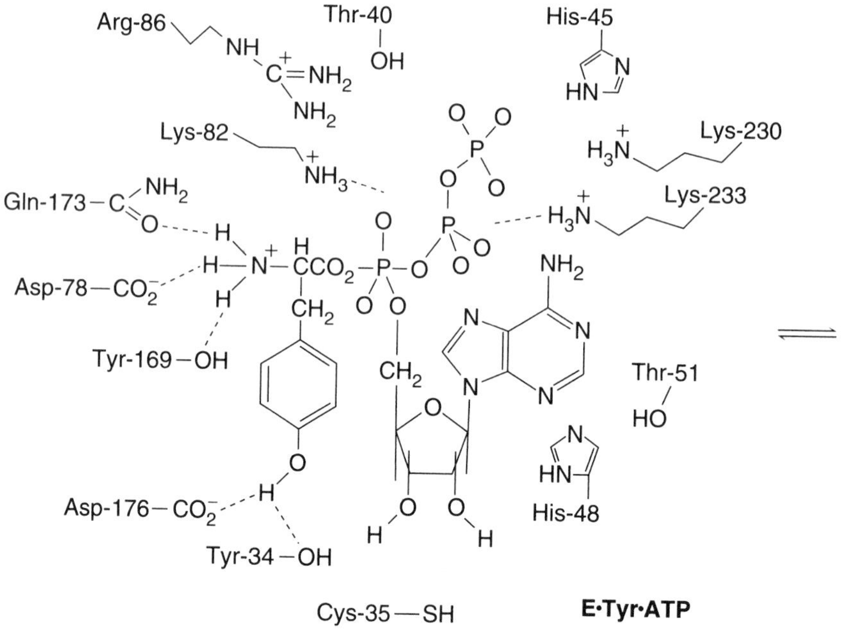

E·Tyr·ATP

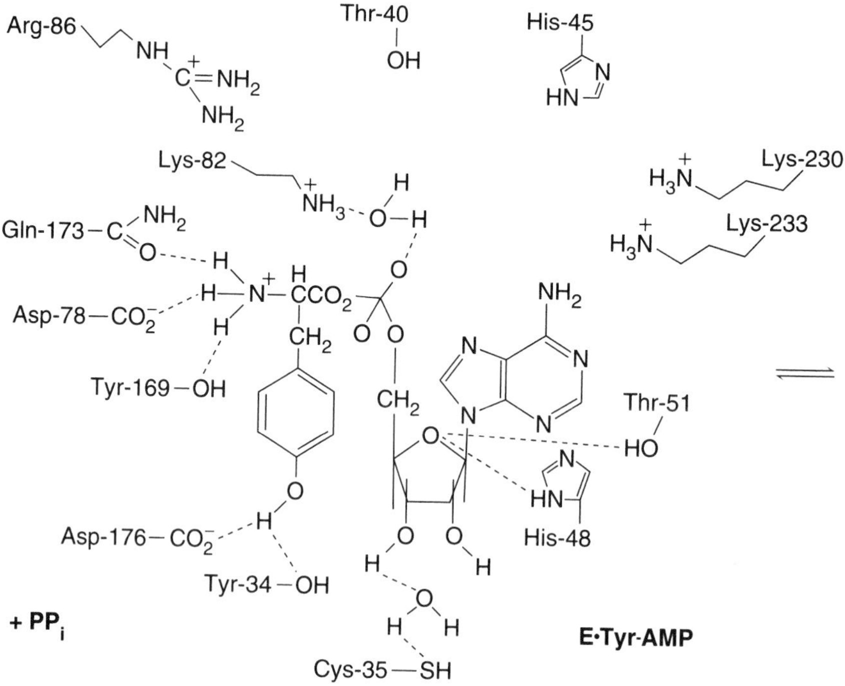

+ PP_i

E·Tyr·AMP

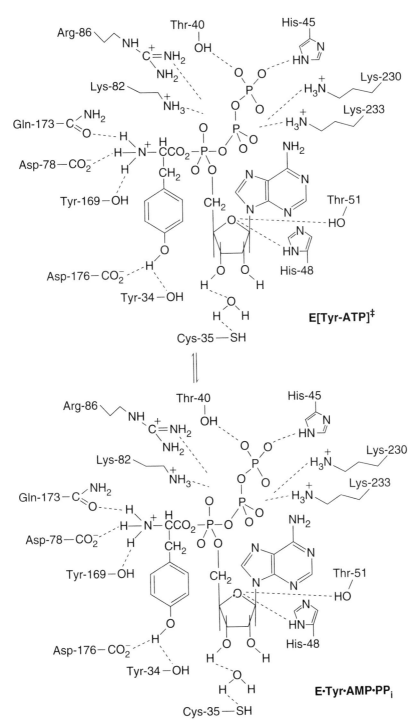

Figure 15.7 Hydrogen-bonding interactions and the movement of the loop on the formation of E·Tyr·AMP·PP$_i$ from E·Tyr·ATP.

Figure 15.8 Cartoons of types of cavities induced on mutation.

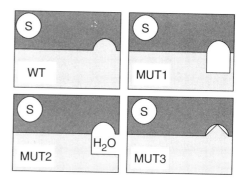

The simplest case is when there is open access of water to the site of mutation (MUT2 in Figure 15.8), because this approximates to the incremental binding energy. A cavity-inducing mutation that blocks access of water (MUT1 in Figure 15.7) is fine, provided that it does not lead to the desolvation of a charged residue (e.g., an Asp or an Lys). If it does, then the apparent binding energy will be dominated by a term involving the desolvation energy of the charged group, and the enzyme will almost certainly rearrange to solvate the charge with other groups. Introduction of a new interaction is very difficult to analyze.

Note that although the apparent binding energies do not necessarily equal the true binding energies, the *changes* in apparent binding energies can equal the *changes* in true binding energies. This holds most often with cavity-inducing mutations in which an interaction is removed from the wild-type complex. Thus, the difference energy diagrams for such deletion mutants do report back on the energy changes during the reaction of wild-type enzyme.

H. Probing evolution: "Reverse genetics" [20,21]

The ability to make small alterations in structure allows us to explore the types of events that occur in the final stages of evolution and that fine-tune an enzyme to its physiological environment. Starting from the final structure, which is presumably optimal for the native environment, we can work backward by inducing mutations that make the enzyme less active. We can first detect the types of changes in activity on mutation.

1. Differential and uniform binding changes

Residues Cys-35 and His-48 are said to exhibit *differential binding*, because they contribute different binding energies with the substrates, transition states, and products (Figure 15.5). The contributions of Thr-40 and His-45 (Figure 15.4) are extreme examples of differential binding, because they do not bind the substrates or intermediate but greatly stabilize the transition state. Differential changes can lower the energy of the transition state with respect to ground state and hence increase the value of k_{cat}. Other residues make the same contribution to the binding throughout the reaction. This is described as *uniform binding*.[22] A good example (Figure 15.9) is the binding of the hydroxyl of Tyr-169 to the —NH_3^+ group of substrate tyrosine (Figure 15.1). Gln-173, which also binds to that —NH_3^+, has approximately uniform binding from the E·Tyr·ATP complex onward.

These types of binding energy changes are precisely those that were discussed in Chapter 12 (see Figure 12.2). Uniform increases in binding energy lower the energies of each of the states equally and so do not alter the rate constants for their interconversion (k_{cat} is unaffected), although the value of k_{cat}/K_M is increased. When [S] $\ll K_M$, the rate is given by $v = (k_{cat}/K_M)[E]_0[S]$. A uniform increase in binding energy raises the value of k_{cat}/K_M, and so the rate increases. When [S] $\gg K_M$, the rate is given by $v = k_{cat}[E]_0$. Thus, uniform binding changes do not increase rate at saturating concentrations of substrate because they do not increase k_{cat}. Only differential binding helps when [S] $\gg K_M$.

2. Fine-tuning activity of tyrosyl-tRNA synthetase toward [ATP]

These effects are nicely illustrated by changes at position 51.[20,21] The tyrosyl-tRNA synthetase from *B. stearothermophilus* has Thr at this position. The tyrosyl-tRNA synthetase from *Bacillus caldotenax* differs by just four amino acid residues from that from *B. stearothermophilus*, but one of the changes is an Ala at position 51.[23] The enzyme from *E. coli* is only 50% identical in sequence but has Pro at

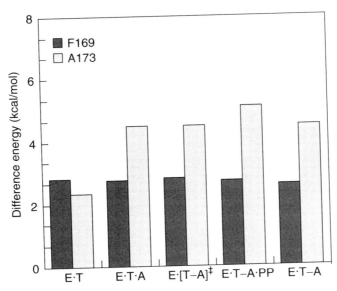

Figure 15.9 Mutation of residues that bind the —NH_3^+ of tyrosine (Tyr → Phe-169, Gln → Ala-173).

position 51. Mutating Thr $\rightarrow$ Pro-51 greatly increases the rate of activation of tyrosine, and other mutations at this position cause significant increases (Table 15.2).[20] Most of the binding energy that is gained on mutating Thr-51 is used to increase the rate of activation of tyrosine and does not increase the affinity of the enzyme for ATP. This classic case of differential binding energy in activation does not, however, increase the value of k_{cat} for the overall aminoacylation reaction. The kinetics of aminoacylation is complex, being similar to that for hydrolysis of esters by chymotrypsin, where it was seen that speeding up the first chemical step does not increase the value of the overall k_{cat} but lowers the overall K_M. The net result is that a family of enzymes is generated, each of which is optimized for different ambient concentrations of ATP (Figure 15.10). The reasons for the lowering of the overall value of k_{cat} on increasing the rate constant for activation are most illuminating and are discussed next.

3. Optimizing rate in a multistep reaction[24]

Mutation of position 51 leads to a lowering of energy of the transition state $[E \cdot Tyr\text{-}AMP]^{\ddagger}$ and a larger lowering of the energy of the E·Tyr-AMP complex (Figure 15.11). This may be viewed as a consequence of the phenomenon of enzyme–intermediate complementarity (section F2). The enzyme stabilizes the intermediate and, in doing so, lowers the energy of the transition state immediately preceding it. In this case, however, there is the consequence that the rate of the following step is slowed down because, as discussed in Chapter 12, the intermediate becomes too stable, as illustrated in Figure 15.12. The enzyme has to juggle the energy level of the intermediate so that it is stable enough to form but is not too stable to decay. This has happened in practice (Table 15.3). The energy level of the wild-type complex is poised to be optimal for both steps, whereas the energy level of Ala-51 and Pro-51 mutants is too stable. The *E. coli* enzyme that has a proline at position 51 has other mutations that compensate for its effect on stability.

Table 15.2 *Fine-tuning the activity of tyrosyl-tRNA synthetase toward [ATP]*

Enzyme	Activation of Tyr		Aminoacylation of tRNA	
	k_3 (s^{-1})	K'_A (ATP) (mM)	k_{cat} (s^{-1})	K_M (ATP) (mM)
Thr-51 (wild type)	38	4.7	4.7	2.5
Ala-51	75	4.7	4.0	1.2
Pro-51	~700	~2	1.8	0.019

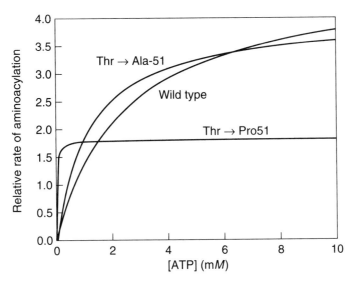

Figure 15.10 Activities on mutation at position 51.

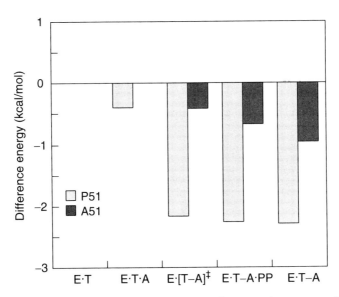

Figure 15.11 Difference energy diagrams (energy of mutant minus energy of wild type) for mutation of residue Thr-51.

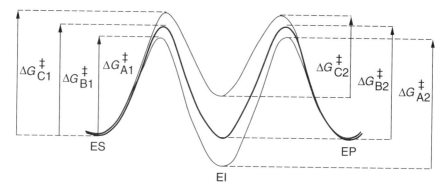

Figure 15.12 Free energy diagram for intermediate flanked by two transition states.

Table 15.3 *Balancing a two-step reaction*

Enzyme	Activation of Tyr k_3 (s^{-1})	Transfer to tRNA k_4 (s^{-1})
Thr-51 (wild type)	38	31
Ala-51	75	16
Pro-51	~700	5

I. Linear free energy relationships in binding energies[25]

The energy changes in Figure 15.12 are very reminiscent of those in Chapter 2 for linear free energy relationships and the Hammond postulate. There is, in fact, a series of good linear free energy relationships relating the rate constant for the activation of tyrosine to various rate and equilibrium constants. These relate changes in binding energies in transition states relative to those in ground or other states. The first relationship that was discovered is the basic plot of log k_3 versus log k_3/k_{-3} for the interconversion of the two ternary complexes in scheme 2 (Figure 15.13a). The residues mutated are primarily those that bind the ribose ring illustrated in Figure 15.1. Those residues bind the ribose ring far more tightly in the E·Tyr-AMP complex than in the E·Tyr·ATP complex. The linear free energy relationship has a slope of 0.8, showing that apparently 80% of the difference in binding energy of those side chains between the E·Tyr-AMP·PP$_i$ and E·Tyr·ATP complexes is realized in the [E·Tyr·ATP]‡ transition state. An objection was raised against this classic plot because log k_3 occurs on both sides of the equation of log k_3 versus log k_3/k_{-3}, and so the two sides of the equation are not truly in-

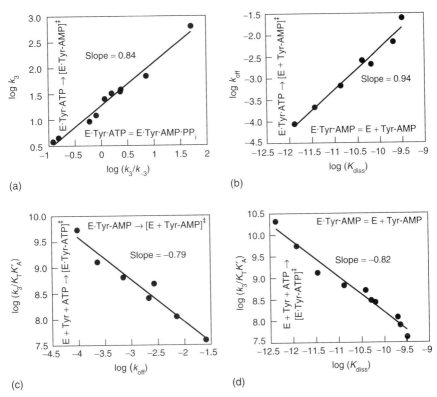

Figure 15.13 LFERs for the formation of tyrosyl adenylate. (a) $\log k_3$ versus $\log k_3/k_{-3}$; (b) $\log k_{off}$ versus K_{diss}; (c) $\log (k_3/K_T K'_A)$ verses $\log k_{off}$; (d) $\log (k_3/K_T K'_A)$ versus $\log K_{diss}$. (Data from A. R. Fersht, R. J. Leatherbarrow, and T. N. C. Wells, *Biochemistry* **26**, 6030 (1987) and T. N. C. Wells and A. R. Fersht, *Biochemistry* **28**, 9201 (1989).)

dependent.[26] This objection brings out an important point, but the linear free energy relationship was shown to be valid both by statistical analysis and by replotting the equation in another form, as follows.[26] The Brønsted plot is given by

$$\log k_3 = \beta \log \frac{k_3}{k_{-3}} + \alpha \tag{15.6}$$

Equation 15.6 may be rearranged to give

$$\log k_3 = \frac{\beta}{\beta - 1} \log k_{-3} + \frac{\alpha}{\beta - 1} \tag{15.7}$$

There are no statistical objections to using equation 15.7 because both sides have independent variables. The experimental validity of linear free energy relationships in the system is demonstrated in the other three panels of Figure 15.13 (see also reference 27). There is a very nice relationship between the observed rate constant for the dissociation of Tyr-AMP (k_{off} in scheme 4) and its equilibrium dissociation constant ($K_{diss} = [E][Tyr\text{-}AMP]/[E\cdot Tyr\text{-}AMP]$).[27] The slope (Figure 15.13b) is 1.0, showing that all the discrimination in binding occurs in the dissociation rate constant (k_{off}). There is also a very nice plot of log ($k_3/K_T K'_A$) against log K_{diss} (Figure 15.13d). The compound rate constant $k_3/K_T K'_A$ is the apparent third-order rate constant for ATP + Tyr + E → E·Tyr-AMP. According to the transition state theory analysis of Chapter 12, section A1, the log of $k_3/K_T K'_A$ is thus proportional to the binding energy of the reagents in the transition state $[E\cdot Tyr\text{-}ATP]^{\ddagger}$. The plot of log ($k_3/K_T K'_A$) against log K_{diss} has a slope β of -0.8, showing directly that the binding energy of the side chains in that transition state is 80% of that when binding Tyr-AMP. A plot of log ($k_3/K_T K'_A$) against log k_{off} also has a slope of -0.8, showing that the mutated side chains interacting with the transition state for the formation of Tyr-AMP have only 80% of the energy as in the transition state for binding Tyr-AMP.

$$E \underset{K_T}{\rightleftharpoons} \overset{Tyr}{E\cdot Tyr} \underset{K'_A}{\overset{ATP}{\rightleftharpoons}} E\cdot Tyr\cdot ATP \underset{k_{-3}}{\overset{k_3}{\rightleftharpoons}} E\cdot Tyr\text{-}AMP\cdot PP_i \underset{K_{PP}}{\rightleftharpoons} E\cdot Tyr\text{-}AMP + PP_i$$

$$K_{diss} \Big\Vert k_{off}$$

$$E + Tyr\text{-}AMP$$

Scheme 4

Brønsted plots for changes in covalent bonds in simple organic chemistry generally have β values between 0 and 1. Those in binding interactions are not so restricted. In Figure 15.14, a plot of log k_3 against the equilibrium constant between E·Tyr·ATP and E + Tyr-AMP for mutants of Thr-40 and His-45 has a slope of ~80. These residues bind to ATP only in the transition state and do not affect the binding of ATP or Tyr-AMP. Thus, they do not affect the equilibrium constant, and so β tends to infinity.

Difference energy diagrams are the bread and butter of mutagenesis studies on binding energies. These linear free energy relationships are a bonus that shows that the reaction responds uniformly to perturbations in its binding energy at certain, but not all, positions.

J. Probing the gross structure and symmetry of the enzyme by mutagenesis[9,28]

In addition to its usefulness in probing the role of individual residues in catalysis, site-directed mutagenesis has proved invaluable in dissecting the gross nature of enzymes. The tyrosyl-tRNA synthetase has some interesting properties

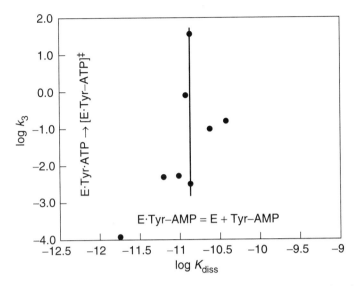

Figure 15.14 LFER for a reaction in which mutations affect only interactions in the transition state. (Data from R. J. Leatherbarrow, and A. R. Fersht, *Biochemistry* **26**, 2524 (1987).)

connected with its symmetry. Although it is a homodimer consisting of two identical chains and appears to be symmetrical in the crystal, it binds only one mole of tyrosine and forms only one mole of tyrosyl adenylate per mole of dimer. Further, it binds tightly only one mole of tyrosine. By using protein engineering, it was possible to resolve the classic dilemma of whether this half-of-the sites reactivity results from ligand-induced asymmetry or from pre-existing asymmetry.

1. Domain structure of the enzyme

Mutagenesis may be conducted at a gross level. Instead of changing a single amino acid residue, whole sections of a protein may be removed by excising a stretch of its gene. This method is called *deletion mutagenesis.* By such means, it was shown that the enzyme is organized into two major domains: the N-terminal domain is solely responsible for the activation step, and the C-terminal is primarily responsible for binding tRNA.[29] Only the N-terminal 319 amino acids are seen in the x-ray map of the enzyme. The 100 residues of the C-terminus are too disordered to give discernible electron density. The N-terminal fragment of the tyrosyl-tRNA synthetase was constructed by deletion mutagenesis and found to be kinetically identical to wild-type enzyme in the activation step but unable to bind or aminoacylate tRNA (Figure 15.15). The crystal structure of the truncated enzyme is very similar to that of the N-terminal domain of wild-type enzyme.

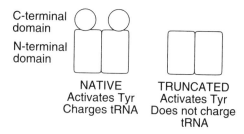

Figure 15.15 Cartoon of full-length and truncated tyrosyl-tRNA synthetase.

C-terminal domain

N-terminal domain

NATIVE
Activates Tyr
Charges tRNA

TRUNCATED
Activates Tyr
Does not charge tRNA

2. Construction of heterodimers

Heterodimers containing one full length and one truncated subunit were constructed and separated from parental full-length and truncated homodimers (Figure 15.16). Denaturing and renaturing an equimolar mixture of truncated and full-length enzymes gives a 1:2:1 mixture of full-length homodimer, heterodimer, and truncated homodimer, as expected from the random association of monomers (Figure 15.17).[9,28] These heterodimers gave the means of studying the symmetry of the enzyme and its complexes in solution.

a. Symmetry of tRNA binding from experiments on heterodimers

Heterodimers of full-length and truncated enzyme have the great utility of providing a way to "tag" the individual subunits so that a second mutation can be

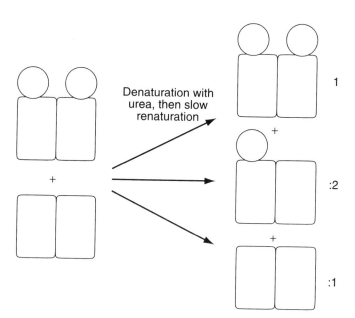

Denaturation with urea, then slow renaturation

1

+

:2

+

:1

Figure 15.16 Formation of different dimers on the denaturation and renaturation of full-length and truncated tyrosyl-tRNA synthetases.

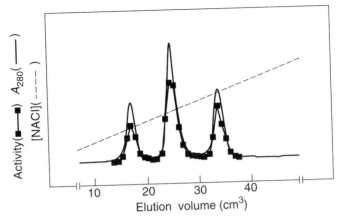

Figure 15.17 Separation of mixture of truncated homodimer, truncated subunit/full-length subunit heterodimer, and full-length homodimer by ion-exchange chromatography with a NaCl gradient.

specifically introduced into one predefined subunit.[30,31] For example (Figure 15.18), the mutation His → Asn-45 was introduced into one subunit to reduce drastically the rate constant for the activation step by that subunit. When Asn-45 is in the large subunit of the heterodimer and the truncated subunit has wild-type site, the rate of aminoacylation of tRNA is half that of wild-type enzyme. But when the heterodimer has Asn-45 in the small subunit, the rate of aminoacylation of tRNA is undetectably low. This shows that, although tRNA binds predominantly to the full-length subunit, amino acid acceptance occurs from the small subunit. tRNA must bind to wild-type enzyme in a similar manner, spanning both subunits (Figure 15.19).

The heterodimer of truncated and full-length enzyme thus provides an asymmetric system for testing the effects of mutations on catalytic and binding sub-

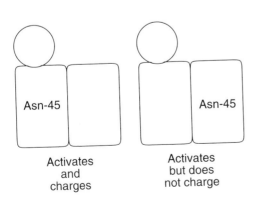

Activates
and
charges

Activates
but does
not charge

Figure 15.18 Positional dependence of the mutation His → Asn-45 on aminoacylation. This mutation greatly slows down activation in its subunit.

Figure 15.19 Cartoon of the mode of binding of tRNA to the dimer.

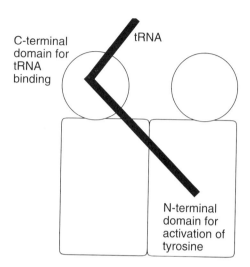

units in aminoacylation. The technique of *surface scanning* was introduced to discover residues that are crucial to either catalysis or binding: lysine and arginine residues were mutated at random to alanine (or hydrophilic residues) and the changes in activity measured. Patches of lysine and arginine residues were implicated in binding tRNA, which enabled a low-resolution model of the enzyme–tRNA complex to be built.[31]

b. Symmetry of enzyme in solution: Mechanism of half-of-the-sites activity

The catalytic asymmetry of heterodimers was used to show that the wild-type enzyme is asymmetrical in solution. The enzyme is frozen into two populations: 50% are active in one subunit and 50% in the other. For example (Figure 15.20), heterodimers containing Asn-45 on one subunit form 0.5 mol of Tyr-AMP per mole of dimer rapidly at wild-type rate ($t_{1/2} = 20$ ms) and a further 0.5 mol four orders of magnitude more slowly ($t_{1/2} = 200$ s) at the rate expected for mutant. If the half-of-the-sites activity were induced by the formation of the first mole of tyrosyl adenylate, then the wild-type site would be the one that is occupied. Because this does not happen, there must be frozen-in asymmetry that is randomly distributed between active wild-type subunit/inactive mutant and inactive wild-type site/active mutant. Any interconversion of active and inactive subunits is on a much slower time scale than the half-life of several minutes required for formation of E·Tyr-AMP at the mutated site (Asn-45). There is no evidence that the behavior of the heterodimers is different from that of wild-type enzyme.

The half-of-the-sites activity is the result of a pre-existing asymmetry in the enzyme in solution. Thus, there are important differences between the crystal and solution structures in terms of both overall symmetry and positions of mobile side chains, such as that containing Lys-230 and Lys-233.

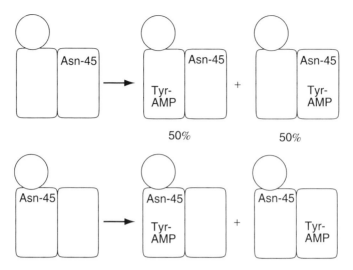

Figure 15.20 Distribution of products on the formation of Tyr-AMP by the heterodimer.

K. Measuring the free energy of hydrolysis of Tyr–AMP[32]

Aminoacyl adenylates have long been known to be "high energy" compounds, but their free energies of hydrolysis had not been accurately measured. This was accomplished for tyrosyl adenylate using the Haldane approach (Chapter 3, section H) and mutants of the tyrosyl-tRNA synthetase. The equilibrium constant for the formation of tyrosyl adenylate in solution ($K_{eq(solution)}$ = [Tyr-AMP][PP$_i$]/ [Tyr][ATP]) is related to the rate and equilibrium constants for the enzymatic reaction illustrated in Figure 15.21 by equation 15.8.

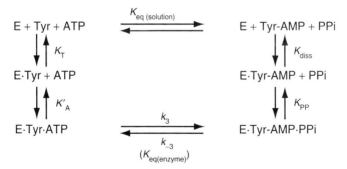

Figure 15.21 Thermodynamic cycle that relates the equilibrium between Tyr and Tyr-AMP + PP$_1$ in solution with the analogous equilibrium between enzyme-bound reagents (the "internal equilibrium").

$$K_{eq(solution)} = \frac{k_3/K_T K_A'}{k_{-3}/K_{PP}K_{diss}} \qquad (15.8)$$

The dissociation constants K_T, K_A', and K_{PP}, and the rate constants k_3 and k_{-3} (defined in scheme 2), were measured as described under scheme 2 by routine procedures. The dissociation constant of the E·Tyr-AMP complex, K_{diss} (scheme 4) is difficult to measure for wild-type enzyme because the value is too low. However, K_{diss} was measured directly using mutant enzymes that had values in an experimentally accessible range.

PART 2

Redesigning an enzyme: Subtilisin

A. Subtilisin

Much of our understanding of basic enzymology comes from studies on the serine proteases (Chapter 1, sections D2 and F2), notably on chymotrypsin and trypsin, because of their ready isolation in large quantities from pancreases from slaughterhouses. But protein engineering pushed the more humble subtilisin BPN' from *Bacillus amyloliquefaciens* to the fore for two reasons. First, because it is a bacterial protein without disulfide crosslinks, it can be easily expressed in very high yields from *E. coli*, whereas the pancreatic enzymes are much more difficult to handle. Second, subtilisin is manufactured by the ton because it is used in soap powders and so is commercially important. Consequently, much of the research on subtilisin has been performed in biotech companies, and subtilisin has been used to pioneer many of the procedures currently used in protein engineering. The properties of the single-chain protease of $M_r 27\,500$ were introduced in Chapter 1. This protease was cloned in and expressed from *B. subtilis* in 1983, with a yield of about 1 gram per liter.[33] It has an active-site triad consisting of Asp-32, His-64, and Ser-221 (Figure 15.22), and it catalyzes the hydrolysis of peptides and esters by the same acylenzyme mechanism as for chymotrypsin (Chapter 1, section F2; Chapter 16, section B). It is important to note that subtilisin differs from the pancreatic enzymes by having a shallow binding groove on the surface, rather than the deep binding pocket of the pancreatic enzymes, to which it is related by convergent evolution (Chapter 1, section D3).

B. Dissection of the catalytic triad and the oxyanion binding site

There are formidable problems to be overcome in analyzing mutants of subtilisin that have had their active sites disrupted. Subtilisin is synthesized as a membrane-associated precursor, preprosubtilisin, that is released by autoproteolytic cleavage.

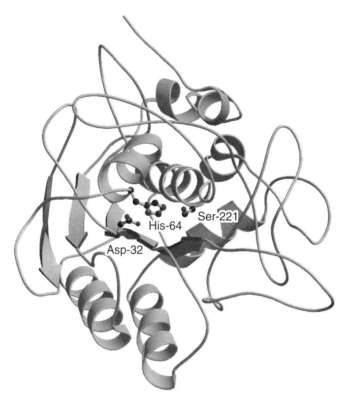

Figure 15.22 Subtilisin, illustrating the position of the active-site triad (drawn with MolScript).

Weakly active mutants were successfully expressed by co-culturing the mutants with a small amount of a "helper" of *B. subtilis* containing wild-type protease.[34] The small amounts of highly active wild-type protease, plus any other endogenous proteases, will mask residual activity in debilitated mutants. This problem was overcome by introducing a surface mutation, Ser → Cys-24. The thiol group of the Cys binds reversibly to an activated thiol sepharose column, and so mutants may be selectively fished out from contaminants.[34]

Wild-type subtilisin BPN′ with the mutation Ser → Cys-24 has a k_{cat} value of 59 s^{-1} and a K_M value of 200 μM with the synthetic substrate N-succinyl-Ala-Ala-Pro-Phe *p*-nitroanilide, compared with a rate constant of 1.1×10^{-8} s^{-1} for its spontaneous hydrolysis under the same conditions. Replacement of Asp-32, His-64, and Ser-221 one at a time by alanine reduced the value of k_{cat} by factors of 3×10^4, 2×10^6, and 2×10^6, respectively. Converting all three to alanine also decreases activity by 2×10^6. The value of K_M increases only by a factor of two on all these mutations.[34] It is unlikely that the residual activity results from the presence of a small amount of wild-type active site in the thiol mutants because of errors in protein synthesis, since they are resistant to inhibition by phenylmethanesulfonyl fluoride, an irreversible inhibitor of wild-type protein. There is thus a residual activity of k_{cat} (3×10^{-5} s^{-1}) that is about a factor of 10^3 above background on disruption of the triad.

Such data must be interpreted with caution for the reasons discussed in Part 1, section G, which are amplified as follows. Removal of the hydroxyl of Ser-221 implies that the reaction mechanism must change from the acylenzyme mechanism to the direct nucleophilic attack of water on the substrate. Thus, the residual activity of the Ser → Ala-221 and triple mutants implies that differential binding interactions between the ground and transition states are worth *at least* a factor of 10^3 in rate. This is the lowest estimate, because it is possible that the presence of the remaining groups in the active site after removal of the serine —OH group inhibits the attack of water on the substrate relative to the situation where there is unrestricted access of the nucleophilic water molecule and any other solvent molecules that could act as general acid-base catalysts. Conversely, the 2×10^6-fold loss of activity on mutation of Ser → Ala-221 or in the triple mutant is an *upper* limit of the importance of the catalytic triad because the resultant mutants may have features that inhibit the attack of water. The hydroxyl of Ser-221 is retained on mutation of Asp → Ala-32 or His → Ala-64 and so the acylenzyme mechanism probably still operates. The factor of 3×10^4 on Asp → Ala-32 is also an *upper* estimate of the intrinsic effect of Asp-32 on catalysis relative to the situation where there is complete freedom of access of water to solvate the imidazole of His-64 as it picks up a proton and becomes positively charged during catalysis.

The oxyanion of the tetrahedral intermediate formed by the attack of Ser-221 on the substrate is solvated by two NH groups of the enzyme, as described for chymotrypsin in Chapter 1, section F2. Whereas the pancreatic enzymes use backbone groups, one of the NH groups used by subtilisin is the —CONH$_2$ side chain of Asn-155, which can, of course be mutated. There is a loss of activity of 10^2- to 10^3-fold on mutation.[35,36] These data are again difficult to interpret, as explained in Chapter 11, section C1e, and Part 1, section G, of this chapter: buried charges that are unsolvated are very unstable, and so it is the instability of the mutant that dominates the energetics. Unless there is access of water to the site of mutation, structures tend to rearrange to allow solvation of buried charges.

C. Redesigning specificity

1. Subsites

a. *Primary specificity site*

The major determinant for specificity is the S$_1$ subsite (see Schechter-Berger notation of Chapter 1, Figure 1.27), which is hydrophobic and binds the side chain of residue P$_1$ of the substrate. The highest activity of wild-type subtilisin is toward tyrosine as P_1, with phenylalanine, methionine, and alanine progressively less active.[37] Gly-166 resides at the back of the pocket for the side chain of P_1. Progressively increasing the size of the side chain of residue 166 progressively decreases the activity toward tyrosine in P_1. Conversely, the activity toward smaller P_1 side chains increases, peaks, and then decreases with increasing size of residue 166.[37]

b. S_4 specificity site

Substrates of the form succinyl-X-Ala-Pro-Phe p-nitroanilide, where P_4 (= X) is a hydrophobic residue, have similar values of k_{cat}/K_M. The residue at the bottom of the S_4 site is a large hydrophobic, Ile-107. Mutating Ile → Gly-107 to increase the size of the pocket decreases k_{cat}/K_M for X = Ala some 200-fold, with only small effects for X = Phe, Ile, or Leu. The larger side chains complement the lesion caused by introducing a cavity in the protein.[38,39]

c. Converting the specificity of subtilisin BPN' to that of furin

Mutations in the S_1 and S_4 subsites have been combined in a modular manner to provide a dramatic change of specificity. Subtilisin BPN' was engineered to cleave proteins after a run of three basic amino acids in a manner resembling the substrate specificity of furin,[40] one of the mammalian subtilisin homologs that processes prohormones.[41] The double mutant of subtilisin BPN' (N62D/G166D) has substantial preference for cleaving after sequences having basic residues at the P_1 and P_2 positions. Additional specificity for basic residues was engineered at the P_4 position by introducing residues that occur in furin into the S_4 subsite (Tyr-104, Ile-107, and Leu-126) of the double mutant of subtilisin.[41] The product, called "furilisin," efficiently catalyzes the hydrolysis of succinyl-Arg-Ala-Lys-Arg p-nitroanilide or succinyl-Arg-Ala-Lys-Arg p-nitroanilide. It discriminates in favor of Arg versus Ala at the P_4 position by a factor of 360. The tribasic substrates are cleaved 60,000 times faster than succinyl-Ala-Ala-Pro-Phe p-nitroanilide. The engineered enzyme may be useful as a protein processing tool.

2. Subtiloligase

One of the achievements of classical protein chemistry was to convert the active-site serine residues of chymotrypsin[42] and subtilisin[43] to cysteines by a series of chemical reactions. Thiolsubtilisin had been used as a peptide ligase to link peptides,[44] using the chemistry described in Chapter 7, section B3: a peptide with an esterified C-terminal forms an acylenzyme with the protease that partitions between hydrolysis and aminolysis products in the presence of another peptide. A double mutant of subtilisin, Ser → Cys-221, Pro → Ala-225, prepared by protein engineering,[45] has 10-fold higher peptide ligase activity and at least 100-fold lower peptidase activity than the singly mutated thiolsubtilisin. The Pro → Ala-225 mutation partly relieves the steric crowding from the Ser → Cys-221 substitution. Further mutations were made in the S_1 site to give a family of enzymes of differing specificity. These have been successfully used to synthesize ribonuclease A.[46]

D. Engineering of stability and other properties

The general problem of engineering stability of proteins is discussed in Chapter 17. Subtilisin has been a test bed for some of the procedures.[47] One important procedure that was pioneered on subtilisin involves examining the structure of homologous thermophilic enzymes and transplanting residues from the thermophile into the mesophile. By this means, a large increase was engineered into subtilisin BPN′.[48] But subtilisin has specific problems in its use as a detergent enzyme. Washing powders tend to contain bleach, which inactivates wild-type subtilisin by oxidation. One susceptible residue is Met-222, adjacent to the active site Ser-221. Substitution by other amino acids led to less active but more oxidation-resistant enzymes.[49] Another problem is subtilism's susceptibility to autolysis. The primary autolysis site in subtilisin J was identified and mutated to render it less susceptible to self-cleavage.[50,51]

The surface charges of subtilisin have also been engineered to alter its pH dependence (described in Chapter 5, Figure 5.4)[52] and specificity toward charged substrates[53] in a rational manner. These data have been used to benchmark the calculation of long-range electrostatic interactions.[54,55]

Further examples of the useful redesign of specificity and the use of mutagenesis to study mechanism are given in Chapter 16. An ingenious example of the engineering of a novel activity into an existing protein has been the conversion of a peptidyl-proline isomerase, cyclophilin, into a protease.[56] These proteins catalyze the *cis-trans* isomerization of peptidyl-proline bonds (see Chapter 17 and Chapter 19, section I1). The mechanism does not involve acid-base or nucleophilic catalysis but probably results from the enzyme being complementary in structure to a peptide bond that has been twisted so that it has lost its delocalization energy (Chapter 1, section C1a).[57] A peptide bond that has lost its delocalization energy would also be activated to nucleophilic attack. Engineering an Asp-His-Ser triad into cyclophilin gave a protease, "cyproase 1," that has a k_{cat} value of 4 s^{-1}—10^8 times higher than the spontaneous rate constant for the hydrolysis of the substrate L-Ala-L-Pro—and a K_M value of 2.4 mM.[56] This is a very respectable protease.

References

1. S. J. Gardell, C. S. Craik, D. Hilvert, M. S. Urdea, and W. J. Rutter, *Nature, Lond.* **317**, 551 (1985).
2. P. T. Jones, P. H. Dear, J. Foote, M. S. Neuberger, and G. Winter, *Nature, Lond* **321**, 522 (1986).
3. G. Winter, A. R. Fersht, A. J. Wilkinson, M. Zoller, and M. Smith, *Nature, Lond* **299**, 756 (1982).

4. E. A. First, in *Comprehensive biological catalysis*, vol. 1, M. Sinnott, ed., Academic Press, 573 (1998).
5. T. N. Bhat, D. M. Blow, P. Brick, and J. Nyborg, *J. Molec. Biol.* **158**, 699 (1982).
6. A. R. Fersht and R. Jakes, *Biochemistry* **14**, 3350 (1975).
7. G. Lowe and G. Tansley, *Tetrahedron* **40**, 113 (1984).
8. P. Brick, T. N. Bhat, and D. M Blow, *J. Molec. Biol.* **208**, 83 (1989).
9. W. Ward and A. R. Fersht, *Biochemistry* **27**, 5525 (1988).
10. P. Schimmel, *Accts Chem. Res.* **22**, 232 (1989).
11. A. R. Fersht, A. Matouschek, and L. Serrano, *J. Molec. Biol.* **224,** 771 (1992).
12. T. N. C. Wells and A. R. Fersht, *Biochemistry* **25**, 1881 (1986).
13. R. J. Leatherbarrow, A. R. Fersht, and G. Winter, *Proc. Natl. Acad. Sci. USA* **82**, 7840 (1985).
14. A. R. Fersht, J. W. Knill-Jones, H. Bedouelle, and G. Winter, *Biochemistry* **27**, 1581 (1988).
15. E. A. First and A. R. Fersht, *Biochemistry* **34**, 5030 (1995).
16. A. R. Fersht, *Biochemistry* **14**, 5 (1975).
17. J. J. Perona, M. A. Rould, and T. A. Steitz, *Biochemistry* **32**, 8758 (1993).
18. G. Eriani, J. Cavarelli, F. Martin, L. Ador, B. Rees, J. C. Thierry, J. Gangloff, and D. Moras, *J. Molec. Evoln.* **40**, 499 (1995).
19. A. R. Fersht, *Biochemistry* **27**, 1577 (1988).
20. A. R. Fersht, A. J. Wilkinson, P. Carter, and G. Winter, *Biochemistry* **24**, 5858 (1985).
21. C. K. Ho and A. R. Fersht, *Biochemistry* **25**, 1891 (1986).
22. W. J. Albery and J. R. Knowles, *Biochemistry* **15**, 5631 (1976).
23. M. D. Jones, D. M. Lowe, T. Borgford, and A. R. Fersht, *Biochemistry* **25**, 1887 (1986).
24. J. M. Avis and A. R. Fersht, *Biochemistry* **32**, 5321 (1993).
25. A. R. Fersht, R. J. Leatherbarrow, and T. N. C. Wells, *Nature* **322**, 284 (1986).
26. D. A. Estell and A. R. Fersht, *Protein Engineering* **6,** 441, 442, 445, 446 (1988).
27. T. N. C. Wells and A. R. Fersht, *Biochemistry* **28**, 9201 (1989).
28. W. Ward and A. R. Fersht, *Biochemistry* **27**, 1041 (1988).
29. M. Waye, G. Winter, A. J. Wilkinson, and A. R. Fersht, *EMBO J.* **2**, 1827 (1983).
30. P. Carter, H. Bedouelle, and G. Winter, *Proc. Natl. Acad. Sci. USA* **83**, 1189 (1986).
31. H. Bedouelle and G. Winter, *Nature, Lond.* **320**, 371 (1986).
32. T. N. C. Wells, C. K. Ho, and A. R. Fersht, *Biochemistry* **25**, 6603 (1986).
33. J. A. Wells, E. Ferrari, D. J. Henner, D. A. Estell, and E. Y. Chen, *Nucleic Acids Research* **11**, 7911 (1983).
34. P. Carter and J. A. Wells, *Nature, Lond.* **332**, 564 (1988).
35. J. A. Wells, B. C. Cunningham, T. P. Graycar, and D. A. Estell, *Phil. Trans. R. Soc. A* **317**, 415 (1986).
36. P. Bryan, M. W. Pantoliano, S. G. Quill, H. Y. Hsiao, and T. Poulos, *Proc. Natl. Acad. Sci. USA* **83**, 3743 (1986).
37. D. A. Estell, T. P. Graycar, J. V. Miller, D. B. Powers, J. P. Burnier, P. G. Ng, and J. A. Wells, *Science* **233**, 659 (1986).
38. M. Rheinnecker, G. Baker, J. Eder, and A. R. Fersht, *Biochemistry* **32**, 1199 (1993).
39. M. Rheinnecker, J. Eder, P. S. Pandey, and A. R. Fersht, *Biochemistry* **33**, 221 (1994).

40. D. J. Matthews, L. J. Goodman, C. M. Gorman, and J. A. Wells, *Protein Science* **3**, 1197 (1994).
41. M. D. Ballinger, J. Tom, and J. A. Wells, *Biochemistry* **35**, 13579 (1996).
42. L. Polgar and M. L. Bender, *J. Am. Chem. Soc.* **88**, (1966).
43. K. Neet and D. Koshland, *Proc. Natl. Acad. Sci. USA* **56**, 1606 (1966).
44. T. Nakatsuka, T. Sasaki, and E. T. Kaiser, *J. Am. Chem. Soc.* **109**, 3808 (1987).
45. L. Abrahmsen, J. Tom, J. Burnier, K. A. Butcher, A. Kossiakoff, and J. A. Wells, *Biochemistry* **30**, 4151 (1991).
46. D. Y. Jackson, J. Burnier, C. Quan, M. Stanley, J. Tom, and J. A. Wells, *Science* **266**, 243 (1994).
47. M. L. Rollence, D. Filpula, M. W. Pantoliano, and P. N. Bryan, *Critical Reviews in Biotechnology* **8**, 217 (1988).
48. M. W. Pantoliano, M. Whitlow, J. F. Wood, S. W. Dodd, K. D. Hardman, M. L. Rollence, and P. N. Bryan, *Biochemistry* **28**, 7205 (1989).
49. D. A. Estell, T. P. Graycar, and J. A. Wells, *J. Biol. Chem.* **260**, 6518 (1985).
50. J. S. Jang, D. K. Park, M. J. Chun, and S. M. Byun, *Biochim. Biophys. Acta* **1162**, 1 (1993).
51. K. H. Bae, J. S. Jang, K. S. Park, S. H. Lee, and S. M. Byun, *Biochem. Biophys. Res. Comm.* **207**, 20 (1995).
52. P. G. Thomas, A. J. Russell, and A. R. Fersht, *Nature, Lond.* **318**, 375 (1986).
53. A. J. Russell and A. R. Fersht, *Nature, Lond.* **328**, 496 (1987).
54. M. J. E. Sternberg, F. R. Hayes, A. J. Russell, P. G. Thomas, and A. R. Fersht, *Nature* **330**, 86 (1987).
55. M. K. Gilson and B. A. Honig, *Nature, Lond.* **330**, 184 (1987).
56. E. Quéméneur, M. Moutiez, J.-P. Charbonnier, and A. Ménez, *Nature, Lond.* **391**, 301 (1998).
57. A. Galat and S. M. Metcalfe, *Prog. Biophys. Mol. Biol.* **63**, 69 (1995).

Case Studies of Enzyme Structure and Mechanism

16

Twenty years ago, it was possible to present a detailed analysis of the structures and mechanisms of all the enzymes whose structures had been solved at high resolution by x-ray crystallography. This is no longer possible in a short text, and the reader is referred to *Comprehensive Biological Catalysis* (M. Sinnott, ed., Academic Press, San Diego, 1998), which attempts to do this in four large volumes, with each section written by experts. It is not just the vastness of structural and kinetic detail that prevents a short overview—there are unresolved controversies over individual mechanisms. These controversies tend to be about enzymes that do not involve detectable intermediates in their reaction pathways. It can be very difficult to distinguish between a mechanism that involves a transient covalent intermediate and one that does not. Further, when covalent intermediates are not involved, there are usually several mechanisms that fit the available kinetic data. Here, we examine a few selected examples that have been of seminal importance in the development of modern enzymology and whose mechanisms are relatively uncontroversial.

Classically, an enzyme mechanism has been developed using the following procedure. The protein was isolated; its kinetic properties were studied to work out the basic chemical mechanism and the groups involved in catalysis; the amino acid sequence was determined; and the crystal structure was solved, crowning the years of intense work and providing unexpected twists. Subsequently, the gene was cloned, which enabled site-directed mutagenesis and biophysical measurements on wild-type protein and mutants to add the final touches. Apart from sequencing the protein via its gene, this is still the procedure for many complex proteins that are discovered from biological function and are not readily crystallized. Two classes of proteins are used to exemplify the classical procedures in depth: the nucleotide-dependent dehydrogenases and the serine proteases. The mechanisms of some other proteases are briefly described to illustrate that nature has many ways of achieving the same ends. The ribonucleases are dealt with in less detail.

The preferred modern procedure for investigating an enzyme mechanism is to identify a protein in a genome sequence, clone and express it, solve the crystal and/or solution structures, guess the mechanism, and then confirm it by site-directed mutagenesis and kinetics. The mechanism of lysozyme almost illustrates that procedure. Although the classical protein chemistry on lysozyme predated the molecular genetics, the kinetic studies were stimulated by the x-ray crystallography, and the previously unknown mechanism almost leapt from the crystal structure.

A. The dehydrogenases[1]

The dehydrogenases discussed in this section catalyze the oxidation of alcohols to carbonyl compounds. They utilize either NAD^+ or $NADP^+$ as coenzymes. The complex of the enzyme and coenzyme is termed the *holo*enzyme; the free enzyme is called the *apo*enzyme. Some dehydrogenases are specific for just one of the coenzymes; a few use both. The reactions are readily reversible, so that carbonyl compounds may be reduced by NADH or NADPH. The rates of reaction in either direction are conveniently measured by the appearance or disappearance of the reduced coenzyme, since it has a characteristic ultraviolet absorbance at 340 nm. The reduced coenzymes also fluoresce when they are excited at 340 nm, which provides an even more sensitive means of assay.

The chemistry of the reduction of NAD^+ has been solved most elegantly (Chapter 8, section B1).[2] Oxidation of the alcohol involves the removal of two hydrogen atoms. One is transferred directly to the 4 position of the nicotinamide ring of the NAD^+, and the other is released as a proton (equation 16.1).[3,4] It is generally thought that the hydrogen is transferred as a hydride ion H^-, but a radical intermediate cannot be ruled out. For convenience, we shall assume that the mechanism is the hydride transfer.

$$(16.1)$$

The transfer is also stereospecific. Through the use of deuterated substrates, it has been found that some dehydrogenases will transfer to one side of the ring, and other enzymes to the opposite side (structures 16.2).

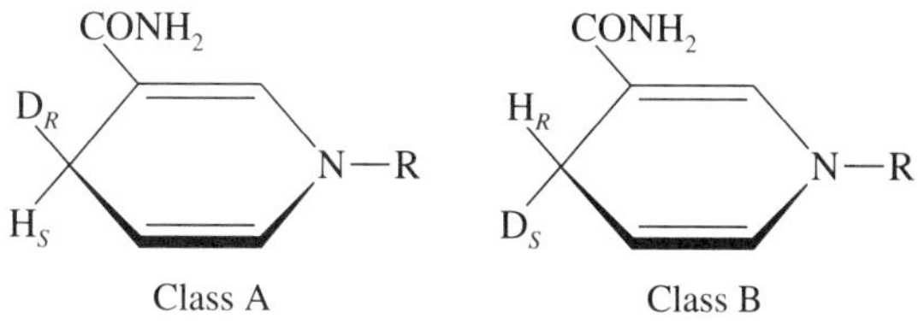

$$\text{(16.2)}$$

Class A   Class B

The enzymes have been classified as "A" or "B" on this basis (Table 16.1). Class A transfers the pro-R hydrogen from NADH, whereas Class B transfers the pro-S. The crystal structures of dehydrogenases solved so far show that NAD$^+$ is bound to the Class A enzymes with the nicotinamide ring in the anti conformation about the glycosidic bond, as is consistent with NMR studies in solution,[5] and that the Class B enzymes have the ring bound in the syn conformation.

A rationale for the two classes has been proposed, based on the observation that Class A enzymes catalyze the reduction of the more reactive carbonyls, while Class B is associated with the less reactive ones.[6-8] NADH is a weaker reducing agent when it is in the syn conformation (Class A) than when it is anti (Class B). This would tend to balance the equilibrium constant between enzyme-bound oxidized and reduced reagents, i.e., the internal equilibrium constant of Chapter 12, section E2, with the consequent rate advantages discussed in that section.

The structures of several dehydrogenases have now been solved. The work on these has been reviewed in depth in the literature, as have their physical and kinetic properties.[1,9,10] Some generalizations can be made. As was discussed in Chapter 1, section D6, the subunits may be divided into two domains: a catalytic domain, which can be quite variable in structure, and a nucleotide-binding domain, which is formed from a similar overall folding of the polypeptide chain for all the dehydrogenases. The detailed geometry of the nucleotide-binding

Table 16.1   *The coenzyme specificity of some dehydrogenases*

Dehydrogenase	Coenzyme required	Stereospecificity class
Glutamate	NAD$^+$ or NADP$^+$	B
Glucose 6-phosphate	NADP$^+$	B
3-Glycerol phosphate	NAD$^+$	B
Glyceraldehyde 3-phosphate	NAD$^+$	B
Malate (soluble)	NAD$^+$	A
Alcohol	NAD$^+$	A
Lactate	NAD$^+$	A
Isocitrate	NADP$^+$	A

domain varies considerably from one enzyme to another. However, the coenzyme binds in a similar extended, open conformation in all cases (Figure 16.1). The most significant variation concerns which side of the nicotinamide ring faces the substrate, with the side in the Class A enzymes being opposite to that in the Class B.

1. The alcohol dehydrogenases[1,10]

The alcohol dehydrogenases are zinc metalloenzymes of broad specificity. They oxidize a wide range of aliphatic and aromatic alcohols to their corresponding aldehydes and ketones, using NAD$^+$ as a coenzyme (see equation 16.1). The two most studied enzymes are those from yeast and horse liver. The crystal structures of the horse liver apo- and holoenzymes were solved at 2.4 and 2.9 Å (0.24 and 0.29 nm), respectively.[11,12] The molecule is a symmetrical dimer, composed of two identical chains of M_r 40 000. Each chain contains one binding site for NAD$^+$ but two sites for Zn^{2+}. Only one of the zinc ions is directly concerned

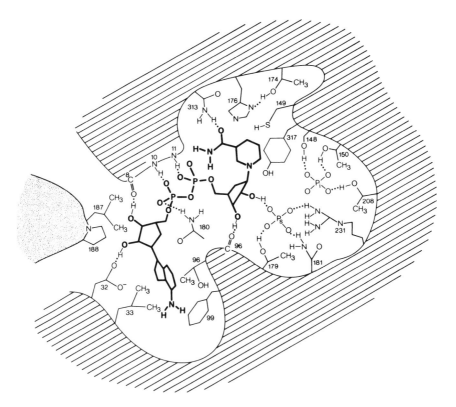

Figure 16.1 The binding of NAD$^+$ to glyceraldehyde 3-phosphate dehydrogenase from *Bacillus stearothermophilus*. [From G. Biesecker, J. I. Harris, J. C. Thierry, J. E. Walker, and A. J. Wonacott, *Nature, Lond.* **266**, 328 (1977).]

with catalysis. The yeast enzyme, on the other hand, is a tetramer of M_r 145 000, and each chain binds one NAD^+ and one Zn^{2+}. Despite these differences, sequence data indicate a good deal of identity between the two enzymes.[13] It is often assumed that the same overall reaction mechanism holds for both enzymes, although details, such as the rate-determining step and the pH dependence, differ between the two.

The liver enzyme exists in two forms, E and S, which differ only by some six residues in their amino acid sequence.[13] Only the S is active toward 3-β-hydroxysteroids, but both forms are active toward ethanol. None of the known amino acid differences is located in the subunit interfaces. Accordingly, E and S chains combine in statistical ratios to form EE, SS, and ES dimers. These different species are termed *isozymes,* which means that they are multiple molecular forms of the same enzyme. When it is isolated from liver, the enzyme consists of about 40 to 60% of the EE dimer, the remainder being SS and ES.

a.   The structure of the active site of liver alcohol dehydrogenase[14]

The Zn^{2+} ion sits at the bottom of a hydrophobic pocket formed at the junction of the catalytic and nucleotide-binding domains (Figure 16.2). It is ligated by the sulfur atoms of Cys-46 and Cys-174, and by a nitrogen atom of His-67. The fourth ligand is an ionizable water molecule that is hydrogen-bonded to the hydroxyl group of Ser-48. It is known from the pH dependence of the binding of NAD^+,[15] from the direct determination of the proton release on the binding of NAD^+,[16] and from other kinetic procedures[17] that the apoenzyme has a functional group of $pK_a \sim 9.2$ that is perturbed to a pK_a of ~ 7.6 in the holoenzyme.

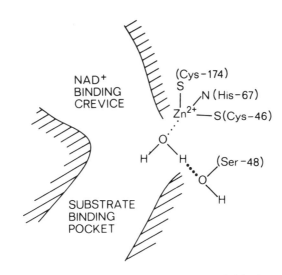

Figure 16.2   Sketch of the active site of horse liver alcohol dehydrogenase. [Courtesy of C.-I. Brändén.]

Crystallographic studies suggest that this is the ionization of the zinc-bound water molecule. The nicotinamide ring of the NAD^+ is bound close to the zinc ion at the bottom of the pocket. The 2′-hydroxyl of the ribose ring is located between the hydroxyl of Ser-48 and the imidazole ring of His-51, within hydrogen-bonding distance of both.

b. The structure of the enzyme–substrate (ternary) complex

The structure of the *reactive* ternary complex composed of the enzyme NAD^+, and 4-bromobenzyl alcohol was solved at 2.9-Å resolution.[14] This was possible because there is a favorable equilibrium between this complex and the enzyme-bound ternary complex reaction product of NADH and 4-bromo-benzaldehyde. The structure of the unreactive product-like complex composed of the enzyme, the coenzyme analogue H_2NADH (i.e., NAD^+ in which the nicotinamide ring has been reduced to 1,4,5,6-tetrahydronicotinamide), and *trans*-4-(*N,N*-dimethylamino)-cinnamaldehyde (DACA) was also solved at the same resolution.[18]

H₂NADH   DACA
                (chromophoric)

In both examples, the relevant oxygen atom of the substrate is directly coordinated to the zinc ion, displacing the bound water molecule so that the metal remains tetracoordinated (Figure 16.3). The hydrophobic side chain of the substrate binds in the hydrophobic "barrel" of the pocket. Both observations are consistent with earlier solution studies using model building,[19] spectroscopy,[20] and kinetics:[21] the binding of the carbonyl oxygen of the chromophoric cinnamaldehyde (DACA) to a positively charged center in the enzyme was indicated by a characteristic shift in the spectrum. (The binding of substituted benzaldehydes is enhanced by electron-donating substituents.)

A major point of contention is the ionization state of the bound alcohol: Is it bound as the alcoholate anion[7,22] or as the neutral alcohol? The position of the proton cannot, of course, be located in the x-ray diffraction studies. Evidence in support of the alcoholate anion being bound comes from the pH dependence of the binding of substituted alcohols.[23] The complexes of the holoenzyme with various alcohols have the following pK_a values: trifluorethanol, 4.3; 2,2-dichloroethanol, 4.5; 2-chloroethanol, 5.4; ethanol, 6.4. These values are some 8 to 9 units below the values of the alcohols in solution. If they represent the ionization of the alcohols in the ternary complexes, then complexing to the enzyme has dramatically lowered the pK_a's. This does seem to have happened, since there is a very good correlation of the pK_a's of the ternary complexes with those of the alcohols in solution (the Brønsted coefficient = 0.6). Kinetic isotope effects are also consistent with the hydride ion transfer taking place from an alcoholate ion (equation 16.3).[24]

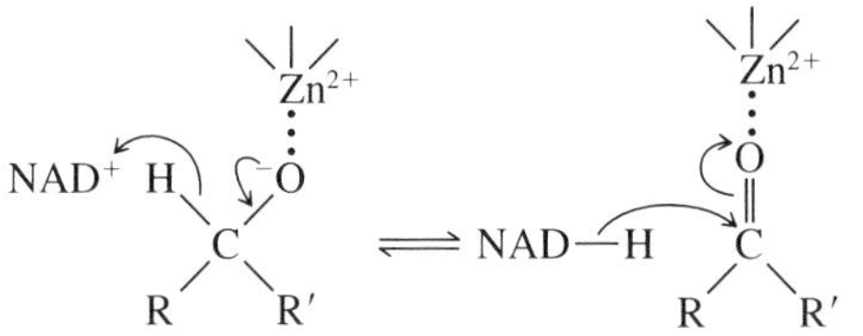

$$(16.3)$$

Interestingly, although the substrate is bound in the pocket in a water-free environment, there is a system of hydrogen bonds that can shuttle the proton from the bound alcohol to the surface (equation 16.4).[17]

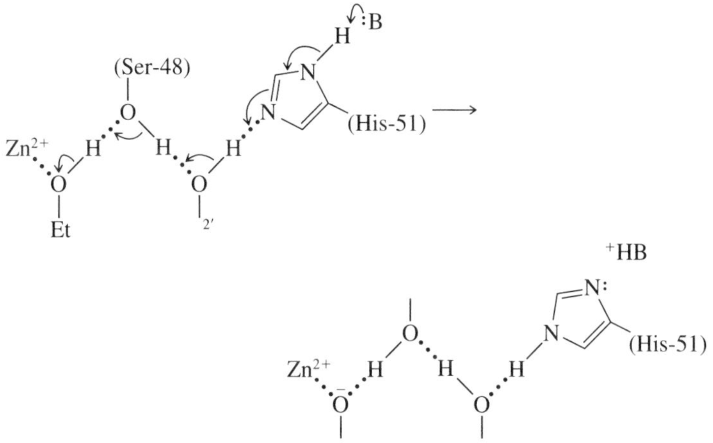

$$(16.4)$$

This is composed of the previously mentioned residues of Ser-48 and His-51 (which is on the surface), and the bridging $2'$-hydroxyl of the coenzyme ribose ring.[25]

c. The kinetic mechanism

The steady state and stopped-flow kinetic studies on the horse liver enzyme are now considered "classic" experiments. They have shown that the oxidation of alcohols is an ordered mechanism, with the coenzyme binding first and the dissociation of the enzyme-NADH complex being rate-determining.[15,26,27] Both the transient state and steady state methods have detected that the initially formed enzyme-NAD$^+$ complex isomerizes to a second complex.[27,28] In the reverse reaction, the reduction of aromatic aldehydes involves rate-determining dissociation of the enzyme–alcohol complex,[27,29] whereas the reduction of acetaldehyde is limited by the chemical step of hydride transfer.

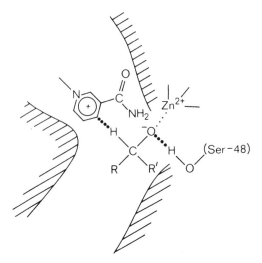

Figure 16.3 A proposed model for the productively bound ternary complex of horse liver alcohol dehydrogenase. It is suggested that the ionized alcohol displaces the zinc-bound water molecule shown in Figure 16.2. [Courtesy of C.-I. Brändén.]

A possible structural basis for the ordered kinetic mechanism of coenzyme binding before alcohol is the need for the $2'$-hydroxyl of the ribose in the proton shuttle of equation 16.4.[14,30] Comparison of the structures of apo- and holoenzymes reveals extensive conformational changes on coenzyme binding.[12] The induced conformational changes are probably related to the kinetically detected isomerization, and may be another factor that contributes to ordered binding.

The enzyme–product complexes of the yeast enzyme dissociate rapidly so that the chemical steps are rate-determining.[31] This permits the measurement of kinetic isotope effects on the chemical steps of this reaction from the steady state kinetics. It is found that the oxidation of deuterated alcohols RCD_2OH and the reduction of benzaldehydes by deuterated NADH (i.e., NADD) are significantly slower than the reactions with the normal isotope ($k_H/k_D = 3$ to 5).[21,31] This shows that hydride (or deuteride) transfer occurs in the rate-determining step of the reaction. The rate constants of the hydride transfer steps for the horse liver enzyme have been measured from pre–steady state kinetics and found to give the same isotope effects.[32,33] Kinetic and kinetic isotope effect data are reviewed in reference 34 and the effects of quantum mechanical tunneling in reference 35.

d. Site-site interactions

Although the liver enzyme binds two moles of NAD^+ with equal affinity, it has been claimed that only one site of the enzyme reacts during the turnover of the enzyme.[29] Structural[14] and NMR[36] work indicate that both subunits are equivalent when substrates are bound. Kinetic data indicate that both sites do react simultaneously.[25,32,37–40] Stopped-flow studies show that the reaction course is biphasic during the turnover of the enzyme, an observation that at first sight appears to confirm that the two sites are nonequivalent. But this is just a natural consequence of the kinetics of an ordered ternary complex mechanism involving a single site:[24,39,40] the appearance or disappearance of the chromophore during the transient state is linked to several equilibria, so the complex kinetics of Chapter 4, sections D2 and D6, apply. This analysis stems from earlier studies on lactate dehydrogenase; these are discussed in the next section.

e.   Changing specificity by protein engineering

The substrate binding pocket of the horse liver enzyme is defined by Ser-48, Leu-57, and Phe-93. The pocket in the yeast enzyme is much narrower because it has the bulkier Thr-48, Trp-57, and Trp-94. Wild-type yeast enzyme is highly active toward the smaller substrate of ethanol and weakly active with hexanol. Mutating Trp $\rightarrow$ Ala-94 reverses the specificity by increasing the activity with hexanol fivefold and decreasing the activity with ethanol 350-fold.[41] Protein engineering has been used to broaden the specificity of the yeast enzyme. For example, comparison with the horse liver enzyme suggests that Trp-54 in the homologous yeast alcohol dehydrogenase prevents it from efficiently catalyzing the oxidation of long-chain primary alcohols that are branched at the 4 position. Mutating Trp $\rightarrow$ Leu-54 yields an enzyme that serves as an effective catalyst for both longer straight-chain primary alcohols and branched-chain alcohols.[42]

A neat use of protein engineering for mechanistic studies was the alteration of the substrate affinity of the horse liver enzyme to study kinetic isotope effects (Chapter 2, section G).[43] The release of product benzaldehyde is slow and partly rate-determining. Placing bulky residues in the binding pocket speeds up release because of steric repulsion. The kinetic isotope effects on the chemical steps could be measured in the mutant, and hydrogen tunneling was thus "unmasked."[43]

2.   L-Lactate and L-malate dehydrogenases[1,6,8,10,44,45]

L-Lactate dehydrogenase catalyzes the reversible oxidation of L-lactate to pyruvate, using NAD^+ as a coenzyme (equation 16.5).

$$
\begin{array}{ccc}
\text{CH}_3 & & \text{CH}_3 \\
| & & | \\
\text{H} - \text{C} - \text{OH} + \text{NAD}^+ \rightleftharpoons & \text{C} = \text{O} + \text{NADH} + \text{H}^+ \\
| & & | \\
\text{CO}_2^- & & \text{CO}_2^-
\end{array}
\qquad (16.5)
$$

The enzyme, which has been isolated from many species, is a tetramer of M_r 140 000. The two forms of the enzyme, the H_4, predominating in heart muscle, and the M_4, predominating in skeletal muscle, give rise to a family of isozymes.[46,47] The amino acid composition of the polypeptide chain that constitutes the M_4 form is significantly different from that of the H_4 form, and the two have different kinetic properties. Despite this, the sites for the association of the subunits had to be very similar, since the hybrids M_3H, M_2H_2, and MH_3 occur in the expected statistical distribution.[48] There are no subunit interactions in catalysis, so the kinetic properties of, say, the H_3M form are identical to those expected for a $3:1$ mixture of H_4 and M_4. The crystal structure of the apoenzyme from dogfish was solved at 2.0-Å resolution, and its complex with substrate analogues at 2.7 to 2.8 Å.[47–51] The molecule is symmetrical and its subunits are structurally equivalent.

L-Malate dehydrogenase catalyzes the reversible oxidation of malate to oxaloacetate, using NAD^+ as a coenzyme (equation 16.6).

$$
\begin{array}{c}
CO_2^- \\
|\\
H-C-OH \\
|\\
CH_2CO_2^-
\end{array}
\ +\ NAD^+ \ \rightleftharpoons\
\begin{array}{c}
CO_2^- \\
|\\
C=O \\
|\\
CH_2CO_2^-
\end{array}
\ +\ NADH \ +\ H^+
\tag{16.6}
$$

Malate dehydrogenase is structurally and mechanistically very similar to the lactate enzyme.

a. The structure of the enzyme–substrate complex

The structures of the ternary complexes of lactate dehydrogenase were deduced from crystallographic studies of the binding of NAD-pyruvate,[50] a covalent analogue of pyruvate plus NADH,[52] and, subsequently, from the binding of the more realistic analogue, S-lac-NAD$^+$.[51]

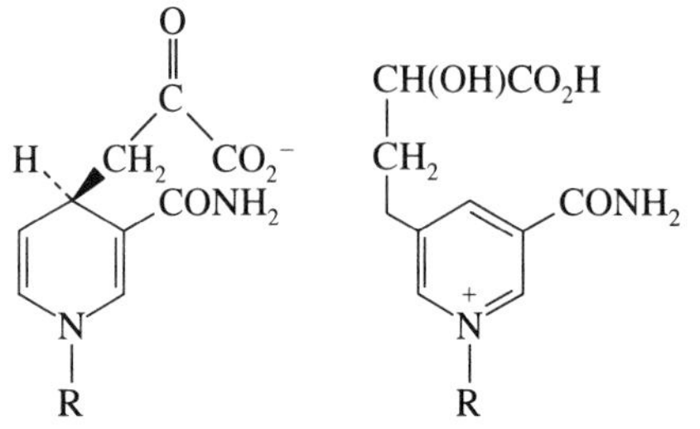

$$\tag{16.7}$$

NAD-pyruvate S-Lac-NAD$^+$

It seems most likely that the active ternary complexes are of the form shown in mechanism 16.8. The carboxylate group of the substrate forms a salt bridge with the side chain of Arg-171. The hydroxyl group of lactate forms a hydrogen bond with the unprotonated imidazole ring of His-195. (When pyruvate is the substrate, its carbonyl group forms a hydrogen bond with the protonated imidazole.) As well as orienting the substrate, His-195 acts as an acid-base catalyst, removing the proton from lactate during oxidation (mechanism 16.8). (In the reaction with pyruvate, His-195 stabilizes the negative charge that develops on the oxygen of pyruvate during reduction.) Mutagenesis of Arg $\rightarrow$ Lys-171 weakens binding by 21 kJ/mol (5 kcal/mol).[53]

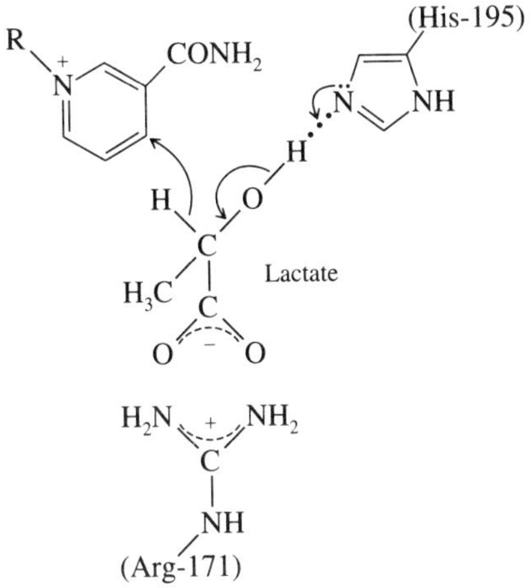

(16.8)

Further mutagenesis studies show that the hydrophobic nature of the nicotinamide-binding pocket causes tighter binding of the neutral NADH than of NADH$^+$, a factor that helps to stabilize the ketoacid form of the ternary complex.[44,45]

The NAD-pyruvate is formed by the enzyme from NAD$^+$ and the enol form of pyruvate[54] (the keto tautomer is the substrate for the normal reaction). The mechanism for the reaction is presumably catalyzed by His-195 in a manner similar to that of the oxidation of lactate (mechanism 16.9).

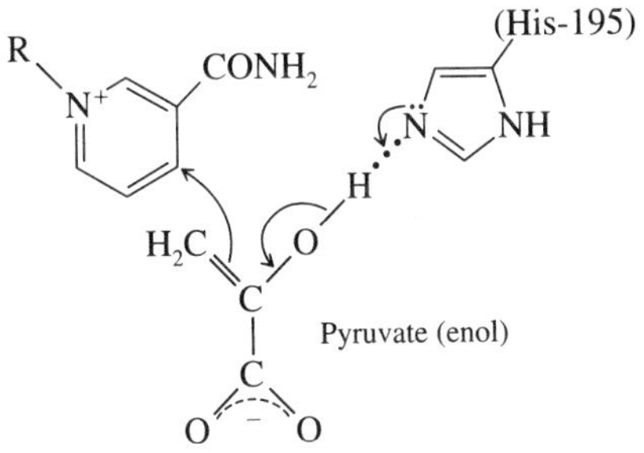

(16.9)

b. The kinetic mechanism[1,55]

The enzyme binds lactate or pyruvate only in the presence of coenzyme.[56] Therefore, the mechanism is ordered, with the coenzyme binding first. Pertinent to this is the observation from the crystallographic studies that coenzyme binding induces a conformational change in which residues 98 to 110 of the chain move through a relatively large distance to close over the active site in the ternary complex.[49,50] Rapid reaction studies show that the mechanism differs somewhat from that of horse liver alcohol dehydrogenase, although the dissociation of the enzyme–NADH complex is rate-determining in both reactions.[55,57] With alcohol dehydrogenase, the dissociation of the aldehyde from the ternary complex E·RCHO·NADH is faster than the rate of hydride transfer, so that the two ternary complexes do not have time to equilibrate. However, the hydride transfer steps are very rapid in the reactions of lactate dehydrogenase and the dissociation of pyruvate is slow, so that the two ternary complexes equilibrate. Further, the equilibrium position favors lactate and NAD$^+$ at neutral pH. When lactate is mixed with the holoenzyme at pH 7, 20% of the bound NAD$^+$ is reduced during the first millisecond of reaction, as the equilibrium between [E·NAD$^+$·lactate] and [E·NADH·pyruvate] is rapidly attained. As the pyruvate dissociates, the equilibrium is displaced toward products, so that all four bound NAD$^+$ molecules are reduced. There is then the slower dissociation of NADH, the rate-limiting step in the steady state (equation 16.10).

$$\text{E·NAD}^+\text{·lactate} \xrightleftharpoons{\text{fast}} \text{E·NADH·pyruvate} \xrightarrow[\text{Pyruvate}]{\text{slow}} \text{E·NADH} \xrightarrow[\text{NADH}]{\text{slowest}} \text{E} \qquad (16.10)$$

This behavior could easily be mistaken for half-of-the-sites reactivity, but all four sites appear to be independent.[55,57] Coenzymes also bind independently to each site. In contrast, the enzyme from *Bacillus stearothermophilus* is allosterically regulated.[44,45] A tetrameric form is stabilized by the binding of two effector molecules of fructose 1,6-bisphosphate and binds pyruvate 50 times more tightly than the dimeric form does. The turnover numbers of tetramer and dimer are the same, and so they form a *V*-system (Chapter 8). The build up of the glycolytic intermediate fructose 1,6-bisphosphate under anaerobic conditions thus stimulates the regeneration of NAD$^+$.

The catalytically important His-195 is unusually reactive toward diethylpyrocarbonate. This enabled the pK_a in both the apo and holoenzymes to be determined directly from the pH dependence of the rate of modification (the pK_a = 6.7).[58] There is evidence that lactate binds preferentially to the holoenzyme containing the un-ionized histidine, whereas pyruvate binds preferentially to the enzyme–NADH complex containing protonated histidine.

c. Changing specificity by protein engineering

A key structural and mechanistic feature of lactate and malate dehydrogenases is the active site loop, residues 98–110 of the lactate enzyme, which was seen in the crystal structure to close over the reagents in the ternary complex.[49,50] The loop has two functions: it carries Arg-109, which helps to stabilize the transition state during hydride transfer; and contacts around 101–103 are the main determinants of specificity. Tryptophan residues were placed in various parts of lactate dehydrogenase to monitor conformational changes during catalysis.[54,59,60] Loop closure is the slowest of the motions.

All malate dehydrogenases have an arginine residue at position 102, whereas most lactate dehydrogenases have a glutamine at this position. Mutation of Gln → Arg-102 in lactate dehydrogenase increases the activity toward oxaloacetate by a factor of 10^3 and decreases the activity toward pyruvate by 10^4; i.e., a single mutation converts lactate into malate dehydrogenase.[61] Such an extreme switch of specificity by a single mutation is rare; in the next section, it will be seen how difficult this is with the serine proteases.

A series of mutations were made in the loop in *B. stearothermophilus* lactate dehydrogenase to convert it into a phenyllactate dehydrogenase.[61] Loops of various lengths and sequences were used and the activities screened. It was found for one longer loop that activity with pyruvate was reduced 10^6-fold but that activity with phenylpyruvate was largely unaltered, to give a 400,000-fold switch in specificity. The 1700:1 selectivity of the mutant for phenylpyruvate over pyruvate may be of diagnostic use for monitoring phenylpyruvate in the urine of patients with phenylketonuria.

3. Glyceraldehyde 3-phosphate dehydrogenase[62,63]

Glyceraldehyde 3-phosphate dehydrogenase, a tetrameric enzyme of M_r 150 000 containing four identical chains, catalyzes the reversible oxidative phosphorylation of glyceraldehyde 3-phosphate to 1,3-diphosphoglycerate, using NAD^+ as a coenzyme (equation 16.11).

$$\underset{\overset{\displaystyle |}{\underset{\displaystyle CH_2OPO_3^{2-}}{H-\overset{\displaystyle |}{C}-OH}}}{\overset{\displaystyle H}{\diagdown}\overset{\displaystyle }{\underset{\displaystyle C}{\diagup}\!\!\overset{\displaystyle O}{}} + NAD^+ + HPO_4^{2-} \rightleftharpoons \underset{\overset{\displaystyle |}{\underset{\displaystyle CH_2OPO_3^{2-}}{H-\overset{\displaystyle |}{C}-OH}}}{\overset{\displaystyle O}{\diagdown}\overset{\displaystyle OPO_3^{2-}}{\underset{\displaystyle C}{\diagup}}} + NADH + H^+ \qquad (16.11)$$

The reaction pathway consists of a series of reactions. The currently accepted mechanism (equations 16.12 to 16.16), which was first proposed in 1953,[64] is supported by extensive pre-steady state[65] and steady state[66,67] kinetic studies.

$$NAD^+ \cdot E-SH + RCHO \rightleftharpoons NAD^+ \cdot E-S-\overset{\overset{\displaystyle OH}{\displaystyle |}}{\underset{\underset{\displaystyle H}{\displaystyle |}}{C}}-R \qquad (16.12)$$

$$NAD^+ \cdot E-S-\overset{\overset{\displaystyle OH}{\displaystyle |}}{\underset{\underset{\displaystyle H}{\displaystyle |}}{C}}-R \rightleftharpoons NADH \cdot E-S-\overset{\displaystyle O}{\underset{\displaystyle R}{C}} + H^+ \qquad (16.13)$$

$$NADH \cdot E-S-\overset{\displaystyle O}{\underset{\displaystyle R}{C}} \rightleftharpoons E-S-\overset{\displaystyle O}{\underset{\displaystyle R}{C}} + NADH \qquad (16.14)$$

$$E-S-\overset{\displaystyle O}{\underset{\displaystyle R}{C}} + NAD^+ \rightleftharpoons NAD^+ \cdot E-S-\overset{\displaystyle O}{\underset{\displaystyle R}{C}} \qquad (16.15)$$

$$\text{NAD}^+ \cdot \text{E}-\text{S}-\text{C} \underset{R}{\overset{O}{\diagup}} + \text{HPO}_4^{2-} \rightleftharpoons \text{NAD}^+ \cdot \text{E}-\text{SH} + \text{RC} \underset{O}{\overset{OPO_3^{2-}}{\diagup}} \tag{16.16}$$

The enzyme has a reactive cysteine residue that is readily acylated by acyl phosphates to form a thioester (the reverse of reaction 16.16).[68] The first step in the reaction sequence is the formation of a hemithioacetal between the cysteine and the substrate. This has the effect of converting the carbonyl group, which is not easy to oxidize directly, into an alcohol that is readily dehydrogenated by the usual procedure (reaction 16.13). The thioester that is formed in reaction 16.13 reacts with orthophosphate to give the acylphosphate (16.16). However, the acyl transfer is very slow unless NAD^+ is bound to the enzyme.[69,70] Therefore, the replacement of NADH by NAD^+ in reactions 16.14 and 16.15 is a necessary part of the reaction sequence. It is of interest that the dissociation of the complex of the acylenzyme and NADH (16.14) is the rate-determining step in the sequence at saturating reagent concentrations at high pH.[69] A consequence of this replacement of NADH by NAD^+ before the release of acylphosphate is that the free apoenzyme does not take part in the reaction. Also, because acylation of the enzyme by the diphosphate is activated by NAD^+, the holoenzyme initiates the reductive dephosphorylation of 1,3-diphosphoglycerate.

The Michaelis complexes of the holoenzyme with the aldehyde or the diphosphoglycerate, and the acylenzyme with orthophosphate, are not included in the scheme because their dissociation constants are too high for their accumulation.

a. The structures of the enzyme–substrate complexes

The crystal structures of the holoenzymes from lobster[71] and *Bacillus stearothermophilus*[72] were solved at 2.9 and 2.7 Å, respectively; the human holoenzyme[73] was solved at low resolution. The lobster apoenzyme was solved at 3.0-Å resolution.[74] The structures of the enzyme–substrate complexes were deduced from model-building experiments on the lobster and bacterial enzymes.[71,72] The following description is a composite of these; we use the specific details of the bacterial enzyme for convenience.

Two binding sites for sulfate ions were identified at the active site of the enzyme that had been crystallized from ammonium sulfate (Figure 16.1).[75] A chemically and stereochemically reasonable model for the course of the reaction may be constructed by assuming that these are the binding sites for the phosphate residue of the substrate and the nucleophilic phosphate in the deacylation reaction (16.16). The aldehyde group of the substrate can form a hemithioacetal with Cys-149 when the 3-phosphate is placed to make hydrogen bonds with the hydroxyl of Thr-179, the positively charged side chain of Arg-231, and the 2′-hydroxyl of the ribose ring that is attached to the nicotinamide of NAD^+ (Figure 16.1). The C-2 hydroxyl of the substrate can then form a hydrogen bond with Ser-148, while the C-1 hydroxyl forms one with a nitrogen of His-176. These interactions orient the substrate so that the C-1 hydrogen points toward the C-4 position of the nicotinamide ring, less than 3 Å away. In this mode of binding, the dehydrogenation reaction may take place as described earlier for lactate dehydrogenase, with His-176 as the general-base catalyst (reaction 16.17).

$$(16.17)$$

The transition state for the attack of orthophosphate on the thioester can be stabilized by hydrogen bonds to the attacking phosphate from the hydroxyl of Ser-148, the hydroxyl of Thr-150, and the C-2 hydroxyl of the substrate, and also from the amido nitrogens of Cys-149 and Thr-150. The presence of this specific binding site for phosphate explains why the thioester is phosphorolyzed rather than hydrolyzed. The sulfur atom of the thioester is close enough to the C-4 carbon of the nicotinamide ring of NAD^+ to be polarized by its positive charge. This perhaps explains the activation of the acyl transfer reactions on NAD^+ binding.

b. The symmetry of the enzyme and the cooperativity of ligand binding

There was considerable controversy in the literature about the symmetry of the dehydrogenase, the cooperativity of ligand binding, and half-of-the-sites versus full-site reactivity.[76-83] The binding of NAD^+ to the enzyme is definitely cooperative; there is strong negative cooperativity in the binding to the rabbit muscle and bacterial enzymes,[76,80] although there is positive cooperativity in the binding to the yeast enzyme at some temperatures.[84] Glyceraldehyde 3-phosphate, on the other hand, binds independently to all four subunits.[83] Half-of-the-sites reactivity is found for the reactions of artificial substrates only; 1,3-diphosphoglycerate, for example, acylates all four reactive cysteines with a single rate constant. All four subunits of the enzyme from *B. stearothermophilus* are structurally identical, and the enzyme has precise 222 symmetry.[72] The bacterial apoenzyme also has this symmetry.[72] The crystals of the *B. stearothermophilus enzyme* in which just one mole of NAD^+ is bound per mole of tetramer have the NAD^+ bound in the same site in each tetramer.[85] That subunit is in the holoenzyme conformation, whereas the other three are in apo-like conformations. Given the symmetry in the fully occupied holoenzyme, this strongly implies an induced-fit mechanism.

4. Some generalizations about dehydrogenases

The structural studies have given a clear and chemically satisfying description of the stereochemical and catalytic requirements of the hydride transfer reaction. In three of the examples, there is an acid-base catalyst that forms a hydrogen bond with the carbonyl or alcohol group of the substrate, helps orient it correctly, and stabilizes the transition state for the reaction (equation 16.18).

$$
\begin{array}{ccc}
\text{B} & \text{B}^{\delta+} & \text{B}^{+} \\
\vdots & \vdots & | \\
\text{H} & \text{H} & \text{H} \\
| & \vdots & \vdots \\
\text{O} & \text{O} & \text{O} \\
| & \| & \| \\
\text{C} & \text{C} & \text{C} \\
\text{NAD}^{+} + \text{H} & [\text{NAD}^{\delta+}\text{---H}]_{\text{TS}} & \text{NADH} +
\end{array}
\tag{16.18}
$$

Liver alcohol dehydrogenase is similar, with the His-51/Ser-48 shuttle and/or the Zn^{2+} ion taking the place of BH^+ in stabilization of the transition state and orientation of the substrate.

A consequence of the direction of the hydrogen bonding is that the alcohol binds preferentially to the basic form (B) of the catalyst, whereas the aldehyde binds preferentially to the acidic form (BH^+). The pK_a of B is lowered in the $E \cdot NAD^+ \cdot RCH_2OH$ ternary complex and raised in the $E \cdot NADH \cdot RCHO$ complex. This means that the proton that is produced during the oxidation does not leave the ternary complex but is taken up by the catalytic group, and vice versa. The proton escapes into solution only when there is a change in substrate binding.[28,55,78,86]

The specificities of the enzymes are also nicely explained: The enantiomers of the substrates of L-lactate and D-glyceraldehyde 3-phosphate dehydrogenases cannot be productively bound; the hydrophobic pocket of alcohol dehydrogenase will not bind the charged side chains of lactate; etc. However, we do not know if conformational changes occur during catalysis or if there is strain.

A general kinetic feature is that NADH usually binds more tightly than NAD^+. The structural features responsible for this are not clear, although the charged nicotinamide ring is clearly more hydrophilic than the reduced form in NADH. The tight binding causes the dissociation of the enzyme–NADH complexes to be largely rate-determining at saturating concentrations of reagents at physiological pH. Further, although the equilibrium constant for the oxidation reaction in solution greatly favors NAD^+ and alcohol, the tighter binding of the NADH causes the equilibrium constant for the enzyme-bound reagents to be more favorable: it was seen that the equilibrium constant between the two ternary complexes in the reactions of lactate dehydrogenase is not far from 1.

B. The proteases

The proteases may be conveniently classified according to their activities and functional groups. The serine proteases are endopeptidases that have a reactive serine residue and pH optima around neutrality. The carboxyl or aspartyl (formerly called acid) proteases are endopeptidases that have catalytically important carboxylates and pH optima at low pH (apart from chymosin, whose activity extends to neutral pH). The thiol proteases are endopeptidases that differ from the serine proteases by having reactive cysteine residues. The zinc proteases are metalloenzymes that function at neutral pH. Many of these proteases are small monomeric enzymes of M_r 15 000 to 35 000, readily amenable to kinetic and structural study. Because of this, they are among the best-understood enzymes. Although the different classes catalyze the same reaction, they utilize different mechanisms. Some are well understood and have chemical models; the others are more obscure.

The proteases were the first enzymes to be discovered because of their abundance in the digestive system. But they have a very wide range of important biological activities because they are involved in processing proteins in general. For example, the activities of some hormones are modulated by proteolysis; e.g., the angiotensin converting enzyme, which is a metalloprotease. The blood clotting cascade involves a series of serine proteases. Activation of the immune system also uses serine proteases. Metalloproteases and thiol proteases are involved in the destruction of cells. Proteases are used to punch holes in cell walls in processes varying from fertilization to the formation of metastases. Viruses encode proteases that act in their maturation: adenoviruses encode a cysteine protease; picornaviruses, a serine protease; and retroviruses, such as HIV, an aspartyl (acid protease).[87] Consequently, proteases are the targets for many drugs, some of which are mechanism-based inhibitors (Chapter 9).

The notation of A. Berger and I. Schechter (Chapter 1, Figure 1.24) is used throughout this section to describe the binding subsites. (The scissile bond of the peptide substrate sits across the S_1 and S_1' subsites with its C-terminal side occupying the S_1' to S_n' subsites, and its N-terminal side occupying S_1 to S_n.)

1. The serine proteases[88–90]

These enzymes have been discussed in various parts of this text. Some major topics are: the enzymes as a family, specificity (Chapter 1, section D2); the structures of the active site, the enzyme–substrate complex, the acylenzyme, and the enzyme–product complex (Chapter 1, section F2); proof of the reaction pathway, reaction kinetics (Chapter 7, section B); the pH dependence of catalysis and the state of ionization of the active site (Chapter 5, sections F and G2a); the utilization of binding energy to increase k_{cat} (Chapter 12, section B); transition state stabilization, specific solvation of the transition state (Chapter 12, section D5c); possible stereoelectronic effects (Chapter 8, section F2). More than 500 serine proteases have been sequenced and more than 20 different structures determined.[89,90] They all have an identical fold, consisting of two β barrels, with the catalytic triad Asp-102, His-57, and Ser-195 at the interface of the two domains (see Chapter 1, section D).

The hydrolysis of ester or amide substrates catalyzed by the serine proteases involves an acylenzyme intermediate in which the hydroxyl group of Ser-195 is acylated by the substrate. The formation of the acylenzyme is the slow step in the reaction of amide substrates at saturating concentrations, but the acylenzyme often accumulates in the hydrolysis of esters. The attack of Ser-195 on the carboxyl group of the substrate probably forms a high-energy tetrahedral intermediate (equation 16.19), but there is no direct evidence for it.

$$E-OH\cdot R-\overset{\overset{\displaystyle O}{\|}}{C}-X \longrightarrow E-O-\overset{\overset{\displaystyle O^-}{|}}{\underset{\underset{\displaystyle R}{|}}{C}}-X \searrow E-O-\overset{\overset{\displaystyle O}{\|}}{C}-R \xrightarrow{H_2O}$$

$$HX$$

$$E-O-\overset{\overset{\displaystyle O^-}{|}}{\underset{\underset{\displaystyle OH}{|}}{C}}-R \longrightarrow E-OH + R-CO_2H \qquad (16.19)$$

There is probably more direct experimental evidence about the mechanism of catalysis and the structures of the intermediates in the reactions of the serine proteases than there is about any other enzyme or family of enzymes. One of the major reasons for the structural knowledge is that it is possible to solve the crystal structures of the co-crystallized complexes of trypsin and some naturally occurring polypeptide inhibitors that mimic substrates (Chapter 1, section F). We know from these studies that the active site of the enzyme is complementary in structure to the transition state of the reaction, a structure that is very close to the tetrahedral adduct of Ser-195 and the carbonyl carbon of the substrate. Further, the structure of the enzyme is not distorted when it binds the substrate. NMR studies on the binding of small peptides show that these are also not distorted on being bound. (The high-resolution study on the crystal structure of the complex between the pancreatic trypsin inhibitor and trypsin shows clearly that the reactive peptide bond is distorted toward its structure in the tetrahedral intermediate. However, this bond is distorted *before* the binding to the enzyme, the inhibitor being "designed" to bind as tightly as possible to the enzyme; i.e., it is a natural transition state analogue.)

a.   The hydrogen-bond network at the active site

It has long been thought that the imidazole base of His-57 increases the nucleophilicity of the hydroxyl of Ser-195 by acting as a general-base catalyst (mechanism 16.20).

$$HN\overset{\frown}{\underset{\underset{\diagdown\_\diagup}{}}{}}N\colon\overset{\frown}{}H-O\diagup \qquad\qquad (16.20)$$

The activity falls off at low pH according to the ionization of a base of $\sim pK_a$ 7, a characteristic value for a histidine residue; His-57 is modified by the affinity label *tos*-L-phenylalanine chloromethyl ketone, with an irreversible loss of enzymatic activity (Chapter 7, section G).[91] It came as a complete surprise when the crystallographers found that the carboxylate of Asp-102 is also involved at the active site to give a catalytic triad (structure 16.21),

(16.21)

which was dubbed the "charge relay system."[92] Although the carboxyl group is completely buried in the interior of the protein, it is surrounded by polar residues and buried water molecules. Similar hydrogen-bond networks have subsequently been found at the active sites of all other serine proteases. In protease A from *Streptomyces griseus*, for example, the buried carboxylate ion is the recipient of four hydrogen bonds in an environment that has been described as very polar.[93]

A controversy arose about which residue is responsible for the ionization of the hydrogen-bond network. Kinetic data suggested that the group ionizing with a pK_a of 6.8 in chymotrypsin is the histidine,[94] and that the buried aspartate is ionized at neutrality, with a pK_a of < 2.[95,96] Subsequent NMR and IR evidence purported to show that the group ionizing with pK_a 6.8 was Asp-102. However, additional NMR work[97,98] and the direct determination by neutron diffraction[99] of the position of the proton between Asp-102 and His-57 proved finally that the histidine is the ionizing residue and that Asp-102 has an unusually low pK_a.

The contribution of the triad to catalysis was quantified experimentally by protein engineering experiments on subtilisin.[100] This is discussed in detail in Chapter 15, Part 2, section B. Replacement of the equivalent of Asp-102, His-57, and Ser-195 one at a time by alanine reduced the value of k_{cat} by factors of 3×10^4, 2×10^6, and 2×10^6, respectively. Converting all three to alanine also decreased activity by 2×10^6.

The importance of the hydrogen bond between Asp-102 and His-57 continues to cause controversy, with some of the most distinguished enzymologists taking opposing views. One camp[101,102] proposes that this is a low-barrier hydrogen bond (Chapter 11, section A3) that causes a significant rate enhancement by stabilizing the transition state, whereas the other side[103,104] argues equally forcibly that it is not.

Another query concerns whether or not there is a hydrogen bond between His-57 (N^ϵ) and Ser-195 (O^γ). The x-ray crystallographic criterion for the existence of a hydrogen bond between two such atoms is that their interatomic distance is about 2.8 to 3 Å (Chapter 11, section A3). The distance between the two relevant atoms is 3.7 Å in subtilisin[105,106] and 3.8 Å in γ-chymotrypsin[107] (see the next section for the different forms of chymotrypsin), distances that are far too long. But in subsequent studies, the distance between the O_γ and the N_ϵ was found to be 3.0 Å in trypsin and 2.9 Å in kallikrein.[108] In elastase, the binding of an inhibitor induces the histidine and serine to be aligned so that the distance is 2.9 Å.[109]

b. The contribution of the oxyanion binding site to catalysis

The protein engineering experiments on subtilisin indicated that, after mutagenesis of all three catalytic residues to alanine, there is a residual activity about 10^3 times higher than the spontaneous hydrolysis.[100] This must result from enzyme–transition state complementarity. One likely candidate for the complementarity is the oxyanion of the tetrahedral intermediate that is formed by the attack of Ser-195 on the substrate and that is solvated by two backbone NH groups (Chapter 1, section F2). This is difficult to study by protein engineering (Chapter 15, Part 2, section B), but vibrational spectroscopy has given clues.[110] Resonance Raman spectra of acylenzymes of serine proteases have a signal at about 1700 cm^{-1} that is attributable to the stretching of its C=O group. Mutagenesis of one of the residues in the oxyanion binding site of subtilisin, Asn-155 → Leu (see Chapter 15, Part 2, section B) removed the spectral features associated with the polarized C=O group of the acylenzyme.[110] A component of the stretching mode is the contribution of the valence bond structures:

$$> C = O \longleftrightarrow > C^+ - O^-$$

(16.22)

Features that specifically stabilize the $> C^+ - O^-$ state should lower the stretching frequency of the C=O because the transition to $> C^+ - O^-$ should be of lower energy. Resonance Raman experiments on a series of substrates of chymotrypsin found that the more reactive the acylenzyme, the lower the stretching frequency.[111] These changes parallel the changes in stretching frequency of carbonyl groups in small organic compounds where the C=O bond length is known from x-ray diffraction studies to increase because of strain and other effects. The spectral properties of the acylenzymes were attributed to ground state strain in the acylenzyme: a 16,000-fold increase in deacylation rate constant from the least to the most reactive acylenzyme was accompanied by a shift in carbonyl stretching of -54 cm^{-1}, which corresponds to a carbonyl bond length increase of 0.025 Å in the model compounds.[112]

c. The structure and the reactivity of the substrate

The structural requirements for a substrate to be reactive have been determined by measuring the values of k_{cat} and K_M for a wide range of ester substrates, and the association constants of reversible inhibitors.[113] The inherently high reactivity of esters causes relatively poor ester substrates to be hydrolyzed at a measurable rate. Thus esters have been most useful for working out the steric requirements of the acyl portion of the substrate. Amides and peptides are so unreactive that the only ones amenable to study are the derivatives of the specific substrates phenylalanine, tyrosine, and tryptophan. The kinetic studies may now be combined with those from x-ray diffraction.

1. The deacylation step.
Listed in Table 16.2 are data for the deacylation of various acylenzymes. (Further values for amino acids were given in Table 7.3). The most reactive derivative is that of acetyl-L-phenylalanine. As was discussed

Table 16.2 *Structural requirements in the deacylation of acylchymotrypsins (at 25°C)*[a]

Acylchymotrypsin R — (RCO$_2$E)	k_{cat} (s^{-1}) (for deacylation)	k_{OH^-} (s^{-1} M^{-1}) (for hydrolysis of RCO$_2$CH$_3$)
CH$_3$—	0.01	0.19
C$_6$H$_5$CH$_2$CH$_2$—	0.178	0.15
CH$_2$(NHCOCH$_3$)—	0.12	2.48
L-C$_6$H$_5$CH$_2$CH(NHCOCH$_3$)—	111	1.94

[a] From A. Dupaix, J.-J. Bechet, and C. Roucous, *Biochem. Biophys. Res. Commun.* **41**, 464 (1970); I. V. Berezin, N. F. Kazanskaya, and A. A. Klyosov, *FEBS Lett.* **15**, 121 (1971) (see Table 7.3).

in Chapter 1, chymotrypsin has a well-defined binding pocket for the aromatic side chain of the amino acid, and a hydrogen-bonding site (the C=O of Ser-214) for the NH of the CH$_3$CONH— of the substrate (Chapter 1, Figure 1.18). When the C$_6$H$_5$CH$_2$— and CH$_3$CONH— groups of acetyl-phenylalanine are replaced by hydrogen atoms to give the simple acetyl group, the deacylation rate drops by a factor of 10^4 (although 10 units of this factor is caused by the inductive effect of the CH$_3$CONH— group, as shown by the hydroxide ion–catalyzed rate constants listed in the last column of Table 16.2). Interestingly, it is seen in Table 16.2 that *both* the aromatic ring *and* the acylamino group are required for high reactivity. Acetylglycine-chymotrypsin deacylates only 12 times faster than acetyl-chymotrypsin, and the increase is seen from the hydroxide ion–catalyzed rate constants to be caused solely by the inductive effect of the acylamino group rather than by any binding effect. Similarly, β-phenylpropionyl-chymotrypsin deacylates only 17.8 times faster than acetyl-chymotrypsin. The fact that both the aromatic ring and the acylamino group are required for high reactivity has been nicely accounted for by x-ray diffraction studies. As was described in Chapter 1 (Figure 1.18), the carbonyl oxygen of a polypeptide substrate sits between the backbone NH groups of Ser-195 and Gly-193. This mode of binding has been found for the specific acylenzyme carbobenzoxy-L-alanine-elastase.[114] However, the carbonyl oxygen of the nonspecific acylenzyme indolylacryloyl-chymotrypsin is not productively bound in this manner.[115] Instead, there is a water molecule forming a hydrogen-bonded bridge between the carbonyl oxygen and the catalytic nitrogen atom of His-57 (Figure 16.4). For reaction to occur, the carbonyl oxygen must swing into the hydrogen-bonding site between Ser-195 and Gly-193, and the bound water molecule must attack the carbonyl carbon. Thus, the acylamino portion and the aromatic ring are together required to anchor the carbonyl group in the productive mode. If either of the anchors is missing, the carbonyl oxygen takes up a nonproductive binding mode.

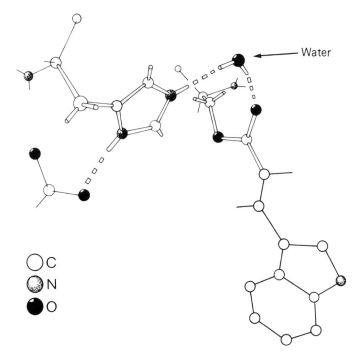

Figure 16.4 The crystal structure of indolylacryloyl-chymotrypsin. [From R. Henderson, *J. Molec. Biol.* **54**, 341 (1970).] Note that the carbonyl oxygen of this nonspecific acylenzyme is not bound between the NH groups of Ser-195 and Gly-193, but is nonproductively linked to His-57 by a hydrogen-bounded water molecule. This is the acylenzyme that was found to deacylate at the same rate in solution and in the crystal (Chapter 1). [G. L. Rossi and S. A. Bernhard, *J. Molec. Biol.* **49**, 85 (1970).]

2. *The acylation step.* It was pointed out in Chapter 12 that the binding energies of the S_2, S_3, S_4, and S_5 subsites often increase k_{cat} for the hydrolysis of polypeptide substrates rather than lower K_M. The reason is not known. The binding energy of the S_1' site is also used to increase k_{cat} rather than give tighter binding.[116] This appears to result from a binding mode that cannot be fully occupied in the enzyme–substrate complex, but that is accessible after the bond changes on formation of the unstable tetrahedral intermediate.[117] It is also likely that the carbonyl oxygen atom of the substrate forms better hydrogen bonds with the backbone $>$ NH groups of Ser-195 and Gly-193 (Figure 1.18) after reaction has occurred to form the tetrahedral intermediate. Evidence supporting this hypothesis has come from studies on the hydrolysis of thionesters (substrates in which the carbonyl oxygen atom is replaced by a sulfur atom). Thionesters are far less reactive than normal esters in the enzymatic reactions. This is attributed to the fact that the $C = S$ bond length and the optimal $S \cdots HN <$ bond distances for the thionesters are longer than those for the ester counterparts. The differential

hydrogen bonding of the substrate and tetrahedral intermediate is thus altered and the transition state stabilization is upset.[118,119]

d. The tetrahedral intermediate

The tetrahedral intermediate does not accumulate during the course of the reaction.[96,120] The seemingly overwhelming mass of evidence that had been gathered to "prove" that the intermediate accumulates was shown convincingly to be an interesting set of artifacts.[121] Nor does the intermediate accumulate in the complexes of serine proteases with their naturally occurring polypeptide inhibitors.[122,123] The evidence for the intermediate is thus circumstantial. An analogous tetrahedral structure has been observed, however, in the complexes of serine proteases with inhibitors that contain an aldehyde group instead of the peptide or ester moiety. The *S. griseus* A protease forms such a covalent complex with Ac-Pro-Ala-Pro-phenylalaninal[124] (equation 16.23) and with the naturally occurring inhibitor chymostatin.[125]

$$\text{Ac-Pro-Ala-Pro-NHCH(CH}_2\text{-Phe)CHO} + \text{HO}-(\text{Ser-195}) \longrightarrow$$

$$\underset{\overset{|}{\text{H}}}{\overset{\overset{\text{OH}}{\diagup}}{\text{Ac-Pro-Ala-Pro-NHCH(CH}_2\text{-Phe)C}}}-\text{O}-(\text{Ser-195}) \qquad (16.23)$$

The oxygen atom is seen to occupy its proposed position between the $>$ NH groups of the residues equivalent to Gly-193 and Ser-195. The lack of accumulation of the intermediate (i.e., its presumed existence at steady state levels only) in the hydrolysis of peptides lends further justification to the assumption that the transition state for the reaction resembles the intermediate.

e. A description of the reaction mechanism

The kinetic and structural data may be combined to give the following qualitative description of the mechanism of acylation of chymotrypsin by a good polypeptide substrate.[116]

The substrate binds in the specificity pocket of the enzyme, with the *N*-acylamino hydrogen binding to the carbonyl group of Ser-214. Any further residues in the *N*-acylamino chain bind in the subsites that are available. The reactive carbonyl group sits with its oxygen between the backbone NH groups of Ser-195 and Gly-193 (Figure 1.18). However, it is possible that the hydrogen bond between the oxygen and Gly-193 is long and weak. The first chemical step in the reaction is the attack of the hydroxyl of Ser-195 on the carbonyl carbon of the substrate to form the tetrahedral intermediate. During this, the proton on the hydroxyl is transferred to the imidazole of His-57. As the bond between Ser-195 and the carbonyl carbon is formed, the C=O bond lengthens to become a single bond. The oxygen, bearing a negative charge, moves closer to the NH of Gly-193, forming a shorter and stronger hydrogen bond. The transition state is stabilized relative to the Michaelis complex because of the better interactions of the leaving group with the S'_1 subsite and the better hydrogen bond with Gly-193. The tetrahedral intermediate collapses to form the acylenzyme and expel the leaving group. The leaving group cannot bind in the S'_1 site in the acylenzyme, as this would force the amino group to be too close to the carbonyl carbon. Thus, in the reverse reaction—the attack of the leaving group on the acylenzyme—the energy of binding to the S'_1 site is realized only in the transition state. Deacylation takes place by the charge relay system activating the attack of water. Another tetrahedral intermediate is formed, and then it collapses to expel Ser-195 and give the enzyme–product complex. The mechanism is sketched in Figure 2.11.

Despite this detailed knowledge, many important questions still remain unanswered. For example, we do not know how the binding energies of the subsites for the N-acylamino chain are sometimes used to increase k_{cat} rather than decrease K_M (Table 12.1).

f. The zymogens

Some of the serine proteases are stored in the pancreas as inactive precursors that may be activated by proteolysis. Trypsinogen, for example, is converted to trypsin by the removal of the N-terminal hexapeptide on the cleavage of the bond between Lys-6 and Ile-7 by enterokinase. Chymotrypsinogen is activated by the tryptic cleavage of the bond between Arg-15 and Ile-16. (In this case, further proteolysis by the chymotrypsin that is released during the activation leads to the different forms of the enzyme—Figure 16.5.)

The mechanism of the activation and the reasons for the inactivity of chymotrypsinogen have been nicely explained by comparison of the crystal structures of the enzyme and the zymogen.[126-128] The zymogen has the charge relay system, and it ionizes in the same manner as it does in the enzyme.[94,129] However, the activity of the zymogen is extremely low, being devoid of proteolytic activity, and is only as reactive toward synthetic substrates as a solution of imidazole.[130,131] The reason for this is that the substrate-binding pocket is not properly formed in the zymogen, and the important NH group of Gly-193 points in the wrong direction for forming a hydrogen bond with the substrate.[128] This is an important lesson about enzyme catalysis. Enzyme catalysis often depends not on the presence of an unusually reactive catalytic group on the enzyme, but rather on the correct alignment of the substrate and ordinary catalytic groups.

The conformational change that creates the binding pocket and rotates Gly-193 results from a movement of Ile-16 as its α-ammonium group forms a salt bridge with the buried carboxylate of Asp-194. The activation process may be mimicked and studied by the effects of pH on the salt bridge.[132] This deprotonates at high pH and is destabilized so that the enzyme takes up a zymogen-like conformation. The energy difference between the two conformations is small and their equilibrium is delicately balanced.[132]

In trypsinogen, a region of the protein at the binding pocket is disordered, indicating conformational flexibility.[133,134] On activation or on the addition of a small peptide that can bind to the buried Asp-194, this region takes on a well-defined structure.

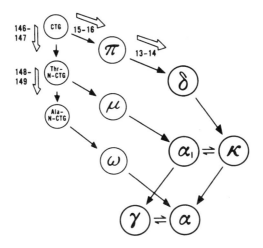

Figure 16.5 Activation of chymotrypsinogen (CTG). It was seen in Chapter 1, Figure 1.1, that CTG is susceptible to tryptic cleavage at the (Arg-15)—(Ile-16) bond, and to chymotryptic cleavages at (Tyr-146)—(Thr-147), (Ile-13)—(Ser-14), and (Asn-148)—(Ala-149) (the latter two because chymotrypsin has relatively broad specificity). In the rapid activation of chymotrypsinogen ([CTG]/[trypsin] $\approx$ 30), there is sufficient trypsin present to activate all the CTG before the accumulated chymotrypsin autolyzes or cleaves the CTG. The pathway of activation is CTG $\rightarrow$ π-chymotrypsin $\rightarrow$ δ-chymotrypsin $\rightarrow$ κ-chymotrypsin $(+\alpha_1-) \rightarrow$ α-chymotrypsin $(+ \gamma$-). The κ and α_1 forms are two different conformational states of the same primary structure, as are the α and γ. During the slow activation ([CTG]/[trypsin] $\approx 10^4$), the small fraction of trypsin activates the zymogen slowly, allowing the chymotrypsin that is initially formed to cleave the unactivated zymogen to form neochymotrypsinogens (N-CTG's). Alpha$_1$- and α-chymotrypsin are produced from the N-CTG's via the generation of μ- and ω-chymotrypsin by tryptic cleavages. [From S. K. Sharma and T. R. Hopkins, *Biochemistry* **18**, 1008 (1979); *Bioorganic Chem.* **10**, 357 (1981).]

g. Redesigning the specificity by protein engineering[99]

The elementary textbook view of chymotrypsin and trypsin is that the primary specificity is determined just by the nature of residue 189 at the bottom of the binding pocket: in trypsin, it is an aspartate that forms a salt bridge with the side chain of arginine or lysine. In elastase, the specificity appears at first sight to result just from two bulky groups at positions 216 and 226 at the mouth of the pocket blocking the access of larger substrates (Chapter 1, section D2). Substitution of the elastase residues at these positions in trypsin failed to confer elastase specificity.[135] Mutation of Asp-189 in trypsin to Ser-189 of chymotrypsin or to other amino acids did not alter the specificity.[136-138] Even mutation of all the amino acid residues in the S$_1$ site of trypsin to their counterparts in chymotrypsin failed to transfer specificity.[139] Conversion to the specificity of chymotrypsin was successful only when surface loops 185 to 188 and 221 to 225 of chymotrypsin were used in addition to replace the analogous loops of trypsin.[139]

These loops are not part of the S_1 binding site or the other direct substrate binding subsites. A further mutation of Tyr-172 → Trp in a third distal loop was required for an additional 50-fold enhancement of binding to give a mutant with about 10% of the k_{cat}/K_M of chymotrypsin.[140]

2. The cysteine proteases[141]

The thiol proteases are widely distributed in nature. The plant enzymes papain (from papaya), ficin (from figs), bromelain (from pineapple), and actinidin (from kiwi fruit or Chinese gooseberry) are members of a structurally homologous family and have been the most widely studied. There are five additional superfamilies that have convergently evolved to having an active-site cysteine: calpain, clostripain, picornavirus proteinase, streptococcal proteinase, and interleukin-1-β-converting enzyme. These have recently been reviewed in depth.[141] Their common feature is the presence of a cysteine and a histidine residue that act in concert. It was originally thought that the mechanism of these enzymes was virtually identical to that of the serine proteases: a histidine-general-base-catalyzed attack of cysteine to form an acylenzyme, followed by deacylation. But there is no evidence from chemical models that thiol nucleophiles are activated by general base catalysis.[142] Denoting the imidazole of the histidine by "Im" and the cysteine as RSH, it was suggested that the reactive form is [RS$^-$·ImH$^+$], which is kinetically equivalent to [RSH·Im].[142] Spectral probe evidence is consistent with this in the active site of papain and related protease.[141] The currently accepted mechanism is shown in equation 16.24.

$$
\begin{bmatrix}
\text{E} \diagup \text{SH} \\
\quad \diagdown \text{Im} \\
\quad\quad \Updownarrow \\
\text{E} \diagup \text{S}^- \\
\quad \diagdown \text{HIm}^+
\end{bmatrix}
\;
\xrightarrow{\;RC\underset{X}{\overset{O}{\diagup}}\;}
\;
\text{E}\diagup\overset{O}{\underset{\text{Im}}{\text{S}-\text{C}}}\diagdown R
\;+\; HX
\;\xrightarrow{\;H_2O\;}\;
RCO_2H \;+\;
\begin{bmatrix}
\text{E} \diagup \text{SH} \\
\quad \diagdown \text{Im} \\
\quad\quad \Updownarrow \\
\text{E} \diagup \text{S}^- \\
\quad \diagdown \text{HIm}^+
\end{bmatrix}
\qquad (16.24)
$$

3. The zinc proteases[143]

There are five distinct families of zinc proteases, classified by the nature of the zinc binding site. These families, and their variously proposed mechanisms, have recently been reviewed in depth.[143] The most studied member is the digestive enzyme bovine pancreatic carboxypeptidase A, which is a metalloenzyme containing one atom of zinc bound to its single polypeptide chain of 307 amino acids and M_r 34 472. It is an exopeptidase, which catalyzes the hydrolysis of C-terminal amino acids from polypeptide substrates, and is specific for the large hydrophobic amino acids such as phenylalanine. The closely related carboxypeptidase B catalyzes the hydrolysis of C-terminal lysine and arginine residues. The two enzymes are structurally almost identical, except that the B form has an aspartate residue that binds the positively charged side chain of the substrate.

Other well-known zinc proteases are: collagenase; angiotensin-converting enzyme (important in regulating blood pressure); thermolysin, a bacterial endopeptidase of M_r 34 600 containing 316 residues in its single polypeptide chain;[144] and the Zn^{2+} G protease from *Streptomyces albus*, a D-alanyl-D-alanine carboxypeptidase that catalyzes carboxypeptidation and transpeptidation in cell wall metabolism.

Matrixin is implicated in arthritis and invasive cancer. There are also aminopeptidases, specific for the N-termini of proteins, that are zinc proteases. One example, leucine aminopeptidase, has a complex hexameric structure, which is unusual for a protease.[145]

The pancreatic carboxypeptidases are not homologous with thermolysin (and the G protease), but their active sites have evolved convergently to having striking similarities.

The chemical mechanism of the zinc proteases is controversial because there is no direct evidence for intermediates, and the Zn^{2+} ion can act as an electrophile to polarize the $>C{=}O$ group of a substrate and/or as a source of metal-bound nucleophilic ^-OH ions (see Chapter 2, section B7).

a. The structure of the active site of carboxypeptidase A

The crystal structure of the enzyme was solved very early at 2.0-Å resolution.[146] The active site consists of a shallow groove on the surface of the enzyme leading to a deep pocket, lined with aliphatic and polar side chains and parts of the polypeptide chain, for binding the C-terminal amino acid. The catalytically important zinc ion is ligated by the basic side chains of Glu-72, His-196, and His-69. In about 20% of the molecules, the phenolic oxygen of Tyr-248 is a fourth ligand. The side chain of this residue is conformationally very mobile, a matter that led to much discussion and debate in the literature about the similarities of the solution and the crystal structure. The phenolic side chain may rotate about its $C_\alpha{-}C_\beta$ bond, and $\sim 80\%$ of its electron density in the map of the crystal structure is found in an orientation on the surface of the molecule pointing into solution.[147] This residue is also so mobile in carboxypeptidase B that its position cannot be determined in the crystal structure.[148] The fourth ligand of the zinc ion is normally a water molecule or a hydroxide ion.

A model was proposed for the structure of a productively bound enzyme–substrate complex, extrapolated from the structures of inert complexes (Figure 16.6). Tyr-248 had been implicated in catalysis from early chemical modification experiments. But one of the first protein engineering experiments, the mutation Tyr-248 → Phe, showed that it is not involved.[149]

b. The structure of the active site of thermolysin

Thermolysin differs from the carboxypeptidases in being an endopeptidase rather than an exopeptidase. This is manifested in the nature of the binding site:

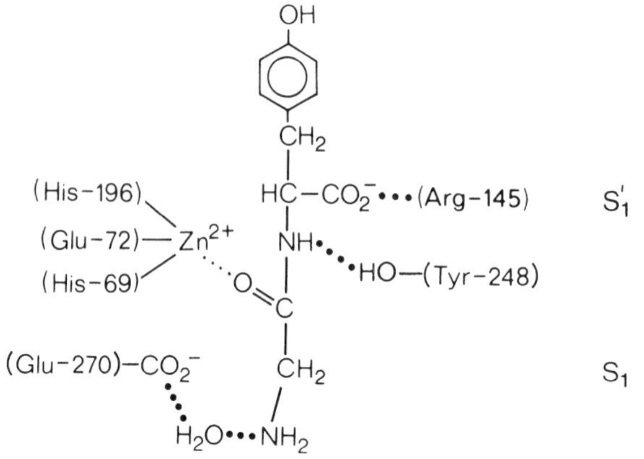

Figure 16.6 The postulated productively bound complex of carboxypeptidase A and a polypeptide substrate. [Courtesy of W. N. Lipscomb.]

instead of the deep pocket of the carboxypeptidases, thermolysin has an open extended cleft that can bind the polypeptide chain of the substrate on both sides of the scissile bond. There are also important differences in the catalytic apparatus:

1. The catalytic glutamate residue (Glu-143) is located at the bottom of a narrow cleft, where it is bound to a water molecule and cannot approach the substrate as readily as can Glu-270 in carboxypeptidase.

2. There is no tyrosine residue corresponding to Tyr-248. Instead, a histidine residue, His-231, is suitably located to be a general-acid-base catalyst in the reaction.

The crystal structure of the complex between the enzyme and a tight-binding inhibitor, phosphoramidon (structure 16.25), has been solved.[149]

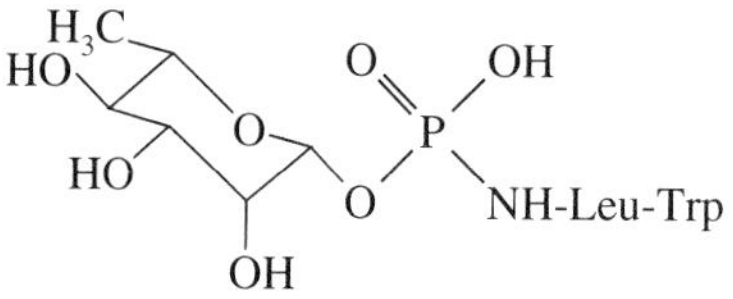

$$(16.25)$$

The phosphoryl group mimics a carbonyl group to which a water molecule has been added to generate a tetrahedral intermediate. A phosphoryl oxygen displaces the molecule of water that is bound to the Zn^{2+} ion and becomes its fourth ligand, at a distance of ~ 2.0 Å. The hydroxyl oxygen bonded to the phosphorus atom is 2.6 Å from an oxygen atom of Glu-143, and is thus hydrogen-bonded to it. The phosphoramide nitrogen is 4.1 Å from a nitrogen atom of His-231.

A likely reaction mechanism is immediately suggested by this structure, since it appears to be a transition state analogue for the Glu-143 general-base-catalyzed attack of a water molecule on a carbonyl group that has been polarized by binding to the zinc ion. Further, His-231 is suitably placed to be a proton donor to the leaving group (equation 16.26).

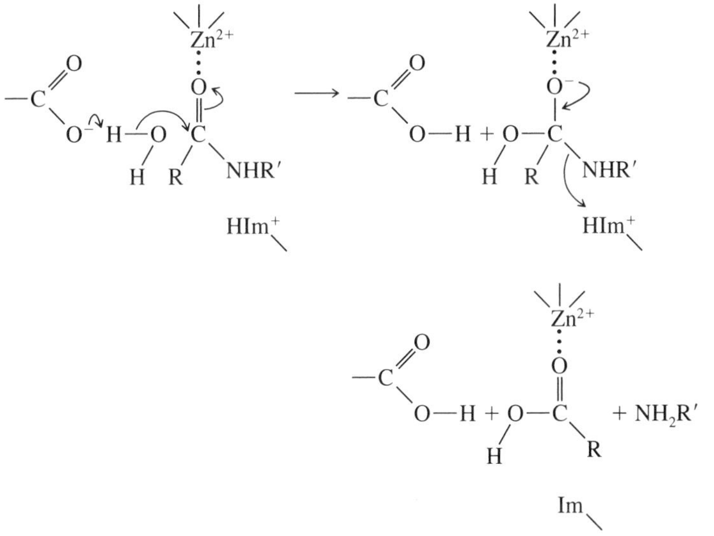

$$\text{(16.26)}$$

An alternative catalytic role for Glu-143 is that it acts as a nucleophile (equation 16.27). Indeed, it is possible to alkylate it with active-site-directed α-chloro-ketone irreversible inhibitors.[150] But the steric inaccessibility of the carboxylate group leads to such distortion of the enzyme on the formation of covalent bonds that nucleophilic attack on a substrate is most unlikely.[150]

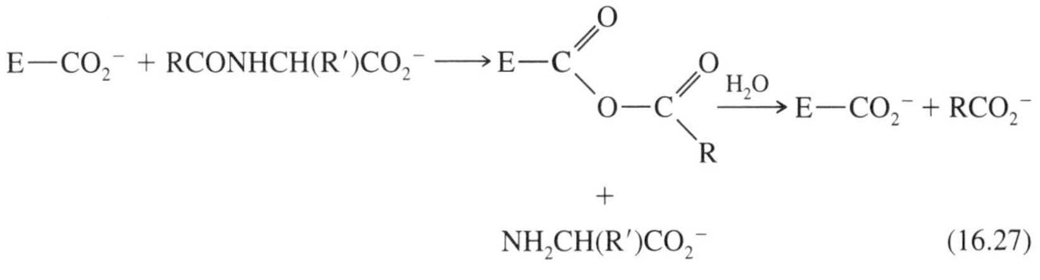

$$\text{(16.27)}$$

c. The reaction mechanism of carboxypeptidase

An important difference between thermolysin and carboxypeptidase leads to the major uncertainty in the mechanism of carboxypeptidase. This difference is that the catalytic carboxylate of carboxypeptidase is far more sterically accessible. The crucial question is whether or not the carboxypeptidase-catalyzed hydrolysis of peptides proceeds via general-base catalysis, as in equation 16.26, or via nucleophilic catalysis, as in 16.27. Early kinetic work concentrated on establishing the participation of the various groups in catalysis.

There is little doubt that the zinc ion acts as an electrophilic catalyst to polarize the carbonyl group and stabilize the negative charge that develops on the oxygen (Chapter 2, section B7).[151] The ionized carboxylate of Glu-270 is implicated in catalysis from the pH-rate profile.[152]

Evidence against the covalent mechanism has been summarized by Mock, who has also proposed alternative general-base-catalyzed mechanisms for thermolysin and carboxypeptidase A.[143] He suggests that His-231 is the general base for thermolysin and the carboxy-terminal carboxylate for carboxypeptidase A. The one common feature of all the proposed mechanisms is the Zn^{2+} functioning as a Lewis acid to polarize the substrate.

d. The zymogen

Procarboxypeptidase A is activated by the removal of a peptide of some 64 residues from the N-terminus by trypsin.[153] This zymogen has significant catalytic activity. As well as catalyzing the hydrolysis of small esters and peptides, procarboxypeptidase removes the C-terminal leucine from lysozyme only seven times more slowly than does carboxypeptidase. Also, the zymogen hydrolyzes Bz-Gly-L-Phe with $k_{cat} = 3$ s^{-1} and $K_M = 2.7$ mM, compared with values of 120 s^{-1} and 1.9 mM for the reaction of the enzyme.[154] In contrast to the situation in chymotrypsinogen, the binding site clearly pre-exists in procarboxypeptidase, and the catalytic apparatus must be nearly complete.

4. The carboxyl (aspartyl) proteases[155]

The carboxyl proteases are so called because they have two catalytically essential aspartate residues. They were formerly called acid proteases because most of them are active at low pH. The best-known member of the family is pepsin, which has the distinction of being the first enzyme to be named (in 1825, by T. Schwann). Other members are chymosin (rennin); cathepsin D; *Rhizopus*-pepsin (from *Rhizopus chinensis*); penicillinopepsin (from *Penicillium janthinellum*); the enzyme from *Endothia parasitica*; and renin, which is involved in the regulation of blood pressure. These constitute a homologous family, and all have an M_r of about 35 000. The aspartyl proteases have been thrown into prominence by the discovery of a retroviral subfamily, including one from HIV that is the target of therapy for AIDS. These are homodimers of subunits of about 100 residues.[156,157] All the aspartyl proteases contain the two essential aspartyl residues. Their reaction mechanism is the most obscure of all the proteases, and there are no simple chemical models for guidance.

a. Pepsin

Pepsin consists of a single polypeptide chain of molecular weight 34 644 and 327 amino acid residues. Ser-68 is phosphorylated, but this phosphate may be removed without significantly altering the catalytic properties of the enzyme. As in other acid proteases, the active site is an extended area that can accommodate at least four or five, and maybe as many as seven, substrate residues.[158,159] The enzyme has a preference for hydrophobic amino acids on either side of the scissile bond. A statistical survey of the bond cleavages in proteins shows that there is a specificity for leucine, phenylalanine, tryptophan, and glutamate (!) in the S_1 site, and for tryptophan, tyrosine, isoleucine, and phenylalanine in the S_1' site.[159] Pepsin rarely catalyzes the hydrolysis of esters, the exceptions being esters of L-β-phenyllactic acid and some sulfite esters.

There are two catalytically active residues in pepsin: Asp-32 and Asp-215. Their ionizations are seen in the pH-activity profile, which has an optimum at pH 2 to 3, and which depends upon the acidic form of a group of $pK_a \sim 4.5$ and the basic form of a group of $pK_a \sim 1.1$.[160,161] The pK_a values have been assigned from the reactions of irreversible inhibitors that are designed to react specifically with ionized or un-ionized carboxyl groups. Diazo compounds—such as N-diazoacetyl-L-phenylalanine methyl ester, which reacts with un-ionized carboxyls—react specifically with Asp-215 up to pH 5 or so (equation 16.28).[162-164] Epoxides, which react specifically with ionized carboxyls, modify Asp-32 (equation 16.29).

$$(\text{Asp-215})-CO_2H + N_2CHCONHCH(Ph)CO_2CH_3 \longrightarrow$$

$$(\text{Asp-215})-CO_2CH_2CONHCH(Ph)CO_2CH_3 + N_2 \quad (16.28)$$

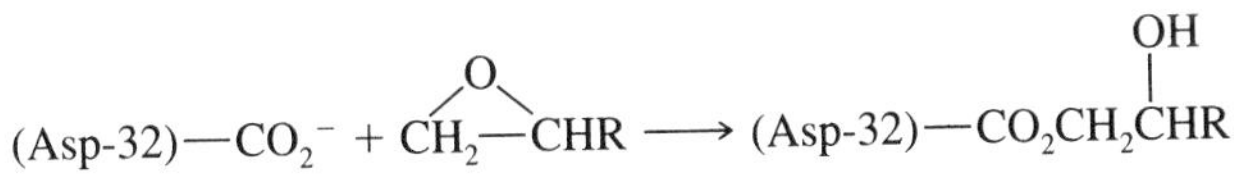

$$(\text{Asp-32})-CO_2^- + \overset{O}{\overset{\diagup\diagdown}{CH_2-CHR}} \longrightarrow (\text{Asp-32})-CO_2CH_2\overset{OH}{\underset{|}{C}HR} \quad (16.29)$$

The pH dependence of the rate of modification shows that the pK_a of Asp-32 is less than 3.[165] It is seen in the high-resolution crystal structures that the carboxyl groups of the two aspartate residues are hydrogen-bonded to each other. This is similar to the ionization of maleic acid, which has pK_a values of 1.9 and 6.2 (equation 16.30).

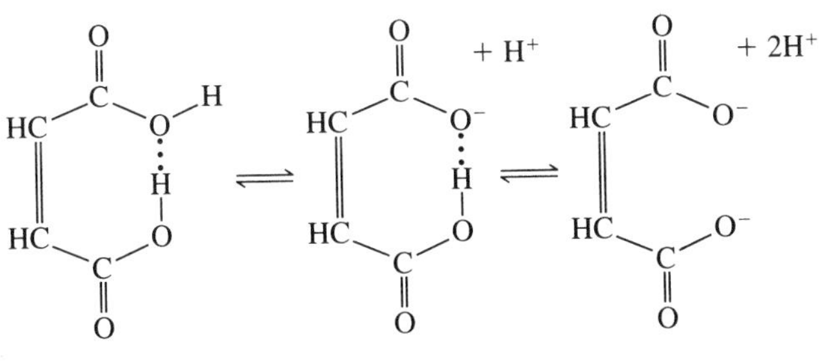

$$(16.30)$$

1. The chemical reaction mechanism of pepsin. Evidence accumulated over the years appeared to support the idea that both an *acyl*enzyme and an *amino*enzyme were formed during the reaction. Some of the evidence, such as the incorporation of ^{18}O into the enzyme during the reaction performed in $H_2^{18}O$, was simply incorrect, but other evidence, principally from transpeptidation studies, may be reinterpreted and shown to be consistent with the *absence* of these intermediates. The interpretation of these experiments is quite instructive and will be discussed next. Current evidence favors general-acid-base catalysis with no intermediates.

Transpeptidation experiments. The pepsin-catalyzed hydrolysis of Leu-Tyr-Leu gives the product Leu-Leu, which can be formed from the acyl transfer shown in equation 16.31.[166,167]

$$E—OH + Leu^*\text{-}Tyr\text{-}Leu \longrightarrow (Leu^*\text{-}CO)—O—E \xrightarrow{Leu^*\text{-}Tyr\text{-}Leu}$$

$$Leu^*\text{-}Leu^*\text{-}Tyr\text{-}Leu + E—OH \qquad (16.31)$$

$$Leu^*\text{-}Leu^* + Tyr\text{-}Leu$$

This experiment has been extended by using the double-labeled substrate [^{14}C]Leu-Tyr-[^{3}H]Leu to show that simultaneous amino and acyl transfer could take place. Both [^{3}H]Leu-[^{3}H]Leu and [^{14}C]Leu-[^{14}C]Leu are formed.[168] The ^{14}C-labeled product, which predominates by a factor of 3 or 4, could come from the acyl transfer route, whereas the ^{3}H-labeled product could arise from the [^{14}C]Leu-Tyr-[^{3}H]Leu-[^{3}H]Leu produced from an aminoenzyme by mechanism 16.32, where E-[^{3}H](NH-Leu) is [^{3}H]Leu bound to E by the NH group of the amino acid.

$$E—OH + [^{14}C]Leu\text{-}Tyr\text{-}[^3H]Leu \longrightarrow E—[^3H](NH\text{-}Leu) \xrightarrow{Leu\text{-}Tyr\text{-}Leu}$$

$$E—OH + [^{14}C]Leu\text{-}Tyr\text{-}[^3H]Leu\text{-}[^3H]Leu \qquad (16.32)$$

$$[^{14}C]Leu\text{-}Tyr + [^3H]Leu\text{-}[^3H]Leu$$

It is clear that this mechanistic interpretation is not unique, because there is no direct evidence for the proposed intermediates. For example, alternative possible explanations are: (1) The amino group in the "aminoenzyme" may not be covalently bound, but may merely be "activated" by the enzyme; and (2) similarly, the acyl transfer reaction of equation 16.31 could occur by the direct attack of Leu-Tyr-Leu on the enzyme-bound Leu-Tyr-Leu. However, M. S. Silver and S. L. T. James[169,170] have proposed a further interpretation, based on the observation that small peptides stimulate the pepsin-catalyzed hydrolysis of other peptides by being first synthesized into larger peptides in a condensation reaction that is the reverse of the hydrolytic step; e.g., equations 16.33. The idea of the condensation of two small peptides to give a larger peptide at a rate that is relatively fast compared with hydrolysis of the small peptides is quite reasonable

Z-Ala-Leu + Phe-Trp-NH$_2$ $\rightleftharpoons$ Z-Ala-Leu-Phe-Trp-NH$_2$

Z-Ala-Leu-Phe-Trp-NH$_2$ $\longrightarrow$ Trp-NH$_2$ + Z-Ala-Leu-Phe $\qquad$ (16.33)

Z-Ala-Leu-Phe $\longrightarrow$ Z-Ala-Leu + Phe

when the following is considered: pepsin has an extended active site so that long peptides react far more rapidly than short ones; and the equilibrium constant for the hydrolysis of a peptide bond is close to 1 M. Therefore, at high concentrations of small peptides that are the reaction products of a reactive larger peptide, the rate of condensation of the small peptides should be comparable with the rate of hydrolysis of the larger peptide, and faster than the rate of hydrolysis of the small peptides. The reaction products in equations 16.31 and 16.32 may be generated by the prior condensation of two molecules of Leu-Tyr-Leu, followed by various modes of hydrolysis and condensation.[170]

Isotope incorporation experiments have provided direct evidence against covalent intermediates between a carboxyl group in the enzyme and the substrate.[171,172] An aminoenzyme can definitely be excluded: if a carboxyl group in the enzyme forms an amide bond with an NH group in a substrate to give E—CO—NHR, then hydrolysis of the aminoenzyme in H$_2$^{18}O will lead to incorporation of ^{18}O into that carboxyl group—and no such incorporation is found. The following experiments on the transpeptidation reaction of Leu-Tyr-NH$_2$ to give Leu-Leu-Tyr-NH$_2$ argue against an acylenzyme. If transpeptidation occurs, as in equation 16.31, by the formation of an acylenzyme that then reacts with the amino group of Leu-Tyr-NH$_2$, there will be no incorporation of ^{18}O from solvent H$_2$^{18}O into the resultant larger peptide. But it is found that such incorporation does occur at a high rate. This is most readily explained by the reaction involving the general-base-catalyzed attack of solvent water on the Leu-Tyr-NH$_2$ to generate Leu-CO^{18}OH, which acts as the acceptor. Further, there is also exchange of some ^{18}O into unreacted Leu-Tyr-NH$_2$. This could result from the formation of a tetrahedral intermediate in which the two oxygens become equivalent and a fraction of the intermediate reverts to starting material. Clearly, however, there are complications from the transpeptidation possibilities suggested by Silver and James.[169,170] For example, products such as Leu-Leu could have arisen from secondary reactions.[173]

The ^{18}O-incorporation experiment was repeated for a reaction in which transpeptidation and secondary cleavage are not observed: the HIV protease–catalyzed cleavage of Ac-Ser-Gln-Asn-Tyr-Pro-Val-Val-NH$_2$ at the Tyr-Pro bond.[174] The substrate was subjected to hydrolysis in 78% enriched H$_2$^{18}O and the remaining substrate isolated after > 80% had converted to products. ^{18}O was found in the Tyr-Pro carbonyl group. The incorporation had to have arisen from the direct attack of H$_2$^{18}O on the substrate to generate a tetrahedral adduct.

The naturally occurring inhibitor pepstatin (structure 16.34) binds very tightly to carboxyl proteases: K_i with porcine pepsin is 4.5×10^{-11} M.[175] The statine residue has a tetrahedral carbon replacing the normal carboxyl carbon, and so perhaps is an analogue of a tetrahedral intermediate that the protein binds tightly.[176] The crystal structure of pepstatin bound to the *Rhizopus chinensis* enzyme has been solved, and it is found that the [middle] statine binds next to the active-site carboxyls.[177] A pepstatin analogue containing a keto group in the reactive position is found from ^{13}C NMR experiments to be bound as a tetrahedral adduct.[178] It is possible that the enzyme binds the hydrated form, $=C(OH)_2$, since this is a transition state analogue of the general-base-catalyzed attack of a water molecule on a carboxyl group (cf. phosphoramidon, structure 16.25).

$$ \text{Valine} \qquad \text{Valine} \qquad \text{Statine} \qquad \text{Alanine} \qquad \text{Statine} $$

$$(16.34)$$

The mechanism in Figure 16.7 fits the structural and kinetic data. There is formation of a terahedral intermediate in which both oxygen atoms become protonated. Such an intermediate has to be formed so that there is a chance that the two oxygen atoms can become equivalent and sufficiently long-lived that exchange can take place in the solvent exchange experiments. Asp-25 (in HIV protease nomenclature) acts as a general acid and Asp-125 as a general base. On the basis of ^{15}N-kinetic isotope effects on the hydrolysis of Ac-Ser-Gln-Asn-Tyr-Pro-Val-Val-NH$_2$, it has been suggested that the first step in this reaction may be the isomerization of the peptide bond (Figure 16.7, bottom).[179]

2. *The zymogen.* Pepsin is formed from pepsinogen by the proteolysis of 44 residues from the N-terminus. The zymogen is stable at neutral pH, but below pH 5 it rapidly and spontaneously activates. The activation process takes place by two separate routes, a pepsin-catalyzed and an intramolecularly catalyzed process. There is much evidence that pepsinogen may activate itself in a unimolecular process, the active site cleaving the N-terminus of its own polypeptide chain.[180–186] Perhaps the neatest demonstration of this intriguing phenomenon is the autoactivation of pepsinogen that is covalently bound to a sepharose resin.[182] The molecules are immobilized and, in general, are not in contact with each other. Yet on exposure to pH 2, the pepsinogen spontaneously activates. The two routes for activation compete, with the bimolecular activation dominating at higher zymogen concentrations and above pH 2.5, and the intramolecular activation dominating at low pH. The result of this spontaneous activation of the pure zymogen is that the majority of pepsinogen molecules will be active 10 s after mixing with the hydrochloric acid in the stomach.

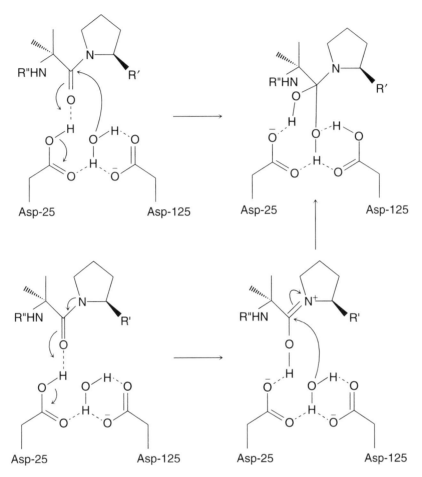

Figure 16.7 Mechanism of aspartyl proteases involving general acid-base catalysis and the formation of a protonated terahedral intermediate. *Bottom:* Proposal by T. J. Rodriguez. T. A. Angeles, and T. D. Meek, *Biochemistry* **32**, 12380 (1993), that the first step is peptide bond isomerization. This accounts for the observed *inverse* $^{15}N/^{14}N$ kinetic isotope effect, which implies that bonding with the N atom becomes stiffer in the transition state.

C. Ribonucleases

Bovine pancreatic ribonuclease catalyzes the hydrolysis of RNA by a two-step process in which a cyclic phosphate intermediate is formed (equation 16.35). The cyclization step is usually much faster than the subsequent hydrolysis, so the intermediate may be readily isolated. DNA is not hydrolyzed, as it lacks the 2′-hydroxyl group that is essential for this reaction. There is a strong specificity for the base B on the 3′ side of the substrate to be a pyrimidine — uracil or cytosine.

$$(16.35)$$

The enzyme consists of a single polypeptide chain of M_r 13 680 and 124 amino acid residues.[187,188] The bond between Ala-20 and Ser-21 may be cleaved by subtilisin. Interestingly, the peptide remains attached to the rest of the protein by noncovalent bonds. The modified protein, called ribonuclease S, and the native protein, now termed ribonuclease A, have identical catalytic activities. Because of its small size, its availability, and its ruggedness, ribonuclease is very amenable to physical and chemical study. It was the first enzyme to be sequenced.[187] The crystal structures of both forms of the enzyme were solved at 2.0-Å resolution several years ago.[189,190] Subsequently, crystal structures of many complexes of the enzyme with substrate and transition analogues and products have been solved at very high resolution.[191] Further, because the catalytic activity depends on the ionizations of two histidine residues, the enzyme has been extensively studied by NMR (the imidazole rings of histidines are easily studied by this method — see Chapter 5).

Bovine pancreatic RNase A is a member of a homologous superfamily. In addition, there is a separate family of guanine-specific microbial RNases that have evolved to have a similar active site.[192,193] Ribonuclease T1 from *Aspergillus oryzae* and the 110-residue barnase from *Bacillus amyloliquefaciens* of M_r 12 392 (see Chapter 19) are the best known examples. One of the histidine residues is replaced by a glutamate in these enzymes. The microbial enzymes are much more amenable to study by protein engineering.

The currently accepted chemical mechanism for the reaction of RNase A was deduced by an inspired piece of chemical intuition before the crystal structure was solved.[194] It was found that the pH-activity curve is bell-shaped, with optimal rates around neutrality. The pH dependence of k_{cat}/K_M shows that the rate depends upon the ionization of a base of pK_a 5.22 and an acid of pK_a 6.78 in the free enzyme, whereas the pH dependence of k_{cat} shows that these are perturbed to pK_a values of 6.3 and 8.1 in the enzyme–substrate complex. It was proposed that the reaction is catalyzed by concerted general-acid–general-base catalysis by two histidine residues, later identified as His-12 and His-119 (reactions 16.36 and 16.37).

In the cyclization step, His-12 acts as a general-base catalyst and His-119 acts as a general acid to protonate the leaving group. Their catalytic roles are reversed in the hydrolysis step: His-119 activates the attack of water by general-base catalysis and His-12 is the acid catalyst, protonating the leaving group. This reversal of roles is quite logical. Reaction 16.37 is essentially the reverse of 16.36, except that HOH replaces ROH. It is expected from the principle of microscopic reversibility that a group reacting as a general acid in one direction will react as a general base in the opposite direction.

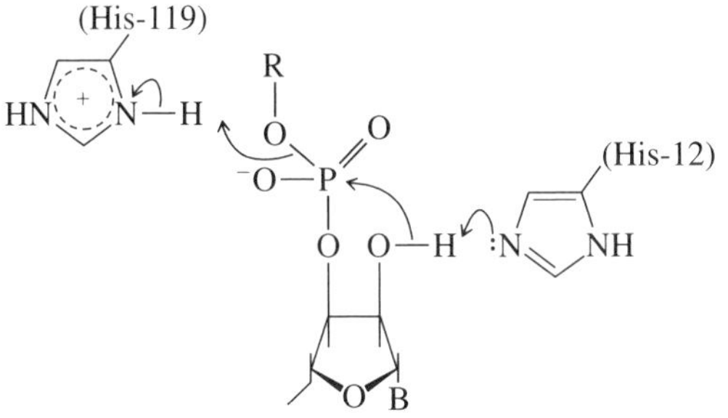

(16.36)

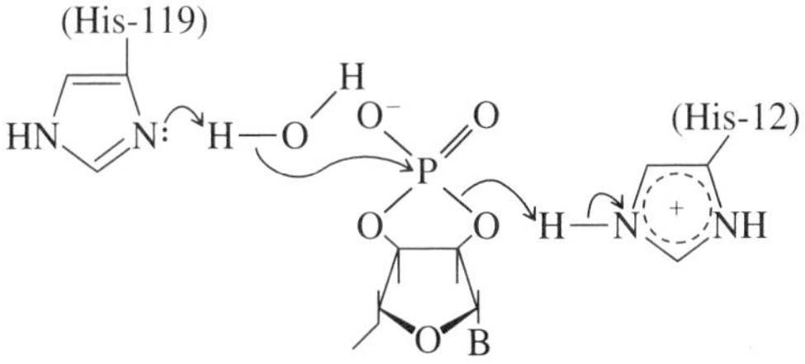

(16.37)

The chemistry and stereochemistry of the reactions were extensively discussed in Chapter 8, sections E1 and E3. There is an in-line mechanism that generates a pentacovalent intermediate or transition state, with the attacking nucleophile and leaving group occupying the apical positions of the trigonal bipyramid.

In the mechanism of barnase, His-119 is replaced by His-102 and His-12 by Glu-73.

1. The structures of ribonuclease A and its complexes

Ribonuclease has a well-defined binding cleft for the substrate. In it are located His-12, His-119, and the side chains of Lys-7, Lys-41, and Lys-66. The structure of the enzyme–substrate complex for the cyclization step was first deduced from the crystal structure of the enzyme and the substrate analogue UpcA (structure 16.38),[195] the phosphonate analogue of UpA. It is a very good analogue, differing from the real substrate only in that a —CH$_2$— group replaces an oxygen atom, so that the structure of its complex with ribonuclease should be close to that of a productively bound enzyme–substrate complex. It was found that His-119 is within hydrogen-bonding distance of the leaving group, and His-12 within hydrogen-bonding distance of the 2'-hydroxyl of the pyrimidine ribose. The pK$_a$ values of His-12 and His-119 have been determined by NMR measurements to be 5.8 and

6.2, respectively, at 40° C.[196] A considerable fraction of each is in the suitable ionic state at physiological pH for the general-acid–general-base catalysis shown in equations 16.36 and 16.37 to occur.

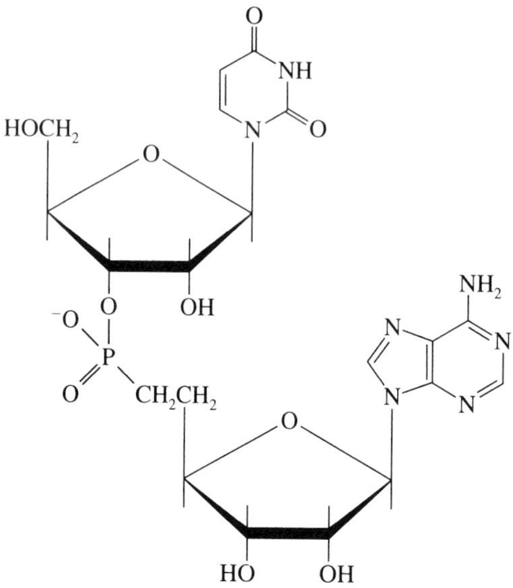

(16.38)

The structures of the native enzyme and its complexes with several inhibitors have since been obtained at higher resolution in other laboratories, to afford a more complete description of the enzyme–substrate interactions.[191] Particularly noteworthy are the lysine residues 7, 41, and 66. That these are an important part of the catalytic machinery has been deduced from their conservation in evolution (they have been found in all homologous ribonucleases that have been se-quenced), and from their loss of activity when they are acetylated. Lys-41 is par-ticularly important. The lysine side chains are very mobile in the free enzyme, but their mobilities are much decreased on the binding of nucleotide substrate analogues. Lys-41 interacts directly with the phosphate moiety and is thought to stabilize the pentacovalent intermediate. Another residue that has been con-served through evolution is Thr-45. This residue is responsible for the specificity of the enzyme for pyrimidines (on the 3′ side). Its backbone NH group and —OH side chain are able to form complementary hydrogen bonds with both uracil and cytosine (structures 16.39).

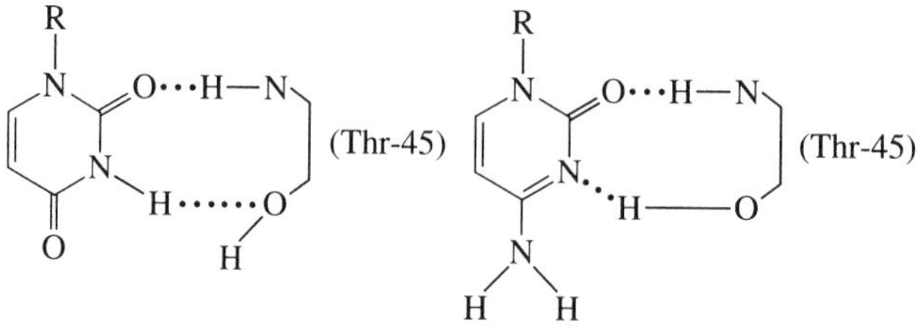

(16.39)

Potentially the most interesting crystal structure solved is that of the complex with uridine vanadate (16.40),

$$(16.40)$$

by a combined neutron and x-ray diffraction study,[197] and most recently refined to 1.3 Å resolution.[198] Uridine vanadate is a transition state analogue, mimicking the pentacovalent intermediate or transition state. An apical and an equatorial oxygen are donated by the ribose, and an apical water molecule is the sixth ligand. Unfortunately, Lys-41 and His-12 are transposed from what is expected from the in-line mechanism of equation 16.36: His-12 is coordinated to an equatorial rather than an apical oxygen and Lys-41 to an apical rather than an axial. Some NMR evidence suggests that His-12 is in a different position in the solution structure from that in the crystal.[199] A quantum mechanical analysis of the structure of the uridine vanadate complex suggests that His-12 may have an additional role in stabilizing the pentacovalent intermediate (or transition state) by acting as a conduit for proton transfers.[200] The subtleties of the mechanism, including the role of Lys-41 and whether the pentacovalent state is an intermediate or transition state, are discussed in reference 200.

2. Mechanism of barnase

Kinetic experiments on substrates of varying length revealed that barnase has subsites that contribute to binding and catalysis (Table 16.3).[201] The subsites are denoted by a subscript to the phosphate 5' to the base: e.g., $5'-Zp_0Gp_1Xp_2Y-3'$ for a substrate with G in the primary binding site; X, Y, and Z other nucleosides; and cleavage of the p_1X bond. The most important subsite is p_2; just the presence of the phosphate increases k_{cat} by a factor of 100, and p_2X increases k_{cat} 3000-fold, with little improvement in K_M. The fragment of DNA, d(CGAC) (i.e., $5'-Cp_0Gp_1Ap_2C-3'$, lacking all 2'-OH groups), binds tightly to barnase, and the crystal structure was solved at 1.76 Å to give a good substrate analogue structure from which the transition state can be modeled (Figure 16.8).[202] Lys-27 takes the place of Lys-41 of ribonuclease A. Mutation of Lys-27 → Ala reduces activity 70-fold.[203] Mutation of His-102 → Ala inactivates the enzyme beyond experimental detection. Mutation of the general base for the first step, Glu-73, to Ala lowers activity 500-fold.[203] Interestingly, mutation of Glu-73 to larger side chains gives larger decreases in rate, consistent with water being the effective base in the Glu-73 → Ala mutant.[204] T1 nuclease has also been subjected to extensive analysis by protein engineering.[205]

Table 16.3 *Kinetic constants for transesterification of oligonucleotide substrates by barnase[a]*

Substrate	K_M (μM)	k_{cat} (s^{-1})	k_{cat}/K_M ($M^{-1} s^{-1}$)
GpA	63	0.73	1.2×10^4
GpG	40	0.25	6.3×10^3
GpC	220	0.34	1.5×10^3
GpU	240	0.55	2.3×10^3
GpUp	19	51	2.7×10^6
GpApC	39	3.6×10^3	9.2×10^7
GpUpC	37	1.7×10^3	4.4×10^7
GpApA	15	2.9×10^3	2.0×10^8
CpGpApC	32	3.2×10^3	9.9×10^7

[a] From A. G. Day, D. Parsonage, S. Ebel, T. Brown, and A. R. Fersht, *Biochemistry* **31**, 6390 (1992).

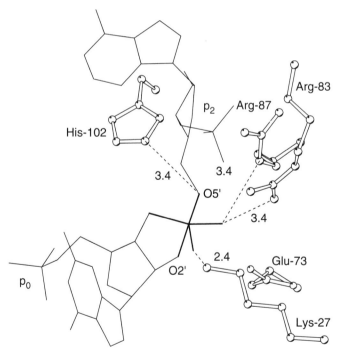

Figure 16.8 The transition state for CGAC bound to barnase, modeled from the structure of the d(CGAC) complex. Arg-83, Arg-87, and Lys-27 can form bonds with equatorial phosphoryl oxygen atoms, whereas His-102 can protonate the leaving group oxygen. [Adapted from A. M. Buckle and A. R. Fersht, *Biochemistry* **33**, 1644 (1994).]

D. Lysozyme[206]

Hen egg white lysozyme is a small protein of M_r 14 500 and 129 amino acid residues. This enzyme was introduced in Chapter 1, where it was pointed out that examination of the crystal structure of the enzyme stimulated most of the solution studies. Hen egg white lysozyme has the distinction of being the first enzyme to have had its structure solved by x-ray crystallography.[207] It is an atypical member of the hexosaminidase class of glycosyl transfer enzymes. It catalyzes the hydrolysis of substrates with retention of stereochemistry. T4 lysozyme was for many years thought to have the same fold and mechanism of lysozyme, despite there being no sequence homology. But it has now been found that the T4 enzyme has inversion of configuration and so operates by a different mechanism.[208,209] A mechanism proposed for the enzymatic reaction was based on the structure of the active site and on ideas from physical organic chemistry.[207,210] This mechanism consisted of the following points:

1. There are six subsites—labeled A, B, C, D, E, and F—for binding the glucopyranose rings of the substrate.
2. The scissile bond lies between sites D and E, close to the carboxyl groups of Glu-35 and Asp-52.
3. The reaction proceeds via an oxocarbenium ion intermediate that is stabilized by the ionized carboxylate of Asp-52 (equation 16.41).

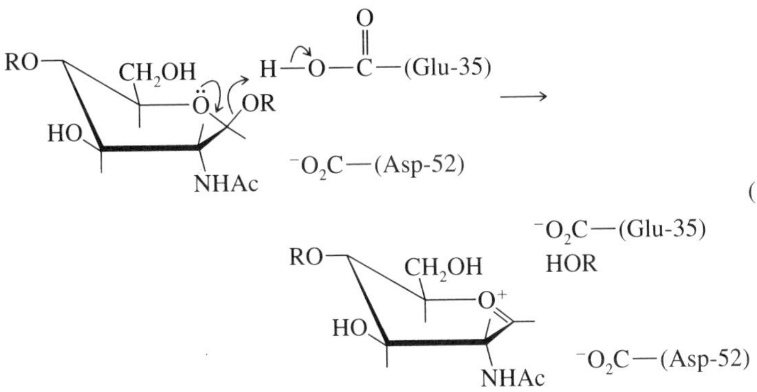

$$(16.41)$$

4. The expulsion of the alcohol is general-acid-catalyzed by the un-ionized carboxyl of Glu-35.
5. Further, as discussed in Chapter 12, sections B2a and D5c, the sugar ring in site D is distorted to the sofa conformation expected for a carbonium ion.
6. Small polysaccharides avoid the strain in the D subsite by binding the A, B, and C sites.

We shall now see how all of these points have been experimentally verified. (The conformation in D was originally called a "half-chair," but "sofa" is more appropriate.)[211]

1. The oxocarbenium ion

Alternatives to the oxocarbenium ion mechanism are the direct attack of water on the substrate, and the nucleophilic attack of Asp-52 on the C-1 carbon to give an ester intermediate. The single-displacement reaction has been ruled out by showing that the reaction proceeds with retention of configuration (Chapter 7, section C3; Chapter 8, section C2).[212–214] The oxocarbenium ion or S_N1 mechanism has been substantiated by secondary isotope effects, using substrates containing either deuterium or tritium instead of hydrogen attached to the C-1 carbon.[215,216] For example, k_H/k_D for structures 16.42 is 1.11,

$$(16.42)$$

compared with values of 1.14 found for a carbonium ion reaction and 1.03 for a bimolecular displacement (S_N2) in simple chemical models. The $^{16}O/^{18}O$ kinetic isotope effect for the leaving group oxygen atom indicates high C—O bond fission in the transition state.[217] It is very difficult to distinguish whether there is a true oxocarbenium ion intermediate or there is an oxocarbenium-like transition state. In free solution, the ion could not exist in the presence of a nucleophile. The question is whether the enzyme can sequester the ion.

2. Electrostatic and general-acid catalysis

The pH dependence of k_{cat}/K_M shows that the reaction rate is dependent on an acid of $pK_a \sim 6$ and a base of $pK_a \sim 4$ in the free enzyme.[218] The pK_a's of all the carboxyls in hen egg white and turkey lysozymes have been measured by two-dimensional ^{1}H-NMR.[219] The values for Glu-35 and As-52 are 6.2 and 3.7, respectively, consistent with the pH dependence of catalysis and their roles as general acids and bases, respectively.

An observation consistent with general-acid catalysis by Glu-35 comes from a study of the reverse reaction. It is found that the rate of reaction of alcohols with the carbonium ion intermediate is virtually independent of their pK_a's. This is consistent with the general-base-catalyzed attack of the alcohol on the ion; hence, by the principle of microscopic reversibility, the expulsion of the alcoholate ion from the glycoside is general-acid-catalyzed.[220]

It was emphasized by C. A. Vernon in the initial formulation of the lysozyme mechanism that the electrostatic stabilization of the oxocarbenium ion by Asp-52 is the most important catalytic factor.[208] Theoretical calculations suggest that the activation energy for k_{cat} is lowered by ~ 32 kJ/mol (~ 8 kcal/mol) by electrostatic stabilization.[221] A high value for this is indicated by experiments showing that the chemical conversion of the carboxyl group of Asp-52 to —CH_2OH abolishes the enzymatic activity.[222] Site-directed mutagenesis of Glu-35 and Asp-52 also abolishes activity.[223,224]

3. Binding energies of the subsites

It was originally proposed that nonbonded interactions between the enzyme and the sugar ring in site D distort it to the sofa conformation of the oxocarbenium ion. Various workers have searched for weak binding in this site by estimating the binding energies of the individual sites from binding and kinetic measurements.[225,226] Some estimates for these are given in Table 16.4. Although the values are not precise, it is clear that there is a repulsive energy against the binding of NAM in site D. Also, as $(NAG)_4$ is found to bind about equally in sites A, B, and C, on the one hand, and A, B, C, and D, on the other, there is no net binding energy for NAG in site D.[227] The positions of binding of small substrates have been located from their interactions with probes, such as the dye Biebrich Scarlet[228] and the lanthanide ion,[227] which have been bound in the cleft; these substrates have been shown to be predominantly nonproductively bound.

The more recent structural evidence supports the role of substrate distortion. A trisaccharide in the BCD sites of crystalline hen egg white lysozyme is distorted to the half-chair,[229] as is a substrate bound to the Asp-52 → Ser mutant studied in solution by NMR.[230] Protein engineering studies on the residues involved support the conclusions.[231] A chitobiose bound to a crystalline chitobiase has the equivalent sugar distorted to a sofa.[232]

The deductions from this remarkable example of x-ray crystallography have not only stood the test of time, but have been neatly confirmed by solution

Table 16.4 *Binding energies of subsites in hen egg white lysozyme[a]*

		Binding energy	
Site	Residue binding[b]	kJ/mol	kcal/mol
A	NAG	−8	−2
B	NAG	−12	−3
	NAM	−16	−4
C	NAG	−20	−5
D	NAM	+12	+3
	NAG	0	0
E	NAG	−16	−4
F	NAG	−8	−2

[a] See also M. Schindler, Y. Assaf, N. Sharon, and D. M. Chipman, *Biochemistry* **16**, 423 (1977).
[b] NAG = *N*-acetylglucosamine; NAM = *N*-acetylmuramic acid.

studies, except that the emphasis of the strain mechanism is now on the electrostatic component rather than on that due to distortion.

E.   Some generalizations

Enzymes can invoke different mechanisms to catalyze the same type of reaction. The different classes of proteases are seen to invoke radically different mechanisms. Glycosyl and phosphoryl transfer can involve retention or inversion of configuration, showing that there are at least two classes of mechanisms for each reaction, retention implying a double displacement (Chapter 8, section C). The involvement of different neighboring groups on different sugar substrates in glycosyl transfer can give rise to unique chemical mechanisms for each substrate.[206] The basic principles of catalysis by enzymes are general, but the details of the covalent chemistry can vary.

It is difficult to attribute quantitatively by experiment the rate enhancements of the different factors contributing to catalysis. Protein engineering can get close to accurate answers when dealing with nonpolar interactions, especially in subsites. But analysis of mutation is at its weakest when altering residues that interact with charges (Chapter 15). The next development must be in improved methods of computer simulation. Controversies arise when there are no intermediates in the reaction because the kinetics can fit more than one mechanism. Again, computer simulation will provide the ultimate answers.

References

1. A. R. Clarke and T. R. Dafforn, in *Comprehensive biological catalysis*, Vol. 3, M. Sinnott, ed., Academic Press, (1998).
2. G. Popják, *The Enzymes* **2**, 115 (1970).
3. H. F. Fisher, E. E. Conn, B. Vennesland, and F. H. Westheimer, *J. Biol. Chem.* **202**, 687 (1953).
4. M. E. Pullman, A. San Pietro, and S. P. Colowick, *J. Biol. Chem.* **206**, 129 (1954).
5. A. M. Gronenborn and G. M. Clore, *J. Molec. Biol.* **157**, 155 (1982).
6. S. A. Benner, *Experientia* **38**, 633 (1982).
7. N. J. Oppenheimer, *J. Am. Chem. Soc.* **106**, 3032 (1984).
8. S. A. Benner, K. P. Nambiar, and G. K. Chambers, *J. Am. Chem. Soc.* **107**, 5513 (1985).
9. H. Eklund and C.-I. Brändén, in *Pyridine nucleotide coenzymes, Part A*, D. Dolphin and R. Poulson, eds., John Wiley, 253 (1987).
10. A. Ohno and K. Ushio, in *Pyridine nucleotide coenzymes, Part B*, D. Dolphin and R. Poulson, eds., John Wiley, 275 (1987).
11. H. Eklund, B. Nordstrom, E. Zeppezauer, G. Söderlund, I. Ohlsson, T. Boiwe, B.-O. Söderberg, O. Tapia, C.-I. Brändén, and Å. Åkeson, *J. Molec. Biol.* **102**, 27 (1976).
12. H. Eklund, J.-P. Samama, L. Wallen, C.-I. Brändén, Å, Åkersen, and T. A. Jones, *J. Molec. Biol.* **146**, 561 (1981).

13. H. Eklund, C.-I. Brändén, and H. Jornvall, *J. Molec. Biol.* **102**, 61 (1976).
14. H. Eklund, B. V. Plapp, J.-P. Samama, and C.-I. Brändén, *J. Biol. Chem.* **257**, 14349 (1982).
15. K. Dalziel, *J. Biol. Chem.* **238**, 2850 (1963).
16. J. D. Shore, H. Gutfreund, R. L. Brooks, D. Santiago, and P. Santiago, *Biochemistry* **13**, 4185 (1974).
17. P. Andersson, J. Kvassman, A. Linström, B. Olden, and G. Pettersson, *Eur. J. Biochem.* **113**, 425 (1981).
18. E. Cedergren-Zeppezauer, J.-P. Samama, and H. Eklund, *Biochemistry* **21**, 4895 (1982).
19. H. Dutler and C.-I. Brändén, *Bioorg. Chem.* **10**, 1 (1981).
20. M. F. Dunn, F. F. Biellman, and G. Bruylant, *Biochemistry* **14**, 3176 (1975).
21. J. P. Klinman, *Biochemistry* **15**, 2018 (1976).
22. J. Kvassman and G. Pettersson, *Eur. J. Biochem.* **103**, 557, 565 (1980).
23. J. Kvassman, A. Larsson, and G. Pettersson, *Eur. J. Biochem.* **114**, 555 (1981).
24. P. F. Cook and W. W. Cleland, *Biochemistry* **20**, 1805 (1981).
25. T. Ehrig, T. D. Hurley, H. J. Edenberg, and W. F. Bosron, *Biochemistry* **30**, 1062 (1991).
26. H. Theorell and B. Chance, *Acta Chem. Scand.* **5**, 1127 (1951).
27. C. C. Wratten and W. W. Cleland, *Biochemistry* **2**, 935 (1963); **4**, 2442 (1965).
28. J. D. Shore, H. Gutfreund, and D. Yates, *J. Biol. Chem.* **250**, 5276 (1975).
29. S. A. Bernhard, M. F. Dunn, P. L. Luisi, and P. Schack, *Biochemistry* **9**, 185 (1970).
30. M. Hennecke and B. V. Plapp, *Biochemistry* **22**, 3721 (1983).
31. J. P. Klinman, *J. Biol. Chem.* **247**, 7977 (1972).
32. J. D. Shore and H. Gutfreund, *Biochemistry* **9**, 4655 (1970).
33. R. L. Brooks and J. D. Shore, *Biochemistry* **10**, 3855 (1971).
34. J. P. Klinman, *C. R. C. Crit. Rev. Biochem.* **10**, 39 (1981).
35. B. J. Bahnson and J. P. Klinman, *Methods in Enzymology* **249**, 373 (1995).
36. D. C. Anderson and F. W. Dahlquist, *Biochemistry* **21**, 3569, 3578 (1982).
37. M. Hadorn, V. A. John, F. K. Meier, and H. Dutler, *Eur. J. Biochem.* **54**, 65 (1975).
38. R. J. Kordal and S. M. Parsons, *Archs. Biochem. Biophys.* **104**, 439 (1979).
39. J. Kvassman and G. Pettersson, *Eur. J. Biochem.* **69**, 279 (1976).
40. C. F. Weidig, H. R. Halvorson, and J. D. Shore, *Biochemistry* **16**, 2916 (1977).
41. D. W. Green, H. W. Sun, and B. V. Plapp, *J. Biol. Chem.* **268**, 7792 (1993).
42. E. G. Weinhold and S. A. Benner, *Protein Engineering* **8**, 457 (1995).
43. B. J. Bahnson, D. H. Park, K. Kim, B. V. Plapp, and J. P. Klinman, *Biochemistry* **32**, 5503 (1993).
44. A. R. Clarke, T. Atkinson, and J. J. Holbrook, *Trends. Biochem. Sci.* **14**, 101 (1989).
45. A. R. Clarke, T. Atkinson, and J. J. Holbrook, *Trends. Biochem. Sci.* **14**, 145 (1989).
46. C. L. Markert and F. Moller, *Proc. Natl. Acad. Sci. USA* **45**, 753 (1959).
47. I. Fine, N. O. Kaplan, and D. Kuftinec *Biochemistry* **2**, 116 (1963).
48. O. P. Chilson, L. A. Costello, and N. O. Kaplan, *Biochemistry* **4**, 271 (1965).
49. J. L. White, M. L. Hackert, M. Buehner, M. J. Adams, G. C. Ford, P. J. Lentz, Jr., I. E. Smiley, S. J. Steindel, and M. G. Rossmann, *J. Molec. Biol.* **102**, 759 (1976).
50. M. J. Adams, M. Buehner, K. Chandrasekhar, G. C. Ford, M. L. Hackert, A. Liljas,

M. G. Rossmann, I. E. Smiley, W. S. Allison, J. Everse, N. O. Kaplan, and S. S. Taylor, *Proc. Natl. Acad. Sci. USA* **70**, 1968 (1973).

51. U. M. Grau, W. E. Trommer, and M. G. Rossmann, *J. Molec. Biol.* **150**, 289 (1981).
52. J. Everse, R. E. Barnett, C. J. R. Thorne, and N. O. Kaplan, *Archs. Biochem. Biophys.* **143**, 444 (1971).
53. K. W. Hart, A. R. Clarke, D. B. Wigley, A. Waldman, W. N. Chia, D. A. Barstow, T. Atkinson, J. B. Jones, and J. J. Holbrook, *Biochim. Biophys. Acta* **914**, 294 (1987).
54. C. J. Coulson and B. R. Rabin, *FEBS Lett.* **3**, 333 (1969).
55. J. J. Holbrook and H. Gutfreund, *FEBS Lett.* **31**, 157 (1973).
56. W. B. Novoa and G. W. Schwert, *J. Biol. Chem.* **236**, 2150 (1961).
57. J. R. Whitaker, D. W. Yates, N. G. Bennett, J. J. Holbrook, and H. Gutfreund, *Biochem. J.* **139**, 677 (1974).
58. J. J. Holbrook and V. A. Ingram, *Biochem. J.* **131**, 729 (1973).
59. A. Waldman, A. R. Clarke, D. B. Wigley, K. W. Hart, W. N. Chia, D. Barstow, T. Atkinson, I. Munro, and J. J. Holbrook, *Biochim. Biophys. Acta* **913**, 66 (1987).
60. H. M. Wilks, K. W. Hart, R. Feeney, C. R. Dunn, H. Muirhead, W. N. Chia, D. A. Barstow, T. Atkinson, A. R. Clarke, and J. J. Holbrook, *Science* **242**, 1541 (1988).
61. H. M. Wilks, K. M. Moreton, D. J. Halsall, K. W. Hart, R. D. Sessions, A. R. Clarke, and J. J. Holbrook, *Biochemistry* **31**, 7802 (1992).
62. J. I. Harris and M. Waters, *The Enzymes* **13**, 1 (1976).
63. K. Dalziel, N. V. McFerran, and A. J. Wonacott, *Phil. Trans. R. Soc.* **B293**, 105 (1981).
64. H. L. Segal and P. D. Boyer, *J. Biol. Chem.* **204**, 265 (1953).
65. P. J. Harrigan and D. R. Trentham, *Biochem, J.* **143**, 353 (1974).
66. R. G. Duggleby and D. T. Dennis, *J. Biol. Chem.* **249**, 167 (1974).
67. J.-C. Meunier and K. Dalziel, *Eur. J. Biochem.* **82**, 483 (1978).
68. I. Krimsky and E. Racker, *Science, N.Y.* **122**, 319 (1955).
69. D. R. Trentham, *Biochem. J.* **122**, 59, 71 (1971).
70. L. D. Byers and D. E Koshland, *Biochemistry* **14**, 3661 (1975).
71. D. Moras, K. W. Olsen, M. N. Sabesan, M. Buehner, G. C. Ford, and M. G. Rossmann, *J. Biol. Chem.* **250**, 9137 (1975).
72. G. Biesecker, J. I. Harris, J. C. Thierry, J. E. Walker, and A. J. Wonacott, *Nature, Lond.* **266**, 328 (1977).
73. H. C. Watson, E. Duée, and W. D. Mercer, *Nature New Biology, Lond.* **240**, 130 (1972).
74. M. R. N. Murthy, R. M. Garavito, J. E. Johnson, and M. G. Rossmann, *J. Molec. Biol.* **138**, 859 (1980).
75. M. Buehner, G. C. Ford, D. Moras, K. W. Olsen, and M. G. Rossmann, *J. Molec. Biol.* **90**, 25 (1974).
76. A. Conway and D. E. Koshland, Jr., *Biochemistry* **7**, 4011 (1968).
77. B. D. Peczon and H. O. Spivey, *Biochemistry* **11**, 2209 (1972).
78. P. J. Harrigan and D. R. Trentham, *Biochem. J.* **135**, 695 (1973).
79. F. Seydoux, S. A. Bernhard, O. Pfenninger, M. Payne, and O. P. Malhotra, *Biochemistry* **12**, 4290 (1973).
80. J. Schlessinger and A. Levitzki, *J. Molec. Biol.* **82**, 547 (1974).
81. A. Levitzki, *J. Molec. Biol.* **90**, 451 (1974).

82. F. Seydoux and S. A. Bernhard, *Bioorg. Chem.* **1**, 161 (1974).
83. N. Kelemen, N. Kellershohn, and F. Seydoux, *Eur. J. Biochem.* **57**, 69 (1975).
84. L. S. Gennis, *Proc. Natl. Acad. Sci. USA* **73**, 3928 (1976).
85. A. Leslie and A. J. Wonacott, *J. Molec. Biol.* **165**, 375 (1983)
86. M. Dunn, *Biochemistry* **13**, 1146 (1974).
87. L. M. Babé and C. S. Craik, *Cell* **91**, 427 (1997).
88. C. W. Wharton, in *Comprehensive biological catalysis*, vol. 1, M. Sinnott, ed., Academic Press, 345 (1998).
89. J. J. Perona and C. S. Craik, *Protein Science* **4**, 337 (1995).
90. J. J. Perona and C. S. Craik, *J. Biol. Chem.* **272**, 29987 (1997).
91. G. Schoellman and E. Shaw, *Biochemistry* **2**, 252 (1963).
92. D. M. Blow, J. J. Birktoft, and B. S. Hartley, *Nature, Lond.* **221**, 337 (1970).
93. G. D. Brayer, L. T. J. Delbaere, and M. N. G. James, *J. Molec. Biol.* **124**, 261 (1978).
94. T. Inagami and J. M. Sturtevant, *Biochim. Biophys. Acta* **38**, 64 (1960).
95. A. R. Fersht and J. Sperling, *J. Molec. Biol.* **74**, 137 (1973).
96. A. R. Fersht and M. Renard, *Biochemistry* **13**, 1416 (1974).
97. M. A. Porubcan, W. M. Westler, I. B. Ibañez, and J. L. Markley, *Biochemistry* **18**, 4108 (1979).
98. W. W. Bachovchin, R. Kaiser, J. H. Richards, and J. D. Roberts, *Proc. Natl. Acad. Sci. USA* **78**, 7323 (1981).
99. A. A. Kossiakoff and S. A. Spencer, *Biochemistry* **20**, 6462 (1981).
100. P. Carter and J. A. Wells, *Nature, Lond.* **332**, 564 (1988).
101. P. A. Frey, S. A. Whitt, and J. B. Tobin, *Science* **264**, 1927 (1994).
102. C. S. Cassidy, J. Lin, and P. A. Frey, *Biochemistry* **36**, 4576 (1997).
103. A. Warshel and A. Papazyan, *Proc. Natl. Acad. Sci. USA* **93**, 13665 (1996).
104. E. L. Ash, J. L. Sudmeier, E. C. DeFabo, and W. W. Bachovchin, *Science* **278**, 1128 (1997).
105. D. A. Matthews, R. A. Alden, J. J. Birktoft, S. T. Freer, and T. Kraut, *J. Biol. Chem.* **252**, 8875 (1977).
106. J. Kraut, *Ann. Rev. Biochem.* **46**, 331 (1977).
107. G. H. Cohen, E. W. Silverton, and D. R. Davies, *J. Molec. Biol.* **148**, 449 (1981).
108. W. Bode, Z. Chen, K. Bartels, C. Kutzbach, G. Schmidt-Kastner, and H. Bartunik, *J. Molec. Biol.* **164**, 237 (1983).
109. D. L. Hughes, L. C. Sieker, J. Bieth, and J.-L. Dimicoli, *J. Molec. Biol.* **162**, 645 (1982).
110. P. R. Carey and P. J. Tonge, *Acct. Chem. Res.* **28**, 8 (1995).
111. P. J. Tonge and P. R. Carey, *Biochemistry* **29**, 10723 (1990).
112. P. J. Tonge and P. R. Carey, *Biochemistry* **31**, 9122 (1992).
113. G. E. Hein and C. Niemann, *J. Am. Chem. Soc.* **84**, 4495 (1962).
114. T. Alber, G. A. Petsko, and D. Tsernoglou, *Nature, Lond,* **263**, 297 (1976).
115. R. Henderson, *J. Molec. Biol.* **54**, 341 (1970).
116. A. R. Fersht, D. M. Blow, and J. Fastrez, *Biochemistry* **12**, 2035 (1973).
117. S. A. Bizzozero, W. K. Baumann, and H. Dutler, *Eur. J. Biochem.* **122**, 251 (1982).
118. P. Campbell, N. T. Nashed, B. A. Lapinskas, and J. Gurrieri, *J. Biol. Chem.* **258**, 59 (1983).
119. B. Asboth and L. Polgar, *Biochemistry* **22**, 117 (1983).
120. J. Fastrez and A. R. Fersht, *Biochemistry* **12**, 1067 (1973).

121. J. L. Markley, F. Travers, and C. Balny, *Eur. J. Biochem.* **120**, 477 (1981).
122. M. Hunkapiller, M. D. Forgac, E. H. Yu, and J. H. Richards, *Biochem. Biophys. Res. Commun.* **87**, 25 (1979).
123. M. Fujinaga, R. J. Read, A. Sielecki, W. Ardelt, M. Laskowski, Jr., and M. N. G. James, *Proc. Natl. Acad. Sci. USA* **79**, 4868 (1982).
124. G. D. Brayer, L. T. J. Delbaere, M. N. G. James, C.-A. Bauer, and R. C. Thompson, *Proc. Natl. Acad. Sci. USA* **76**, 96 (1979).
125. L. T. J. Delbaere and G. D. Brayer, *J. Molec. Biol.* **139**, 45 (1980).
126. S. T. Freer, J. Kraut, J. D. Robertus, H. T. Wright, and Ng. H. Xuong, *Biochemistry* **9**, 1997 (1970).
127. H. T. Wright, *J. Molec. Biol.* **79**, 1, 13 (1973).
128. J. J. Birktoft, J. Kraut, and S. T. Freer, *Biochemistry* **15**, 4481 (1976).
129. G. Robillard and R. G. Shulman, *J. Molec. Biol.* **86**, 519 (1974).
130. A. R. Fersht, *FEBS Lett.* **29**, 283 (1973).
131. A. Gertler, K. A. Walsh, and H. Neurath, *Biochemistry* **13**, 1302 (1974).
132. A. R. Fersht, *J. Molec. Biol.* **64**, 497 (1972).
133. H. Fehlhammer, W. Bode, and R. Huber, *J. Molec. Biol.* **111**, 415 (1977).
134. W. Bode, *J. Molec. Biol.* **127**, 357 (1979).
135. C. S. Craik, C. Largman, T. Fletcher, S. Roczniak, P. J. Barr, R Fletterick, and W. J. Rutter, *Science* **228**, 291 (1985).
136. L. Graf, C. S. Craik, A. Patthy, S. Roczniak, R. J. Fletterick, and W. J. Rutter, *Biochemistry* **26**, 2616 (1987).
137. L. B. Evnin, J. R. Vasquez, and C. S. Craik, *Proc. Natl. Acad. Sci. USA* **87**, 6659 (1990).
138. W. S. Willett, S. A. Gillmor, J. J. Perona, R. J. Fletterick, and C. S. Craik, *Biochemistry* **34**, 2172 (1995).
139. L. Hedstrom, L. Szilagyi, and W. J. Rutter, *Science* **255**, 1249 (1992).
140. L. Hedstrom, J. J. Perona, and W. J. Rutter, *Biochemistry* **33**, 8757 (1994).
141. K. Brocklehurst, A. B. Watts, M. Patel, and E. W. Thomas, in *Comprehensive biological catalysis*, vol. 1, M. Sinnott, ed., Academic Press, 381 (1998).
142. A. R. Fersht, *J. Am. Chem. Soc.* **93**, 3504 (1971).
143. W. L. Mock, in *Comprehensive biological catalysis*, vol. 1, M. Sinnott, ed., Academic Press, 425 (1998).
144. B. W. Matthews, *Acc. Chem. Res.* **21**, 333 (1988).
145. S. K. Burley, P. R. David, R. M. Sweet, M. Taylor, and W. N. Lipscomb, *J. Molec. Biol.* **224**, 113 (1992).
146. G. N. Reeke, J. A. Hartsuck, M. L. Ludwig, F. A. Quiocho, T. A. Steitz, and W. N. Lipscomb, *Proc. Natl. Acad. Sci. USA* **58**, 2220 (1967).
147. W. N. Lipscomb, *Proc. Natl. Acad. Sci. USA* **70**, 3797 (1973).
148. M. F. Schmid and J. R. Herriott, *J. Molec. Biol.* **103**, 175 (1976).
149. S. J. Gardell, C. S. Craik, D. Hilvert, M. S. Urdea, and W. J. Rutter, *Nature, Lond.* **317**, 551 (1985).
150. L. H. Weaver, W. R. Kester, and B. W. Matthews, *J. Molec. Biol.* **114**, 119 (1977).
151. M. A. Holmes, D. E. Tronrud, and B. W. Matthews, *Biochemistry* **22**, 236 (1983).
152. B. L. Vallee, J. F. Riordan, and J. E. Coleman, *Proc. Natl. Acad. Sci. USA* **49**, 109 (1963).
153. J. H. Freisheim, K. A. Walsh, and H. Neurath, *Biochemistry* **6**, 3010, 3020 (1967).
154. J. R. Uren and H. Neurath, *Biochemistry* **13**, 3512 (1974).

155. T. D. Meek, in *Comprehensive biological catalysis*, vol. 1, M. Sinnott, ed., Academic Press, 327 (1998).
156. J. K. M. Rao, J. W. Erickson, and A. Wlodawer, *Biochemistry* **30**, 4663 (1991).
157. A. Wlodawer and J. W. Erickson, *Ann. Rev. Biochem.* **62**, 543 (1993).
158. P. S. Sampath-Kumar and J. S. Fruton, *Proc. Natl. Acad. Sci. USA* **71**, 1070 (1974).
159. A. A. Zinchenko, L. D. Rumsh, and V. K. Antonov, *Bioorg. Chem. (USSR)* **2**, 803 (1976).
160. J. L. Denburg, R. Nelson, and M. S. Silver, *J. Am. Chem. Soc.* **90**, 479 (1968).
161. A. J. Cornish-Bowden and J. R. Knowles, *Biochem. J.* **113**, 353 (1969).
162. G. R. Delpierre and J. S. Fruton, *Proc. Natl. Acad. Sci. USA* **54**, 1161 (1965); **56**, 1817 (1966).
163. R. L. Lundblad and W. H. Stein, *J. Biol. Chem.* **244**, 154 (1969).
164. R. S. Bayliss, J. R. Knowles, and G. B. Wybrandt, *Biochem. J.* **113**, 377 (1969).
165. J. A. Hartsuck and J. Tang, *J. Biol. Chem.* **247**, 2575 (1972).
166. M. Takahashi and T. Hofmann, *Biochem. J.* **127**, 35P (1972); **147**, 549 (1975).
167. T. T. Wang and T. Hofmann, *Biochem. J.* **153**, 691 (1976).
168. A. K. Newmark and J. R. Knowles, *J. Am. Chem. Soc.* **97**, 3557 (1975).
169. M. S. Silver and S. L. T. James, *Biochemistry* **20**, 3177 (1980).
170. M. S. Silver and S. L. T. James, *Biochemistry* **20**, 3183 (1980).
171. V. K. Antonov, L. M. Ginodman, Y. V. Kapitannikov, T. N. Barshevskaya, A. G. Gurova, and L. D. Rumsh, *FEBS Lett.* **88**, 87 (1978).
172. V. K. Antonov, L. M. Ginodman, L. E. Rumsh, Y. V. Kapitannikov, T. N. Barshevskaya, L. B. Yavashev, A. G. Gurova, and L. I. Volkova, *Eur. J. Biochem.* **117**, 195 (1981).
173. T. Hofman and A. L. Fink, *Biochemistry* **23**, 5247 (1984).
174. L. J. Hyland, T. A. Tomaszek, Jr., G. D. Roberts, S. A. Carr, V. W. Magaard, H. L. Bryan, S. A. Fakhoury, M. L. Moore, M. D. Minnich, J. S. Culp, R. L. Desjarlais, and T. D. Meek, *Biochemistry* **30**, 8441 (1991).
175. R. J. Workman and D. W. Burkitt, *Archs. Biochem. Biophys.* **194**, 157 (1979).
176. J. Marciniszyn, J. A Hartsuck, and J. Tang, *J. Biol. Chem.* **251**, 7088 (1976).
177. R. Bott, E. Subramanian, and D. R. Davies, *Biochemistry* **21**, 6956 (1982).
178. D. H. Rich, M. S. Bernatowicz, and P. G. Schmidt, *J. Am. Chem. Soc.* **104**, 3535 (1982).
179. T. J. Rodriguez, T. A. Angeles, and T. D. Meek, *Biochemistry* **32**, 12380 (1993).
180. M. Bustin, M. C. Lin, W. H. Stein, and S. Moore, *J. Biol. Chem.* **245**, 846 (1970).
181. J. Tang, *Biochem. Biophys. Res. Commun.* **41**, 697 (1970).
182. M. Bustin and A. Conway-Jacobs, *J. Biol. Chem.* **246**, 615 (1971).
183. J. Al-Janabi, J. A. Hartsuck, and J. Tang, *J. Biol. Chem.* **247**, 4628 (1972).
184. P. McPhie, *J. Biol. Chem.* **247**, 4277 (1972).
185. C. G. Sanny, J. A. Hartsuck, and J. Tang, *J. Biol. Chem.* **250**, 2635 (1975).
186. C. W. Dykes and J. Kay, *Biochem. J.* **153**, 141 (1976).
187. C. H. W. Hirs, S. Moore, and W. H. Stein, *J. Biol. Chem.* **235**, 633 (1960).
188. D. G. Smyth, W. H. Stein, and S. Moore, *J. Biol. Chem.* **238**, 227 (1963).
189. G. Kartha, J. Bello, and D. Harker, *Nature, Lond.* **213**, 862 (1967).
190. H. W. Wyckoff, D. Tsernoglou, A. W. Hanson, J. R. Knox, B. Lee, and F. M. Richards, *J. Biol. Chem.* **245**, 305 (1970).
191. G. L. Gilliland, in *Ribonucleases: structure and function*, G. D'Alessio and J. Riordan, eds., Academic Press (1997).

192. Y. Mauguen, R. W. Hartley, E. J. Dodson, G. G. Dodson, G. Bricogne, C. Chothia, and A. Jack, *Nature, Lond.* **297**, 162 (1982).
193. C. Hill, G. Dodson, U. Heinemann, W. Saenger, Y. Mitsui, K. Nakamura, S. Borisov, G. Tischenko, K. Polyakov, and S. Pavlovsky, *Trends Biochem. Sci.* **8**, 364 (1983).
194. D. Findlay, D. G. Herries, A. P. Mathias, B. R. Rabin, and C. A. Ross, *Nature, Lond.* **190**, 781 (1961).
195. F. M. Richards, H. W. Wyckoff, W. D. Carlson, N. M. Allewell, B. Lee, and Y. Mitsui, *Cold Spring Harbor Symp. Quant. Biol.* **36**, 35 (1971).
196. J. L. Markley, *Biochemistry* **14**, 3546 (1975).
197. B. Borah, C. W. Chen, W. Egan, M. Miller, A. Wlodawer, and J. S. Cohen, *Biochemistry* **24**, 2058 (1985).
198. J. E. Ladner, B. D. Wladkowski, L. A. Svensson, L. Sjolin, and G. L. Gilliland, *Acta Crystallog. Section D Biological Crystallography* **53**, 290 (1997).
199. T. D. Veenstra and L. Lee, *Biophysical Journal* **67**, 331 (1994).
200. B. D. Wladkowski, L. A. Svensson, L. Sjolin, J. E. Ladner, and G. L. Gilliland, *J. Amer. Chem. Soc.* **120**, 5488 (1998).
201. A. G. Day, D. Parsonage, S. Ebel, T. Brown, and A. R. Fersht, *Biochemistry* **31**, 6390 (1992).
202. A. M. Buckle and A. R. Fersht, *Biochemistry* **33**, 1644 (1994).
203. D. E. Mossakowska, K. Nyberg, and A. R. Fersht, *Biochemistry* **28**, 3843 (1989).
204. G. Schreiber, C. Frisch, and A. R. Fersht, *J. Molec. Biol.* **270**, 111 (1997).
205. J. Steyaert, *Europ. J. Biochem.* **247**, 1 (1997).
206. G. Davies, M. L. Sinnot, and S. G. Withers, in *Comprehensive biological catalysis*, M. Sinnott, ed., Academic Press, 119 (1998).
207. C. C. F. Blake, L. N. Johnson, G. A. Mair, A. C. T. North, D. C. Phillips, and V. R. Sarma, *Proc. R. Soc.* **B167**, 378 (1967).
208. R. Kuroki, L. H. Weaver, and B. W. Matthews, *Science* **262**, 2030 (1993).
209. R. Kuroki, L. H. Weaver, and B. W. Matthews, *Nature Structural Biology* **2**, 1007 (1995).
210. C. A. Vernon, *Proc. R. Soc.* **B167**, 389 (1967).
211. L. O. Ford, L. N. Johnson, P. A. Machin, D. C. Phillips, and R. Tijian, *J. Molec. Biol.* **88**, 349 (1974).
212. J. A. Rupley and V. Gates, *Proc. Natl. Acad. Sci. USA* **57**, 496 (1967).
213. M. A. Raftery and T. Rand-Meir, *Biochemistry* **7**, 3281 (1968).
214. U. Zehavi, J. J. Pollock, V. I. Teichberg, and N. Sharon, *Nature, Lond.* **219**, 1152 (1968).
215. F. W. Dahlquist, T. Rand-Meir, and M. A. Raftery, *Proc. Natl. Acad. Sci. USA* **61**, 119 (1968).
216. L. E. H. Smith, L. H. Mohr, and M. A. Raftery, *J. Am. Chem. Soc.* **95**, 7497 (1973).
217. S. Rosenberg and J. F. Kirsch, *Biochemistry* **20**, 3196 (1981).
218. J. J. Pollock, D. M. Chipman, and N. Sharon, *Archs. Biochem. Biophys.* **120**, 235 (1967).
219. K. Bartik, C. Redfield, and C. M. Dobson, *Biophysical Journal* **66**, 1180 (1994).
220. J. A. Rupley, V. Gates, and R. Bilbrey, *J. Am. Chem. Soc.* **90**, 5633 (1968).
221. A. Warshel, *Biochemistry* **20**, 3167 (1981)
222. Y. Eshdat, A. Dunn, and N. Sharon, *Proc. Natl. Acad. Sci. USA* **71**, 1658 (1974)
223. B. A. Malcolm, S. Rosenberg, M. J. Corey, J. S. Allen, A. Debaetselier, and J. F. Kirsch, *Proc. Natl. Acad. Sci. USA* **86**, 133 (1989).

224. I. Matsumura and J. F. Kirsch, *Biochemistry* **35**, 1881 (1996).
225. B. Dunn and T. C. Bruice, *Adv. Enzymol.* **37**, 1 (1973).
226. D. M. Chipman and N. Sharon, *Science, N.Y.* **165**, 454 (1969).
227. I. I. Secemski and G. E. Lienhard, *J. Biol. Chem.* **249**, 2932 (1974).
228. E. Holler, J. A. Rupley, and G. P. Hess, *Biochemistry* **14**, 1088, 2377 (1975).
229. N. Strynadka and M. James, *J. Molec. Biol.* **220**, 401 (1991).
230. A. T. Hadfield, D. J. Harvey, D. B. Archer, D. A. Mackenzie, D. J. Jeenes, S. E. Radford, G. Lowe, C. M. Dobson, and L. N. Johnson, *J. Molec. Biol.* **243**, 856 (1994).
231. I. Matsumura and J. F. Kirsch, *Biochemistry* **35**, 1890 (1996).
232. I. Tews, A. Perrakis, A. Oppenheim, Z. Dauter, K. S. Wilson, and C. E. Vorgias, *Nature Structural Biology* **3**, 638 (1996).

Protein Stability

17

One of the great unsolved problems of science is the prediction of the three-dimensional structure of a protein from its amino acid sequence: the "folding problem." An even more elusive goal is the prediction of the catalytic activity of an enzyme from its amino acid sequence. There are two reasons that the folding problem is so important. First, the acquisition of sequence data by DNA sequencing is relatively quick, and vast quantities of data have become available through international efforts such as the human genome project and other genome sequencing projects (Chapter 1). The acquisition of three-dimensional data is still slow and is limited to proteins that either crystallize in a suitable form or are sufficiently small and soluble to be solved by NMR in solution. Algorithms are thus required to translate the linear information into spatial information. Second, we are now able to synthesize proteins by way of their genes (Chapter 14), and so the production of new enzymes with specified catalytic activities is a challenging prospect. Producing such new enzymes requires five underpinning and interrelated abilities: (1) the ability to predict the most stable fold of a particular sequence; (2) the ability to design a novel fold; (3) the ability to predict whether the desired fold is kinetically accessible; (4) the ability to design the precise features for specific binding in the fold; and (5) the ability to design the precise orientation of groups in the protein for efficient catalytic function. Each of these five prerequisites is beyond the current capabilities of theory. To appreciate some of the difficulties facing theoreticians, we must consider the physical nature of protein folding. A denatured protein makes many interactions with solvent water. As the protein folds up, it exchanges those noncovalent interactions with others that it makes within itself: its hydrophobic side chains tend to pack with one another, and many of its hydrogen bond donors and acceptors pair with each other, especially those in the polypeptide backbone that form the hydrogen-bonded networks in helixes and sheets. Each interaction energy is small, but because of their large numbers, the total interaction energies in the native and denatured states are huge, being some thousands of kilocalories per mole, depending on the size of the protein. Yet proteins are only marginally stable, their free energies of unfolding ranging from 5 to 15 kcal/mol (20–60 kJ/mol). This tiny amount of energy is the difference between the free energies in the native and denatured states, and those states are almost balanced in energy. Thus, whether a protein folds depends crucially on a balance between two large numbers, each of which is very difficult to calculate with precision. To predict the stability of a protein, we have to calculate not only the interaction energy between any two atoms within a protein, but also this energy relative to the interactions that the individual atoms make with water in the denatured state. Current potential functions are not sufficiently accurate for this purpose. But we can use protein engineering and other experimental procedures to make changes in existing proteins. The protein engineering experiments have provided an experimental route into determining quantitatively the factors that govern stability. In turn, the information derived from the initial experiments has provided the basis for designing rational changes in protein stability and allows the modest redesign of proteins. Such data are also being used to benchmark the computer calculations that will eventually be used for design.

A. Protein denaturation

Proteins can be denatured by changing their physical or chemical environments. The most common methods are by heating; adding a chemical denaturant, such as urea or guanidinium chloride (GdmCl); changing the pH to acidic or alkaline; or applying high pressure. Many proteins, especially small ones, denature reversibly; they regain their native structures spontaneously when returned to conditions that favor folding. Many larger proteins denature irreversibly; they often aggregate and precipitate after denaturant has been removed or hot solutions are cooled. Even small, readily renaturable proteins tend to precipitate at high concentrations of their denatured states. Sometimes renaturation may be induced by a slow return to favorable conditions, perhaps by dialysis against decreasing concentrations of denaturant. The reversible equilibria for denaturation of proteins can be treated by thermodynamics in the same way as simple chemical equilibria, except that there are some added twists because of the multitude of weak noncovalent interactions.

1. Thermodynamics of protein folding[1]

The denatured state has considerable conformational freedom: it is not a rigid structure but individual segments of polypeptide chain can move relative to one another, and groups can rotate about single bonds (see section B). The denatured state has an inherently high configurational entropy from the simple Boltzmann formula $S = k \ln W$, where W is the number of accessible states and k is the Boltzmann constant (R divided by Avogadro's number). Conversely, the native state is very conformationally restricted and has a low entropy. Thus, as a protein folds, it loses considerable entropy, which must be balanced by a gain of enthalpy for the free energy to favor folding. The enthalpy of packing of side chains in the native state is favorable and compensates, just barely, for its low entropy. The classic thermodynamic description of protein folding is that it has large negative values of ΔS and ΔH. We know (Chapter 11) that it is not just the thermodynamics of the polypeptide chain *per se* that contributes to the observed values of ΔH and ΔS but also the thermodynamics of solvent water: the entropy and enthalpy of water must be added to the entropy and enthalpy of the protein to give the gross thermodynamic properties of the denatured or native states. The contribution of solvent has two important consequences.

a. Entropy of release of water on folding

The native state of a protein has many of its hydrophobic side chains shielded from water because they are packed in hydrophobic cores. Conversely, the denatured state has many of its hydrophobic side chains exposed to solvent. The water molecules stack around these in "icebergs" as they maximize their hydrogen bonds with one another (Chapter 11). This lowers the entropy of water, because the individual molecules have less freedom of movement, and lowers the enthalpy because more hydrogen bonds are made.[2] Similarly, the hydrogen bond donors and acceptors in the polypeptide backbone of the denatured protein are largely exposed to solvent and tie down more water molecules.[3] These water molecules are released as the protein folds, and the gain in entropy of water compensates considerably for the loss of conformational entropy.

b. The specific heat of unfolding

The solvent makes another important contribution to the thermodynamics of protein folding; to the specific heat,[4] C_p (subscript "p" is the value at constant pressure, the usual conditions used for biology). The specific heat is the energy required to raise 1 mole through 1°C (or 1 K). The change in specific heat, ΔC_p, for most simple chemical reactions is very small because the reactants and products have similar values of C_p. However, the specific heat of denatured states is particularly high, mainly because of the icebergs of solvent around the exposed hydrophobic groups. As the temperature is raised, the icebergs melt and their latent heat of fusion contributes significantly to C_p of the denatured state. Also, if there are weakly structured regions in the denatured state, they can make further contributions to C_p as they melt. To a rough approximation, the value of ΔC_p of unfolding is about 12 cal/deg/mol

(50 J/deg/mol) per residue of protein.[5] The value does become distinctly lower at higher temperatures, perhaps because the icebergs have melted more. The high value of ΔC_p has several important consequences.

The first is that the enthalpy of unfolding of a protein changes significantly with temperature. Consider that the protein denatures reversibly according to a simple two-state transition in which the denatured state D is in equilibrium with the native structure N. Then, according to classical thermodynamics, the value of ΔH_{D-N} at a temperature T_2 (i.e., $\Delta H_{D-N(T_2)}$) is related to that at another temperature T_1 ($\Delta H_{D-N(T_1)}$) by

$$\Delta H_{D-N(T_2)} = \Delta H_{D-N(T_1)} + \Delta C_p(T_2 - T_1) \tag{17.1}$$

where $\Delta H_{D-N(T_1)}$ is the enthalpy of unfolding at temperature T_1 and $\Delta H_{D-N(T_2)}$ the enthalpy of unfolding at temperature T_2. Because ΔC_p is positive, the enthalpy of unfolding becomes more positive (that is, favors the native state) as the temperature increases. The physical interpretation of this fact is that the enthalpy of the unfolded state increases as the temperature increases because the icebergs surrounding the hydrophobic side chains melt progressively and have fewer hydrogen bonds with one another.

Entropy is related to C_p by $\delta S = C_p \delta T / T$, so that $S_{T_2} = S_{T_1} + \int_{T_1}^{T_2} C_p dT$. The entropy of unfolding thus changes according to the formula

$$\Delta S_{D-N(T_2)} = \Delta S_{D-N(T_1)} + \Delta C_p \ln \frac{T_2}{T_1} \tag{17.2}$$

The entropy of unfolding becomes higher (favoring the denatured state) with increasing temperature because fewer water molecules are tied down in the icebergs that surround the denatured state.

c. Free energy of unfolding

The free energy of unfolding, $\Delta G_{D-N}, = \Delta H_{D-N} - T\Delta S_{D-N}$, is obtained from equations 17.1 and 17.2.

$$\Delta G_{D-N(T_2)} = \Delta H_{D-N(T_1)} + \Delta C_p(T_2 - T_1) - T_2(\Delta S_{D-N(T_1)} + \Delta C_p \ln (T_2/T_1)) \tag{17.3}$$

As the temperature increases, the term $T\Delta S_{D-N}$ increases and causes the protein to unfold when it becomes greater than ΔH_{D-N}; i.e., when ΔG_{D-N} becomes negative. Thus, thermal unfolding is caused by the entropy of the unfolded state dominating at high temperature.

d. Cold unfolding

Intriguingly, the high value of ΔC_p leads to the phenomenon of cold denaturation: as the temperature becomes lower, the enthalpy of unfolding decreases, according to equation 17.1, and the free energy term becomes dominated again by the entropy component. The melting temperature for cold denaturation, T_c, is given by

$$T_c = \frac{T_m^2}{T_m + 2(\Delta H_{D-N}/\Delta C_p)} \approx T_m - \frac{2(\Delta H_{D-N})}{\Delta C_p} \tag{17.4}$$

where T_c is the temperature at which 50% of the protein is cold denatured and ΔH_{D-N} is the enthalpy of unfolding at the melting temperature, T_m. Consequently, proteins with a small numerical value of ΔH_{D-N} and a large ΔC_p are expected to cold unfold at higher temperatures. In most cases, however, the temperature is too far below the freezing point of water for cold unfolding to be detected. But addition of the denaturant urea decreases T_m, decreases ΔH_{D-N}, and increases the ΔC_p term so that cold unfolding may be observed more readily.

e. Measuring thermal denaturation

The thermal unfolding of proteins is best measured by differential scanning calorimetry, which measures the heat absorbed by a protein as it is slowly heated through its melting transition (Figure 17.1). A solution of about 1 mg of protein in 1 mL of buffer and a separate reference sample of buffer alone are heated electrically.[6] The additional current required to heat the protein solution is recorded. As the protein denatures, there is a large uptake of heat because the process is highly endothermic. The temperature at the maximum of the peak is

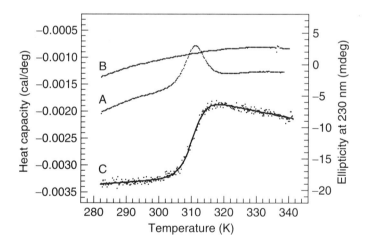

Figure 17.1 Thermal unfolding of barnase measured by calorimetry and spectroscopy. The heat capacity of barnase (trace A) was measured using differential scanning calorimetry with a baseline (trace B) of buffer versus buffer: T_m is 310.9 ± 0.01 K, $\Delta H_{D-N(cal)} = 98.4 \pm 0.2$ kcal/mol, and $\Delta H_{D-N(vh)} = 98.1 \pm 0.3$ kcal/mol. The ellipticity at 230 nm in the circular dichroism (trace C) under identical conditions fits $T_m = 310.5 \pm 0.1$ K, and $\Delta H_{D-N(vh)} = 93 \pm 3$ kcal/mol (equation 17.5).

the T_m, the melting temperature. The area under the curve, after subtracting the baselines, is the enthalpy of unfolding, ΔH_{D-N}. This direct measurement of the enthalpy is usually given the subscript "cal"; i.e. $\Delta H_{D-N(cal)}$. The slope of the baseline at temperatures above the T_m is the product of the specific heat of the denatured state and its concentration. At the T_m, ΔG_{D-N} is zero because there are equal concentrations of D and N, and $\Delta G_{D-N} = -RT \ln ([D]/[N])$. Thus, $\Delta H_{D-N(Tm)} = T_m \Delta S_{D-N(Tm)}$. Thus, the value of T_m for a particular class of protein is not necessarily a measure of its stability, but is merely equal to the ratio of its enthalpy and entropy of denaturation.

The equation for ΔG_{D-N} may be rearranged and ΔH_{D-N} and ΔS_{D-N} substituted to give

$$\frac{[D]}{([D] + [N])} = \frac{\exp(\{T\Delta S_{D-N} - \Delta H_{D-N}\}/RT)}{1 + \exp(\{T\Delta S_{D-N} - \Delta H_{D-N}\}/RT)} \tag{17.5}$$

The fraction unfolded, $[D]/([D] + [N])$, may be determined by spectroscopy (for example, absorbance, fluorescence, or circular dichroism) and fitted to equation 17.5. The enthalpy term derived from these experiments is referred to as the van't Hoff value, $\Delta H_{D-N(vh)}$. (Note that the specific heat terms of equations 17.1 and 17.2 have to be substituted into the equation.)

f. Measurement of ΔC_p from change of ΔH_{D-N} with pH

It is difficult to measure ΔC_p with precision from an individual thermogram. A convenient alternative procedure is to measure $\Delta H_{D-N(cal)}$ in increasingly acidic solutions. Proteins usually become progressively less stable with decreasing pH, and the T_m becomes progressively lower. The decrease in stability results from the protonation of carboxylic acid groups, and these are known to have small enthalpies of ionization. It is found that $\Delta H_{D-N(cal)}$ decreases with decreasing pH at acidic pH, and a plot of $\Delta H_{D-N(cal)}$ against T_m for each pH is linear. The slope of the plot is ΔC_p.

2. Solvent denaturation

Why denaturants such as urea and GdmCl cause proteins to denature may be considered empirically. Those denaturants solubilize all the constituent parts of a protein, from its polypeptide backbone to its hydrophobic side chains. To a first approximation, the free energy of transfer of the side chains and polypeptide backbone from water to solutions of denaturant is linearly proportional to the concentration of denaturant.[7,8] Because the denatured state is more exposed to solvent than the native state, the denatured state is preferentially stabilized by denaturant. Thus, the free energy of denaturation at any particular concentration of denaturant is given by

$$\Delta G_{D-N} = \Delta G_{D-N}^{H_2O} - m_{D-N}[\text{denaturant}] \tag{17.6}$$

where $\Delta G_{\text{D-N}}^{\text{H}_2\text{O}}$ is the value in water and $m_{\text{D-N}}$ is a constant of proportionality ($= -\partial(\Delta G_{\text{D-N}})/\partial[\text{denaturant}]$) and has the dimensions of cal/mol/M or J/mol/M.[9] Typical denaturation curves are illustrated in Figure 17.2.

The term $m_{\text{D-N}}$ contains some interesting information. Suppose each group, i, in the protein has a free energy of transfer of δg_i cal/mol from water to a 1-M so-

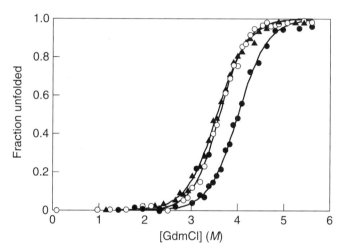

Figure 17.2 Denaturation of wild-type and two destabilized mutants of chymotrypsin inhibitor 2 (CI2) induced by guanidinium chloride. Substituting equation 17.6 into equation 17.5 gives

$$[D] = \frac{([D] + [N])\exp(\{m_{\text{D-N}}[\text{denaturant}] - \Delta G_{\text{D-N}}^{\text{H}_2\text{O}}\}RT)}{1 + \exp(\{m_{\text{D-N}}[\text{denaturant}] - \Delta G_{\text{D-N}}^{\text{H}_2\text{O}}\}/RT)}$$

In practice, spectroscopic data have sloping baselines because the signals of the denatured and native states change approximately linearly with [denaturant], and so these factors must be introduced into the equation for fitting data (M. M. Santoro and D. Bolen, *Biochemistry* **27**, 8063 (1988)). The author prefers the use of the equation

$$F = $$

$$\frac{(\alpha_N + \beta_N[\text{denaturant}]) + (\alpha_D + \beta_D[\text{denaturant}])\exp[m_{\text{D-N}}([\text{denaturant}] - [D]^{50\%})/RT]}{1 + \exp[m_{\text{D-N}}([\text{denaturant}] - [D]^{50\%})/RT]}$$

where F is the observed spectroscopic signal; α_N is the signal at the native state at 0-M denaturant and $\beta_N = d\alpha_N/d[\text{denaturant}]$ (i.e., the slope); α_D and β_D are the corresponding quantities for the denatured state; and $[D]^{50\%}$ is the denaturant concentration at which the protein is 50% unfolded (J. Clarke and A. R. Fersht, *Biochemistry* **32**, 4322 (1993)). Solving the equation in this form by curve-fitting procedures gives $[D]^{50\%}$ and $m_{\text{D-N}}$ with their standard errors. These quantities are used in comparing mutants. Note that (1) the denaturation curve is less sharp than it is for larger proteins, because CI2 is very small, just 64 residues; and (2) the destabilized mutants unfold at lower concentrations of GdmCl.

lution of urea when it is fully exposed to solvent. If it does not become fully exposed in the denatured state, but only increases its exposure by a fraction α_i, then the effective free energy change for the group on denaturation is $\alpha_i \delta g_i$. Thus,

$$m_{D-N} = \sum_i \alpha_i \delta g_i \qquad (17.7)$$

The value of $\sum_i \delta g_i$ can be calculated from standard values[8] and α from the structure of the protein. A low value of m_{D-N}, compared with that calculated, indicates that the protein does not become highly unfolded on denaturation. The value of m_{D-N} thus provides a test for the degree of unfolding. A low value of m_{D-N} may also indicate that the denaturation process is occurring stepwise rather than in a single cooperative transition.

Because m_{D-N} is proportional to the number of groups in the protein, large proteins are more sensitive to solvent denaturation than small ones. It is a common mistake to attribute resistance to solvent denaturation of small proteins to high stability; in fact, it is an inherent characteristic of a small change in surface area on denaturation. Also, proteins that are elongated have higher solvent exposure in the native state and correspondingly lower values of m_{D-N}.

Equation 17.6 can be tested over only a very narrow range of [denaturant] because the fraction of denatured state switches from being immeasurably small to being indistinguishable from 100% on going from, for example, 3- to 5-M urea for a typical small protein. Good linearity is observed over this range. But there are difficulties in estimating $\Delta G_{D-N}^{H_2O}$ from equation 17.6 because of the long extrapolation back to 0-M denaturant. However, equation 17.6 has been tested over a wide range of denaturant (0- to 5.5-M urea) by combining differential scanning calorimetry and urea denaturation on barnase, which has a midpoint for unfolding of 4.6-M urea at 25°C (Figure 17.3).[10] The plot is noticeably curved, and the linear extrapolation of the data from 3.5- to 5.5-M urea gives a value of 9.2 kcal/mol for $\Delta G_{D-N}^{H_2O}$ compared with a value of 10.5 kcal/mol from calorimetry. Nevertheless, the linear assumption is good enough for most practical purposes.

There are two mechanisms by which the denaturant can stabilize groups on the protein: indirectly by altering the solvent properties of water and directly by specifically interacting with groups on the protein.[11] Specifically bound molecules of denaturant have been observed in crystal structures, and calculation suggests direct binding to certain side chains, such as the aromatics.[12] The polypeptide backbone should certainly make hydrogen bonds with urea. A model based on specific binding gives

$$\Delta G_{D-N} = \Delta G_{D-N}^{H_2O} - \Delta n\, RT \ln (1 + [\text{denaturant}]K_{ass}) \qquad (17.8)$$

where Δn is the number of urea molecules binding upon unfolding and K_{ass} is the average association constant at each site.[11] Equation 17.8 predicts the curvature found for the urea denaturation of barnase (Figure 17.3), but the curvature does not prove this model.

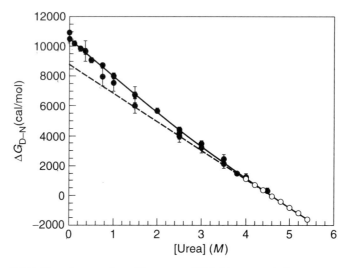

Figure 17.3 Urea denaturation of barnase at 25°C. The open circles were measured directly at 25°C from the ratios [D]/[N], which were determined spectroscopically. The solid circles are values extrapolated to 25°C from thermal unfolding experiments at lower concentrations of urea and higher temperatures.

3. Acid- or base-induced denaturation

Most proteins denature when the pH is changed to very low or very high values. The primary reason for extremes of pH causing denaturation is that proteins usually have buried groups that have highly perturbed pK_a's (see Chapter 5). Suppose, for example, that a glutamate side chain has a pK_a of 4.5 in the denatured state but that the pK_a is lowered to 1.5 in the native state because it makes a buried salt bridge with a lysine or arginine side chain. At pH 7, the side chain is ionized in both states. At pH 4.5, the denatured state is 50% protonated, and this protonation displaces the denaturation equilibrium by a factor of 2 toward denaturation. At pH 3.5, the denatured side chain is one unit below its pK_a and so displaces the equilibrium toward unfolding by a factor of 10, and so on. This situation may be formally analysed by a thermodynamic cycle (Figure 17.4), as described in Chapter 3, section L. If the dissociation constant of the group in the native state is K_a^N and in the denatured state is K_a^D, and the equilibrium constant

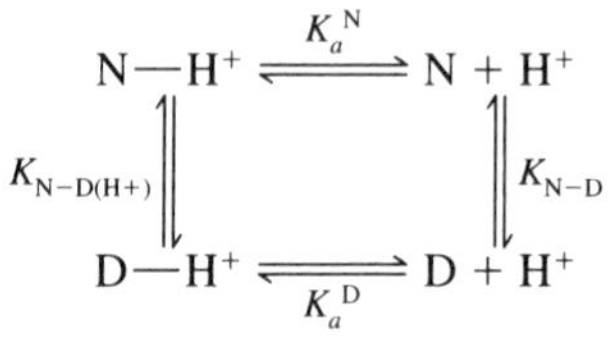

Figure 17.4 Thermodynamic cycle for the mechanism of acid-induced denaturation by a group that has a lower pK_a in the native state than in the denatured state.

$$
\begin{array}{ccc}
N\!-\!H^+ & \overset{K_a^N}{\rightleftharpoons} & N + H^+ \\
K_{N-D(H+)} \big\updownarrow & & \big\updownarrow K_{N-D} \\
D\!-\!H^+ & \underset{K_a^D}{\rightleftharpoons} & D + H^+
\end{array}
$$

for folding in the unprotonated state is K_{N-D} and in the protonated state is $K_{N-D(H^+)}$, then the equilibrium constants are related by

$$\frac{K_{N-D(H^+)}}{K_{N-D}} = \frac{K_a^N}{K_a^D} \tag{17.9}$$

The simple thermodynamic cycle must be expanded for use with proteins because there are so many ionizing groups with overlapping pK_a's. If $Q_D(pH)$ and $Q_N(pH)$ are the number of protons bound to the denatured state and the native state, respectively, then

$$\frac{\partial(\Delta G_{D-N})}{\partial(pH)} = 2.3RT[Q_D(pH) - Q_N(pH)] = 2.3RT \cdot \Delta Q_{D-N}(pH) \tag{17.10}$$

where ΔG_{D-N} is the difference in free energy between the native and the denatured state.[7,8]

A further reason for the change in stability is that at low pH, the protein becomes highly positively charged as the aspartates and glutamates become protonated; and at high pH, the protein becomes negatively charged as the lysine, tyrosine, and, eventually, arginine residues become deprotonated. Electrostatic repulsion then causes destabilization.

4. Two-state versus multistate transitions

The equations so far apply to a simple two-state equilibrium of $D \rightleftharpoons N$. That is, protein denaturation is a cooperative (or all-or-none) transition in which denaturation occurs in a single step without any intermediates accumulating. How can we tell that this is so?

a. Spectroscopic evidence for two-state transitions

A two-state transition is usually identified by all spectroscopic probes changing simultaneously as the equilibrium changes. The far ultraviolet circular dichroism signals, which are a measure of secondary structure, should change in parallel with the near ultraviolet, which are a measure of tertiary structure. Fluorescence and near ultraviolet absorbance spectra also probe tertiary structure and should change in parallel with each other and the circular dichroism spectra. Ideally, there should be isosbestic or isodichroic points where spectra converge.

b. Enthalpy test for a two-state transition

The calorimetric measurement *per se* can detect additional intermediates during transitions. There are clear-cut cases where two or more phases are seen in melting curves, or thermograms, showing that intermediates are appearing. The difficult case is when only one transition is seen. Whether or not there is a single transition can be tested by measuring $\Delta H_{D-N(cal)}$ and $\Delta H_{D-H(vh)}$ for the transition.[13] (This does not require two separate experiments because the bell-shaped curve of the thermogram can be readily fitted to the van't Hoff equation.) $\Delta H_{D-N(cal)}$ measures the total change in enthalpy for a reaction, no matter the number of steps. $\Delta H_{D-N(vh)}$ is identical to $\Delta H_{D-N(cal)}$ for a two-state reaction. If there are more steps that are close together and appear to merge into one, the van't Hoff plots become smeared out and give an artifactually low value. The ratio $\Delta H_{D-N(vh)}/\Delta H_{D-N(cal)}$ is the diagnostic test for the two-state mechanism. If the ratio is 1, then no intermediates accumulate; there may be intermediates present, but they are at high energy. If $\Delta H_{D-N(vh)}/\Delta H_{D-N(cal)} < 1$, then there is an intermediate present. Occasionally, $\Delta H_{D-N(vh)}/\Delta H_{D-N(cal)}$ is found to be > 1. This indicates that aggregation or another artifact is occurring.

c. Apparent two-state transitions

Intermediates that are present at equilibrium at low concentration will not be detected by most spectroscopic methods. (One procedure that may do this is $^1H/^2H$-exchange at equilibrium, described in Chapter 18, section E). Further, they will not be detected by calorimetry, either, if they do not build up during the thermal transition.

There are many proteins in which there are stable intermediates present in water, but these are not observed during the denaturation transition. For barnase, for example (Figure 17.5), there is a folding intermediate (I) that is more stable than the highly denatured state (D) at low concentrations of denaturant.[14] But as the concentration of the denaturant is increased, D becomes progressively more stable than I. The stability of I varies with denaturant in a manner analogous to equation 17.6, in which m_{I-N} replaces the term m_{D-N}. Because intermediates are more compact than D, $m_{D-N} > m_{I-N}$ (see equation 17.7). It is found in the transition region for barnase, as for many proteins, that I is the highest energy species and so is not observed at equilibrium. The same is true during thermal denaturation.

d. Extrapolated thermodynamic quantities
from apparent two-state transitions

There are two important consequences from Figure 17.5. First, many proteins appear to follow two-state denaturation around the midpoint of folding, despite intermediates accumulating under more renaturing conditions. Second, the measurements made under these conditions give information on the transition between D and N, and not on D and I. The extrapolation of equilibrium data from the transition region for denaturation in a solution of denaturant will give the thermodynamic properties of the denatured state. The same is true for data from a calorimetric experiment.

B. Structure of the denatured state

We have used the notation "D" for the denatured state so far rather than "U" for unfolded, for good reason. The denatured state of a protein is not a single fixed

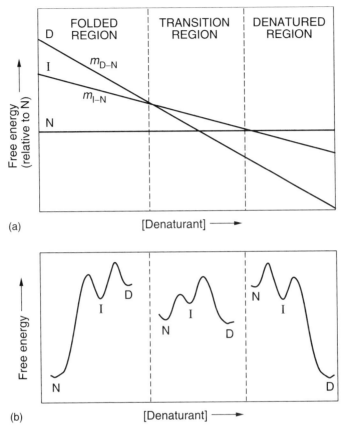

Figure 17.5 Apparent two-state transitions in multistate reactions. (a) Free energies of D and I are plotted against [denaturant], relative to that of N, assuming that simple linear equations such as 17.6 are followed. Typically, in the absence of denaturant, D is the most unstable, and I the next most unstable. The free energy of D decreases most rapidly with increasing denaturant because D has the greatest exposed surface area. Consequently, the relative energies of the three states change with increasing [denaturant]. In the transition region, the equilibrium is effectively between D and N, because I is present at low concentration. (b) The relative energies of D, I, and N.

state. First, it is a collection of states of very similar energies that are in very rapid equilibrium with one another (a "distribution function" of states).[15,16] Second, this collection of states is very sensitive to its physical and chemical environment. The denatured state is not easy to characterize. Gross properties such as hydrodynamic volume and radius of gyration may be readily measured to indicate the average compactness. Infrared and circular dichroism spectra give indications of residual secondary structure. NMR cannot solve the structure at the level of individual residues in the same detail as for a native protein because of the conformational heterogeneity, the lack of NOEs, and the poor dispersion of chemical shifts.

1. The denatured state under denaturing conditions, U

The denatured state of barnase is highly unfolded under denaturing conditions. The most highly unfolded state is the urea-denatured state, followed by the temperature- and acid-denatured state and then the acid-denatured state alone.[17] This pattern of unfolding seems typical. There is evidence for some flickering helical structure in the regions that are helical in the native state of barnase,[18] this being most pronounced in the acid-denatured state. There is also some residual structure in the center of the β sheet. The pK_a values of acidic groups in the acid-denatured state are perturbed downward 0.4 unit at low ionic strength, indicating compactness and electrostatic interactions with positively charged residues.[19] (Further details of specific interactions are dealt with in Chapter 18). We term the highly unfolded state in concentrated denaturant as U, the unfolded state, because this most resembles the random coil.

2. The denatured state at physiological conditions, D^{Phys}

The starting state for protein folding studies *in vitro* is the one that is present under physiological conditions. Refolding is initiated by restoring a protein from denaturing to folding conditions. The unfolded state U rearranges in the dead-time (< 1 ms) of a stopped-flow mixing experiment. The properties of the starting state are difficult to measure directly because it rapidly reacts to give products. But its properties can be inferred from kinetic studies and shown to be an increasingly compact state with increasing size of the protein. Such states are usually termed folding intermediates. But because the intermediate is more stable than U, the intermediate is, in fact, the most stable of the non-native states under physiological conditions. We can also term the intermediate, D^{Phys}, the denatured state under physiological conditions.

D^{Phys} may be related to the so-called *molten globule* states. These are compact, partly folded states of proteins that can sometimes be isolated under mildly denaturing conditions or when a cofactor or metal ion that is essential for stability has been removed.[20] They are characterized by having few tertiary interactions, some secondary structure, and a fluctuating hydrophobic core and by being separated from the native state by a high activation energy. Their hydrophobic nature is indicated by binding to the dye ANS, which has an affinity for mobile hydrophobic regions. Other attempts to model D^{Phys} use fragments of larger proteins.[21] Detailed deductions about D^{Phys} from measurements relative to the transition state are described in Chapter 18. Stopped-flow data on barnase indicate that D^{Phys} is 3 kcal/mol (12 kJ/mol) more stable than U.[22]

Very small proteins, such as CI2, can have a very expanded D^{Phys} (see Chapter 19, section B4). The IgG binding domain of streptococcal protein L (62 residues) is of similar size to CI2 and also folds by two-state kinetics (see Tables

18.1 and 19.2). The radius of gyration of the denatured state in GdmCl (> 4-*M*) is 26 Å and that of the native state is 16.5 Å, determined by small-angle x-ray scattering. The radius in 1.4-*M* GdmCl, immediately after dilution from concentrated GdmCl under conditions which favor folding, is 25.9 Å, determined from stopped-flow measurements of small-angle x-ray scattering.[23] The larger the protein, the more likely that D^{Phys} has compact regions.

3. First- and second-order transitions

Most classical chemical reactions of "A" in equilibrium with "B" proceed by a *first-order transition*. That is, A and B are always present as two distinct chemical species that interconvert with first-order kinetics. The equilibrium constant may change with conditions, but not the chemical nature of A and B. In a *second-order transition*, A would slowly convert into B with changing reaction conditions.[15] There would be a single species present under all conditions: at one extreme it would have the structure of A, at the other it would be B, and in between it would have a structure intermediate between those of A and B. The formation of B from A would not be a cooperative process. This could happen in protein denaturation if the denatured state that is present at high concentration of denaturant slowly interconverted into D^{Phys} with decreasing concentration of denaturant (Figure 17.6). Careful experiments on the folding intermediate of barnase show that it does interconvert cooperatively with U,[22] as do experiments on some molten globules.[24] But less compact states are likely to interconvert non-cooperatively.

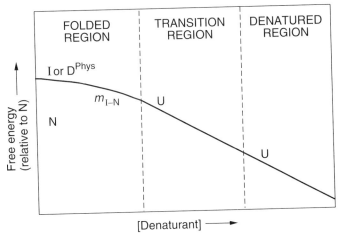

Figure 17.6 Free energies of N and U against [denaturant] where there is a second-order transition between U and D^{Phys} (or I).

C. Measurement of changes in stability

The important contribution of protein engineering to the study of protein stability is that defined structural changes can be made in a protein and the change in stability measured. In this way, the relationship between structure and stability can be established experimentally. Fortunately, it is easier to measure changes in stability accurately than it is absolute values of stability.

1. Thermal denaturation

Absolute values of ΔH_{D-N} are very high, typically on the order of 100 kcal/mol (400 kJ/mol). The calculation of $\Delta H_{D-N(cal)}$ requires determining the absolute value of the protein concentration, which is difficult to measure to $\pm 1\%$. Realistic errors in ΔH_{D-N} are ± 1 kcal/mol (± 4 kJ/mol) and greater, even though statistical analyses of data suggest higher precision. An error of ± 1 kcal/mol is unacceptable because the free energy of unfolding of proteins is only $5-15$ kcal/mol and changes of only $0.5-5$ kcal/mol in enthalpy are measured on mutation. However, the errors in $\Delta H_{D-N(cal)}$ and in ΔC_p tend to cancel out in calculating changes in ΔG_{D-N}.[25] If the error in determining $\Delta H_{D-N(cal)}$ at the T_m is $\delta \Delta H_m$ and the error in determining ΔC_p is $\delta \Delta C_p$, then the error in calculating ΔG_{D-N} at any temperature T is

$$\delta \Delta G_{D-N} = \left(1 - \frac{T}{T_m}\right)\delta \Delta H_m + \left[T - T_m - T\ln\frac{T}{T_m}\right]\delta \Delta C_p \tag{17.11}$$

The errors are greatly attenuated: a large extrapolation of 25 K from a T_m of 350 K leads to an error of only $0.07\,\delta \Delta H_m + \delta \Delta C_p$. Thus, the free energies of unfolding of the wild-type protein and a mutant protein may be extrapolated to a common temperature and the difference determined with precision.[25]

The value of T_m is determined with good precision, even from spectroscopic measurements of unfolding and the van't Hoff plot. A change in stability causes a change in T_m, ΔT_m, and the change in stability may be calculated as follows.[26] Differentiating equation 17.3 with respect to temperature gives $\partial(\Delta G_{D-N})/\partial T = \Delta S_{D-N}$. Therefore, the change in free energy on mutation, $\Delta\Delta G_{D-N}$, is given by

$$\Delta\Delta G_{D-N} \approx \Delta T_m \Delta S_{D-N} \approx \Delta H_{D-N(T_m)}\frac{\Delta T_m}{T_m} \tag{17.12}$$

This equation holds very well in practice and provides a useful rule for calculating the change in T_m from changes in ΔG_{D-N} when ΔS_{D-N} or $\Delta H_{D-N(T_m)}$ is known.

2. Solvent denaturation

The determination of ΔG_{D-N} from equation 17.6 is too crude because the extrapolation to 0-M denaturant is too long and there may be curvature in the plots.

The determination of $[D]_{50\%}$, the concentration of denaturant at which 50% of the protein is denatured, is particularly reproducible. The change in stability on mutation can be measured by the change in $[D]_{50\%}$, $\Delta[D]_{50\%}$, on mutation using equation 17.13, where m'_{D-N} is the value for the mutant and $\Delta\Delta G_{D-N}^{D_{50\%}}$ is the value of $\Delta\Delta G_{D-N}$ at the mean value of $[D]_{50\%}$ for the two proteins.[27]

$$\Delta\Delta G_{D-N}^{D_{50\%}} = 0.5(m_{D-N} + m'_{D-N})\Delta[D]_{50\%} \tag{17.13}$$

The presence of denaturant at a concentration of $[D]_{50\%}$ changes the overall free energy of folding by only some 10 kcal/mol (40 kJ/mol), which averages at only ~ 0.1 kcal/mol per individual residue, which is within the experimental error. The value of m_{D-N} changes only insignificantly on mutation of proteins that unfold extensively in denaturant. One advantage of equation 17.13 is that it holds even when there is curvature in the plots of $\Delta\Delta G_{D-N}$ versus $[D]$, provided the curvature is the same for wild type and mutant.[25] There is very good agreement between values of $\Delta\Delta G_{D-N}^{D_{50\%}}$ and values of $\Delta\Delta G_{D-N}$ determined from calorimetry and equation 17.12.[25] However, high concentrations of GdmCl mask electrostatic interactions because of its ionic strength.

D. Energetics of formation of structure

The standard experimental procedure for determining the energetics of interactions in proteins is to use protein engineering to make a series of mutations in a protein and then determine the changes in its stability[28] (see Chapter 15, Part 1, section D for a similar discussion on mutagenesis of enzymes). These changes give the *relative* effects on stability of the different amino acids at the site of mutation. In general, these data must not be equated directly with the absolute values of the interaction energies. But they are precisely the empirical values that are required for use in protein design and for calculation of the specificity of interactions. Alanine is a good choice as the reference amino acid in calculating relative energetics because it has the smallest side chain of the L-amino acids and does not make electrostatic interactions. Glycine is unique because it can adopt dihedral angles that are not allowed for other residues (see Chapter 1, Figure 1.11). The strategy of analysis is at two levels. The first is a purely empirical comparison of the relative effects of the different amino acids at particular positions in a structural theme. The second is to gain insight about the specific types of interactions. This requires a logical choice of mutation, usually achieved by deleting the moieties of side chains that are involved in the desired interaction.

1. α Helixes

A disordered polypeptide chain in solution has considerable conformational freedom and a high entropy. The side chains are free to rotate, which gives them maximal entropy. The side chains are able to avoid unfavorable steric contacts with other side chains. If the side chains have hydrogen-bonding groups, then these groups make bonds with the polypeptide backbone, if favorable, or are highly exposed to solvent. On forming a helix, much of this conformational freedom is lost, although the intrahelical hydrogen bonds are favored entropically because of the release of water. Part of the side chain is buried in the helix, relative to its being highly exposed in solution, even in the surface solvent exposed face. The partial burial of hydrophobic side chains provides a considerable driving force for helix formation.[29]

It is important to recall from Chapter 1, section C1b, that an α helix is not a uniform structure. The hydrophobic face of an amphipathic helix is buried in a hydrophobic region of the protein, and the importance of mutations in this face is described below (section D4). Further, the first and last four residues differ from the interior ones because they cannot make the intrahelical hydrogen bonds between the backbone C=O groups of one turn and the NH groups of residues in the next.[30,31] Because of this, the ends of helixes require hydrogen bonds to be made with either solvent or groups in the protein. The helix is an electrostatic dipole with the positive end at the N-terminus.[32] The following discussion refers principally to the solvent exposed regions of helixes.

a. Helix-forming propensities of amino acids

Glycine versus alanine. The relative merits of Gly and Ala depend on their position in a helix and on their neighbors. Three principles are involved: The first is that Gly is inherently more destabilizing than Ala because Gly has more conformational freedom in the unfolded state[33]—see the Ramachandran diagram (Chapter 1, section C2)—but this should be a relatively constant factor within a protein. The second is that Ala buries more solvent-accessible hydrophobic surface area on folding from an extended conformation than Gly does.[29,34] The amount buried depends on the nature of the surrounding side chains. There is an empirical equation for the relative effects of Ala and Gly on stability ($\Delta\Delta G_{Gly\rightarrow Ala}$) at solvent-exposed positions that are not at the N- or C-caps:

$$\Delta\Delta G_{Gly\rightarrow Ala} = 1.85 - 0.055\Delta A_{HP} \text{ (kcal/mol, or kJ/4.184 mol)} \qquad (17.14)$$

where ΔA_{HP} is the difference in the solvent-exposed hydrophobic surface area of the helix containing Ala minus that containing Gly, measured in Å^2 units (1 Å = 0.1 nm) (Figure 17.7).[34,35]

The third principle is that the side chain of Ala can hinder the solvation of NH and CO groups that need to make hydrogen bonds with solvent.[35] This is important for the C- and N-termini of the helixes.[36] There is an empirical correlation that relates $\Delta\Delta G_{Gly\rightarrow Ala}$ to both ΔA_{HP} and the difference in solvent-accessible surface area of NH and CO groups in the helix that require solvation (ΔA_{HB}, = area of Ala-containing helix − area of Gly-containing helix):

$$\Delta\Delta G_{Gly\rightarrow Ala} = 1.61 - 0.045\Delta A_{HP} + 0.19\Delta A_{HB} \text{ (kcal/mol, or kJ/4.184 mol)} \quad (17.15)$$

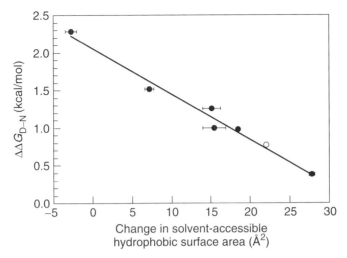

Figure 17.7 Change in ΔG_{D-N} on mutation of Ala to Gly in the solvent-exposed (noncap positions) of two helixes in barnase plotted against the change in solvent-exposed hydrophobic surface area on mutation (area of Ala-containing protein − area of Gly-containing protein). The data point at the top left is for an Ala that is surrounded by other hydrophobic side chains. When the denatured state of this protein folds, it buries near-maximal hydrophobic surface area. Mutation to Gly is thus particularly destabilizing. At the bottom right is an Ala that is highly solvent-exposed. Little surface area is buried on folding, so mutation to Gly is hardly destabilizing. The linearity of the plot with surface area suggests that the classical hydrophobic bond is operating here (Chapter 11). The open circle is for a model peptide. [Modified from L. Serrano, J. L. Neira, J. Sancho, and A. R. Fersht, *Nature, Lond.* **356**, 453 (1992)].

In general, Ala is more stable in the middle positions of an α helix because it buries more solvent-accessible surface area, but Gly is preferred at the ends because it allows greater exposure for solvation. This is more noticeable at the N-terminus because the side chain points backward along a helix.

A more advanced general equation uses A_{HP}, A_{HB}, and a third term involving the dihedral angles of the residue at the mutated position. This last term extends the calculation to any conformation, not only α helixes.[37]

Residues at C-caps. Mutational experiments on the two C-caps of the ribonuclease barnase (Chapter 1, Figure 1.6) give the rank order of preference of Gly ≫ Arg,His > Asn,Lys > Ala,Ser,Thr > Asp > Val (Table 17.1).[35,36] This order may be rationalized by three basic principles. The first is that amino acid side chains that can make hydrogen bonds with the exposed CO groups are stabilizing with respect to Ala (see Table 17.1). For example, the C-cap residue His-18 makes a hydrogen bond with the backbone CO of Gln-15. All things being equal, an intramolecular hydrogen bond is preferred over an intermolecular one, because there is a greater loss of entropy on tying down an independent

Table 17.1   *Change in stability on mutation of Ala at the C-caps of barnase*

| Mutation | $\Delta\Delta G_{D-N}{}^a$ | |
	kcal/mol	kJ/mol
Ala→Gly	−2.23	−9.3
Ala→His	−0.67	−2.8
Ala→Asn	−0.47	−2.0
Ala→Arg	−0.47	−2.0
Ala→Lys	−0.01	−0.04
Ala→Ser	0.16	0.7
Ala→Asp	0.27	1.1

^{a} $\Delta\Delta G_{D-N}$ is the free energy of unfolding of the Ala mutant minus that of the other mutant. Negative sign = increase in stability on mutation. The values are the average of mutations at positions 18 and 34 of barnase.

solvent molecule. This is an important general principle that is discussed in Chapter 11. The second reason is that positively charged side chains can interact favorably with the helix dipole, whereas negative ones are repelled. The third is that large hydrophobic side chains hinder the solvation of the CO groups of the backbone in the last turn of the helix. The fourth results from the conformational preferences in the Ramachandran diagram. The His at the C-cap of helix$_1$ and the Gly at the C-cap of helix$_2$ in barnase are in the α_L conformation with positive ϕ and ψ angles (α_L is the left-handed helical conformation in Figure 1.11). This is a common motif of helixes: some 30% of helixes have their C-cap in such a conformation and 60% of these are a Gly. This can readily be rationalized by the empirical relationship (equation 17.13) between stabilization energy and the degree of solvent exposure of groups that require hydrogen bonding with solvent. An Ala residue in an α_L conformation at the C-terminus of a model *poly*-alanine helix exposes 32 Å^2 more solvent-accessible hydrophilic surface area and buries 26 Å^2 more hydrophobic surface area than when in the regular α_R conformation. This could contribute 3.4 kcal/mol (14 kJ/mol) of energy to compensate for the unfavorable conformation. Gly is the most favorable residue at the C-cap of a helix because the lack of a side chain allows it to adopt an α_L conformation without strain, and it also allows the greatest solvent exposure of the C-terminal of a helix.

Residues at N-caps. The rank order of preference at the N-caps of barnase is Asp,Thr,Ser > Asn,Gly > Glu,Gln > His > Ala > Val (Table 17.2).[35,36] The basic principles behind these preferences are similar to those at the C-cap. (1) Hydrogen-bond acceptors are particularly effective at stabilizing the NH groups

Table 17.2 *Change in stability on mutation of Ala at the N-cap of barnase*

	$\Delta\Delta G_{D-N}{}^a$	
Mutation	(kcal/mol)	(kJ/mol)
Ala→Asp	−2.02	−8.4
Ala→Thr	−2.05	−8.6
Ala→Ser	−1.64	−6.9
Ala→Asn	−0.86	−3.6
Ala→Gly	−0.69	−2.9
Ala→Gln	−0.42	−1.8
Ala→Glu	−0.25	−1.0
Ala→His	−0.16	−0.7
Ala→Val	0.15	0.6
Ala→Pro	0.87	3.6

^{a} $\Delta\Delta G_{D-N}$ is the free energy of unfolding of the Ala mutant minus that of the other mutant. Negative sign = increase in stability on mutation. The values are the average of mutations at positions 6 and 26 of barnase.

by forming intramolecular hydrogen bonds (mainly to residue N-cap + 3; e.g., Thr-6 in barnase forms, typically, a good hydrogen bond with the backbone NH of residue 9, a poorer one with residue 8, and a bad one with residue 7 (Figure 1.5). (2) Gly allows the greatest exposure of the unpaired NH groups to solvent, whereas bulky hydrophobics inhibit solvation. (3) Negative charges can interact favorably with the helix dipole. A water molecule that solvates the N-terminal of the helix of chymotrypsin inhibitor 2 is clearly seen in the crystal structure of a mutant with Gly at the N-cap. There is a good $H_2O\cdots HN$ bond with N-cap + 3; $O\cdots N = 2.7$ Å. Mutation of Gly → Ala causes the water to move and inhibits the solvation; $O\cdots N = 3.4$ Å. Mutation to Ser leads to its side chain making the hydrogen bond with N-cap + 3; $O\cdots N = 2.6$ Å.

The negatively charged carboxylate of Asp that acts as a hydrogen-bond acceptor at the N-cap stabilizes barnase and T4 lysozyme by some 0.6 to 1.3 kcal/mol (2.5 to 5.5 kJ/mol) more than does the carboxyamide of Asn. Asp residues at positions N + 1 and N + 2 also stabilize the protein with respect to Asn residues by some 0.6 to 0.8 kcal/mol (2.5 to 3.5 kJ/mol). The interaction of a single charge at the end of the helix with its dipole appears to be worth some 0.5–1.5 kcal/mol (2–6 kJ/mol).

Amino acid residues at an internal position. The changes in stability of barnase on making all 19 substitutions of Ala-32 in the second helix[38] are in reasonable agreement with equivalent changes in internal positions in model

Table 17.3 *α-Helix- and β-sheet stabilizing effects of amino acids relative to alanine[a]*

Amino acid	Relative helix stability[a] (kcal/mol)	(kJ/mol)	Agadir[b] (kcal/mol)	(kJ/mol)	Relative β-sheet stability[c] (kcal/mol)	(kJ/mol)
Ala	0.00	0.0	0.00	0.00	0.00	0.00
Arg	0.17	0.7	0.06	0.25	−0.40	−1.7
Leu	0.17	0.7	0.19	0.79	−0.45	−1.9
Glu (0)[d]	0.17	0.7	—	—	—	—
Met	0.19	0.8	0.21	0.88	−0.90	−3.8
Lys	0.31	1.3	0.15	0.63	−0.35	−1.5
Trp	0.31	1.3	0.47	1.97	−1.04	−4.2
Gln	0.33	1.4	0.32	1.34	−0.38	−1.6
Ser	0.44	1.8	0.52	2.18	−0.87	−3.6
Ile	0.43	1.8	0.35	1.46	−1.25	−5.2
Phe	0.47	2.0	0.47	1.97	−1.08	−4.5
Cys	0.54	2.3	0.61	2.55	−0.78	−3.3
Asp (0)[d]	0.54	2.3	—	—	—	—
Glu (−)	0.56	2.3	0.34	1.42	−0.23	−1.0
Tyr	0.56	2.4	0.47	1.97	−1.63	−6.8
Asn	0.61	2.6	0.60	2.51	−0.52	−2.2
Thr	0.61	2.6	0.57	2.38	−1.36	−5.7
Val	0.63	2.7	0.51	2.13	−0.94	−3.9
His (0)	0.65	2.7	0.62	2.59	−0.37	−2.4
Asp (−)[d]	0.68	2.8	0.59	2.47	0.85	3.6
His (+)	0.88	3.7	—	—	—	—
Gly	0.90	3.8	1.11	4.64	1.21	5.1
Pro	3.47	14.5	2.72	11.4	>5	>20

helixes[39,40] and other proteins[41,42] (Table 17.3). The model peptides and barnase are similar to a *poly*-alanine helix at the site of mutation. Apart from Pro, which is very destabilizing because of the loss of two intrahelical hydrogen bonds, there is a spread of only 1 kcal/mol (4 kJ/mol); Ala is the most stabilizing residue, and Gly destabilizes the most. Factors that contribute to the differences in the intrinsic effect of different amino acids at internal positions on α-helix stability are (1) burial of hydrophobic surfaces and van der Waals contacts; (2) differences in the conformational restriction of the different amino acids in the helix relative to that in the unfolded proteins that may account for most of the effects by contributing to the entropy of folding; and (3) different solvation effects.[38] Individual reasons for the differences in stability are tabulated (Table 17.4). Briefly, the straight-chain amino acids are the most stabilizing because they can be incorporated into the helix with the least loss of rotational freedom, and they bury hydrophobic surface area. Side chains that are branched at the C^β atom destabilize α helices because they are conformationally restricted: the *trans* conformation of the dihedral angle χ_1 (C^α—C^β rotation) becomes increasingly the only allowed conformation with increasing size, whereas the side chain has three allowed conformations, g^-, g^+, and *trans*, in the random coil state. Residues with side chains that can hydrogen-bond with the main chain in the unfolded state will tend to be helix destabilizing; e.g., His, Tyr, Ser, Cys, Gln, and Asn and possibly Asp, Glu, Lys, and Arg. The burial of the —OH group of the side chain of Ser is correlated with destabilization because of the restrictions on its hydrogen bonding with solvent.[38]

---

[a] Amino acids are at solvent-exposed internal positions in helixes that mostly approximate to *poly*-alanine near the site of substitution. All substitutions are destabilizing relative to alanine. The values are the medians from six sets of experiments: mutations in barnase (A. Horovitz, J. M. Matthews, and A. R. Fersht *J. Molec. Biol.* **227**, 560–568 (1992)); T_4 lysozyme (M. Blaber, X. J. Zhang, and B. W. Matthews, *Science* **260**, 1637 (1993)); RNase T_1 (J. K. Myers, C. N. Pace, and J. M. Scholtz, *Biochemistry* **36**, 10923 (1997)); a model helix that is probably helical in its central positions (P. C. Lyu, M. I. Liff, L. A. Marky, and N. R. Kallenbach, *Science* **250**, 669 (1990)); a model peptide that dimerizes to give a stable helix (K. T. O'Neil and W. F. DeGrado, *Science* **250**, 646–651 (1990)); and a peptide from RNase T_1, equivalent to that in which the mutations were made (J. K. Myers, C. N. Pace, and J. M. Scholtz, *Biochemistry* **36**, 10923 (1997)). [(0) = neutral form of an ionizable side chain, (+) = positively charged, and (−) = negatively charged].

[b] Used as the intrinsic values in the program Agadir (V. Muñoz and L. Serrano, *Nature Struct. Biol.* **1**, 399 (1994)).

[c] These values are highly context dependent. These data are for substitutions of Thr-53 in the 56-residue B1 domain of staphylococcal IgG binding protein G (C. K. Smith, J. M. Withka, and L. Regan, *Biochemistry* **33**, 5510 (1994)). Thr-53 has nearest neighbors of Ile-16, Thr-44, Thr-51, and Thr-55, with side chains in the same face of the sheet. Pairwise interaction energies are up to 1.8 kcal/mol (8 kJ/mol), which are comparable with the intrinsic energies (C. K. Smith and L. Regan, *Science* **270**, 980 (1995)).

[d] From relatively small data sets.

Table 17.4 Structural properties of amino acids in α helixes

Amino acid	Effects relative to Ala
1. N-cap	
Gly	Stabilizing. Exposes terminal NH groups to solvent.
Ser, Thr	Stabilizing. Hydrogen bonds with $N + 3$.
Asp	Stabilizing. Hydrogen bonds with $N + 3$. Interacts with helix dipole.
Asn	Stabilizing if the dihedral angles are appropriate. Hydrogen bonds with $N + 3$.
Gln	Weakly stabilizing. Hydrogen bonds with $N + 3$. Loss of entropy.
Glu	Weakly stabilizing. Hydrogen bonds with $N + 3$. Loss of entropy. Interacts with helix dipole.
His	Weakly stabilizing. Hydrogen bonds with $N + 3$. Loss of entropy.
His(+), Lys, Arg	Destabilizing. Unfavorable interaction with helix dipole. Cannot form hydrogen bonds with terminal NH groups. Shield terminal NH groups from solvent.
Val and larger hydrophobics	Destabilizing. Shield terminal NH groups from solvent.
2. C-cap	
Gly	Stabilizing. Exposes terminal CO groups to solvent. Can take up α_L conformation and expose more hydrophilic surface area.
Asn	Stabilizing. Makes hydrogen bonds with terminal CO groups.
His(+), Lys, Arg	Weakly stabilizing. Make hydrogen bonds with terminal CO groups and interact favorably with helix dipole but lose entropy, especially in the case of Lys and Arg. More effective when side chains also interact elsewhere.
Ser	Weakly destabilizing. Poor geometry for making hydrogen bonds with terminal CO groups.

Surface salt bridges. A salt bridge is the interaction between a positively charged side chain of one amino acid with the negatively charged side chain of another in which the charged groups are within hydrogen-bonding distance (e.g., the $-CO_2^-$ of an Asp and the $-NH_3^+$ of a lysine). Salt bridges that can occur between residues i and $i + 4$ in helixes have been thought to be important in ther-

Amino acid	Effects relative to Ala
Asp	Weakly destabilizing. Unfavorable interaction with helix dipole, cannot make hydrogen bonds with CO groups.
Large hydrophobics	Destabilizing? Shield terminal CO groups.

3. Solvent-exposed internal positions (not N + 1 to N + 3 or C − 1 to C − 3)

Met, Leu, Arg, Lys	Mildly destabilizing. Intrahelical interactions may be made.
Ser	Destabilizing. γOH group partly shielded from solvent.
Cys	Destabilizing. The γSH group prevents hydrogen bonding of water to the main chain CO at position $i − 3$ and does not make as good a hydrogen bond as the γOH group of Ser. There are discrepancies in the results from barnase and the results obtained from model peptides.
Gln, Glu	Destabilizing. May form hydrogen bonds with polypeptide backbone in unfolded state.
Thr, Val, Ile	Destabilizing because β-branching allows only the *trans* conformation in folded state.
Asp, Asn	Destabilizing. May form hydrogen bonds with polypeptide backbone in unfolded state, and there could be a reduction of the conformational freedom and solvation of the side chain.
Phe, Tyr, His, Trp	Destabilizing. β-substitution reduces conformational freedom. His and Tyr may also form hydrogen bonds with polypeptide backbone in unfolded state.
Gly	Very destabilizing. Buries less hydrophobic surface area, makes less van der Waals contacts, and loses more conformational entropy on folding.
Pro	Very destabilizing. Loses two intrahelical hydrogen bonds.

mostability. However, experiments on several proteins and model systems show that a single solvent-exposed salt bridge contributes only 0.1 to 0.5 kcal/mol (0.4 to 2 kJ/mol) to stability.[43]·The low values result from two factors. First, the energy of the salt bridge must be balanced against the better solvation of the ions when they are completely surrounded by water on breaking the bridge. Second, there is a loss of entropy on fixing mobile side chains of the two charged residues, although this may be partly compensated by any extraneous hydrophobic and van der Waals interaction that a side chain can make with the protein. In double salt bridges, however, there can be synergy between the components because of the sharing of the entropy cost between the different interactions.

Prediction of helix stability. There is not a unique helix-forming propensity for each individual amino acid residue. Contacts elsewhere are all important. In solvent-exposed positions, contacts of residue i with $i + 3$ and $i + 4$ are important. A more general treatment of helix-forming propensities must have pairwise interaction energy terms included because side chains interact, especially residue i with $i + 3$ and $i + 4$. The program AGADIR incorporates these terms and is successful at predicting the stability of model helical peptides,[44] as do other procedures.[45,46] Within a protein, however, interactions with the body of the protein can stabilize or destabilize a helix greatly.

Helix stability and protein stability. We can predict the stability of helixes more reliably than we can any other element of protein structure. This provides a means for increasing the stability of proteins, because naturally occurring helixes are not always optimized for stability. If we make a mutation in the face of a helix that is exposed to solvent, and the mutation does not affect interactions elsewhere in the protein, then the overall free energy of folding of the protein generally changes by the same amount as the stability of the helix.[47] This rule breaks down if we overstabilize the helix: if the helix becomes so stable that it is still present as a helix in the denatured state, then increasing its stability further does not increase the stability of the protein, because both the native and denatured states are increased equally in stability.

2. β-Sheet propensities

The formation of β sheet is driven primarily by hydrophobic interactions between side chains in the strands. The calculation of β-sheet propensities of different amino acids is far more complicated than for the α helix because they depend on the context of both secondary and tertiary structure, and the local interactions can far outweigh intrinsic propensities.[48-52] There are also differences between the edge and central strands. But there is a broad agreement between experimental scales determined for different systems, suggesting that there are secondary structural preferences for β strands, although they are sometimes masked by context effects (Table 17.3). Generally, β-branched residues, such as Thr and Val, are favored because they have more degrees of freedom in the Ramachandran plot for extended strands. Gly is disfavored, just as it is in helixes, because it loses more entropy than any other amino acid on being restricted.

3. The hydrophobic core

The importance of the hydrophobic core is most easily probed by using mutations that create small deletions in side chains without changing the geometry. Changes that cause geometric clashes by increasing the size of a side chain or changing its stereochemistry are far more difficult to analyze. The truncation of the side chains of Ile to Val, Ala, and Gly; truncation of Leu to Ala and Gly; truncation of Val to Ala and Gly; and so on decrease stability by up to 2.5 kcal/mol (10 kJ/mol) per methylene group deleted, and typically 1.5 ± 0.5 kcal/mol (6 ± 2 kJ/mol). Mutation of Ile $\rightarrow$ Val in barnase just leaves a hole in the hydrophobic core with little, if any, rearrangement of the surrounding residues.[53] Mutation of Leu and Ile $\rightarrow$ Ala in the hydrophobic cores of T4 lysozyme[54] and barnase[55] leaves cavities, some of which partly collapse as the surrounding side chains move in. The largest losses in energy occur where the mutant has the full-sized cavity. As the cavity collapses due to neighboring side chains moving in, there are correspondingly smaller changes in energy. Rearrangement of structure on mutation must always compensate the loss of missing interactions.

There is a correlation between the observed change in energy and the number of $-CH_2-$ and CH_3- groups in a sphere of radius 6 Å surrounding each methylene group that is deleted from hydrophobic side chains in barnase[27] and chymotrypsin inhibitor 2[56] (Figure 17.8). The higher the packing density, the stronger the van der Waals interactions with the target methylene and also among the surrounding side chains, so that they are less likely to collapse into the cavity formed on mutation.

Because the hydrophobic interactions are cumulative, the effects of removing several methylene groups cause the largest decreases in stability; for example, Ile or Leu mutated to Ala can cost 5 to 7 kcal/mol (20 to 30 kJ/mol). The hydrophobic core is surprisingly tolerant to substitution of hydrophobic side chains by others of different stereochemistry from what would be expected, and radical changes have been made in cores with the retention of activity.[57-59]

a. Are the cavities empty?

There is no crystallographic evidence for water molecules occupying the cavities produced on mutation of the hydrophobic core of T4 lysozyme.[54] But x-ray crystallography can detect solvent molecules only if they are extremely precisely located; otherwise, the electron density is smeared out. There is a bound water in a small cavity in barnase, which makes a single hydrogen bond with a backbone peptide atom.[55] All cavities produced on mutation of hydrophobic cores are lined by a proportion of about 12% of polar atoms, and so purely hydrophobic cavities are very rarely produced by mutation.[55]

b. Nature of the hydrophobic effect in cores

The magnitude of the classical hydrophobic effect is proportional to the amount of hydrophobic surface area buried on formation of the hydrophobic interaction, ΔA_{HP} (Chapter 11). Changes in energy on mutation of Ala $\rightarrow$ Gly in the solvent exposed surfaces of α helixes are proportional to ΔA_{HP}, and so these energetics are presumably driven by hydrophobicity.[34,35] On the other hand, the correlation of energy changes in the core with packing density indicates that the interactions are dominated by the van der Waals term, because van der Waals interactions are

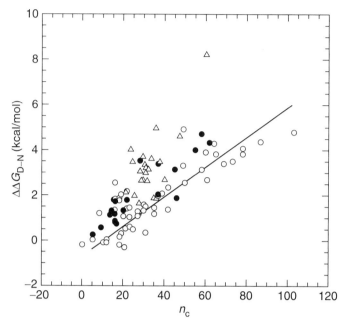

Figure 17.8 Correlation of $\Delta\Delta G_{D-N}$ for deletion mutations in the hydrophobic cores of CI2 [open circles, from D. E. Otzen and A. R. Fersht, *Biochemistry* **34**, 13051 (1995)], barnase [filled circles, from L. Serrano, J. T. Kellis, P. Cann, A. Matouschek, and A. R. Fersht, *J. Molec. Biol.* **224**, 783 (1992)], and T4 lysozyme [triangles, from J. Xu, W. A. Baase, E. Baldwin, and B. W. Matthews, *Protein Sci.* **7**, 158 (1998)] with packing density, n_c (= number of methyl and methylene groups within a 6 Å sphere of each deleted methylene group). The line is the best fit to the CI2 and barnase data. This correlation indicates the importance of van der Waals interactions in stabilizing the core, since they are additive and depend on atom density. Although the size of the deleted chain tends to increase along the series, a good correlation with n_c is also found for the restricted series of Ile $\rightarrow$ Val mutations in CI2 and barnase, showing that the correlation does not depend on increasing size of side chain [S. E. Jackson, M. Moracci, N. Elmasry, C. M. Johnson, and A. R. Fersht, *Biochemistry* **32**, 11259 (1993)].

additive, and so the higher the packing density, the stronger the van der Waals interactions. Mutation of partly exposed hydrophobic patches in proteins gives a mixture of correlations among energy, surface area, and packing density. The hydrophobic effect appears to be greater in the interactions within proteins than it is in simple model reactions. The reasons are discussed in Chapter 11.

4. Disulfide crosslinks

Disulfide bridges are thought to stabilize proteins primarily by destabilizing the denatured state by reducing its conformational entropy. They can contribute up to 4 kcal/mol (17 kJ/mol) per disulfide bridge to the stability, and similar values have been found on introducing bridges into proteins that lack them.[60–63] Stabilization occurs only when the bridge can be introduced with the correct stereochemistry; otherwise, destabilization may occur.

The change in configurational entropy, ΔS_{config}, on forming a loop of n residues from an open random coil has been estimated from simple polymer theory and the probability that the two points where the loop in the random coil is linked lie simultaneously within a defined volume element V. Thus, equations of the form

$$\Delta S_{\text{config}} = a - \frac{3}{2} R \ln n \tag{17.16}$$

can be derived. The value of a for disulfide-linked loops is estimated to be between -2 and -8 cal/mol/deg (-8 and -34 J/mol/deg), depending on the choice of V.[64,65] The factor of 3/2 assumes a complete lack of defined structure in the residues in the absence of the link and no additional interactions of the loop with the rest of the protein. The presence of structure alters this term. In practice, adventitious interactions within real proteins outweigh these effects.

5. Relationship between statistical surveys and measured energetics

The results of statistical surveys may be compared with detailed energetic studies. If the probability of finding a particular residue, AA, at a particular position is P_{AA}, and if this represents true thermodynamic preference relative to, say, Ala (P_{Ala}), then a free energy may be calculated from the statistics ($\Delta\Delta G_{\text{stat}}$):

$$\Delta\Delta G_{\text{stat}} = -RT \ln \frac{P_{\text{AA}}}{P_{\text{Ala}}} \tag{17.17}$$

Equation 17.17 is based on the general equation that the probability of occupying a state of energy ΔG varies as $\exp(-\Delta G/RT)$. There is, in fact, a fair correlation between the statistical and observed experimental data, which indicates that the main factor in determining the statistical preferences is, indeed, thermodynamic.[28]

6. Additivity of binding energy changes

The effects of mutations in residues that are not in direct contact are generally additive. For example, barnase has a close homolog, binase, which differs by just 17 mutations that are scattered throughout the sequence. All single mutations between the two have been made, as have many of their combinations.[66] The changes in stability are strictly additive in combination. Barnase and binase have similar net stabilities, because six of the changes in binase are stabilizing, six are destabilizing, and the remainder are neutral. The stability of barnase is increased by 3.4 kcal/mol (14 kJ/mol) when all six stabilizing mutations are incorporated.

This appears to be quite a general phenomenon and is a way of empirically stabilizing proteins.[66–68]

E. Stability – activity trade-off?

Some enzymes from thermophiles that are very stable at normal temperatures have low activities at the lower temperatures. It has been speculated that there is a trade-off between stability and activity. But barnase that has been stabilized by 3.4 kcal/mol (14 kJ/mol) by a combination of six mutations that are external to its active site has no loss of activity.[66] However, there is a compromise between stability and activity in the structure of the active site.[69] There are several positions in the active site of barnase that can be mutated to give more stable but less active protein. Activity can even be increased further at an unacceptable expense to stability. Some of these changes can be rationalized. The active site of the ribonuclease is highly positively charged because it has to stabilize the highly negatively charged transition state generated at the pentacoordinate phosphate. The lysine, arginine, and histidine at the active site are mutually repulsive. Other regions of the active site have to be mobile, and so some hydrogen bonds are weak.

Active sites of enzymes and binding sites of proteins are a general source of instability because they contain groups that are exposed to solvent in order to bind substrates and ligands and so are not paired with their normal types of partners. Stability – activity trade-off is also seen with residues in the natural polypeptide inhibitor of barnase, barstar, that has evolved to bind as rapidly as possible to barnase,[70] and also in the active site of T4 lysozyme.[71]

F. Prediction of three-dimensional structure from primary structure

The *de novo* prediction of the three-dimensional structure of a protein from its primary sequence involves two distinct tasks: description of the native state and knowledge of the kinetic pathway of folding to determine if the protein will fold. These tasks are beyond present methods of computation both now and in the foreseeable future. Direct computation of protein structure from interaction energies is feasible for small perturbations from a *known* three-dimensional structure. If a structure is known with a fair degree of precision from protein crystallography or NMR, then the coordinates of the atoms may be refined by calculation using energy functions. This is now a standard technique in structure refinement. Further, fluctuations in the structure caused by the vibrations of the atoms about their mean positions may be calculated from the energy functions and Newton's equations of motion; this is the technique of *molecular dynamics*. Thus, a possible strategy for structure prediction is the derivation of methods for predicting approximate three-dimensional structure, followed by computer refinement. Unfortunately, it is still not yet possible to obtain even a crude starting structure *de novo* by computer calculation, for the reasons discussed at the very beginning of this chapter.

The alternative procedure is to ignore detailed energetics and prediction *de novo* and use, instead, modeling by *homology*. This procedure may be applied to proteins that already exist in nature and thus must form a stable folded structure. For example, trypsin was known to be homologous with chymotrypsin, whose structure had been solved with about 50% identity of sequence. It was possible some 30 years ago to build a reasonable structure for trypsin by building the amino acid sequence of trypsin into the known three-dimensional structure of chymotrypsin. This approach works with varying degrees of success with homologous proteins that have a high degree of sequence identity. More sophisticated procedures have been developed to analyze proteins, or parts of proteins, in which the homologies are not obvious. One procedure uses *neural networks*, whereby computer programs train themselves to recognize patterns. By using a database of all proteins of known structures, it is possible to predict to some extent the secondary structure of proteins that do not have close homologs. An important method of improving the accuracy of prediction for an unknown protein is to increase the amount of information by finding the sequences of the same protein from different organisms. The prediction algorithm is then applied to the whole class of unknown structures, with a consequent increase in reliability of prediction. Despite much effort, however, the algorithms have been developed only to the stage where at most 60–70% of the residues in a protein may be successfully predicted to belong to a particular secondary structure that forms an α helix, a β sheet, or a β bend. The identification of segments of structure (i.e., sheet or helix) is more successful.

An extension of homology modeling is to include empirical energy functions. For example, tables may be prepared of the relative frequency with which certain amino acid residues are in proximity to one another, and an energy can be calculated from the probability by equation 17.17. The sequence of the unknown protein may be "threaded" through all known three-dimensional structures, and its energy can be calculated each time the sequence is moved along by one amino acid. In this way, it has been possible to find known structures that are homologous with the unknown structures.

References

1. G. I. Makhatadze and P. L. Privalov, *Adv. Prot. Chem.* **47**, 307 (1995).
2. P. L. Privalov and G. I. Makhatadze, *J. Molec. Biol.* **232**, 660 (1993).
3. G. I. Makhatadze and P. L. Privalov, *Protein Science* **5**, 507 (1996).
4. G. I. Makhatadze and P. L. Privalov, *J. Molec. Biol.* **213**, 375 (1990).
5. P. L. Privalov and G. I. Makhatadze, *J. Molec. Biol.* **213**, 385 (1990).
6. P. L. Privalov, E. I. Tiktopulo, S. Y. Venyaminov, Y. V. Griko, G. I. Makhatadze, and N. N. Khechinashvili, *J. Molec. Biol.* **205**, 737 (1989).

7. C. Tanford, *Adv. Prot. Chem.* **23**, 121 (1968).
8. C. Tanford, *Adv. Prot. Chem.* **24**, 1 (1970).
9. C. N. Pace, *Methods Enzymol.* **131**, 266 (1986).
10. C. M. Johnson and A. R. Fersht, *Biochemistry* **34**, 6795 (1995).
11. J. A. Schellman, *Biopolymers* **34**, 1015 (1994).
12. J. Tirado Rives, M. Orozco, and W. L. Jorgensen, *Biochemistry* **36**, 7313 (1997).
13. P. L. Privalov, *Adv. Prot. Chem.* **33**, 167 (1979).
14. J. M. Sanz and A. R. Fersht, *Biochemistry* **32**, 13584 (1993).
15. K. A. Dill and D. Shortle, *Ann. Rev. Biochem.* **60**, 795 (1991).
16. D. R. Shortle, *Current Opinion in Structural Biology* **6**, 24 (1996).
17. V. L. Arcus, S. Vuilleumier, S. M. V. Freund, M. Bycroft, and A. R. Fersht, *J. Molec. Biol.* **254**, 305 (1995).
18. S. M. V. Freund, K. B. Wong, and A. R. Fersht, *Proc. Natl. Acad. Sci. USA* **93**, 10600 (1996).
19. M. Oliveberg, V. L. Arcus, and A. R. Fersht, *Biochemistry* **34**, 9424 (1995).
20. O. B. Ptitsyn, *Trends in Biochemical Sciences* **20**, 376 (1995).
21. J. R. Gillespie and D. Shortle, *J. Molec. Biol.* **268**, 170 (1997).
22. P. M. Dalby, M. Oliveberg, and A. R. Fersht, *J. Molec. Biol.* **276**, 625 (1998).
23. K. W. Plaxco, I. S. Millett, D. J. Segel, S. Doniach, and D. Baker, personal communication (1998).
24. V. N. Uversky and O. B. Ptitsyn, *Fold Des.* **1**, 117 (1996).
25. A. Matouschek, J. M. Matthews, C. M. Johnson, and A. R. Fersht, *Protein Engineering* **7**, 1089 (1994).
26. W. J. Becktel and J. A. Schellman, *Biopolymers* **26**, 1859 (1987).
27. L. Serrano, J. T. Kellis, P. Cann, A. Matouschek, and A. R. Fersht, *J. Molec. Biol.* **224**, 783 (1992).
28. A. R. Fersht and L. Serrano, *Current Opinion in Structural Biology* **3**, 75 (1993).
29. F. M. Richards and T. Richmond, *Ciba Symp.* **60**, 23 (1978).
30. J. S. Richardson and D. C. Richardson, *Science* **240**, 1648 (1988).
31. L. G. Presta and G. D. Rose, *Science* **240**, 1632 (1988).
32. W. G. J. Hol, *Prog. Biophys. Mol. Biol.* **45**, 149 (1987).
33. B. W. Matthews, H. Nicholson, and W. J. Becktel, *Proc. Natl. Acad. Sci. USA* **84**, 6663 (1987).
34. L. Serrano, J. L. Neira, J. Sancho, and A. R. Fersht, *Nature, Lond.* **356**, 453 (1992).
35. L. Serrano, J. Sancho, M. Hirshberg, and A. R. Fersht, *J. Molec. Biol.* **227**, 544 (1992).
36. L. Serrano and A. R. Fersht, *Nature, Lond.* **342**, 296 (1989).
37. E. Lopez Hernandez and L. Serrano, *Proteins: Structure Function and Genetics* **22**, 40 (1995).
38. A. Horovitz, J. M. Matthews, and A. R. Fersht, *J. Molec. Biol.* **227**, 560 (1992).
39. K. T. O'Neil and W. F. Degrado, *Science* **250**, 646 (1990).
40. P. C. Lyu, M. I. Liff, L. A. Marky, and N. R. Kallenbach, *Science* **250**, 669 (1990).
41. M. Blaber, X. J. Zhang, and B. W. Matthews, *Science* **260**, 1637 (1993).
42. J. K. Myers, J. S. Smith, C. N. Pace, and J. M. Scholtz, *J. Molec. Biol.* **263**, 390 (1996).
43. D. Sali, M. Bycroft, and A. R. Fersht, *J. Molec. Biol.* **220**, 779 (1991).
44. V. Munoz and L. Serrano, *Biopolymers* **41**, 495 (1997).
45. B. J. Stapley, C. A. Rohl, and A. J. Doig, *Protein Sci.* **4**, 2383 (1995).

46. C. A. Rohl, A. Chakrabartty, and R. L. Baldwin, *Protein Sci.* **5**, 2623 (1996).
47. V. Villegas, A. R. Viguera, F. X. Aviles, and L. Serrano, *Fold Des.* **1**, 29 (1996).
48. D. L. Minor and P. S. Kim, *Nature, Lond.* **371**, 264 (1994).
49. C. K. Smith, J. M. Withka, and L. Regan, *Biochemistry* **33**, 5510 (1994).
50. D. E. Otzen and A. R. Fersht, *Biochemistry* **34**, 5718 (1995).
51. M. Ramirez-Alvadao, F. J. Blanco, H. Niemann, and L. Serrano, *J. Molec. Biol.* **273**, 898 (1997).
52. C. K. Smith and L. Regan, *Science* **270**, 980 (1995).
53. A. M. Buckle, K. Henrick, and A. R. Fersht, *J. Molec. Biol.* **234**, 847 (1993).
54. A. E. Eriksson, W. A. Baase, X. J. Zhang, D. W. Heinz, M. Blaber, E. P. Baldwin, and B. W. Matthews, *Science* **255**, 178 (1992).
55. A. M. Buckle, P. Cramer, and A. R. Fersht, *Biochemistry* **35**, 4298 (1996).
56. S. E. Jackson, M. Moracci, N. ElMasry, C. M. Johnson, and A. R. Fersht, *Biochemistry* **32**, 11259 (1993).
57. W. A. Lim, D. C. Farruggio, and R. T. Sauer, *Biochemistry* **31**, 4324 (1992).
58. N. C. Gassner, W. A. Baase, and B. W. Matthews, *Proc. Natl. Acad. Sci. USA* **93**, 12155 (1996).
59. D. D. Axe, N. W. Foster, and A. R Fersht, *Proc. Natl. Acad. Sci. USA* **93**, 5590 (1996).
60. M. W. Pantoliano, R. C. Ladner, P. N. Bryan, M. L. Rollence, J. F. Wood, and T. L. Poulos, *Biochemistry* **26**, 2077 (1987).
61. R. Wetzel, L. J. Perry, W. A. Baase, and W. J. Becktel, *Proc. Natl. Acad. Sci. USA* **85**, 401 (1988).
62. A. Shaw and R. Bott, *Current Opinion in Structural Biology* **6**, 546 (1996).
63. J. Clarke and A. R. Fersht, *Biochemistry* **32**, 4322 (1993).
64. C. N. Pace, G. R. Grimsley, J. A. Thomson, and B. J. Barnett, *J. Biol. Chem.* **263**, 11820 (1988).
65. A. D. Nagi and L. Regan, *Fold Des.* **2**, 67 (1997).
66. L. Serrano, A. G. Day, and A. R. Fersht, *J. Molec. Biol.* **233**, 305 (1993).
67. M. W. Pantoliano, M. Whitlow, J. F. Wood, S. W. Dodd, K. D. Hardman, M. L. Rollence, and P. N. Bryan, *Biochemistry* **28**, 7205 (1989).
68. B. Steipe, B. Schiller, A. Pluckthun, and S. Steinbacher. *J. Molec. Biol.* **240**, 188 (1994).
69. E. M. Meiering, L. Serrano, and A. R. Fersht, *J. Molec. Biol.* **225**, 585 (1992).
70. G. Schreiber, A. M. Buckle, and A. R. Fersht, *Structure* **2**, 945 (1994).
71. B. K. Shoichet, W. A. Baase, R. Kuroki, and B. W. Matthews, *Proc. Natl. Acad. Sci. USA* **92**, 452 (1995).

Kinetics of Protein Folding

18

C. B. Anfinsen discovered that the small proteins ribonuclease A and *staphylococcal* nuclease could be reversibly denatured.[1] On removal of a chemical denaturant, such as urea, they spontaneously refold to their native structures after denaturation. Similarly, they spontaneously refold on cooling after thermal denaturation. Not only do the amino acid sequences of these proteins encode their final folded structures, they also encode the information on how to get to the structures. These ideas have, with some modification, stood the test of time. But there are proteins that will not renature after being denatured, sometimes because they have been processed after biosynthesis. For example, subtilisin has a prosequence that is required to direct folding after biosynthesis.[2,3] This prosequence is removed by autolysis after folding. Proteins such as insulin and chymotrypsin have internal segments that are removed by proteolysis, the cleaved chain being held together by disulfide bonds (Chapter 1). In some cases, larger or multimeric proteins do require assistance to fold, which is provided by *molecular chaperones*, which are described in Chapter 19, section I. The kinetics of folding and unfolding of proteins appear to be a very complex process, but this process follows very simple rate laws, governed by a few basic principles. In this chapter, we discuss these basic principles as a prelude to examining the mechanism of protein folding. We also outline two powerful techniques for examining detailed structural events in protein folding, Φ-value analysis and ^{1}H/^{2}H-exchange.

A. Kinetics of folding

1. Basic methods

The folding of proteins is usually studied *in vitro* by first denaturing them in solutions of urea, guanidinium chloride (GdmCl), or acid and then diluting the denaturant. Stopped-flow methods (Chapter 4, section A2) are the most convenient, because they cover a useful time range and are ideal for mixing two reagents. Fluorimetry, following the change in tryptophan fluorescence in the near ultraviolet, is used because of its sensitivity. This technique is generally used to monitor tertiary interactions because the fluorescence yield and emission wavelength of tryptophan are sensitive to its environment. Stopped-flow circular dichroism is useful because it can detect changes in secondary structure in the far ultraviolet. But it has an inherently poor signal-to-noise ratio and requires considerably more protein than does fluorescence.

Helixes form in a few hundred nanoseconds[4] and β turns in a few microseconds in model peptides.[5,6] Short loops in proteins form with an upper limit of about 10^6 s^{-1}.[7,8] Thus, a lower limit for the initial collapse of a denatured protein is about 1 μs.[7] Conventional rapid mixing methods are limited to a time scale of milliseconds or greater, but specialized continuous-flow apparatus has been used for tens of microseconds.[9] Relaxation methods or flash photolysis ("optical triggering"[10]) are necessary for investigating faster reactions (Chapter 4, section C). Temperature-jump procedures are ideal for denaturing proteins by jumping from ambient to elevated temperatures. This method has been adapted to study rapid events in folding on time scales of nanoseconds to microseconds by temperature-jumping a cold-denatured protein.[11,12] Pressure jumps on the 50-μs time scale have some limited application. Another promising technique for the 10- to 100-μs region is the analysis of line-width broadening in the NMR signals from a mixture of rapidly converting native and denatured states.[13]

2. Multiple phases and *cis*-peptidyl-prolyl bonds

a. Effects of peptidyl-proline isomerization on kinetics

Refolding is generally found to proceed by a series of exponential phases. Many of these exponentials are a consequence of *cis-trans* isomerization about peptidyl-prolyl bonds.[14,15] The equilibrium constant for the normal peptide bond in proteins favors the *trans* conformation by a factor of $10^3 - 10^4$ or so. The peptidyl-prolyl bond is an exception that has some 2–20% of *cis* isomer in model peptides (see Chapter 1, Figure 1.3). Further, it is often found as the *cis* isomer in native structures. (Replacement of *cis*-prolines with other amino acids by protein engineering can retain the *cis* stereochemistry.[16]) The interconversion of *cis* to *trans* in solution is quite slow, having half-lives of 10–100 s at room temperature and neutral pH. This has two important consequences. First, a protein that has several prolines all in the *trans* conformation in the native structure will equilibrate when denatured to give a mixture of *cis* and *trans* forms. On refolding, there will be an apparent succession of slow refolding phases as the different isomers fold *in parallel*. The isomerization of the proline bonds can be accelerated by being coupled with folding to have rate constants of $1\ s^{-1}$, or so, which are comparable with those of the folding process itself. Second, proteins that naturally have *cis*-peptidyl prolines equilibrate on denaturation to give denatured states that are predominantly in the wrong conformation. The kinetics of their refolding is dominated by the isomerization events. The proline-related phases in refolding can sometimes be accelerated by peptidyl-prolyl isomerase, if they are accessible to the enzyme.[15,17]

That peptidyl-proline isomerization is a post-unfolding event is readily demonstrated by double-jump experiments.[17] The protein is denatured for several milliseconds and then restored to renaturing conditions before the peptidyl prolines have had time to isomerize. The proline-related phases are then no longer seen.

b. Effects of peptidyl-proline isomerization on equilibria

Equilibrium parameters derived from kinetics will differ from those measured at equilibrium because of the effects of proline isomerization. The additional equilibria between *cis*- and *trans*-peptidyl-proline bonds in the denatured state add to its stability and so favor denaturation (equation 18.1).

$$N_{trans} \rightleftharpoons D_{trans} \rightleftharpoons D_{cis} \tag{18.1}$$

If, for example, there is a peptidyl-proline bond in the denatured state that has an equilibrium constant K_{iso} ($= [D_{cis}]/[D_{trans}]$), then the equilibrium constant for unfolding will be increased by a factor of $1 + K_{iso}$, since $[D_{cis}] + [D_{trans}] = (1 + K_{iso})[D_{trans}]$. Since K_{iso} is generally in the range of 0.05 to 0.2, this is a small factor if there are just three or four *trans*-peptidyl prolines in the native state. But if N contains a *cis*-peptidyl-proline bond, then this leads to very large factors (5 to 20) for each such prolyl residue.

c. Effects of peptidyl-proline isomerization on unfolding kinetics

The proline isomerization phases are normally seen in the folding kinetics. Denaturation of the protein at [denaturant] $\gg$ [denaturant]$_{50\%}$ (the concentration for 50% denaturation) is close to 100%, and only a single exponential phase is observed in the unfolding kinetics for most small proteins. But, at denaturant concentrations of approximately [denaturant]$_{50\%}$, a protein with a *trans*-peptidyl-prolyl bond goes first to D_{trans}, as in equation 18.1, and then there is a second minor phase as $D_{trans} \rightarrow D_{cis}$, displacing the equilibrium. This phase will be a major one if the protein has a *cis* bond. The two phases can, of course, become coupled and give complex kinetics, as described in Chapter 4, section D5.

B. Two-state kinetics

Chymotrypsin inhibitor 2 (CI2) folds rapidly by simple two-state kinetics; that is, $D \rightleftharpoons N$, with a $t_{1/2}$ of 13 ms.[18,19] CI2 is a small 64-residue protein that has all its peptidyl-proline bonds in the favorable *trans* conformation.[20] (There are, of course, additional slow *cis* $\rightarrow$ *trans* peptidyl-prolyl isomerization events, which account for about 20–30% of the refolding amplitudes.) The occurrence of two-state kinetics does not prove that there are no intermediates on the folding pathway; there could be intermediates that are present at high energy and are kinetically undetectable (see section B4). Two-state behavior has subsequently been found for many other small proteins. The simplicity of two-state folding kinetics provides the ideal starting point for the analysis and illumination of the basic principles of folding.

1. Effects of denaturant on unfolding and folding kinetics

A plot of the logarithm of the rate constant for unfolding, k_u, against the concentration of denaturant is found to be close to linear for many proteins at [denaturant] > [denaturant]$_{50\%}$:

$$\ln k_u = \ln k_u^{H_2O} + m_{k_u}[\text{denaturant}] \tag{18.2}$$

where $k_u^{H_2O}$ is the value extrapolated to the absence of denaturant and m_{k_u} is a constant of proportionality.[21,22] This can be recast in terms of activation energies:

$$\Delta G_{\ddagger-N} = \Delta G_{\ddagger-N}^{H_2O} - m_{\ddagger-N}[\text{denaturant}] \tag{18.3}$$

where $\Delta G_{\ddagger-N}^{H_2O}$ is the free energy of activation for unfolding in water, and the subscripts N and $\ddagger$ refer to the native states and transition states, respectively. This equation is analogous to that for equilibrium unfolding, $\Delta G_{D-N} = \Delta G_{D-N}^{H_2O} - m_{D-N}[\text{denaturant}]$ (Chapter 17, equation 17.6). The term $m_{\ddagger-N}$ is analogous to m_{D-N} for equilibrium unfolding.

For CI2 and some other small proteins, but not in general, the folding rate constant, k_f, follows a similar relationship for [denaturant] < [denaturant]$_{50\%}$:

$$\ln k_f = \ln k_f^{H_2O} - m_{k_f}[\text{denaturant}] \tag{18.4}$$

which may also be recast in terms of activation energies:

$$\Delta G_{\ddagger-D} = \Delta G_{\ddagger-D}^{H_2O} + m_{\ddagger-D}[\text{denaturant}] \tag{18.5}$$

The two curves form a V-shaped kinetics curve (Figure 18.1), which is sometimes called a chevron plot, constructed by combining the two rate constants:

$$\ln k_{obs} = \ln (k_f^{H_2O} \exp(-m_{k_f}[\text{denaturant}]) + k_u^{H_2O} \exp(m_{k_u}[\text{denaturant}])) \tag{18.6}$$

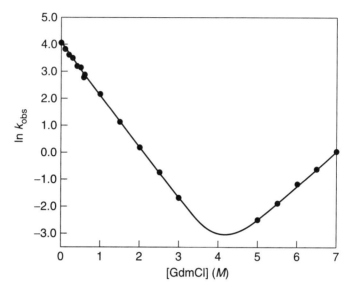

Figure 18.1 Plot of ln k_{obs} against [GdmCl] for the folding of CI2. $k_{obs} = k_f + k_u$ (s^{-1}) (i.e., the sum of forward and reverse rate constants (see Chapter 4, section D1c). k_{obs} approximates to the folding rate constant (k_f) for low [GdmCl], and the unfolding constant (k_u) at high [GdmCl]. ln k_{obs} fits perfectly to a two-state equation (solid line). [Modified from S. E. Jackson and A. R Fersht, *Biochemistry* **30**, 10436 (1991).]

The linearity of the equations holds very well for denaturation and folding of CI2 in GdmCl, but there is some downward curvature for the unfolding of barnase in urea,[23] which follows the equation:

$$\ln k_u = \ln k_u^{H_2O} + m_{k_u}[\text{urea}] - m^*[\text{urea}]^2 \qquad (18.7)$$

Possible reasons for the curvature are discussed in the next chapter.

2. Interpretation of the rate laws for denaturation and folding: The Tanford β value

It was explained in Chapter 17, equation 17.6, that proteins are denatured by urea and GdmCl because all parts of them are more soluble in denaturant, and their free energy of transfer to denaturant is nearly linear with [denaturant].[21,22] The transition state for protein folding and unfolding is intermediate in structure between the denatured and native states. Thus, the transition state is stabilized by denaturant with respect to the native structure but destabilized relative to the denatured state. Accordingly, denaturant lowers the activation energy of denaturation and raises the activation energy of folding. The stabilization energy terms are clearly linear with [denaturant].

Recall from Chapter 17, equation 17.7, that if each group, i, in the protein has a free energy of transfer of δg_i cal/mol from water to a 1-M solution of denaturant and each group becomes exposed by only a factor α_i on denaturation, then the effective free energy change for the group on denaturation is $\alpha_i \delta g_i$. This leads to an m value for unfolding of $m_{D-N} = \Sigma_i \alpha_i \delta g_i$. The same equation applies to the formation of the transition state for unfolding:

$$m_{\ddagger-N} = \sum_i \alpha_{i\ddagger} \delta g_i \qquad (18.8)$$

where $a_{i\ddagger}$ is the fractional degree of exposure in the transition state for denaturation. We can define a quantity, the Tanford β value, β_T:

$$\beta_T = \frac{m_{\ddagger-N}}{m_{D-N}} = \frac{\sum_i \alpha_{i\ddagger} \delta g_i}{\sum_i \alpha_i \delta g_i} \qquad (18.9)$$

β_T is a measure of the average degree of exposure in the transition state relative to that of the denatured state from the native state. It is a useful index of the compactness of the transition state. Because the plots against denaturant may be slightly curved, it is better to define $m_{D-N} = \partial \Delta G_{D-N}/\partial[\text{denaturant}]$ and $m_{\ddagger-N} = \partial \Delta G_{\ddagger-N}/\partial[\text{denaturant}]$ over the ranges of [denaturant] employed in the comparison. Accordingly, $\beta_T = (\partial \Delta G_{\ddagger-N}/\partial[\text{denaturant}])/(\partial \Delta G_{D-N}/\partial[\text{denaturant}])$.[24,25] The Tanford β value is often measured for unfolding kinetics. This is especially convenient for reactions in which there are folding intermediates because unfolding kinetics are usually simple. For two-state systems, the β values for folding and unfolding should add up to 1.

3. Effects of temperature on folding

The rate constant, k, for most elementary chemical reactions follows the Arrhenius equation, $k = A \exp(-E_A/RT)$, where A is a reaction-specific quantity and E_A the activation energy. Because E_A is always positive, the rate constant increases with temperature and gives linear plots of $\ln k$ versus $1/T$. Kinks or curvature are often found in Arrhenius plots for enzymatic reactions and are usually interpreted as resulting from complex kinetics in which there is a change in rate-determining step with temperature or a change in the structure of the protein. The Arrhenius equation is recast by transition state theory (Chapter 3, section A) to

$$k = k_B T \frac{\kappa}{h} \exp \frac{\Delta S^{\ddagger}}{R} \exp \frac{-\Delta H^{\ddagger}}{RT} \qquad (18.10)$$

where k_B is the Boltzmann constant, h is the Planck constant, κ is the transmission coefficient, $\Delta S^{\ddagger}$ is the entropy of activation, and $\Delta H^{\ddagger}$ is the enthalpy of activation. The value of $\Delta H^{\ddagger}$ is usually determined from the slope of a plot of $\ln (k/T)$ versus $1/T$, the Eyring plot. $\Delta H^{\ddagger}$ is always positive for an elementary reaction step because energy is required to create the partly broken bonds of the activated complex. The denaturation of proteins appears to follow equation 18.10. But folding reactions are often distinctly curved, and the kinetics is said to be non-Arrhenius.[26]

a. Non-Arrhenius folding kinetics: Negative enthalpies of activation

The simple two-state kinetics of folding of CI2 has been followed over a wide range of temperature (Figure 18.2).[26,27] The rate first increases with temperature, then goes through a maximum, and finally decreases with increasing temperature. $\Delta H^{\ddagger}$ varies with temperature, being positive at lower temperatures and negative at sufficiently high temperatures.

The simplest interpretation is that $\Delta H^{\ddagger}$ for protein folding has a large ΔC_p term, as described in Chapter 17 for overall denaturation. If the transition state for folding is relatively compact and shields the hydrophobic side chains, then there will be the characteristic ΔC_p term associated with the hydrophobic effect (cf. equation 17.1):

$$\Delta H_{\ddagger-D(T_2)} = \Delta H_{\ddagger-D(T_1)} + \Delta C_{p(\ddagger-D)}(T_2 - T_1) \tag{18.11}$$

where $\Delta C_{p(\ddagger-D)}$ is algebraically negative because it is measured from the denatured state to the transition state.

Interestingly, similar non-Arrhenius kinetics is found in computer simulations of simple models that do not consider interactions with solvent. At high temperature, the model polymer is highly random and has a high configurational entropy

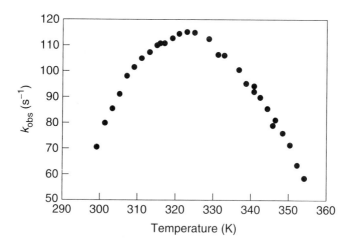

Figure 18.2 Plot of k_{obs} for the folding of CI2 against temperature. Note how the rate constant decreases with increasing temperature above 325 K. The activation energy barrier is purely entropic at this temperature, and the enthalpy of activation becomes negative. [Modified from M. Oliveberg, Y. J. Tan, and A. R. Fersht, *Proc. Natl. Acad. Sci. USA* **92**, 8926 (1995).]

that is lost in the folding reaction, and so the barrier to folding is entropic. At low temperatures, there are local minima in the model that lower the entropy of D and require enthalpic barriers to be broken (that is, there is structure present at low temperature that is melted out at high temperature). The contribution from these effects is probably much less than that from the classical hydrophobic effect.

The free energy of activation, $\Delta G_{\ddagger-D}$, is always positive. At high temperature, the barrier is entropic, with $T\Delta S_{\ddagger-D}$ being the dominant term. At low temperature, the barrier is primarily enthalpic, with $\Delta H_{\ddagger-D}$ being dominant.

Although it is difficult to detect curvature in the Arrhenius plots of the rate constant for unfolding, $\Delta C_{p(\ddagger-D)}$ can be determined by the same trick used for measuring $\Delta C_{p(N-D)}$ in equilibrium unfolding (Chapter 17, section A1F): measure $\Delta H_{\ddagger-D}$ at various values of acid pH and plot log $\Delta H_{\ddagger-D}$ versus $<T>$, the average temperature range for the experiments (Figure 18.3).[27]

The ratio $\Delta C_{p(\ddagger-D)}/\Delta C_{p(N-D)}$ should be a measure of the relative burial of hydrophobic surface areas in the transition and folded states, if the hydrophobic model is correct. The ratio is 0.51 for CI2, compared with a value of β_T of 0.6,[27] which is a measure of the overall change in surface area (equation 18.9).

4. Two-state kinetics and intermediates

a. Apparent two-state kinetics

It was described in Chapter 17, section A4c, that many multistate denaturation equilibria exhibit apparent two-state behavior around the transition region because the intermediates are present at low concentrations. The same is often true for folding and unfolding kinetics. The occurrence of simple two-state kinetics does not prove that there are no intermediates on the reaction pathway.[28] For example, the energy profile in Figure 18.4 has a high energy intermediate. But the rate laws that describe the reaction are identical to those of a simple two-state reaction at all concentrations of denaturant. This is easily demonstrated algebraically, but it also can be seen intuitively, because there is the same transition state, and the ground states are always D and N at all concentrations of denaturant. Figure 18.5 is a more typical situation where the intermediate is more stable than the more highly denatured state D in the absence of denaturant (cf. Figure 17.5 and Chapter 17, section B).

5. Kinetics tests for intermediates

Intermediates are usually detected in reaction pathways by the appearance of additional phases. Most small proteins have only one non-proline-related folding phase in the accessible time range. But the simple kinetics can be deceptive, and there can be hidden intermediates, besides the high energy states discussed in the last section. In Chapter 17, section A4, procedures were described to test for equilibrium two-state transitions; these procedures are also effective for detecting intermediates that may accumulate close to the transition region. Similar tests, such as the simultaneous change of tryptophan fluorescence or near

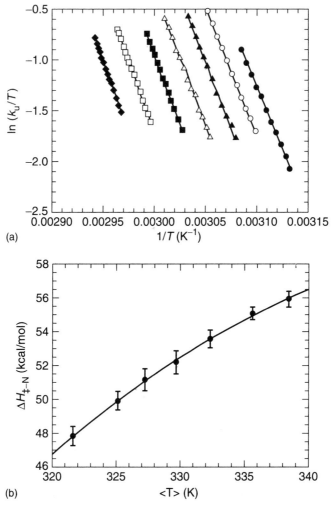

Figure 18.3 (a) Eyring plots of $\ln(k_u/T)$ versus $1/T$ for the unfolding of CI2 at different values of acid pH. Unfolding kinetics was measured over a range of acid values of pH. In this region, the rate of unfolding increases with decreasing pH, and the T_m decreases with decreasing values of pH. This is a consequence of the protonation of carboxylate groups in side chains, which have enthalpies of ionization close to zero. As the pH decreases, the unfolding rates are measured at progressively lower values of temperature. At each value of pH, the deviation from linearity is imperceptible. But the value of $\Delta H_{\ddagger-N}$ decreases with the decreasing average temperature of each set. (b) Determination of $\Delta C_{p,\ddagger-N}$ for unfolding for CI2 from a plot of $\Delta H_{\ddagger-N}$ versus $\langle T \rangle$, the average temperature over which the measurements are made. A linear fit to the data gives a value of 480 cal/mol/deg (2.0 kJ/mol/deg) for ΔC_p. It is more likely that the downward curvature is real and $\Delta C_{p,\ddagger-N} = 6.0 - 0.0168T$ cal/mol/deg (25 − 0.07T J/mol/deg). This implies that $\Delta C_{p,D-N}$ becomes zero at 360 K. [Modified from Y. J. Tan, M. Oliveberg, and A. R. Fersht, *J. Molec. Biol.* **264**, 377 (1996).]

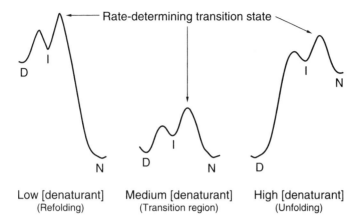

Figure 18.4 Free energy profiles for a folding reaction in which there is an intermediate that is always at higher energy than D or N. The kinetics are indistinguishable from those of a two-state system.

ultraviolet circular dichroism (which detect tertiary structure changes) and far ultraviolet circular dichroism signals (which detect secondary structure changes), can be applied to refolding where intermediates accumulate.

The reaction illustrated in Figure 18.5, however, often leads to kinetics that appears to be two-state: in the absence of denaturant, the transition from D to I is often too fast to be measured, and only the slower transition of I to N is observed. There are powerful kinetic tests for detecting intermediates under these conditions even if only a single transition is observed.[19]

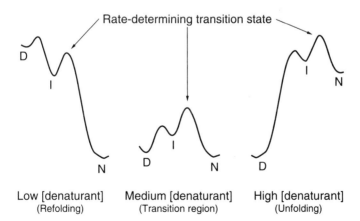

Figure 18.5 Free energy profiles for a folding reaction in which there is an intermediate that is at higher energy than D or N in the transition region but is more stable than D at low [denaturant]. The step D → I is often fast, and the kinetics can be deceptively simple.

The first test is that the ratio of the forward and reverse rate constants for a two-state transition must give the same equilibrium constant as that measured directly by solvent denaturation. Recall from Chapter 17, section A4d, that measurements around the transition region, even for reactions involving intermediates, extrapolate back to pure water to give the properties of the highly denatured state. The ratio of the rate constants for the apparent folding step and for unfolding in water of the three-state reaction in Figure 18.5 gives the equilibrium constant between I and N, and not between D and N.

The second test is that the sum of the forward and reverse m values must equal the m value for equilibrium unfolding for a two-state transition; that is,

$$m_{\ddagger-N} + m_{\ddagger-D} = m_{D-N} \tag{18.12}$$

where all three are algebraically positive, since equations 18.3 and 18.5 must combine to give equation 17.6 ($\Delta G_{D-N} = \Delta G_{D \rightleftharpoons N}^{H_2O} - m_{D-N}[\text{denaturant}]$, Chapter 17).

The third test is deviations from the simple V-shaped curves of rate constants versus [denaturant] that were shown in Figure 18.1 for CI2. An intermediate was detected from the denaturant dependence of the single phase for the folding of barnase (Figure 18.6) because of downward curvature. This is sometimes called "rollover." The rate constant for folding for the two-state transition for barnase can be calculated as a function of [urea] from the equilibrium and rate constants

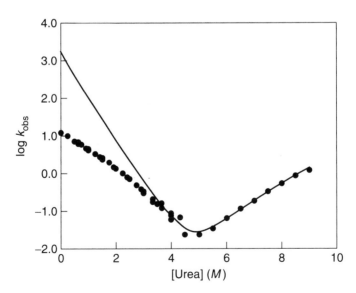

Figure 18.6 Plot of log k_{obs} against [urea] for the folding of barnase. $k_{obs} = k_f + k_u$ (s^{-1}). The solid line is the fit to a two-state equation based on the rate and equilibrium constants for folding. The deviation from the fit shows that an intermediate (on or off the pathway) accumulates at low [urea].

for denaturation. The calculated values deviate significantly from the observed values being much higher as [urea] tends to zero. This behavior is a diagnostic test for the occurrence of an intermediate.

It is essential in kinetic studies to measure the V-shaped curves down to zero molar denaturant to show that there is no deviation from the kinetics in pure aqueous solution. Kinetic measurements of refolding should always be performed over a wide range of protein concentration to check for aggregation, the effects of which should depend on concentration. Folding intermediates and denatured states are usually sticky because of exposed hydrophobic surfaces and so tend to aggregate. This aggregation causes artifacts in kinetics that can easily be mistaken for additional intermediates.[29] Aggregation can often be detected by light scattering. Some examples of two-state folding and β_T values are listed in Table 18.1.

Table 18.1 *Proteins that fold with two-state kinetics[a]*

Protein	Chain length	$\Delta G_{D-N}^{H_2O}$ (kcal/mol)	T (°C)	$k_f^{H_2O}$ (s^{-1})	$k_u^{H_2O}$ (s^{-1})	$^a\beta_T$
α-*Helical proteins*						
Monomeric						
λ-repressor (λ_{6-85})						
Wild type[b]	80	3.0	37	4900[c]	30	0.39
G46AG48A[c]		4.8	37	88000	36	0.83
Acylbinding proteins[d]						
Bovine	86	7.1	5	279	0.0001	0.61
Rat		6.1	5	395	0.0067	0.69
Yeast		7.8	5	4105	0.0015	0.57
Cytochrome c[e]	104	6.9	20	2800	0.017	0.47
Cytochrome c[f]						
Horse, oxidized (Fe^{II})	104	17.7	23	400	ND	0.40
Yeast, oxidized (Fe^{II})	103	14.6	23	15000	ND	0.34
Cold-shock β-*barrel proteins*						
CspB (*Bacillus subtilis*)[g]	67	3.0	25	1070	12	0.85
CspB (*Bacillus subtilis*)[h]	67	2.7	25	689	9.9	0.91
CspB (*Bacillus caldolyticus*)[h]	66	4.8	25	1370	0.64	0.93

(continued)

Table 18.1 *Proteins that fold with two-state kinetics[a] (continued)*

Protein	Chain length	$\Delta G_{D-N}^{H_2O}$ (kcal/mol)	T (°C)	$k_f^{H_2O}$ (s^{-1})	$k_u^{H_2O}$ (s^{-1})	$^a\beta_T$
CspB (*Thermotoga maritima*)[h]	68	6.3	25	565	0.018	0.86
CspA[i]	69	3.0	25	200	4.2	0.94
		2.9	10	188	3.3	0.91
SH3 domains (β barrels)						
α-Spectrin[j]	62	2.9	25	8.1	0.045	0.74
				4.1	0.0067	0.69
Src[k]	64	4.1	22	57	0.1	0.69
PI3 kinase[l]	84	3.4	20	0.35	0.00067	0.60
Fyn[m]	67	6.0	20	94	0.00099	0.68
β-Sandwich domains						
⁹FNIII[n]	90	1.2	25	0.4	ND	0.63
TWIg18[o]	93	3.9	20	1.50	0.000203	0.70
TNfn3 (short form)[p]	90	4.8	20	2.9	0.0028	0.76
TNfn3 (long form)		6.7	25	6.0	0.000072	0.75
CD2[o]	98	8.0	25	19	0.0017	0.65
α/β proteins						
CI2[q]	64	7.0	25	48	0.00018	0.61
Activation domain Procarboxypeptidase A2 (ADAh2)[r]	81	4.1	25	897	0.65	0.74
Arc repressor (single chain)[s]	106	6.3	25	>10^4	1.5	0.81
Ubiquitin[t] V26A	76	3.9	25	102	0.08	0.70
Ubiquitin[t] V26G		3.7	25	300	0.2	0.61
IgG binding domain of streptococcal Protein L[u]	62	4.6	22	60	0.02	0.75
Spliceosomal protein U1A[v]	102	9.3	25	316	0.000063	0.55
Hpr (histidine-containing phospho-carrier protein)[w]	85	4.6	20	14.9	0.0021	0.64
FKBP12[x]	107	5.5	25	4.3	0.00017	0.67
Muscle AcP[y]	98	5.4	28	0.23	0.00011	0.79

[a] Data taken from S. E. Jackson, *Fold. Des.* **3**, R81 (1998). β_T is for folding.
[b] G. S. Huang and T. G. Oas, *Biochemistry* **34**, 3884 (1995).
[c] R. E. Burton, G. S. Huang, M. A. Daugherty, P. W. Fullbright, and T. G. Oas, *J. Molec. Biol.* **263**, 311 (1996).
[d] B. B. Kragelund, C. V. Robinson, J. Knudsen, C. M. Dobson, and F. M. Poulsen, *Biochemistry* **34**,

Table 18.1 *Proteins that fold with two-state kinetics^a (continued)*

7217 (1995); B. B. Kragelund, P. Hojrup, M. S. Jensen, C. K. Scherling, E. Juul, J. Knudsen, and F. M. Poulsen, *J. Molec. Biol.* **256**, 187 (1996).

^{e} C. K. Chan, Y. Hu, S. Takahashi, D. L. Rousseau, W. A. Eaton, and J. Hofrichter, *Proc. Natl. Acad. Sci. USA* **94**, 1779 (1997).

^{f} G. A. Mines, T. Pascher, S. C. Lee, J. R. Winkler, and H. B. Gray, *Chemistry & Biology* **3**, 491 (1996).

^{g} T. Schindler, M. Herrler, M. A. Marahiel, and F. X. Schmid, *Nature Structural Biology* **2**, 663 (1995).

^{h} D. Perl, C. Welker, T. Schindler, K. Schroder, M. A. Marahiel, R. Jaenicke, and F. X. Schmid, *Nature Structural Biology.* **5**, 229 (1998).

^{i} K. L. Reid, H. M. Rodriguez, B. J. Hillier, and L. M. Gregoret, *Protein Science* **7**, 470 (1998).

^{j} A. R. Viguera, J. Martinez, V. Filimonov, P. Mateo, and L. Serrano, *Biochemistry,* **33**, 2142 (1994); A. R. Viguera, L. Serrano, and M. Wilmanns, *Nature Structural Biology* **3**, 874 (1996).

^{k} V. P. Grantcharova and D. Baker, *Biochemistry* **36**, 15685 (1997).

^{l} J. I. Guijarro, C. J. Morton, K. W. Plaxco, I. D. Campbell, and C. M. Dobson, *J. Molec. Biol.* **276**, 657 (1998).

^{m} K. W. Plaxco, J. I. Guijarro, C. J. Morton, M. Pitkealy, I. D. Campbell and C. M. Dobson, *Biochemistry* **37**, 2529 (1998).

^{n} K. W. Plaxco, C. Spitzfaden, I. D. Campbell, and C. M. Dobson, *J. Molec. Biol.* **270**, 763 (1997).

^{o} S. Fong, S. J. Hamill, M. R. Proctor, S. M. V. Freund, G. M. Benian, C. Chothia, M. Bycroft, and J. Clarke, *J. Molec. Biol.* **264**, 624 (1996); J. Clarke, personal communication (1998).

^{p} J. Clarke, S. J. Hamill, and C. M. Johnson, *J. Molec. Biol.* **270**, 771 (1997); S. J. Hamill, A. C. Meekhof, and J. Clarke, *Biochemistry,* **37**, 8071 (1998).

^{q} S. E. Jackson and A. R. Fersht, *Biochemistry* **30**, 10428 (1991).

^{r} V. Villegas, A. Azuuaga, L. Catasus, D. Reverter, P. L. Mateo, F. X. Aviles, and L. Serrano, *Biochemistry* **34**, 15105 (1995).

^{s} C. R. Robinson and R. T. Sauer, *Biochemistry* **35**, 13878 (1996). The single-chain arc repressor has a 15-amino-acid, glycine-rich linker region; the chain length given here is twice that of the monomer.

t S. Khorasanizadeh, I. D. Peters, T. R. Butt and H. Roder, *Biochemistry* **32**, 7054–7063 (1993); S. Khorasanizadeh, I. D. Peters, and H. Roder, *Nature Structural Biology* **3**, 193 (1996). Wild-type ubiquitin folds with three-state kinetics; the mutants V26A and V26G (in the presence of 0.4M Na_2SO_4) fold with two-state kinetics.

^{u} M. L. Scalley, Q. Yi, H. Gu, A. McCormack, J. R. Yates III, and D. Baker, *Biochemistry* **36**, 3373 (1997).

^{v} M. Silow and M. Oliveberg, *Biochemistry* **36**, 7633 (1997).

^{w} N. A. J. van Nuland, W. Meijberg, J. Warner, V. Forge, R. M. Scheek, G. T. Robillard, and C. M. Dobson, *Biochemistry* **37**, 622 (1998).

^{x} E. R. G. Main, and K. F. Fulton, S. E. Jackson, *J. Molec. Biol.,* in press (1999). FKBP12 = FK506 binding protein.

^{y} N. A. J. van Nuland, F. Chiti, N. Taddei, G. Raugei, G. Ramponi, and C. M. Dobson, *J. Molec. Biol.* **283**, 883 (1998). ACP = Acylphosphatase.

C.   Multistate kinetics

If additional kinetic phases are observed in a folding pathway, does this mean that there are folding intermediates?

1.   Are intermediates on or off pathway?

The most difficult question to answer is whether an observed "intermediate," I, is on or off the reaction pathway.[30,31] That is, is there the compulsory order

$D \rightarrow I \rightarrow N$, or is I a nonproductive side product that has to decompose back to D for folding to proceed? The problem is that although we are performing kinetic experiments, there are equilibria involved and some of the key species are effectively at equilibrium. It is a fundamental tenet of kinetic and thermodynamic analysis that measurements of the equilibrium distribution of intermediates give just the relative thermodynamic properties of the intermediates and not the pathway between them. This is a consequence of the first law of thermodynamics: one cannot derive a kinetic mechanism from an equilibrium analysis, because the free energy of a particular molecule at equilibrium depends only on the state of the molecule and is independent of the pathway by which it is formed (Chapter 3, section L). This concept is so important that it is a worthwhile exercise to go through particular mechanisms and discover the ambiguities. So as not to presuppose the order of events, we shall temporarily drop the terminology "I" and assume folding from a highly unfolded state, U.

a. Kinetic consequences of rapid pre-equilibria

Suppose that, as on the left of Figure 18.7, there is a compact intermediate state, C, on the pathway that is a genuine folding intermediate, and so the reaction follows the scheme:

$$U \xrightarrow{K_{C-U}} C \xrightarrow{k_c} N \tag{18.13}$$

where $K_{C-U} = [C]/[U]$ and k_c is the rate constant for C reacting to give N. If U and C are in rapid equilibrium, the fraction of the denatured state, D, that is present as [C] is $K_{C-U}/(1 + K_{C-U})$. As U and C equilibrate rapidly, we can observe only a single rate constant for folding that is given by

$$k_{obs} = \frac{k_c[D]K_{C-U}}{1 + K_{C-U}} \tag{18.14}$$

Since $K_{C-U} \gg 1$ and $[C] \gg [U]$, the observed rate of formation of N is given by

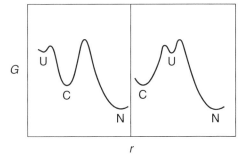

Figure 18.7 Energy profiles for compact states in the folding reaction. *Left*: The fully unfolded state collapses to a genuine intermediate that rearranges to form products. *Right*: The compact state C is off the pathway and has to unfold for productive folding to occur.

$$\frac{d[N]}{dt} = k_c[C] \tag{18.15}$$

Now suppose that C is not on the pathway but is a compact product of a side reaction, as on the right of Figure 18.7; that is,

$$C \xrightleftharpoons{1/K_{C-U}} U \xrightarrow{k_t} N \tag{18.16}$$

The fraction of the denatured state that is present as [U] is $1/(1 + K_{C-U})$. The observed rate constant for folding is given by

$$k'_{obs} = \frac{k_f[D]}{1 + K_{C-U}} \tag{18.17}$$

Since $K_{C-U} \gg 1$ and $[C] \gg [U]$,

$$\frac{d[N]}{dt} = \frac{k_f[C]}{K_{C-U}} \tag{18.18}$$

that is,

$$\frac{d[N]}{dt} = k'_f[C] \tag{18.19}$$

Equations 18.14 and 18.17 (or 18.15 and 18.19) are kinetically equivalent (Chapter 2, section F): the rate of formation of N by both pathways follows a rate constant multiplied by the concentration of C, the dominant species in solution. Thus, the rate constant for formation of N reports back simply on the difference in structural properties between the transition state for the formation of N and the ground state that is C. This means that we cannot determine whether the reaction is $U \rightarrow C \rightarrow N$ or $C \rightarrow U \rightarrow N$ by kinetics. The physical reason for this is that we are examining what is, in effect, a rapidly interconverting system of U and C that is being slowly converted to N; that is,

$$\{U \rightleftharpoons C\} \longrightarrow N \tag{18.20}$$

For example, suppose we examine the effects of increasing concentrations of denaturant, which has two effects: the intrinsic rate constants for folding, k_f and k_c, become lower, and the concentration of U increases relative to C. Initially, the change in concentration is not important while K_{C-U} is greater than 1, and so both mechanisms slow down because of the intrinsic rate constants decreasing. When $[C] = [U]$ at higher concentrations of denaturant, k_{obs} for scheme 18.13 slows down by a further factor of 2 since $K_{C-U}/(1 + K_{C-U}) = 0.5$ (in equation

18.14), and so does k_{obs} for scheme 18.16 since $1/(1 + K_{C-U}) = 0.5$. The same arguments can be applied to the effects of temperature, mutation, and any other factor.

b. Both kinetic steps resolvable

The arguments in the previous section are based on measuring just the single rate constant that is observed for the final step of folding. Can we resolve the situation by directly measuring the rate constants for the pre-equilibrium? In principle, yes, but in practice this is extremely difficult for the most common situation where a highly unfolded denatured protein in concentrated denaturant is diluted into a folding buffer and there is fast formation of C, either on or off the pathway, followed by slower folding. Recall from Chapter 4, sections D3 and D6, that two relaxation times are observed, but measurement of rate constants alone cannot give the order of events because the rate equations are symmetrical in the relaxation times. In theory, the associated magnitudes and signs of the amplitudes of the rate constants can determine the sequence, for the following reasons. If the intermediate is on the pathway, there will be a slight lag in the rate of formation of N. Conversely, if the intermediate is off the pathway, there will be an initial rapid partitioning of U between N and C so that there is a burst of formation of N followed by slower folding as C gives N via U. But the amplitudes of the lag or burst phases in [N] are low when the rate constant for the formation of C is much greater than for N, and so data of exquisite accuracy are required to detect a burst or lag. The detection is further complicated because the sum of all the spectroscopic signals from U, C, and N are collected, and so the absolute values of extinction coefficients, fluorescence yields, etc., must be known. In addition, heterogeneity in the denatured state, resulting naturally from isomerization of peptidyl-proline bonds or from the presence of small amounts of impurities, can further obscure the analysis. (Most protein folding studies concentrate on just the major phases in the kinetics.)

c. Indirect evidence for sequence of events

The structure of a compact state on the folding pathway of barnase has been mapped out in detail at the level of individual residues and is discussed in detail later (Chapter 19) using the Φ-value analysis of the next section.[32] Its structure is clearly between that of the unfolded state and the major transition state for folding. It has the same elements of secondary structure and the same tertiary contacts, but they are all weaker, especially in the hydrophobic core. The structure has all the properties expected of a folding intermediate that is on the pathway and has no indications of alternative structure. This does not prove that it is on the pathway, but the most likely explanation is that it is.

D. Transition states in protein folding

It is pertinent to ask what is the nature of a transition state in protein folding and if there is evidence that transition states do occur.[33]

1. What is a transition state in protein folding?

The conventional description of a transition state is a structure that is at a saddle point in an energy surface, at a maximum of energy along a reaction coordinate and at a minimum of energy perpendicular to it (Figure 18.8). Transition states of simple chemical reactions have only a few bonds being made and broken, and these are very strong, so that quantum theory has to be applied. But protein folding involves the making and breaking of very many weak bonds, which can be described by classic statistical mechanics and Newton's equations. The energy surface can be calculated from these equations, and it will have a saddle point. The transition state is the ensemble of states that differ slightly from one another in energy around the saddle point. This type of energy surface should lead to transition states in protein folding being at very wide and long saddle points, with many small dips in the profile (Figure 18.8). The energy surfaces in protein folding also have very high entropy components because of the large changes in configurational entropy as a protein folds and the changes in entropies of hydration.

The energy landscape around transition states can be probed by structure–reactivity relationships.[33,34] These are movements of the transition state along the

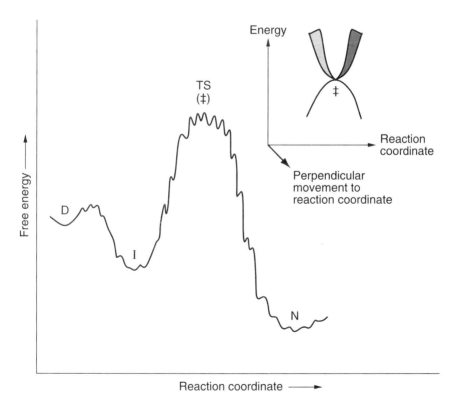

Figure 18.8 Sketch of a reaction coordinate and a saddle point for a folding reaction. [Modified from A. R. Fersht, *Curr. Opin. Struct. Biol.* **5**, 79 (1995).]

reaction coordinate on perturbation[24,25] (the Hammond effect discussed in Chapter 2, section A2, and illustrated in Chapter 19, Figure 19.8) or perpendicular to it ("anti"-Hammond effects[35]). Protein folding studies in mutants have an advantage over those in simple organic reactions because the Tanford β_T value (equation 18.9) measures the position of the transition state along the reaction coordinate. By combining Tanford β_T values and Φ-value analysis, which is described in section E, it has been shown that transition states in protein folding do act as if they are at the bottom of long and wide saddle points (Chapter 19, section B5).

2. Can we apply transition state theory?

Transition state theory (Chapter 2, section A) was derived for chemical bonds that obey quantum theory. An equation analogous to that for transition state theory may be derived even more simply for protein folding because classical low energy interactions are involved and we can use the Boltzmann equation to calculate the fraction of molecules in the transition state; i.e., $= \exp(-\Delta G_{\ddagger-D}/RT)$, where $\Delta G_{\ddagger-D}$ is the mean difference in energy between the conformations at the saddle point of the reaction and the ground state. Then, if v is a characteristic vibration frequency along the reaction coordinate at the saddle point, and κ is a transmission coefficient, then

$$k_f = v\kappa \exp \frac{-\Delta G_{\ddagger-D}}{RT} \tag{18.21}$$

Equation 18.21 is occasionally used to estimate barrier heights assuming that κ is 1.0 and $v = k_B T/h$ (Chapter 2, section A, and the rate constant defined as in equation 18.10), but its main use is the calculation of *changes* in $\Delta G_{\ddagger-D}$—i.e., $\Delta\Delta G_{\ddagger-D}$—for which the terms $v\kappa$ cancel out. For example, if a mutant folds with a rate constant k_f', compared with k_f for wild type, then the change in activation energy on mutation is given by

$$\Delta\Delta G_{\ddagger-D} = RT \ln \frac{k_f'}{k_f} \tag{18.22}$$

The value of $v\kappa$ is probably many orders of magnitude below that found for chemical reactions, but equation 18.22 is perfectly sound and reliable. An empirical estimate for $v\kappa$ is 10^6 s^{-1}.[7]

E. Introduction to Φ-value analysis

We saw in Chapter 15 how the contributions of the binding energy of different side chains of the tyrosyl-tRNA synthetase to catalysis were measured by systematic protein engineering experiments. Apparent binding energies of each side chain were measured by comparing catalytic rate constants of wild-type and mutant enzymes. It was seen how the binding energy developed during the reaction. The structures of transition states and intermediates in folding may be inferred in a similar manner from the changes in the kinetics and equilibria of folding on mutation. The procedure has become more formal, and it is described in terms of Φ-value analysis,[33,34] by analogy with classical Brønsted β-value analysis. A more detailed description of the procedure in terms of individual interactions is given elsewhere,[36] but the following qualitative approach is adequate for most purposes. Φ-value analysis is the only procedure for obtaining the structures of transition states (and some unstable intermediates) at the level of individual residues, at almost atomic resolution. This is because the only way of analyzing transition states experimentally is by kinetics, and the only way of analyzing detailed structure is by structure–activity relationships.

1. Changes in energy levels on mutation

Suppose that we make a mutation Ala → Gly in the solvent-exposed face of a helix in a protein that destabilizes it by 1 kcal/mol (4 kJ/mol) relative to the denatured state (Figure 18.9). Suppose that we then measure the kinetics of folding and unfolding of the protein and find that the transition states for folding and for a folding intermediate are also destabilized by 1 kcal/mol (4 kJ/mol) relative to the denatured state. Then it seems likely that the helix is fully formed at the site of mutation in those two states, because it responds to a destabilizing mutation

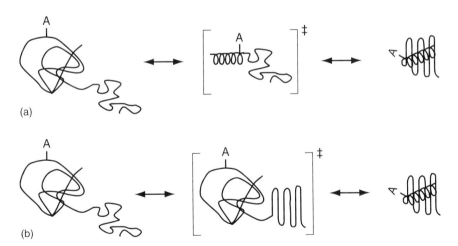

Figure 18.9 Two extreme folding pathways for a protein that has a helix and a sheet. (a) The helix that is present in the final folded structure is also present in the transition state. Mutation of residue A that makes entirely intrahelical interactions thus destabilizes the transition state by the same energy as it does the native structure. (b) The helix that is present in the final folded structure is as unfolded in the transition state as it is in the denatured state. Mutation of residue A thus does not destabilize the transition state at all relative to the ground state.

in exactly the same way as the fully folded structure does. Suppose, in addition, that other mutations at the same position and at neighboring residues also destabilize the transition state and intermediate by the same amount as they do the fully folded structure. Then it is beyond reasonable doubt that the helix is fully formed in those states in the region of mutation. Suppose, conversely, that a mutation destabilizes the native state by 1 kcal/mol (4 kJ/mol) but does not affect the energy of a transition state or intermediate at all, measured relative to the denatured state. Then it seems likely that the helix is fully denatured at the site of mutation in those two states, because it responds to destabilizing mutations in the same way as the denatured state does. Again, if further mutations at the same site or at neighboring sites also do not affect the energy of the intermediate and transition states relative to the denatured state, then it is beyond reasonable doubt that the helix is fully denatured in those states at the region of mutation.

The two situations above do occur for many mutations in barnase. We can simply go around the structure checking "fully native" or "fully denatured" at the relevant regions and so get a detailed residue-by-residue characterization of its folding intermediate and transition state. Each mutation is a reporter group for what is happening at the site of mutation during the folding pathway.

This analysis is formalized in terms of Φ values (Figure 18.10). Suppose that a mutation destabilizes the folded structure by $\Delta\Delta G_{N-D}$ units of energy, measured relative to the unfolded state. Then, if the free energy of a transition state, measured relative to the unfolded state, changes by $\Delta\Delta G_{\ddagger-D}$, Φ is defined by

$$\Phi_F = \frac{\Delta\Delta G_{\ddagger-D}}{\Delta\Delta G_{N-D}} \tag{18.23}$$

A Φ_F value of 0 means that the energy of the transition state is perturbed on mutation by the same amount as the denatured state is perturbed. A Φ_F value of 1.0 means that the energy of the transition state is perturbed on mutation by the same amount as the native state is perturbed. It is inferred from the changes in energy that a Φ_F value of 0 means that the structure is unfolded at the site of mutation as much as it is in the denatured state, and a Φ_F value of 1.0 means that the structure is folded at the site of mutation as much as it is in the native state.

It is possible, of course, that a side chain makes an interaction with a different residue in the transition state from the one in the native structure. This can be checked by the procedure of *double-mutant cycles* (Chapter 3, section L), in which two residues are mutated pairwise and singly. These can be analyzed to measure the coupling energies between pairs of residues.

2. Choice of mutations: Nondisruptive deletions

The principles for making mutations to investigate folding are the same as discussed in Chapter 15, section D, for investigating catalysis by site-directed mu-

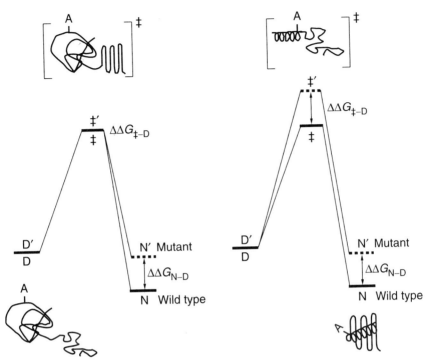

Figure 18.10 Free energy diagrams for the folding of the protein from Figure 18.9. Residue A in the helix is mutated to give a mutant (denoted by $'$). The change on mutation of the free energy of the transition state divided by that of the native protein, $\Delta\Delta G_{\ddagger-D}/\Delta\Delta G_{N-D} = \Phi$. The free energies of D and D$'$ are superposed in the figure, but this does not affect the analysis. A common misconception is that the analysis assumes that the free energy of the denatured state is unaffected by mutation and that the analysis fails if the mutation does affect this state. This is not correct; it can be shown from a formal analysis that the change in the energy of the denatured state cancels out in the equations. The physical reason for this is that all quantities are measured relative to the change in energy of the denatured state. For example, if the denatured state changes in absolute terms by ΔG_D, then the transition state changes in absolute terms by $\Delta G_D + \Delta\Delta G_{\ddagger-D}$ and the native state by $\Delta G_D + \Delta\Delta G_{N-D}$.
$\Phi_F = \{(\Delta G_D + \Delta\Delta G_{\ddagger-D}) - \Delta G_D\}/\{(\Delta G_D + \Delta\Delta G_{N-D}) - \Delta G_D\} = \Delta\Delta G_{\ddagger-D}/\Delta\Delta G_{N-D}$.

tagenesis, but the requirements for folding are more stringent. The ideal mutations are those that remove moieties of side chains that make well-defined interactions. It is important to avoid mutations that add new functional groups that can make extraneous interactions within the protein, mutations that change stereochemistry or can cause stereochemical clashes, and mutations that leave charged groups without solvating partners. The best mutations are those that remove a small number of methylene groups because there is little change in the solvation energy of the denatured state; this causes a complicating term in the equations to be canceled. These hydrophobic mutations can be on the surface or the interior of the protein. Residues on the surface are also promising targets because they do not have to pack as well as in the interior.

The ideal mutations for investigating surface and buried hydrophobic interactions are Ile $\rightarrow$ Val, Ala $\rightarrow$ Gly, and Thr $\rightarrow$ Ser, followed by Val $\rightarrow$ Ala, Ile $\rightarrow$ Ala, and Leu $\rightarrow$ Ala. Phe $\rightarrow$ Ala and Tyr $\rightarrow$ Ala are radical but still work well enough in practice.

A variation is *Ala $\rightarrow$ Gly scanning*:[35] mutate a larger amino acid to Ala and to Gly, and then compare the two. Ala $\rightarrow$ Gly is a very benign mutation and affects the stability of helixes, for example, just by simple surface area terms (Chapter 11, section B1). The scanning procedure is very useful when there is not a convenient mutation of the wild-type residue.

Hydrogen bonding in buried residues places severe constraints on substitutions, but a variety of changes can be made to surface residues. For example, Ser $\rightarrow$ Ala may remove the hydrogen-bonding potential of the —OH group, but water may be able to take its place.

3.   Relationship between Φ and the Brønsted β

Φ_F is similar to, but not identical with, the Brønsted β of physical organic chemistry. Φ_F and β are the same at the two extreme values of 0 and 1: measured in the direction of folding (or bond making), Φ_F (or β) = 0 corresponds to fully denatured (no bond making), and Φ_F (or β) = 1 to fully native (complete bond formation). These are, of course, the values that were easily interpreted for barnase. In general, there need not be a linear relationship between Φ_F and the extent of noncovalent bond formation. The worst situation is for the formation of salt bridges, because of the complexities of the energetics. But, there should be some correlation between Φ_F and the number of van der Waals interactions made in the packing of hydrophobic cores, in particular.[34,36]

4.   Fractional values of Φ

Fractional values of Φ are more difficult to interpret because they can be caused by a variety of factors. First, there is a fundamental ambiguity in that they can result either from genuinely weakened interactions in a single species or from there being a mixture of states, some with the interactions fully formed, others with the interactions fully broken. The mixture could stem from an equilibrium between various states in an intermediate or from parallel pathways in kinetics leading to a mixture of transition states. The folding of the chymotrypsin inhibitor 2, for example, has predominantly fractional values of Φ in its transition state. By making a large number of mutations in the same elements of structure, it was possible to show that the fractional values do not result from parallel pathways with a mixture of fully formed and fully denatured transition states. Instead, the fractional Φ values must represent genuinely weakened interactions[37] (discussed in Chapter 19, section B5c).

Fractional values may also arise if a side chain makes one set of interactions with one element of structure and a further set with a different element, and these elements react in different ways in unfolding. This possibility was noticed by examining the structure of barnase and performing a fine-structure analysis by making systematic mutations in the side chains involved; e.g., Ile $\rightarrow$ Val $\rightarrow$ Ala $\rightarrow$ Gly.[38]

Fractional values of Φ may arise from artifacts because of, say, a distortion of the structure of the folded molecule on mutation that is not the same in the transition or intermediate states; that is, there are various reorganization energies introduced by mutation that do not cancel. Again, analyzing a large number of mutations in the same regions of structure allows the detection of artifacts and the identification of patterns of structure. For example, in the transition state for the unfolding of barnase, there is a beautiful gradation of Φ values from 1.0 for mutation of residues in the center of the β sheet to small fractional values at the edges.

The other major problem is that, in general, there is not a linear relationship between Φ and the extent of formation of structure. All that can be said without sophisticated calculation is that a significant value of Φ means that the structure is basically present but weakened. Despite all these caveats, the method works very well in practice if sensible mutations are made. The important principle is to make as many mutations as possible of the right sort to get as much information as possible.

5. Benchmarking of simulations with Φ values

There are various procedures for simulating the folding and unfolding of proteins using computers (Chapter 19). The interaction of side chains, probed by Φ values, is just what is required for benchmarking the calculations, which in turn check the Φ values. There is generally excellent agreement between experiment and theory,[39-46] and the simulations can be used to convert the energies from kinetics into real structures, as discussed in Chapter 19.

F. ^{1}H/^{2}H-exchange methods

1. ^{1}H/^{2}H-exchange at equilibrium

The protons of backbone $>$NH groups of small peptides dissolved in water are labile and exchange with those of the solvent. The exchange is readily detectable by NMR methods by studying a peptide containing the normal isotope ^{1}H dissolved in deuterium oxide (D_2O or ^2H_2O), or *vice versa*.[47,48] The reaction is acid- and base-catalyzed, following the rate law

$$k_{int} = k_{H_2O} + k_{OH^-}[OH^-] + k_{H^+}[H^+] \tag{18.24}$$

where the observed rate constant for exchange is k_{int} (the "intrinsic" rate constant), the spontaneous value in water is k_{H_2O}, the base-catalyzed second-order rate constant is k_{OH^-}, and the acid-catalyzed second-order rate constant is k_{H^+}. The absolute and relative values of these terms vary with the different amino acids and are affected by the residues immediately preceding and following in the sequence. Values of k_{int} have been tabulated for each of the individual amino acids in different sequences as a function of temperature and pH.[49] The major term for most amino acids above pH 6 is that for the base catalysis, and the other terms may be ignored.

Backbone $>$NH groups of proteins may be highly protected against exchange, by factors of 10^4–10^6 or more by being in stable elements of secondary structure (for example, as in strands of β sheet in Figure 18.11) or by simply being buried in the protein. A protection factor P is defined for any proton by

$$P = \frac{k_{int}}{k_{ex}}$$

(18.25)

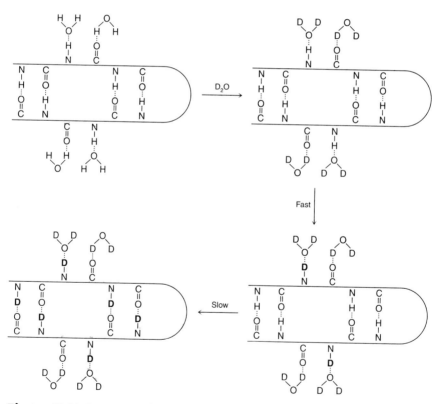

Figure 18.11 Protection of backbone $>$NH protons in strands of β sheet against exchange with solvent D_2O. Surface $>$NH groups exchange at the same high rate as do those in small peptides, but the buried ones exchange slowly, being limited by the rate of structural fluctuations that expose them.

where k_{int} is the intrinsic rate constant for that proton under the reaction conditions and k_{ex} is the observed value for exchange. For hydrogen exchange of buried groups to occur, there must be either small-scale structural fluctuations, termed local unfolding or breathing, involving the breakage of hydrogen bonds and access of solvent to the exchangeable proton, or full unfolding of the protein (18.26);

$$C^H \underset{k_c}{\overset{k_o}{\rightleftharpoons}} O^H \xrightarrow[D_2O]{k_{int}} O^D \underset{k_o}{\overset{k_c}{\rightleftharpoons}} C^D \qquad (18.26)$$

where C and O are either the locally "closed" and "open" conformations or globally folded or unfolded states and k_c and k_o are the rate constants of local (or global) folding and unfolding, respectively. If D_2O is in great excess,

$$k_{ex} = \frac{k_o k_{int}}{k_c + k_o + k_{int}} \qquad (18.27)$$

Under conditions that strongly favor folding, the rate constant of refolding is considerably faster than that of unfolding, so that $k_c \gg k_o$, and k_o can be ignored. Equation 18.27 has two extreme limits. When $k_c \gg k_{int}$, as is usual for most globular proteins under most conditions, the observed rate constant of exchange reduces to

$$k_{ex} = \frac{k_o}{k_c} k_{int} \qquad (18.28)$$

or

$$k_{ex} = K k_{int} \qquad (18.29)$$

where $K = k_o/k_c$, is the effective equilibrium constant between open and closed forms. That is, $K = [O]/[C] = 1/P$. This situation is termed the EX2 limit.

At the other extreme, when $k_c \ll k_{int}$, the rate-determining step is the opening, and

$$k_{ex} = k_o \qquad (18.30)$$

This is termed the EX1 limit.

The mechanism of exchange may be probed by the effects of denaturant,[50] pH, and mutation[51] on k_{ex}. Use of mutation is particularly powerful for determining how much structural fluctuation is required for exchange to take place.[52] Suppose, as is common, that exchange is occurring at the EX2 limit. Then we can calculate the free energy between the open and closed states from equation 18.29, as follows:

$$\Delta G_{ex} = -RT \ln K = RT \ln P \tag{18.31}$$

The value of ΔG_{ex} will differ from the real value of the equilibrium free energy because of uncertainties in the value of k_{int}: the residue in state O may not be as fully accessible as in model peptides. Suppose we make a mutation in the protein that changes its stability by $\Delta\Delta G_{D-N}$. We can then measure the change in ΔG_{ex} on mutation from the change in protection factors (P_{wt} in wild type and P_{mut} in the mutant):

$$\Delta\Delta G_{ex} = -RT \ln \frac{P_{mut}}{P_{wt}} \tag{18.32}$$

Providing that the residues examined are removed from the site of mutation and that the mutation is nondisruptive, all the uncertainties in k_{int} cancel out in equation 18.32, and it is a precise equation.

Mutation of many of the deeply buried residues in barnase gives exactly the same values of $\Delta\Delta G_{ex}$ as $\Delta\Delta G_{D-N}$, showing that those residues exhibit the full change in free energy of unfolding on going from C to O. Those residues thus exchange by global unfolding. Others, especially those close to the surface in helices, have values of $\Delta\Delta G_{ex}$ that are very close to zero, and so they are unaffected by mutations that alter global stability. These residues thus exchange by local fluctuations.

2. Exchange at equilibrium cannot be used to determine pathways

Because of the apparent power of exchange at equilibrium to measure fluctuations in structure, there have been many attempts to use the procedure to determine pathways. We know from section C1 that this is a futile activity: equilibrium measurements just give the relative thermodynamic properties of intermediates and not the pathway between them.[53] The value of ΔG_{ex}, the equilibrium constant between particular open and closed states of a protein derived from hydrogen exchange at equilibrium, is just such a thermodynamic measurement and does not give information about when that state is formed on a pathway. For example, consider the exchange from a reaction sequence:

$$N \underset{}{\overset{K_1}{\rightleftharpoons}} I \underset{}{\overset{K_2}{\rightleftharpoons}} U \tag{18.33}$$
$$\Big\downarrow k_{int} \qquad \Big\downarrow k_{int}$$

where an intermediate I is formed from a native structure, N, with formation constant K_1 ($= [I]/[N]$), and the unfolded state U has formation constant $K_1 K_2$ ($= [U]/[N]$), and I and U are "open" states for exchange. The rate constant for the exchange of a particular residue i from state I is given by $k_{ex}^i = k_{int}^i K_1$ and that of residue j from state U by $k_{ex}^j = k_{int}^j K_1$. Suppose, instead, that I is an off-pathway state (equation 18.34),

$$I \underset{1/K_1}{\overset{\text{}}{\rightleftharpoons}} N \underset{K_1 K_2}{\overset{\text{}}{\rightleftharpoons}} U \qquad (18.34)$$

$$\downarrow k_{\text{int}} \qquad\qquad \downarrow k_{\text{int}}$$

formed with the same equilibrium constants; $K_1 = [I]/[N]$, $K_1 K_2 = [U]/[N]$. Again, the rate constant for the exchange of a particular residue i from state I is given by $k_{\text{ex}}^i = k_{\text{int}}^i K_1$ and that of residue j from state U by $k_{\text{ex}}^j = k_{\text{int}}^j K_1$. Thus, identical values of ΔG_{ex} are obtained for both pathways. This situation is equivalent to mechanisms in equations 18.13 and 18.16.

3. Uses of equilibrium $^1H/^2H$-exchange in folding studies

$^1H/^2H$-exchange at equilibrium comes into its own in studying structure in stable intermediates.[47,48] It is particularly useful for detecting regions of protected structure in partly unfolded states; regions that are protected will have significant protection factors. But this can be obscured by rapid conformational equilibria, when equation 18.26 applies, without the proviso that $k_c \gg k_o$. Suppose, for example, that a state exists in equal amounts in two rapidly interconverting conformations, one of which is highly protected and the other of which is unprotected against exchange. Then the observed protection factor is only two, no matter how high the degree of protection in the protected state, because the observed exchange is occurring from the other state ($k_c = k_o$ in equation 18.27). $^1H/^2H$-exchange at equilibrium is also an invaluable tool for determining the nature of the states of compounds bound to large complexes, such as molecular chaperones (Chapter 19), that are detected by mass spectrometry[54] as well as NMR, and for detecting small amounts of unfolded protein that would otherwise be undetectable.

4. Quenched-flow $^1H/^2H$-exchange

The onset of protection of $>$NH groups as a protein folds may be measured by rapid quenching techniques[55-58] (Figure 18.12). A crude form of the experiment is to denature deuterated protein in ^2H_2O containing deuterated urea. The urea is diluted out with a large volume of ^2H_2O in a quenched-flow machine (see Chapter 4, section A3), and the protein is allowed to refold for varying lengths of time (x). It is then pulsed with ^1H_2O for a defined time at a sufficiently high pH for all unprotected deuterons to exchange nearly completely. It is then allowed to fold to completion. The NMR spectrum is then measured for each sample to determine the fraction of each $>N^2H$ group that has exchanged, and the results are plotted against x.

This is a very useful procedure to determine when protected secondary structure is formed. But it is quite difficult to perform quantitatively in practice. Individual time points, each of not particularly high precision, are measured separately, and a very large number of these points is required to analyze multiphasic kinetics. Also, folding intermediates can be very labile and become unprotected in the labeling pulse, especially at high pH.

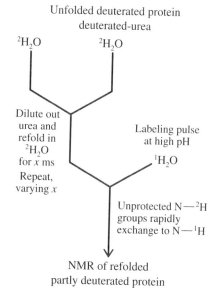

Figure 18.12 Prototype scheme for quenched-flow $^1H/^2H$-exchange experiments.

5. Φ-value analysis *versus* quenched-flow $^1H/^2H$-exchange

The two techniques are complementary in some ways: Φ values measure tertiary interactions and quenched-flow exchange mainly detects secondary interactions. Φ-Value analysis is by far the better technique for kinetics (Table 18.2). It can be used over any time scale accessible to kinetics, whereas quenched-flow exchange is limited to several milliseconds or more. Φ values can be used to analyze transition states and unstable intermediates, whereas exchange is limited to intermediates. Φ values do detect secondary structure by inference and can in principle be used to analyze the fate of virtually every side chain, whereas exchange is limited to analyzing just the relatively few protons that are highly protected in secondary structure. Φ values give energetics as well as structure and tell whether interactions are fully or partly formed. This enables the order of events to be defined with some precision and gives evidence unavailable from any other technique. For example, Φ-value analysis has provided evidence that the folding intermediate of barnase has a structure between that of the transition state for unfolding and the unfolded structure (section C1c). It is seen in the next chapter (section C) how different mechanisms of folding can be distinguished, because some predict fully formed units of structure in the transition state. There is also an inherent ambiguity in the exchange experiments because it is not known if a particular >NH group in an intermediate is protected by the same partners in an intermediate as in the native structure. Evidence about whether or not a set of >NH groups in the same element of secondary structure has coordinated protection is a useful check. This type of ambiguity applies also to Φ-

Table 18.2 *Comparison of protein engineering and pulsed $^1H/^2H$-exchange methods.*

Application to	Protein engineering	Pulsed $^1H/^2H$-exchange
Time scales	>fs	>ms
Transition states	Yes	No
Intermediates	Yes	Yes
Secondary structure	Yes	Yes
Tertiary structure	Yes, nearly all side chains	Very little
Core	Yes	No
Energetics	Yes	No
Ambiguity in partner	Not for double mutant cycles; multiplicity of probes	Yes, unless a coordinated set
Size restrictions	None	<25 kD

value analysis. But the ambiguity can be checked in Φ-value analysis by using double-mutant cycles and using an excess of mutations in each region of structure.

G. Folding of peptides

As a prelude to discussing the mechanism of folding of intact proteins in the next chapter, we end this one with a discussion of the kinetics and mechanism of folding of isolated secondary structural elements of proteins from the random coil conformation.

1. Loops

Theory suggests that the most probable loops in proteins contain 10 residues.[59] Shorter loops become stiff and longer ones lose more conformational entropy — recall from Chapter 17 that $\Delta S_{config} = a - (3/2)R \ln n$. Indeed, the kinetics of formation of longer loops does fall off as $\exp(-3/2) \ln n)$, $= n^{-(3/2)}$; where n is the number of residues in the loop.[60] A 50-residue loop in cytochrome c closes with $t_{1/2}$ of 40 μs, implying a maximum closing rate of $\sim 10^6$ s^{-1}.[7]

2. α Helixes

A simple model for the kinetics of helix formation invokes as the rate-determining step the formation of a nucleus of neighboring residues with ϕ and ψ angles in the α-helical region of the Ramachandran diagram (Chapter 1, section C1 and Figure 1.11).[5] The nucleus is so improbable that there is only a single one formed in each molecule. Once the first hydrogen bond between the $>C=O$ of residue i and the $>NH$ of residue $i + 4$ is formed in the nucleus in a helix that should be stable, the subsequent growth is thermodynamically downhill because the stabilization energy of forming a single extra hydrogen bond is greater than the unfavorable energy of fixing a pair of residues in the α-helical conformation.

3. β Hairpin

The simplest β structure is the β hairpin (Chapter 1, section C3 and Figure 1.12). The β hairpin requires the pairing of hydrophobic side chains to stabilize it. The hairpin has to nucleate at its central turn, unlike the helix, which can nucleate at any residue whose $>C=O$ can form a hydrogen bond with residue $i + 4$. The formation of β structure is inherently slower than the formation of α helices because of the fewer nucleation sites and the requirements to make side-chain interactions.[5] Depending on precise structure, helixes form with half-lives of a few hundred nanoseconds, and β structures form with half-lives 10 times longer.

4. Very fast folding small proteins

The fastest-folding small proteins generally fold on much slower time scales than the time scale of formation of secondary structure. The speed record is currently held by lambda(6-85), a truncated, monomeric form of the N-terminal domain of lambda repressor, which refolds with a half-life of approximately 140 μs. A thermostable lambda(6-85) variant with alanine substituted for glycine residues 46 and 48 in the third helix folds faster in dilute solutions of denaturant, with an extrapolated half-life of less than 10 μs in water.[13] Cold-shock protein CspB from *Bacillus subtilis* folds in about 1 ms.[61] Engineered mutants of the P22 Arc repressor[62] and CI2[63] fold in a fraction of a millisecond.

References

1. C. B. Anfinsen, *Science* **181**, 223 (1973).
2. U. Shinde and M. Inouye, *Trends Biochem. Sci.* **18**, 442 (1993).
3. P. Bryan, L. Wang, J. Hoskins, S. Ruvinov, S. Strausberg, P. Alexander, O. Almog, G. Gilliland, and T. Gallagher, *Biochemistry* **34**, 10310 (1995).
4. S. Williams, T. P. Causgrove, R. Gilmanshin, K. S. Fang, R. H. Callender, W. H. Woodruff, and R. B. Dyer, *Biochemistry* **35**, 691 (1996).
5. V. Munoz, P. A. Thompson, J. Hofrichter, and W. A. Eaton, *Nature, Lond.* **390**, 196 (1997).
6. T. Veitshans, D. Klimov, and D. Thirumalai, *Fold Des.* **2**, 1 (1997).
7. S. J. Hagen, J. Hofrichter, A. Szabo, and W. A. Eaton, *Proc. Natl. Acad. Sci. USA* **93**, 11615 (1996).
8. W. A. Eaton, V. Munoz, P. A. Thompson, C. K. Chan, and J. Hofrichter, *Curr. Opin, Struct. Biol.* **7**, 10 (1997).
9. C. K. Chan, Y. Hu, S. Takahashi, D. L. Rousseau, W. A. Eaton, and J. Hofrichter, *Proc. Natl. Acad. Sci. USA* **94**, 1779 (1997).

10. C. K. Chan, J. Hofrichter, and W. A. Eaton, *Science* **274**, 628 (1996).
11. B. Nolting, R. Golbik, and A. R. Fersht, *Proc. Natl. Acad. Sci. USA* **92**, 10668 (1995).
12. R. M. Ballew, J. Sabelko, and M. Gruebele, *Proc. Natl. Acad. Sci. USA* **93**, 5759 (1996).
13. R. E. Burton, G. S. Huang, M. A. Daugherty, T. L. Calderone, and T. G. Oas, *Nature Struct. Biology* **4**, 305 (1997).
14. J. F. Brandts, H. R. Halvorson, and M. Brennan, *Biochemistry* **14**, 4953 (1975).
15. G. Fischer and F. X. Schmid, *Biochemistry* **29**, 2205 (1990).
16. C. Odefey, L. M. Mayr, and F. X. Schmid, *J. Molec. Biol.* **245**, 69 (1995).
17. F. X. Schmid, *Ann. Rev. Biophys. Biomolec. Struct.* **22**, 123 (1993).
18. S. E. Jackson and A. R. Fersht, *Biochemistry* **30**, 10428 (1991).
19. S. E. Jackson and A. R. Fersht, *Biochemistry* **30**, 10436 (1991).
20. Y. Harpaz, N. Elmasry, A. R. Fersht, and K. Henrick, *Proc. Natl. Acad. Sci. USA* **91**, 311 (1994).
21. C. Tanford, *Adv. Prot. Chem.* **23**, 121 (1968).
22. C. Tanford, *Adv. Prot. Chem.* **24**, 1 (1970).
23. A. Matouschek, J. M. Matthews, C. M. Johnson, and A. R. Fersht, *Protein Engineering* **7**, 1089 (1994).
24. A. Matouschek and A. R. Fersht, *Proc. Natl. Acad. Sci. USA* **90**, 7814 (1993).
25. A. Matouschek, D. E. Otzen, L. S. Itzhaki, S. E. Jackson, and A. R. Fersht, *Biochemistry* **34**, 13656 (1995).
26. M. Oliveberg, Y. J. Tan, and A. R. Fersht, *Proc. Natl. Acad. Sci. USA* **92**, 8926 (1995).
27. Y. J. Tan, M. Oliveberg, and A. R. Fersht, *J. Molec. Biol.* **264**, 377 (1996).
28. H. Roder and W. Colon, *Curr. Opin. Struct. Biol.* **7**, 15 (1997).
29. M. Silow and M. Oliveberg, *Proc. Natl. Acad. Sci. USA* **94**, 6084 (1997).
30. A. Ikai and C. Tanford, *Nature, Lond.* **230**, 100 (1971).
31. R. L. Baldwin, *Fold Des.* **1**, R1 (1996).
32. A. R Fersht, *FEBS Letters* **325**, 5 (1993).
33. A. R. Fersht, *Curr. Opin. Struct. Biol.* **5**, 79 (1995).
34. A. Matouschek, J. T. Kellis Jr., L. Serrano, and A. R. Fersht, *Nature, Lond.* **340**, 122 (1989).
35. J. M. Matthews and A. R. Fersht, *Biochemistry* **34**, 6805 (1995).
36. A. R. Fersht, A. Matouschek, and L. Serrano, *J. Molec. Biol.* **224**, 771 (1992).
37. A. R. Fersht, L. S. Itzhaki, N. elMasry, J. M. Matthews, and D. E. Otzen, *Proc. Natl. Acad. Sci. USA* **91**, 10426 (1994)
38. L. Serrano, A. Matouschek, and A. R. Fersht, *J. Molec. Biol.* **224**, 805 (1992)
39. A. Caflisch and M. Karplus, *Proc. Natl. Acad. Sci. USA* **91**, 1746 (1994).
40. A. Caflisch and M. Karplus, *J. Molec. Biol.* **252**, 672 (1995).
41. T. Lazaridis and M. Karplus, *Science* **278**, 1928 (1997).
42. A. J. Li and V. Daggett, *Proc. Natl. Acad. Sci. USA* **91**, 10430 (1994).
43. A. J. Li and V. Daggett, *J. Molec. Biol.* **257**, 412 (1996).
44. V. Daggett, A. J. Li, L. S. Itzhaki, D. E. Otzen, and A. R. Fersht, *J. Molec. Biol.* **257**, 430 (1996).
45. J. Tirado-Rives, M. Orozco, and W. L. Jorgensen, *Biochemistry* **36**, 7313 (1997).
46. B. A. Shoemaker, J. Wang, and P. G Wolynes, *Proc. Natl. Acad. Sci. USA* **94**, 777 (1997).

47. S. W. Englander and N. R. Kallenbach, *Quart. Rev. Biophys.* **16**, 521 (1984).
48. S. W. Englander, T. R. Sosnick, J. J. Englander, and L. Mayne, *Curr. Opin. Struct. Biol.* **6**, 18 (1996).
49. Y. Bai, J. S. Milne, L. Mayne, and S. W. Englander, *Proteins: Structure, Function and Genetics* **17**, 75 (1993).
50. Y. W. Bai, T. R. Sosnick, L. Mayne, and S. W. Englander, *Science* **269**, 192 (1995).
51. J. Clarke, A. M. Hounslow, M. Bycroft, and A. R. Fersht, *Proc. Natl. Acad. Sci. USA* **90**, 9837 (1993).
52. J. Clarke, L. S. Itzhaki, and A. R. Fersht, *Trends Biochem. Sci.* **22**, 284 (1997).
53. A. R. Clarke and J. P. Waltho, *Curr. Opin. Biotechnol.* **8**, 400 (1997).
54. C. V. Robinson, M. Gross, S. J. Eyles, J. J. Ewbank, M. Mayhew, F. U. Hartl, C. M. Dobson, and S. E. Radford, *Nature, Lond.* **372**, 646 (1994).
55. J. B. Udgaonkar and R. L. Baldwin, *Nature, Lond.* **335**, 700 (1988).
56. H. Roder, G. A. Elöve, and S. W. Englander, *Nature, Lond.* **335**, 694 (1988).
57. M. Bycroft, A. Matouschek, J. T. Kellis, Jr., L. Serrano, and A. R. Fersht, *Nature, Lond.* **346**, 488 (1990).
58. E. W. Chung, E. J. Nettleton, C. J. Morgan, M. Gross, A. Miranker, S E. Radford, C. M. Dobson, and C. V. Robinson, *Protein Sci.* **6**, 1316 (1997).
59. Z. Y. Guo and D. Thirumalai, *Biopolymers* **36**, 83 (1995).
60. A. G. Ladurner and A. R. Fersht, *J. Molec. Biol.* **273**, 330 (1997).
61. T. Schindler and F. X. Schmid, *Biochemistry* **35**, 16833 (1996).
62. C. R. Robinson and R. T. Sauer, *Biochemistry* **35**, 13878 (1996).
63. A. Ladurner, L. S. Itzhaki, V. Daggett, and A. R. Fersht, *Proc. Natl. Acad. Sci. USA* **95**, 8473 (1998).

Folding Pathways and Energy Landscapes

19

Solving how a protein folds up from its denatured state to its native structure poses an intellectual challenge that is far more complex than solving classical chemical mechanisms. In simple chemical mechanisms, there are usually changes in just a small number of bonds as a reaction progresses. Chemical bonds are often strong, and so stable covalent intermediates can sometimes be isolated and characterized. Often, the rules for analyzing mechanisms (Chapter 7, section A) can be applied simply and rigorously. In protein folding, on the other hand, the whole molecule changes in structure. Thousands of weak noncovalent interactions are made or broken, and it is very difficult to trap intermediates because of their unstable nature. An astronomical number of conformations must be considered. The denatured state is particularly difficult to analyze because it is an ensemble of many ill-defined and rapidly fluctuating conformations. But the experimentalist's basic strategy for analyzing the pathway of protein folding must still be the same as that for analyzing a simple chemical mechanism: characterize all the stable and metastable states on (and off) the reaction pathway and the transition states that link them. The role of the experimentalist is to determine the sequence of intermediates and their rates of interconversion and to characterize their structures and energetics as far as possible to provide the essential basis for calculations and simulations by theoreticians.

The essence of modern work on structure and activity of proteins is analysis at the level of individual residues and atoms. Unfortunately, the only state that is readily studied at this level is the native state. The only spectroscopic technique that can be applied in solution with sufficient resolution is NMR, but it works best with stable, compactly folded proteins. New techniques are required for the study of unstable intermediates and transition states in folding. One of the most productive techniques for analyzing simple chemical mechanisms has been the use of structure–activity relationships, in which small changes are made in the reagents and correlations made with changes in activity. The advent of protein engineering has allowed the use of such techniques to tackle the problem of protein folding: the Φ-value method, which was introduced in Chapter 18, section F. Even though there are many differences between noncovalent and covalent chemistry, the fundamental ideas are the same, and so we can adapt the classical methods. The crucial importance of Φ-value analysis is that it is the only method for studying side-chain interactions in transition states, which can then be used to infer secondary and tertiary structural information.

There is also a computer simulation procedure for analyzing folding at atomic resolution: molecular dynamics,[1] which applies Newton's laws of motion to every atom in a protein and to the surrounding solvent. There are three major difficulties of applying the method to protein folding. First, there is the shear scale of the problem: the computer must handle many conformations of the denatured and partly denatured states and then must calculate how they change over a relatively long period of time (calculating time scales of nanoseconds in femtosecond steps). Second, the limited time scale accessible to simulation means that the reactions must be speeded up by, for example, performing them at very high temperatures. The question is then raised of whether this is realistic. Third, there are the same uncertainties in the potential functions that are used to calculate the energetics as there are in calculating the stabilities of the native state, discussed in Chapter 17. Molecular dynamics is very well suited for analyzing protein unfolding, because the calculations start from well-defined crystal and solution structures, solved by x-ray crystallography and NMR. Also unfolding can be speeded up by simulating denaturing conditions, such as high temperature or the presence of simulated denaturants. The folding pathway is then simulated by running unfolding in reverse. There has been a synergy in the development of Φ-value analysis and molecular dynamics simulations because the experimental data from Φ values have been used to benchmark molecular dynamics simulations, which, in turn, flesh out the Φ-value data into real structures and make predictions that can be tested.[2–6] The molecular dynamics simulations have the potential of describing the whole pathway of folding, but they require experimental validation.

The combination of Φ-value analysis, NMR, and molecular dynamics simulation has allowed the description of folding pathways at atomic resolution. The choice of proteins for folding studies depends on the questions being asked. The proteins used to illustrate folding pathways here are chymotrypsin inhibitor 2 (CI2), barnase, and barstar because they have been extensively analyzed by the combination of the three procedures from the denatured to the native state, including studies on fragments. The three proteins each consist of a single polypeptide chain. An important point is that they have no disulfide crosslinks, which renders them much better suited for studying the early stages of folding than proteins that have disulfide crosslinks. Disulfide crosslinks are put in at a late stage in protein biosynthesis and staple the structure together. When such proteins are denatured and retain their crosslinks, their denatured states retain this late stage information, which constrains them, and they may be thought of as being "spring loaded" to snap back into shape. But disulfide-linked proteins can illuminate different stages of folding processes. For example, T. E. Creighton has pioneered the study of the refolding of reduced and denatured proteins, such as bovine pancreatic trypsin inhibitor, and has been able to trap and analyze unambiguously intermediates on the pathway that are covalently linked by —S—S— bond formation.[7,8] Studies on the refolding of disulfide-bridged lysozyme show how a eukaryotic protein refolds after nonreductive denaturation[9] and how amyloid fibrils are formed via partly unfolded intermediates,[10] which is important for understanding protein misfolding and disease.

There is a difference between experimentalists and theoreticians: experimentalists observe the minima and maxima in free energy profiles—the experimental entities of intermediates and transition states—whereas theoreticians wish to calculate the entire energy surface of a reaction. Experimentalists talk about pathways, theoreticians about energy landscapes. Experiment and theory touch base around the ground and transition states that provide the milestones in the energy landscapes for the theoreticians to benchmark their calculations. The two views are reconciled in section G.

A. Levinthal's paradox

For more than a quarter of a century, ideas about protein folding were dominated by two interrelated concepts; the Levinthal paradox[11] and the supposed necessity of folding intermediates. Levinthal argued that because there is an astronomical number of conformations open to the denatured state of a protein, an unbiased search through these conformations would take too long for a protein to fold. It was a short logical step to argue that there must be defined pathways to simplify the choices in folding (Figure 19.1). Three mechanisms emerged to provide these routes. They simplified the search for the folded state by uncoupling the formation of secondary structure from tertiary structure, thus removing the stringent requirement of simultaneous formation of secondary and tertiary structure. The *framework model* proposed that local elements of native local secondary structure could form independently of tertiary structure.[12,13] These elements would diffuse until they collided, successfully adhered, and coalesced to give the tertiary structure *(diffusion-collision model)*.[14,15] The classical *nucleation model* postulated that some neighboring residues of sequence would form native secondary structure that would act as a nucleus from which the native structure would propagate, in a stepwise manner.[16,17] Thus, tertiary structure would form as a necessary consequence of the secondary structure. The *hydrophobic-collapse model* hypothesized that a protein would collapse rapidly around its hydrophobic side chains and then rearrange from the restricted conformational space occupied by the intermediate.[18,19] Here, the secondary structure would be directed by native-like

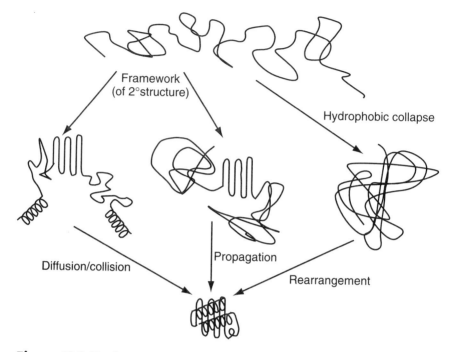

Figure 19.1 The three classical mechanisms for folding

tertiary interactions. A contrary view, the *jigsaw model,* is that each molecule of protein folds by a different distinct path.[20] The hydrophobic-collapse mechanism and the full blown framework model imply the existence of folding intermediates, whereas nucleation does not. Because of the subsequent finding of so many apparent folding intermediates, it was assumed that the presence of intermediates on pathways is an essential requirement for folding, and so nucleation mechanisms dropped out of favor.

The flaw in Levinthal's analysis is that the search is assumed to be *unbiased*: it presupposes that the groups on the protein rotate around their single bonds at random with no stabilization of any particular conformation until they are all in the right orientation, which then clicks into place. If there is a conformational bias in the sequence toward the correct structure, then the paradox disappears.[21,22]

B. Folding of CI2

1. Structure of the native protein

CI2 (Figure 19.2) is a 64-residue polypeptide inhibitor of serine proteases.[23] It has a binding loop (Met-40, which binds in the primary site of chymotrypsin or subtilisin), a single α helix running from residues 12 to 24, and a mixed parallel and antiparallel β sheet. The strands and the amphipathic helix interact to form

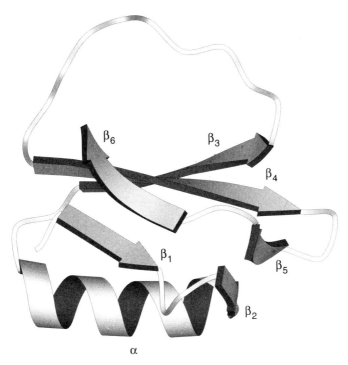

β₆ β₃ β₄ β₁ β₅ β₂ α

Figure 19.2 Structure of CI2 in MolScript.

the small hydrophobic core. CI2 is an α/β structure, the α and β elements being interspersed (Chapter 1, section C4). Its four peptidyl-proline bonds are all in the favored *trans* conformation. CI2 is a single module of structure: the interatomic interactions are quite uniform over the structure, and they do not segregate into regions that make more tertiary interactions within themselves than they do with neighboring regions. As such, CI2 is a basic folding unit or *foldon*.[24] It is, perhaps, a model for a foldon within a larger protein.

2. Folding kinetics

CI2 was used in the last chapter to exemplify the basic kinetics of two-state reactions. The small fraction that folds slowly because of $cis \rightarrow trans$ peptidyl-proline isomerization is ignored in the following discussion. CI2 folds according to first-order kinetics from a relatively open denatured state, with a half-life of some 10 ms.[25] Some mutants fold 10 times faster.

3. Structures of peptide fragments

Fragments of CI2 of increasing length from the N-terminus have been constructed and examined for residual structure in aqueous buffers by NMR and CD spectroscopy.[26,27] Interestingly, the α helix does not form to a significant extent

($<5\%$) until nearly the complete protein is present (residues $1-60$). Very few other interactions are detectable. The amphipathic helix has to make hydrophobic interactions, especially from Ala-16, with residues in the core that are located around positions $50-60$, at the other end of the protein, to be stable.

There are two important lessons from these and other experiments on peptide fragments.

1. In general, isolated elements of secondary structure are relatively unstable without the rest of the protein. Thus, short-range interactions (that is, those within regular secondary structure) must be stabilized by long-range interactions (that is, by tertiary interactions with residues distant in sequence). This emphasizes the necessary balance between local and long-range interactions in stabilizing proteins.

2. Although the native-like secondary structure interactions are weak in the absence of tertiary interactions, residual native structure does exist in some small peptide fragments and can help guide the protein on its way to folding.

4. Structure of the denatured protein

General details about denatured states were discussed in Chapter 17, section B. Specific details are obtained from advanced NMR methods. But any residual native structure in CI2 under denaturing conditions is below the limits of direct structural detection by NMR. The pK_a values of the 10 acidic residues in the native state of CI2 have been measured by NMR to give indirect information.[28] Most of them are $2-3$ units lower than they are in model compounds. The pK_a values in the acid-denatured state are, on average, 0.3 unit lower than the model compound values in pure water, but the difference disappears as ionic strength increases. This indicates some compactness in the denatured state, but it could be induced by electrostatic interactions. By all criteria, the denatured state of CI2 is a relatively expanded state.

5. Structure of the transition state

The structure of the transition state for the folding of CI2 has been mapped by kinetic measurements and Φ-value analysis of more than 100 mutants under a variety of conditions.[29,30] The properties were measured in water from folding kinetics and extrapolated to water from unfolding kinetics as a function of [GdmCl]. The results from both directions agree.

a. Gross features

The average properties of the transition state may be estimated from a variety of criteria. The Tanford β_T value, measured from the relative sensitivities of the folding kinetics and equilibrium constants to [GdmCl] (Chapter 18, section B1, equation 18.9), is 0.6, indicating that about 60% of the surface area of the protein is buried in the transition state for folding, relative to that buried in the native structure. The change in heat capacity of the transition state relative to that of the folded state ($\Delta C_{p(\ddagger-D)}/\Delta C_{p(N-D)}$) is about 0.5, indicating that about 50% of the hydrophobic side chains become buried (see Chapter 17).[31]

Interestingly, a plot of changes in activation energy of unfolding ($\Delta\Delta G_{\ddagger-N}$) against the change of free energy ($\Delta\Delta G_{D-N}$) on mutation fits well to a Brønsted plot of slope 0.7 (Figure 19.3) so that, on average, about 30% of the net stabilization energy of the side chains is present in the transition state.[32] This relatively good fit to the Brønsted equation shows that CI2 unfolds and folds as a single cooperative unit.

b.   Fine details from Φ-value analysis

The Brønsted plot just represents the average change in energy of activation on mutation. The variations are seen from the individual Φ_F values, most of which are between 0.2 and 0.5. The higher values tend to be for the formation of the α helix, especially at its N-terminus. The energetics imply a structure for the transition state that looks like an expanded version of the native structure. But the expansion is not uniform. It is as if there is a center located in the N-terminal part of the helix, and the structure becomes more expanded the farther away it is from the center.[29,30] The structure, fleshed out by molecular dynamics, is described in section 6.

c.   Fractional Φ values = fractional bond formation

Do the fractional Φ values correspond to partial noncovalent bond formation or to a mixture of transition states that arise from parallel pathways? For example, suppose that, as sketched in Figure 19.4, a protein that consists of an α helix and

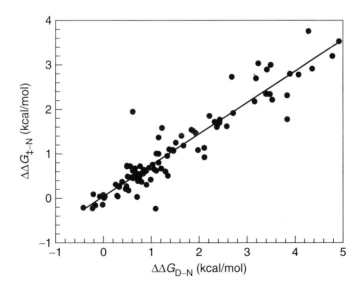

Figure 19.3 Brønsted plot for unfolding of CI2.

a β sheet has two pathways of folding. In one hypothetical pathway (the upper pathway in Figure 19.4), the transition state has the helix fully formed ($\Phi_{F(helix)} = 1$) and the sheet unfolded ($\Phi_{F(sheet)} = 0$). In the other pathway, the helix is fully unfolded ($\Phi_{F(helix)} = 0$) and the sheet is fully formed ($\Phi_{F(sheet)} = 1$). The average structure of the two transition states will thus appear to have partially formed helix and sheet and so have apparent fractional values of Φ_F. We can detect this mechanism from the effects of mutations. Quite simply, if there are two distinct pathways, we can reduce the importance of the one with $\Phi_F = 1$ by destabilizing mutations. The first few destabilizing mutations will lower the importance of this route until it is negligible, then subsequent mutations will have no further effects. This situation would not lead to a linear Brønsted plot but would give a plot of changing slope (Figure 19.5). Multiple mutations can be made in various parts of CI2 that lead to good Brønsted plots over a wide range of changes of energy. For example, the minicore of CI2 (three residues: Leu-32, Val-38, and Phe-50) has been extensively probed by mutation to other hydrophobic residues; a plot of ln k_U versus ln K_{D-N} (Figure 19.6) fits nicely to a Brønsted equation with $\beta_U = 0.64 \pm 0.02$ over a range of nearly 5 kcal/mol (20 kJ/mol) in $\Delta\Delta G_{D-N}$ and a factor of 300 in rate constants (correlation coefficient, $r = 0.99$). (Similarly, the plot of ln k_F versus ln K_{D-N} gives $\beta_F = 0.36$, as expected, since

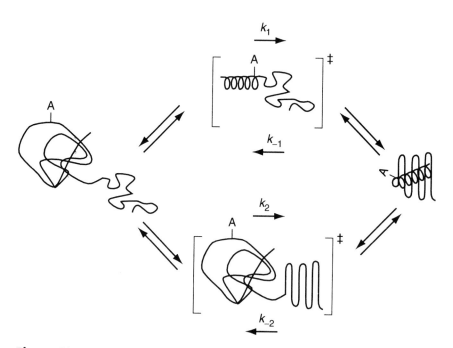

Figure 19.4 A folding mechanism that proceeds along two parallel pathways.

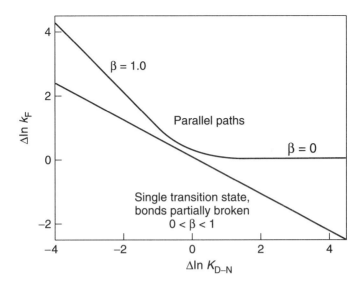

Figure 19.5 Effects of destabilizing mutations on the Brønsted plot for a reaction that proceeds along two parallel pathways (Figure 19.4) and one that has a single expanded transition state.

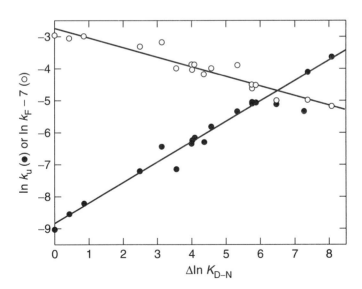

Figure 19.6 Brønsted plot for unfolding and folding of CI2 minicore. k_U is the rate constant for unfolding and k_F that for folding. K_{D-N} is the equilibrium constant for denaturation.

$\beta_F + \beta_U = 1$ for a two-state reaction.) Thus, the data do not fit a model in which there are parallel pathways with largely different transition states but are consistent with a narrow ensemble of structures whose average resembles an expanded transition state.[32]

d. Hammond behavior

Further evidence for a transition state composed of closely spaced energy levels is the finding of Hammond behavior (Chapter 2, section A2): as a reagent is destabilized, the transition state for its reaction moves along the reaction coordinate to become closer in structure to the destabilized state. Plotted in Figure 19.7 is the Tanford β_T value for unfolding $(m_{\ddagger-N}/m_{D-N})$ against the change in the free energy of activation of unfolding. Wild-type protein has a β_T value for unfolding of 0.4, which drops progressively as the protein becomes destabilized to about 0.25 when the free energy of activation falls by 3 kcal/mol (12 kJ/mol). This shows a smooth movement in structure in the transition state for folding, from about 60% burial of surface area in wild type to 75% in the most destabilized mutants.[33]

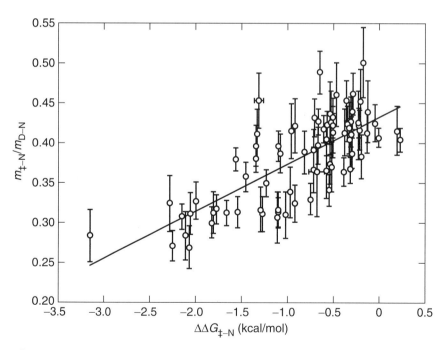

Figure 19.7 Hammond plot for the unfolding of CI2. β_T is a measure of the position of the transition state on the folding/unfolding pathway. The transition state moves closer to the native structure as the activation energy of unfolding, $\Delta G_{\ddagger-N}$, becomes lower ($\Delta G_{\ddagger-N}$ becomes more negative).

6. Molecular dynamics simulations of the transition state

Application of molecular dynamics to the unfolding of CI2[4,5,34,35] gives results in excellent agreement with experiment where they meet up at the transition state. However, whereas the experimentalist looks at the average properties of a very large number of molecules during each folding experiment—typically $> 10^{14}$— simulations look at individual molecules. There are some variations in individual unfolding simulations.[34] By performing a large number of simulations, the energy landscape can be determined by applying statistical mechanics to the distribution of structures. Twenty-four simulations of CI2 unfolding produced a statistical cluster of structures around the transition state that was determined experimentally.[4] A common feature of the different simulations is that the helix is well-formed in the distribution of transition states of CI2. This shows that the Φ_F values of 0.5 to 0.6 in the α helix mean that the structure is basically formed, although the interactions are somewhat weakened. The cluster of structures explains the Hammond data: mutation or change of conditions leads to the transition state moving smoothly around the energy landscape.

The simulations make predictions that can be tested experimentally. For example, as illustrated in Figure 19.8, the C-terminal region of the helix moves away from first β strand in the transition state and breaks a salt bridge between Asp-23 and Lys-2. Now, an Asp is unfavorable energetically in this position in an isolated helix relative to an Ala (see Chapter 17). Mutation of Asp-23 to Ala destabilizes the native protein, as expected from the disruption of the Asp-23$\cdots$Lys-25 salt bridge, but speeds up folding because it stabilizes the transition state in which the C-terminal region resembles an isolated helix.[36] The success of the testing of the simulations lends confidence to their use.

C. The nucleation–condensation mechanism

1. The lessons from CI2 folding

The kinetics of folding of CI2 and the Brønsted and Φ-value analysis of its transition state established important principles.

1. Small proteins can fold rapidly in a first-order reaction from a relatively expanded denatured state.

2. The transition state of CI2 is an expanded structure in which secondary and tertiary structures are formed in parallel.

3. There are no completely fully formed elements of secondary structure in the transition state of folding of CI2; all elements are in the actual process of being formed.

All three of the classical mechanisms of folding (Figure 19.1) are ruled out by these observations, as is the jigsaw model. The framework and nucleation models imply that there are fully formed elements of secondary structure in the transition state of folding. Regions within the classical nucleus or the preformed

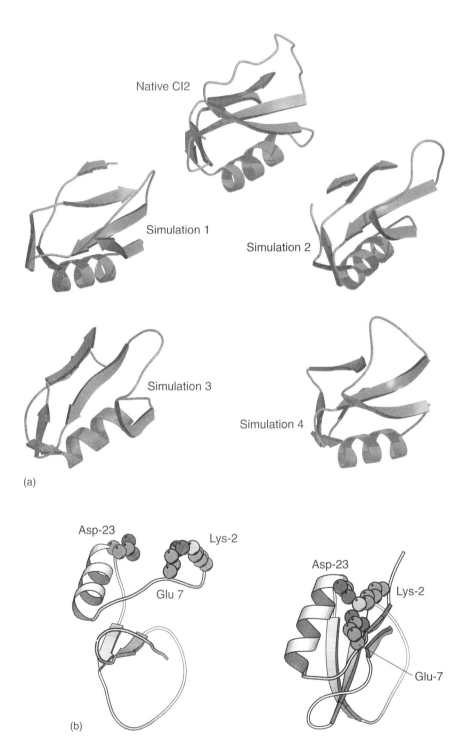

Figure 19.8 (a) Structure of native CI2 plus four simulated transition states for unfolding. [From A. J. Li and V. Daggett, *Proc. Natl. Acad. Sci. USA* **91**, 10430 (1994); A. J. Li and V. Daggett, *J. Molec. Biol.* **257**, 412 (1996)). (b) Residues Lys-2, Glu-7, and Asp-23 form a triple salt bridge in the native structure (*right*) that is broken in the simulated transition states for the unfolding (*left*).

secondary structure would have Φ_F values of 1.0, and these are not seen. The hydrophobic-collapse model is ruled out because there is not a folding intermediate and tertiary and secondary interactions are formed in parallel. Jigsaw and parallel-path mechanisms are ruled out by the Brønsted analysis of the previous section.

2. The nucleation-condensation (or -collapse) mechanism

The transition state is constructed around an extended, delocalized nucleus (Figure 19.9) consisting of the helix (residues 12–24) and the interactions it makes with hydrophobic residues in the core that are located around position 50, especially from Ala-16. Recall from section B3 that these long-range interactions are precisely those required to stabilize the helix in peptide fragments of CI2.

A mechanism called *nucleation-condensation* describes the folding.[30,37,38] (The term *nucleation-collapse* is also used.[39,40]) This mechanism involves a nucleus that consists primarily of adjacent residues (Table 19.1). The nucleus does not form stable structure without assistance from interactions made with residues that are distant in sequence. Formation of the small nucleus cannot be solely rate-determining because a significant fraction of the overall structure

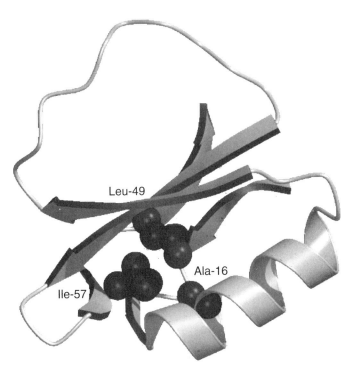

Figure 19.9 Interactions in the nucleus of CI2.

Table 19.1 *Classical nucleation versus nucleation-condensation*	
Classical nucleation	Nucleation-condensation
Strong localized, nucleus; e.g., two or three turns of helix Structure grows from this; growth follows nucleation	Weak local nucleus that is stabilized by long-range interactions to give large extended nucleus Nucleus consolidates as structure forms Consolidation of nucleus and extended structure is concurrent

must be in the approximately correct conformation to provide the long-range contacts to stabilize the nucleus. In CI2, for example, the α helix has a very weak tendency to be formed in a nascent manner, driven by local interactions. The helix remains embryonic until sufficient long-range interactions are built up that it becomes stable. The rest of the protein then condenses around the helix and the other native interactions that are developed during the stabilization of the helix. The onset of stability as the multiple interactions are made is so rapid that the helix is still in the process of being consolidated as the transition state is reached. Thus, formation of the nucleus is coupled with more general formation of structure, and so nucleation is just part of the rate-determining step. The coupling of nucleation and condensation leads to a relatively compact transition state. Thus, the whole molecule is involved in forming the transition state; the nucleus is simply the best formed part of the structure in the transition state.

This nucleation-condensation mechanism has certain characteristics. First, the nucleation site does not need to be extensively preformed in the denatured state. It may or may not be detectable in isolated peptides, but it is essential only in that it is extensively formed in the transition state. Further, it need not be formed completely in the transition state; because of the onrush of cooperative stabilizing interactions, however, it may be in the process of being formed in the transition state. Second, although much of the structure is from local, contiguous residues, there are important stabilizing contributions from long-range interactions; i.e., from contacts with residues that are far removed in sequence.

It must be emphasized that the choice of which residues constitute the nucleus is somewhat arbitrary; they are simply those residues that have the highest Φ values. Thus, destabilization of the nucleus does not prevent folding, but it is the most sensitive region to destabilizing mutations. The wild-type SH3 domain of α-spectrin folds by a nucleation-condensation mechanism. The gene of SH3 has been shuffled so that it is cleaved in its folding nucleus and the former N- and C-termini ligated to give a circularly permuted protein that retains the wild-type fold. The circularly permuted protein finds another nucleus to condense around.[41,42] CI2 can be circularly permuted, leaving the nucleus intact, and it folds around the usual nucleus.

The physicochemical basis of the nucleation-condensation mechanism is that there must be a critical number of interactions made in the transition state for folding.[31,43] This may be understood by a simple argument based on the reasoning that as a protein folds there is an unfavorable large loss of entropy that is compensated by the favorable decrease in enthalpy of the interactions that are formed. After a critical number of interactions are made, the lowering of enthalpy is more rapid than the loss of entropy as further interactions are made (Figure 19.10). The reason the enthalpy is lowered more rapidly is that the stabilizing interactions cluster and so are formed cooperatively. The cluster of favorable interactions constitutes the folding nucleus.

Various theoretical procedures (described in sections G and H) suggest that nucleation mechanisms are optimal for fast folding.[44-47]

3. Direct evidence for nucleation-condensation in assembly of protein fragments

It could be argued that the nucleus is formed in the denatured state of the protein immediately on its transfer from denaturing to renaturing conditions. But nature

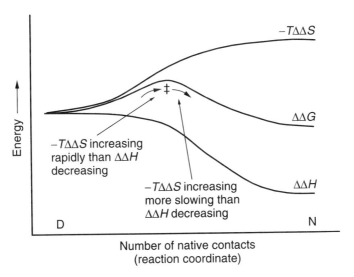

Figure 19.10 A critical number of interactions are required to form a folding nucleus. The figure is oversimplified in that the enthalpy is shown to decrease throughout folding. In practice, there will be some gain in enthalpy as the transition state is reached because of the loss of energy of solvation and the incomplete formation of the interactions present in the native state.

has allowed a trick by which the structures of the starting materials, transition state for folding, and products may all be analyzed under the same set of conditions that favor folding. CI2 can be cleaved in two by the action of CNBr on Met-40 in the binding loop, to give the peptides CI2(1–40) and CI2(41–64).[48] NMR and CD spectroscopy show that they are largely of random structure in water. The few specific interactions that are present are non-native. Yet the two peptides, when added together, associate and fold with a second-order rate constant of $4 \times 10^3 \text{ s}^{-1} M^{-1}$ to give a product that has the wild-type structure, apart from the region immediately around the cleaved bond. The transition state for the association/folding step, analyzed by Φ_F values, has the same nucleus as in the folding of intact protein.[49]

D. Folding of barnase

1. Structure of the native protein

Barnase (Figure 19.11), the 110-residue RNase that is secreted from *Bacillus amyloliquefaciens*, is of $\alpha + \beta$ structure.[50] It has a major α helix running from residues 6 to 18 (α helix$_1$), two smaller helixes in the first half of the sequence, and five strands of antiparallel β sheet in the second half. There is a well-packed hydrophobic core formed by helix$_1$ packing onto one face of the sheet. All the peptidyl-proline bonds are in the *trans* conformation. There are obvious regions

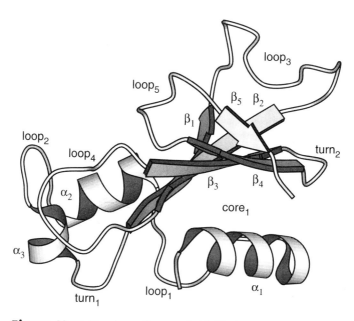

Figure 19.11 Structure of barnase in MolScript.

in the structure that make more interactions between themselves than they do with their neighbors.[51] Thus, barnase is a representative of the small multifoldon proteins.

2. Folding kinetics

Barnase was used in the last chapter to exemplify the basic kinetics of multistate reactions. Barnase has only one observable step for refolding, consisting of first-order kinetics ($t_{1/2} \sim 30$ ms) from a folding intermediate that is the major stable species under conditions that favor folding (equation 19.1).

$$U \rightleftharpoons \quad \dots\dots \quad \rightleftharpoons I \underset{k_u}{\overset{k_{-u}}{\rightleftharpoons}} N \tag{19.1}$$

The kinetics of formation of the intermediate from its more unfolded state is unknown. The small fraction that folds slowly because of *cis* → *trans* peptidyl-proline isomerization is ignored in the following discussion.

3. Structures of peptide fragments

The region corresponding to helix$_1$ of barnase has only about 6% or less helical structure in small fragments.[52] The peptide corresponding to the β hairpin at the center of barnase (equivalent to residues 85–102) does have detectable native and non-native structure, even in urea solutions.[53] The fragment 23–110 appears from its CD spectrum to be largely disordered.[54] This fragment has all the residues necessary for catalysis and was found to have weak catalytic activity that is enhanced on the addition of fragment 1–22. This implies that a small fraction of the larger fragment does exist in the native conformation. The small (1–22) and large (23–110) fragments also bind to each other very rapidly and tightly to give a native-like structure.[54] This behavior is preliminary evidence that barnase can fold in separate parts.

4. Structure of the denatured protein

Advanced NMR methods detect weak residual native structure in helix$_1$ and helix$_2$ of barnase under denaturing conditions.[55,56] There is also some local structure in the central β hairpin of barnase, although it maybe non-native. The denatured state ensemble of barnase has been analyzed by molecular dynamics, using a trick of statistical mechanics. According to the ergodic principle, the behavior of one molecule over a period of time gives the same variations in energies and structures as does the simultaneous behavior of many molecules. The denatured state of barnase, generated by a long molecular dynamics simulation of unfolding, was followed for 10,000 time steps.[57] The ensemble is heterogeneous, but it contains non-random, residual structure with persistent interactions. There is helical structure in the C-terminal portion of the major helix (residues 13–17) and non-native hydrophobic clustering between β strands 3 and 4, consistent with NMR data. In addition, there were tertiary contacts between residues in α1 and the C-terminal portion of the β sheet. As with CI2, long-range interactions are required to stabilize the helix.

5. Structures of the intermediate and transition state for unfolding

Barnase was extensively analyzed by measurements on more than 130 mutants.[58–60] In addition to the Φ_F values for single mutants, the tertiary structure was probed by double mutant cycles that check whether different elements are in contact in the intermediate and transition state. The transition state for unfolding has some regions with Φ_F values of 0, others with values close to 1.0, and a few regions with intermediate values. This heterogeneity is seen in the Brønsted plot for unfolding (Figure 19.12), where many points fit the lines of slopes 0 and 1. This is in contrast to the plot for CI2 (Figure 19.3), which folds and unfolds as a single cooperative unit. Barnase clearly unfolds and folds in parts. The main parts of barnase that are nearly completely folded in the transition state are the β sheet, with Φ_F values of 1.0 at the center and slightly decreasing at the edges, and the α helix$_1$, apart from the first turn. The hydrophobic core, which is formed by the docking of helix$_1$ on the sheet, is weak, with low Φ_F values. The folding intermediate is of similar structure; the Φ_F values that are 1.0 in the transition state are still 1 or only slightly lower. The major difference is in the hydrophobic core, which is very weak. Molecular dynamics simulations are in excellent agreement with these findings.[2,3,6]

The Φ_F values show clearly that the rate-determining step in folding of barnase involves the docking of the preformed α helix$_1$ on the preformed β sheet,

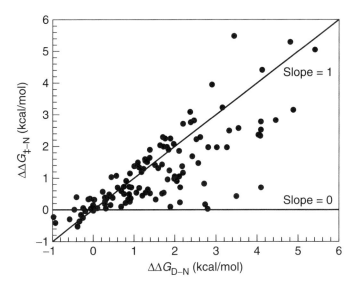

Figure 19.12 Brønsted plot for unfolding of barnase.

with the formation of the hydrophobic core being a significant energetic part of the process.

The structures of the intermediate and transition state can be perturbed by radical mutations that cause significant decreases in stability.[61] Interestingly, there is a mixture of the Hammond effect, as found for CI2 in Figure 19.7, and the very rare anti-Hammond effect (discussed in detail in referance 61). Molecular dynamics simulations are consistent with the experimental data and show that radical mutations in the major helix cause it to be more unfolded in the transition state for unfolding (the anti-Hammond effect) but the β sheet is less unfolded (Hammond Effect).[62]

6. Molecular dynamics, Φ values, and NMR conspire to describe the folding pathway

The Φ values describe the energetics of interactions of side chains and implicate the presence of secondary structure in transition states and intermediates. NMR studies on denatured states give rudimentary information on persistent structure. This is precisely the type of information required to benchmark molecular dynamics simulations, which then flesh out the experimental data into true atomic resolution structures, as described for the transition state for CI2. The consistency between simulation and experiment in describing the denatured state of barnase and the transition states for unfolding of barnase and CI2 inspires confidence in the molecular dynamics simulations. There is excellent agreement between simulated and experimental Φ values. A description of the folding pathway of barnase from the denatured to the native state can be constructed by combining the simulation with experimental data from Φ-value analysis and NMR measurements on the denatured state (Figure 19.13).[57]

E. Folding pathway of barstar at microsecond resolution

The missing link in the folding pathway of barnase is the measurement of the kinetics of formation of its folding intermediate. Studies on barstar have been able to fill in what happens at this stage. Barstar (Figure 19.14) is an 89-residue polypeptide inhibitor of barnase that is produced by *B. amyloliquefaciens* to prevent its RNA from being destroyed by any barnase that is not secreted.[63] It is of α/β structure, consisting of three strands of parallel β sheet, three large helixes, and two small helixes.[64] The distinction between single- and multifoldon proteins can be subjective: CI2 is clearly single and barnase multi, but barstar is borderline between the two. One peptide bond, that to Pro-48, is in the *cis* conformation in the native state. The *trans* $\rightarrow$ *cis* peptidyl-proline isomerization event dominates the folding kinetics of barstar because the *trans* is the major form in the denatured state.

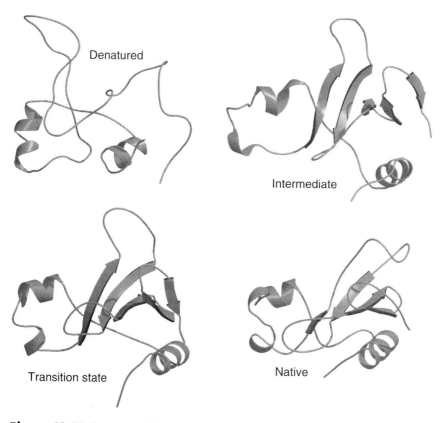

Figure 19.13 Structures of denatured, intermediate, major transition, and native states of barnase from molecular dynamics, Φ values, and NMR. [Coordinates from C. J. Bond, K. B. Wong, J. Clarke, A. R. Fersht, and V. Daggett, *Proc. Natl. Acad. Sci. USA* **94**, 13409 (1997).]

Figure 19.14 Structure of barstar in MolScript.

The early stages of folding of barstar have been measured on the microsecond time scale by temperature jumping its cold-denatured state from 2 to 10°C.[65,66] There is the fast formation of a folding intermediate ($t_{1/2} \sim 200 \ \mu s$) with the peptidyl-proline 48 bond *trans*, followed by the formation with $t_{1/2} \sim 60$ ms of a second intermediate that is highly native-like because it binds to and inhibits barnase. The native-like intermediate then undergoes *trans* → *cis* peptidyl-proline isomerization on the time scale of minutes to give the final native structure (equation 19.2).

$$D \rightleftharpoons I_{trans} \rightleftharpoons N'_{trans} \rightleftharpoons N_{cis} \qquad (19.2)$$

The fragment corresponding to the helix$_1$-loop-helix$_2$ region of barstar has barely detectable helical structure in water. NMR methods detect weak residual native structure in helix$_1$ and helix$_2$ of barstar under denaturing conditions, indicating that long range interactions stabilize secondary structure in the denatured state. Φ_F-Value analysis shows that the first helix becomes substantially consolidated as the intermediate is formed in a few hundred microseconds, as does the second to a lesser extent. The conversion of D to I$_{trans}$ fits the nucleation-condensation mechanism.

F. Unified folding scheme?

It is unlikely that there is a single mechanism for protein folding. Proteins vary so much in structure, size, and properties that there are bound to be many mechanisms. Further, evolution toward a specific function may be at the expense of stability or optimization of folding rate (see Chapter 17, section E, for "stability–activity trade-off"). But the folding mechanisms of CI2, barstar, and barnase point to a unified scheme, variations of which could describe a large number of pathways (Figure 19.15).[30,31,38] The basic hypothesis is that the folding of CI2 is a model for the folding of individual foldons in larger proteins, and the features of the folding of the larger barnase and barstar apply to larger proteins in general. CI2 folds by nucleation-condensation. Barnase folds in a clearly stepwise manner, with rapid formation of the individual foldons, followed by their rate-determining docking and consolidation. Nucleation-condensation happens in the first stage of the folding of barstar, but the formation of secondary and tertiary structure is less coupled than in CI2 and so has elements of the barnase pathway. Recently, several proteins, mainly the small two-state systems of Table 18.1, have been analyzed by Φ values. The majority have a distribution of Φ values similar to that of CI2 (Table 19.2), suggesting that nucleation-condensation is the most common mechanism.

Whether folding is a concerted or stepwise process depends on the stability of individual substructures of the protein when they are considered in isolation or loosely complexed with one another. For example, because the α helix and β sheet of CI2 are unstable without the rest of the protein, they are formed in a

Table 19.2 *Proteins analyzed by Φ values*

Protein	Chain length	Inter-mediates	Number of mutations	Φ Pattern model[a]	Comments
α-Helical proteins					
Monomeric λ repressor[b]	80	0	8	Barnase	Fits diffusion-collision model
Acylbinding protein (bovine)[c]	86	0	26	CI2	Nucleus in hydrophobic core
SH3 domains (β barrels)					
α-spectrin[d]	62	0	10	CI2	
Src[e]	64	0	21	Barnase[f]	
β-sandwich domains					
TNfn3 (long form)[g]	92	0	33	CI2	
[10]FNIII[h]	94	1	41	CI2	
CD2d1[i]	98	1	7	CI2	
α/β proteins					
IgG binding domain of streptococcal protein L[j]	62	0	4	Barnase	Turns analyzed
CI2[k]	64	0	150	CI2	
Activation domain ProcarboxypeptidaseA2 (ADAh2)[l]	81	0	15	CI2	
Barstar[m]	89	2	25	CI2+ Barnase	
FKBP12[n]	107	0	34	CI2	
Barnase[o]	110	1	130	Barnase	
CheY[p]	129	0	34	CI2	One domain = CI2, other is unstructured
PGK[q]	394	0	8	CI2	Double-mutant cycles detected tertiary interactions

(continued)

Table 19.2 *Proteins analyzed by Φ values (Continued)*

Protein	Chain length	Inter-mediates	Number of mutations	Φ Pattern model[a]	Comments
Bimolecular association					
CI2 fragments 1−40 + 41−64[r]	64	0	23	CI2	Concurrent folding and association
Barnase fragments 1−22 + 23−110[s]	110	0	5 (1−22)	Barnase	Docking of fully formed helix of 1−22 and consolidation of 23−110
Arc repressor (dimer)[t]	53	0	44	CI2	Concurrent folding and dimerization

[a] CI2 pattern = mainly fractional, with a few high values. Barnase = mixture of close to 0 and 1 plus some fractional.

[b] R. E. Burton, J. K. Myers, and T. G. Oas, *Biochemistry* **37**, 5338 (1998).

[c] B. B. Kragelund, K. Poulsen, K. V. Andersen, T. Baldursson, J. B. Kroll, T. B. Neergard, J. Jepsen, P. Roepstorff, K. Kristiansen, F. M. Poulsen, and J. Knudsen, *Biochemistry* **38**, 2386−2394 (1999).

[d] A. R. Viguera, L. Serrano, and M. Wilmanns, *Nature Structural Biology* **3**, 874 (1996), and J. C. Martinez, M. T. Pisabarro, and L. Serrano, *Nature Structural Biology* **5**, 721 (1998).

[e] V. P. Grantcharova, D. S. Riddle, J. V. Santiago, and D. R. Baker, *Nature Structural Biology* **5**, 714 (1998).

[f] It has been suggested that this is more like CI2 than barnase (M. Gruebele and P. G. Wolynes, *Nature Structural Biology* **5**, 662 (1998).

[g] S. J. Hamill and J. Clarke, personal communication (1999).

[h] E. Cota and J. Clarke, personal communication (1999).

[i] M. J. Parker and A. R. Clarke, *Biochemistry* **36**, 5786 (1997); and M. Lorch, M. J. Parker and A. R. Clarke, personal communication (1998).

[j] H. Gu, D. Kim, and D. Baker, *J. Molec. Biol* **274**, 588 (1997).

[k] S. E. Jackson, N. elMasry, and A. R. Fersht, *Biochemistry* **32**, 11270 (1993); L. S. Itzhaki, D. E. Otzen, and A. R. Fersht, *J. Molec. Biol.* **254**, 260 (1995).

[l] V. Villegas, F. X. Aviles, and L. Serrano, *J. Molec. Biol.* **283**, 1027 (1998).

[m] B. Nölting, R. Golbik, J.-L. Neira, A. S. Soler-Gonzalez, G. Schreiber, and A. R. Fersht *Proc. Natl. Acad. Sci. USA* **94**, 826 (1997).

[n] K. F. Fulton, E. R. G. Main, and S. E. Jackson, personal communication (1998). FKBP12 = FK506 binding protein.

[o] A. Matouschek, J. T. Kellis, Jr., L. Serrano, and A. R. Fersht, *Nature, Lond.* **340**, 122 (1989); L. Serrano, A. Matouschek, and A. R. Fersht, *J. Molec. Biol.* **224**, 847 (1992).

[p] E. Lopez Hernandez and L. Serrano, *Fold. Des.* **1**, 43 (1996).

[q] M. J. Parker, R. B. Sessions, I. G. Badcoe, and A. R. Clarke, *Fold. Des.* **1**, 145 (1996).

[r] G. de Prat Gay, J. Ruiz-Sanz, B. Davis, and A. R. Fersht *Proc. Natl. Acad. Sci. U.S.A.* **91**, 10943−10946 (1994; J. L. Neira, B. Davis, A. G. Ladurner, A. M. Buckle, G. D. Prat Gay, and A. R. Fersht, *Fold. Des.* **1**, 189 (1996).

[s] A. D. Kippen and A. R. Fersht, *Biochemistry* **34**, 1464 (1995).

[t] M. E. Milla, B. M. Brown, C. D. Waldburger, and R. T. Sauer, *Biochemistry* **34**, 13914 (1995).

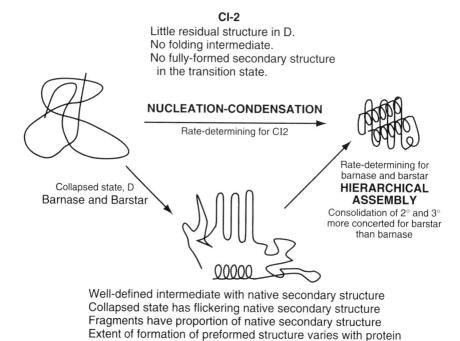

Figure 19.15 Unified nucleation-condensation scheme.

concerted manner. If they were stable in isolation, then the protein could form by a framework or collision/diffusion model, as suggested by experiments on a λ repressor fragment. The same is true for individual foldons within a protein. If a foldon is stable without the other modules, then it could form independently. The more the individual structures are stable, the more stepwise and hierarchical the folding process. One extreme example is the folding of the 129-residue protein CheY, which has two domains. The transition state has one foldon folding by a nucleation-condensation mechanism, whereas the second foldon is unstructured.[67] Just as the mechanism of chymotrypsin appears to change with different substrates because the rate-determining step changes, so the mechanism of folding can appear to change because the balance of rate constants change. Nevertheless, the basic pathway is the same.

The nucleation-condensation mechanism can be accommodated in modified framework and hydrophobic-collapse models: the framework model must be modified so that formation of secondary structure is linked to the formation of tertiary interactions; and the hydrophobic collapse model must have the formation of tertiary interactions linked to the formation of secondary structure. Another variation of concerted structure formation is the "hydrophobic zipper."[68] Whatever the distinctions of names, stable tertiary and secondary structural interactions must form concurrently.

G. Insights from theory

There is an arsenal of theoretical methods for the analysis of folding, ranging from precise atomic analysis by molecular dynamics simulation, through models involving simplified polymer chains, to completely abstract procedures. The most precise procedure of molecular dynamics simulation applies readily to unfolding and can take a protein to the denatured state in solution. Molecular dynamics simulation has not, as yet, been applied to the folding of a completely random chain because of the plethora of possible conformations. In comparison, the more abstract procedures allow simple models for folding to be analyzed directly, and they start from a state prior to the partly collapsed state in solution. These procedures tend to stress general principles. All the methods have given insights into the principles of folding that can be tested against experiment, make predictions, and fill in the entire energy landscape.

1. Lattice simulations

A step down in detail from molecular dynamics simulation is to simplify the protein into a string of beads that is arranged on a cubic lattice; e.g., a 27-mer on a cube of $3 \times 3 \times 3$ points, or a 64-mer on a $4 \times 4 \times 4$ cube.[69] In its simplest form, each bead of the string is spherical, has no side chains, and is assigned a characteristic of being hydrophobic or polar. A simple potential function is assigned to the interaction of the various beads, defining them as being polar or nonpolar. (Off-lattice simulations use similar simplifying assumptions about the protein but do not use the lattice. This is computationally more demanding.) The folding of the random chain can be simulated by Monte Carlo methods ("spinning a roulette wheel" based on the energetics at each step in the folding simulation to see what happens next). Each simulation gives a different pathway that is recorded and analyzed. In effect, a folding experiment is performed in the computer. Once the computer experiment is benchmarked to fit in with reality, then the parameters in the program can be systematically changed to monitor how folding changes. The modeler can "mutate" the protein and watch its properties evolve with changing structure. Insights that have come from this approach include the following:

1. For a protein to be able to fold, there must be a series of structures with decreasing energy levels so that the final state is of distinctly lower energy than structurally unrelated states (an "energy gap").[70]

2. Strings of beads that are modeled on real proteins fold much better than random sequences in the computer.[45]

3. Optimal rates of folding occur when there is not an initial collapse of structure, hydrophobic or otherwise, but when there is a specific and extended nucleus in the transition state; model proteins evolve toward nucleation mechanisms with extended nuclei.[45,47]

4. The folding of larger proteins splits up into domains.[71]

2. Spin glass theory and other abstract methods

Spin glass theory was originally formulated to analyze the orientations of the spins of ferromagnetics that are frozen in a diamagnetic material (e.g., manganese dissolved in molten copper and frozen). The spins want to be antiparallel to each other, but they cannot all be and so are "frustrated." By analogy, a residue in a protein is frustrated if it wants to be in more than one conformation; for example, to be both in a helix and in a sheet.[72] The most stable structure is "minimally frustrated," compared with misfolded states. Another approach is to use "capillarity," a combination of considering the protein as a droplet, with a surface tension, and using ideas of nucleation.[73]

From these and other methods of physics, including classical polymer theory, it is possible to describe protein folding in terms of energy landscapes.[74] A "smooth" energy landscape allows rapid transitions. A "rough" energy landscape has kinetic traps that slow down folding and so is to be avoided. These methods make predictions analogous to those of the more discrete lattice simulations, such as the size of the nucleus, non-Arrhenius kinetics of folding, and the transition from single- to multinucleation. An extended nucleus appears necessary from all these approaches.

3. The folding funnel

a. The golf course analogy

If a golfer was blindfolded, stood at the edge of a very large golf course, and was told to hit the golfball in a random direction, then the chances of sinking the ball in a single hole are infinitesimally small. This is analogous to a protein existing in a large number of non-native states of equal energy and having just one native state of much lower energy. There is no driving force to push the protein in the direction of folding; this is the Levinthal paradox. Suppose, instead, that the golf course sloped down from all directions to the hole. Then gravity would funnel the ball to the hole, and the golfer would always score a hole-in-one. Theoreticians have likened the process of a protein overcoming the Levinthal paradox to a progression down a funnel.[75,76]

b. The funnel

A cross section through a funnel is shown in Figure 19.16. The funnel is a conceptual mechanism for understanding the self-organization of a protein and how it avoids the Levinthal paradox.[76] At the top of the funnel, the protein exists in a large number of random states that have relatively high enthalpy and high entropy. There is a fight between the maximization of entropy keeping the protein as random as possible at the top of the funnel and the minimization of enthalpy trying to drag the protein down the funnel as it folds. Progress down the funnel is accompanied by an increase in native-like structure as folding proceeds. Proteins fold from the random state by collapsing and reconfiguring. The reconfiguring occurs by a Brownian type of motion between conformations that are geometrically similar and follows a general drift from higher energy to lower energy

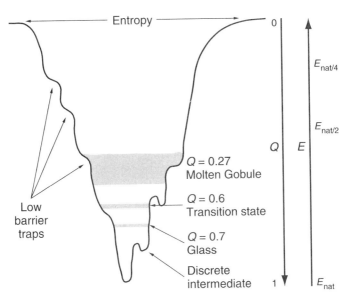

Figure 19.16 Cross section through a folding funnel. E corresponds to free energy.
[Courtesy of P. G. Wolynes]

conformations. The funnel is thus a progressive collection of geometrically simi-lar collapsed structures, one of which is more thermodynamically favorable than the rest. Progress down the funnel can be described by a parameter Q, the frac-tion of correct contacts made in a state i.

$$Q_i = \frac{\text{(number of native pairwise contacts)}_{\text{in state i}}}{\text{(total number of pairwise contacts in native state)}} \qquad (19.3)$$

Folding is driven by an increase in Q as the collection of states rapidly intercon-verts; i.e., by the acquisition of native-like interactions.

The width of the funnel in Figure 19.16 represents the entropy of the system. The depth represents the energy, and also Q (which is not linear with free en-ergy). The molten globule state (a loosely collapsed state with fluctuating ter-tiary interactions and very weak secondary structure) is formed with Q around 0.27. The transition state occurs around $Q = 0.6$. There must be a single unique state of low energy for the protein to fold; otherwise, it gets trapped in the "glass" region with $Q = 0.7$. If the increase in Q is not uniform, but some re-gions increase in Q more rapidly than others, then there is nucleation.

There is an explicit assumption in the funnel mechanism that there is a kinetic flow down through a series of states that are progressively lower in energy. The assumption that kinetic pathways follow thermodynamic energies downhill is normally unwarranted, because there is not a formal link between kinetics and thermodynamics. (The Hammond postulate and linear free energy relationships are *extra*thermodynamic relationships, which occur in practice under a limited set of conditions.) The assumption is justified when the barriers between states are very low ($\sim k_B T$) so that they are all in rapid thermal equilibrium, as in the early stages of folding, but it is not justified at the later stages when there are high energy barriers ($\gg k_B T$). The increase in Q in the early stages of folding is the downward slope in the golf course that solves the Levinthal paradox.

c. Two-dimensional view of the funnel

We can convert the energy landscape funnel to the free energy–reaction coordinate diagram of the experimentalist by lumping together the various states as "ensembles" (Figure 19.17).[77] It is perfectly acceptable to marry the energy landscapes of theory and the folding pathways of experiment by using the idea of ensembles. Conventional two-state folding for a small protein such as CI2 (Figure 19.17a) has energetically "downhill" folding to the denatured state ensemble that consists of a large number of conformations that have flickering native interactions. These then converge to the "transition state ensemble," which is a much more restricted set of states. A more complex protein has downhill folding from the U ensemble via more compact states to a D ensemble that has well-defined elements of structure (Figure 19.17b). The energy landscape is bumpier. If it is sufficiently bumpy, an early molten globule will form that can be isolated. The earliest state that is seen by an experimentalist is the D ensemble. This varies from relatively expanded states of very small proteins to well-structured folding intermediates of larger proteins.

H. Optimization of folding rates

Computer simulation using lattice models and energy landscape theory using abstract models predict that the fastest folding of small proteins should occur without intermediates and by an extended nucleation process. Stable intermediates slow down folding.[43,78,79] This can be seen by analogy with enzyme catalysis (Chapter 12).[37] The fastest reaction occurs when the reagents are at the highest permissible energy level and the transition state at the lowest. The fastest mechanism is nucleation-condensation because it has a starting state with the least possible structure (Figure 19.18).

There is a difference, however, between the evolutionary pressure on rates of folding and on enzyme catalysis. There is a constant evolutionary pressure to increase the efficiency of enzymes because less protein has to be produced and so energy is saved. But a protein has to fold only fast enough to be produced. Any further increase in folding rate is of no apparent advantage to the cell, and so evolutionary pressure has a cutoff. Thus, proteins need not have evolved to the most efficient folding mechanisms. Coupled with the inherent complications of folding in larger proteins, this means that a wide variety of folding mechanisms, not just the optimal ones, are liable to be encountered in practice (see section F). Indeed, intermediates are frequently found in the folding of small proteins.[80] There will not be a unique mechanism of folding, but a range in which mechanisms will blur into one another.

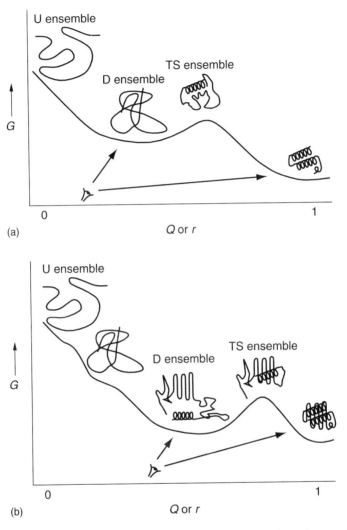

Figure 19.17 Reduction of the folding funnel to a conventional reaction coordinate diagram. This reconciles the classical view of a pathway with the new view of an energy landscape and an ensemble of conformations. [After W. A. Eaton, P. A. Thompson, C. K. Chan, S. Hagen, and J. Hofrichter, *Structure* **4**, 1133 (1996).]

1. Factors that determine rate constants for two-state folding

There is a severe practical problem in looking for correlations between the rate constants for folding of small proteins and their structural or thermodynamic properties—specific structural features can dominate the rate of folding. For example, we know from the protein engineering studies on barnase and CI2 that specific mutations can slow down the rate of folding by several orders of

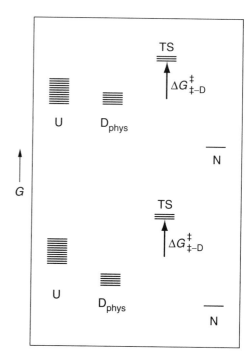

Figure 19.18 Energy level diagrams for protein folding. The conformations of the denatured state ensemble U and D_{phys} are drawn, for convenience, as stacked. Those denatured conformations that have some elements of native structure could be drawn more accurately as being on the reaction pathway and closer to the transition state. *Top*: There is no intermediate, and the average energy of the D_{phys} ensemble, G_{Dphys}, is similar to that of G_U. *Bottom*: As above, but $G_{Dphys} < G_U$ because stabilizing interactions are present in D_{phys}. The activation energy is higher at the bottom.

magnitude and some mutations even speed up folding by more than an order of magnitude. We saw in Table 18.1 that the rate constants for the folding of two-state proteins span only four orders of magnitude. Thus, when comparing different classes of proteins, the "noise" from specific structural features is likely to obscure any "signal" from underlying trends in folding rates. We can eliminate some of the noise by looking for correlations in families of mutant proteins derived from mutagenesis experiments on a single protein. Clearly, a mutation in a region that is unstructured in the transition state will not affect the rate of folding even if it affects stability. For mutations in the single folden protein CI2 that folds by nucleation condensation, there is a Brønsted plot between rate and equilibrium constants for the folding (Figure 19.3), which shows that there is a relationship between stability and folding rate in this limited system. But, for the folding of the multimodular barnase that does have unstructured regions in the transition state, there is not a simple Brønsted plot (Figure 19.12).

There does, however, appear to be a statistically significant correlation between the rate constants for folding of single domain proteins and the average sequence separation between contacting residues in the native state. Proteins that have primarily local contacts (i.e., have a low contact order) tend to fold more rapidly than those that have more non-local interactions (i.e., have a high contact order).[81,82]

I. Molecular chaperones[83-85]

1. Chaperones and heat-shock proteins

Even though their three-dimensional structures are encoded in their sequences, proteins face an obstacle race on their way to successful folding. At sufficiently high concentrations, the denatured states of even the small, fast-folding proteins aggregate and precipitate instead of folding. Denatured states and folding intermediates are generally "sticky" because they have exposed hydrophobic surfaces. These states can also polymerize by forming extended β-sheet structures (this is the root cause of amyloidosis diseases). Aggregation processes are of second or higher order in concentration and so eventually compete with first-order folding kinetics as concentrations increase. Even if aggregation does not occur rapidly, misfolded states or even very slowly reacting intermediates can form and be kinetically trapped so that they aggregate or become degraded in the cell. Proteins with multiple disulfide linkages can have problems in forming the correct ones spontaneously. The biochemist can perform tricks to encourage proteins to fold *in vitro*. Cycles of addition and removal of denaturant help unfold misfolded states, thus allowing further attempts at successful folding. This approach can be combined with redox buffers that allow rapid equilibration of disulfide linkages. The means that nature uses to overcome the problems of folding of real proteins gives insights into mechanistic problems of folding.

The cell uses a series of accessory proteins, collectively termed molecular chaperones, to help some proteins (perhaps ~10%) fold correctly. Some of their functions are obvious. There are protein disulfide isomerases that rapidly catalyze the shuffling of disulfide linkages. Peptidyl-prolyl isomerases (e.g., cyclophilin, FK506 binding protein) have a weak activity in equilibrating the *cis-trans* isomers. The cell does not allow unfolded polypeptide chains of any significant length to be free in solution because of problems of aggregation or proteolysis. There are many chaperones that have evolved to bind denatured or unfolded states at various stages in their development. Trigger factor, for example, binds to nascent polypeptide chains as they are formed on the ribosome.

Many of the chaperones double as heat shock-proteins (Hsp). When a cell is put under stress that can cause proteins to denature, such as too high a temperature, it produces heat-shock proteins. Their names are abbreviated to "Hsp" plus their subunit molecular mass in kDa. Hsp70, for example, is a ubiquitous heat-shock protein in eukaryotes. It is known in *E. coli* as DnaK for historical reasons because it was first discovered from a supposed role in DNA replication. Hsp70 is also important in protein trafficking and the conveying of proteins across membranes, because the denatured state is important in these processes. In protein biosynthesis, the unfolded state of the nascent polypeptide chain is passed on to DnaK, which maintains it in an extended form. The chain, under the influence of ATP and co-chaperones such as DnaJ and GrpE, is handed over to GroEL.

2. GroEL (Hsp60 or Cpn60)

GroEL is the most interesting molecular chaperone in the process of protein folding from the point of view of folding mechanisms. GroEL from *E. coli* is a typical member of the Hsp60 class of molecular chaperones, also known as chaperonins; GroEL is Cpn60 (Cpn = chaperonin). It works with a co-chaperone, GroES, that is a typical member of the Hsp10 or Cpn10 class of chaperonins. Successful folding of protein substrates is coupled *in vivo* to the hydrolysis of ATP, but some substrates work *in vitro* with GroEL in the absence of ATP. GroEL consists of 14 identical subunits, each of 57 kDa, that form a cylinder containing a central cavity. (Note that GroEL usually refers to the 14-mer.) Each subunit consists of three domains. The equatorial domain (residues 6–133 and 409–523) has the binding site for ATP and forms most of the intersubunit contacts. The intermediate domain (residues 134–190 and 377–408) connects the equatorial domain to the apical domain (residues 191–376), which binds polypeptides.[86]

a. GroEL is an allosteric protein that has weak and tight binding states

GroEL undergoes a series of conformational changes on the binding of its ligands, revealed at low resolution by electron microscopy and at high resolution by x-ray crystallography.[87,88] Free GroEL is a short, squat, symmetrical molecule. On the binding of ATP, it elongates (Figure 19.19). GroES binds as a heptamer to the apex of GroEL in the presence of ATP to give a bullet-shaped molecule. In doing so, it causes the cavity in the ring to which it is attached (the *cis*) to become larger. The other end of the *cis* ring is sealed off. Electron microscopy studies on a $GroEL_{14} \cdot GroES_7$ complex that had been frozen in the presence of denatured substrates found that a proportion bind in the *cis* "cage," although a higher proportion is found bound to the apex of the opposite (*trans*) ring.[89] At higher ratios of GroES to GroEL, symmetrical $GroEL_{14} \cdot (GroES_7)_2$ complexes are formed ("footballs").[90]

The hydrolysis of ATP follows classical sigmoid kinetics (Chapter 10, section A), characteristic of an allosteric protein (Chapter 10, section B).[91,92] GroEL undergoes a switch from a tense T state in the absence of ligands to the relaxed R state on the binding of ATP (Figure 19.20). The T state has a predominantly hydrophobic cavity. The conformational changes on the binding of ATP and GroES cause the hydrophobic surfaces to become buried and new hydrophilic ones to be exposed.[88] Most important, the change in state is accompanied by a change in the affinity toward denatured proteins; GroEL binds them very tightly, $GroEL_{14} \cdot GroES_7$ much more weakly.[93]

b. GroEL slows down individual steps in protein folding

Individual steps in the GroEL mechanism were measured from its effects on the refolding of mutants of barnase. GroEL binds barnase at diffusion-controlled rates. Barnase refolds when complexed to GroEL, but it does so 400 times more slowly than when it is free in solution. The $GroEL_{14} \cdot GroES_7 \cdot$ nucleotide complexes retard the rate of refolding only fourfold, paralleling the loss of affin-

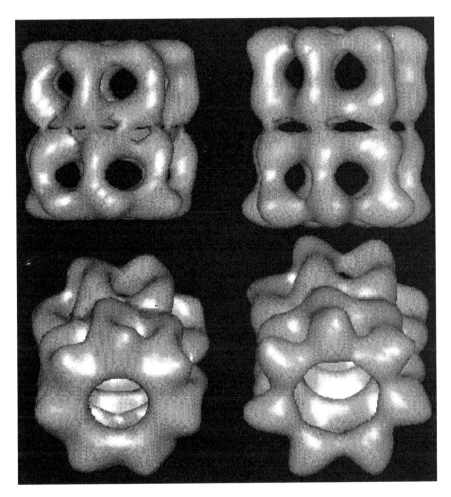

Figure 19.19 Reconstruction of the structure of GroEL from electron microscopy. *Top left*: Side view in the absence of ATP. *Bottom left*: Tilted view of same. *Right*: Elongation and opening of the molecule on the addition of saturating concentrations of ATP. [After A. M. Roseman, S. X. Chen, H. White, K. Braig, and H. R. Saibil, *Cell* **87**, 241 (1996).] Peptides bind to the apical domains. In the absence of ATP, the binding sites from the seven subunits form a nearly continuous ring around the inner neck of the cylinder, seen in the bottom left. On the binding of ATP, the individual binding sites are forced apart, thus weaking the binding of a large substrate that spans more than one subunit. [A. M. Buckle, R. Zahn, and A. R. Fersht, *Proc. Natl. Acad. Sci. USA* **94**, 3571 (1997).]

ity of GroEL for denatured states on the binding of ATP and GroES.[94] GroEL also retards the rate of refolding of CI2. Truncation of any of the hydrophobic side chains of CI2 by mutation leads to both weaker binding and less retardation of folding.[95] (Note that barnase and CI2 do not need GroEL to fold, but they are excellent substrates for studying its kinetics.)

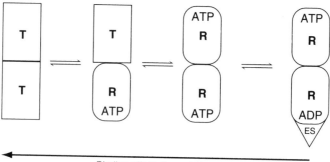

Binding affinity for proteins

Figure 19.20 Allosteric nature of GroEL. The rectangles represent the T state, which has strong affinity for peptides and weak affinity for GroES and ATP. Rectangles with rounded corners represent the R state, which has weak affinity for peptides and strong affinity for GroES and ATP. In the absence of ATP, GroEL exists in the T state. Seven molecules of ATP bind cooperatively to one ring, which flips to the R state. There is negative cooperativity between the two rings, but at higher concentrations of ATP, a symmetrical R state forms. GroES binds to one end of the complex, and there is hydrolysis of ATP. The affinity for denatured states decreases progressively on the binding of ATP and GroES. The hydrolysis of ATP in the *trans* ring causes the concomitant dissociation of GroES.

These experiments show two key features of the GroEL mechanism: (1) weaker binding of denatured states leads to faster folding and (2) GroEL is targeted against denatured and partly denatured states of proteins in general, because it binds hydrophobic side chains, which are the signatures of such states. Folding is slowed down because the bonds between the denatured state and GroEL are broken during folding. Thus, the T state of GroEL is an enzyme that is complementary in structure to the denatured states of proteins. Its role is not to catalyze the steps of folding *per se*, but to act in the direction of unfolding to increase the yield of products.

c. GroEL has an unfolding activity

Under physiological conditions, the T state of GroEL catalyzes the ^{1}H/^{2}H-exchange of buried >NH groups of barnase that must be fully unfolded for exchange to happen. GroEL can thus cause stable states to become transiently unfolded.[96] The weak binding R state has much weaker activity.

d. The active site: Minichaperones show the basic activity of GroEL

A peptide of some 150 residues of GroEL (e.g., residues 191–345) folds stably into a monomer that is functionally active as a chaperone *in vitro*.[97] Further, it can be covalently attached to agarose and other solid supports where the monodispersed fragment is extremely active as a chaperone.[98] Crystal structures of recombinant "minichaperones" reveal that the active site is a flexible hydrophobic patch.[99] It fits best to extended β strands. The basic function of GroEL is to provide a surface for binding exposed hydrophobic patches of denatured states and folding intermediates, which prevents self-aggregation. The process of sequestration is then enhanced by the binding site providing a "strain" mechanism for disrupting patches into separate β strands.

e. The annealing mechanism

The essence of the mechanism of GroEL is first to prevent hydrophobic folding intermediates from sticking to one another and then, when necessary, to allow a misfolded state to unfold and have a fresh start for productive folding (Figure 19.21).[101–103] This is analogous to an annealing activity, whereby flaws in a metal, or mispairing in DNA, can be corrected by annealing: successive cycles of heating and cooling provide the activation energy for escaping from a kinetic trap to reach thermodynamic equilibrium. Every time the unfolded state in Figure 19.21 reacts, it partitions to give a proportion $k_{fold}/k_{misfold}$ of correctly folded native state. Successive rounds of annealing and refolding decrease the amount of misfolded product by a fraction of $k_{fold}/k_{misfold}$ at each step.

f. Role of ATP hydrolysis: Energy transduction and the annealing mechanism

GroEL is essential for the growth of *E. coli* cells. There are temperature sensitive mutants of *E. coli* that are viable at 37°C but do not grow at 43°C because they have a mutation in GroEL that causes it to be nonfunctional at the higher temperatures. Viability can be restored by expressing a small minichaperone from a plasmid introduced into the cell. But, the complete absence of GroEL cannot be replaced.[105] Single-ring constructs are also functional in vivo.[106] The basic chaperone activity of the 150-residue active-site peptide of GroEL is thus enhanced by the allosteric properties of the large GroEL·GroES complex, which is regulated by the hydrolysis of ATP.

The first role of ATP hydrolysis is to keep out unwanted proteins. Every cycle of the GroEL reaction consumes 7 mol of ATP. In order not to waste ATP or saturate GroEL with all the nascent chains in the cell, nature uses the time constant for ATP hydrolysis to act as a filter or gatekeeper (Figure 19.22).[94,104] The half-

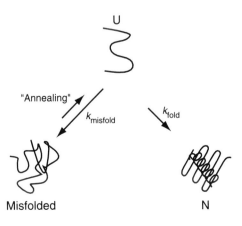

Figure 19.21 Annealing mechanism for GroEL. A misfolded state that does not fold productively is formed more rapidly from the initial unfolded state (U) than the correctly folded state (N) is formed. Unfolding the misfolded state (annealing) allows further attempts at productive folding. At each attempt, the proportion of N formed is $k_{fold}/(k_{misfold} + k_{fold})$.

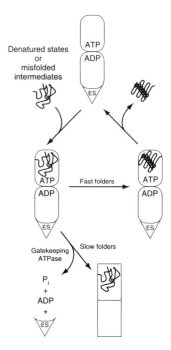

Figure 19.22 The gatekeeping or filtering activity of the GroEL ATPase. The turnover number for the ATPase reaction ($0.05-0.1$ s^{-1}) is far slower than the refolding rate constant of $2-2.5$ s^{-1} for GroEL-bound barnase. Only slowly folding proteins bind long enough to enter the chaperoning cycles of Figure 19.23.

life for the hydrolysis of ATP in the GroEL$_{14}$·GroES$_7$·nucleotide complex is some 20 s. This is 60 times longer than that for the refolding of barnase when bound. Proteins that fold as fast as barnase and do not need a chaperone bind to the GroEL$_{14}$·GroES$_7$·nucleotide *holo*enzyme and refold before significant ATP is hydrolyzed. But slower folding proteins that require help reside long enough to be channeled into the annealing process.

The second role of ATP hydrolysis in the GroEL mechanism is to regulate the affinity toward different substrates and to shuttle the protein between the tight-binding/slow-folding T state and the weak-binding fast-folding R state. The natural resting state of GroEL in the cell is the GroEL$_{14}$·GroES$_7$·nucleotide *holo*enzyme R state. To generate the T state, ATP must be hydrolyzed, which leads to the concomitant expulsion of GroES. The shuttling of a bound protein between the R and T states would put it in a hydrophobic unfolding environment at one extreme and a hydrophilic refolding environment at the other. This accomplishes two results: it enhances the annealing mechanism, and it allows tightly bound states to escape from the R state.

The complex mechanism shown in Figure 19.23 summarizes these steps. In addition, some proteins may need to be sequestered inside the cage-like structure of GroEL, which can probably accommodate proteins up to 70 kDa.[88] The

Figure 19.23 The GroEL cycle, compiled from Figures 19.21 and 19.22. A denatured state binds to ①, the GroEL·GroES·ATP·ADP complex. Fast folders, such as barnase, fold rapidly to give ③ and release folded protein. (Fast-folding parts of larger proteins may also fold rapidly to give partly folded proteins on ③, where they may translocate or dissociate.) For slow folders, GroES is expelled from ② on ATP hydrolysis to give ④, the T state. There is an equilibrium with the R state ⑥, induced by the binding of ATP. GroES does not directly give ② on binding to ⑥. The T-state ④ has the potential for unfolding a compact denatured state. The top cycle ((① → ② → ③)) is a selection step that uses ATP hydrolysis as a gatekeeping action to allow access of slow folders to the bottom cycle, which encompasses rounds of annealing, if necessary, and any folding or translocation. This cycle uses ATP hydrolysis to pump the conversion of R state to T state. ⑥ can be shunted back to ② via ⑩. [After F. J. Corrales and A. R. Fersht, *Proc. Natl. Acad. Sci. USA* **93**, 4509 (1996); and *Fold. Des.* **1**, 265 (1996).]

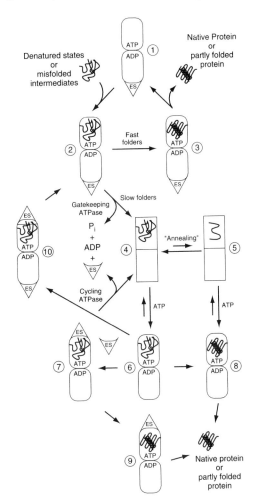

minichaperones or GroEL by itself are capable of effecting the refolding of many proteins, but the full machinery of the *holo*enzyme complex is required for those proteins that are particularly difficult to refold.[107]

3. A real folding funnel

The initial stages of folding presented a puzzle, the Levinthal paradox, that was solved by the theoreticians using a folding funnel. The final stages of folding present a problem for the experimentalists, especially the biotechnol-

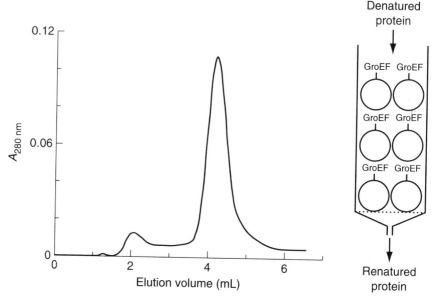

Figure 19.24 A real folding funnel. Refolding chromatography of IGPS (49–252) (indole 3-glycerol phosphate synthase lacking residues 1–48). Denatured IGPS from an inclusion body was dissolved in 8-M urea and diluted onto a 3-mL column of a GroEL minichaperone (GroEF) that was immobilized on agarose. Ninety-six percent of the protein eluted as fully active material.

ogists, because many desirable proteins that are produced by recombinant DNA technology do not fold successfully. Frequently, they precipitate as inclusion bodies. But the minichaperones provide an answer in some cases: attached to agarose, they provide a real folding funnel to which protein is added in denaturant at the top and pours out native at the bottom (Figure 19.24). This procedure works for proteins that have a low refolding yield of 5–10 % or so in the absence of chaperones, but not, as yet, for the more recalcitrant ones.

References

1. M. Karplus and D. C. Weaver, *Nature, Lond.* **260**, 404 (1976).
2. A. Caflisch and M. Karplus, *Proc. Natl. Acad. Sci. USA* **91**, 1746 (1994).
3. A. Caflisch and M. Karplus, *J. Molec. Biol.* **252**, 672 (1995).
4. T. Lazaridis and M. Karplus, *Science* **278**, 1928 (1997).
5. V. Daggett, A. J. Li, L. S. Itzhaki, D. E. Otzen, and A. R. Fersht, *J. Molec. Biol.* **257**, 430 (1996).

6. J. Tirado Rives, M. Orozco, and W. L. Jorgensen, *Biochemistry* **36**, 7313 (1997).

7. T. E. Creighton, *Bioessays* **14**, 195 (1992).

8. T. E. Creighton, *Biol. Chem.* **378**, 731 (1997).

9. C. M. Dobson, P. A. Evans, and S. E. Radford, *Trends Biochem. Sci.* **19**, 31 (1994).

10. D. R. Booth, M. Sunde, V. Bellotti, C. V. Robinson, W. L. Hutchinson, P. E. Fraser, P. N. Hawkins, C. M. Dobson, S. E. Radford, C. C. F. Blake, and M. B. Pepys, *Nature, Lond.* **385**, 787 (1997).

11. C. Levinthal, *J. Chim. Phys.* **85**, 44 (1968).

12. O. B. Ptitsyn, *Dokl. Acad. Nauk. SSR* **210**, (1973).

13. P. S. Kim and R. L. Baldwin, *Ann. Rev. Biochem.* **59**, 631 (1990).

14. D. Bashford, F. E. Cohen, M. Karplus, I. D. Kuntz, and D. L. Weaver, *Proteins: Struct. Func. Genet.* **4**, 211 (1988).

15. M. Karplus and D. L. Weaver, *Protein Sci.* **3**, 650 (1994).

16. D. B. Wetlaufer, *Proc. Natl. Acad. Sci. USA* **70**, 697 (1973).

17. D. B. Wetlaufer, *Trends Biochem. Sci.* **15**, 414 (1990).

18. O. B. Ptitsyn, *Trends Biochem. Sci.* **20**, 376 (1995).

19. K. Kuwajima, *Proteins: Struct. Funct. Genet.* **6**, 87 (1989).

20. S. C. Harrison and R. Durbin, *Proc. Natl. Acad. Sci. USA* **82**, 4028 (1985).

21. A. V. Finkelstein and A. Y. Badretdinov, *Fold. Des.* **2**, 115 (1997).

22. M. Karplus, *Fold. Des.* **2**, 569 (1997).

23. M. Bycroft, S. Ludvigsen, A. R. Fersht, and F. M. Poulsen, *Biochemistry* **30**, 8697 (1991).

24. A. R. Panchenko, Z. Luthey Schulten, and P. G. Wolynes, *Proc. Natl. Acad. Sci. USA* **93**, 2008 (1996).

25. S. E. Jackson and A. R. Fersht, *Biochemistry* **30**, 10436 (1991).

26. G. D. Prat Gay, J. Ruiz-Sanz, J. L. Neira, L. S. Itzhaki, and A. R. Fersht, *Proc. Natl. Acad. Sci. USA* **92**, 3683 (1995).

27. J. L. Neira, L. S. Itzhaki, A. G. Ladurner, B. Davis, G. D. Prat Gay, and A. R. Fersht, *J. Molec. Biol.* **268**, 185 (1997).

28. Y. J. Tan, M. Oliveberg, B. Davis, and A. R. Fersht, *J. Molec. Biol.* **254**, 980 (1995).

29. D. E. Otzen, L. S. Itzhaki, N. F. elMasry, S. E. Jackson, and A. R. Fersht, *Proc. Natl. Acad. Sci. USA* **91**, 10422 (1994).

30. L. S. Itzhaki, D. E. Otzen, and A. R. Fersht, *J. Molec. Biol.* **254**, 260 (1995).

31. Y. J. Tan, M. Oliveberg, and A. R. Fersht, *J. Molec. Biol.* **264**, 377 (1996).

32. A. R. Fersht, L. S. Itzhaki, N. elMasry, J. M. Matthews, and D. E. Otzen, *Proc. Natl. Acad. Sci. USA* **91**, 10426 (1994).

33. A. Matouschek, D. E. Otzen, L. S. Itzhaki, S. E. Jackson, and A. R. Fersht, *Biochemistry* **34**, 13656 (1995).

34. A. J. Li and V. Daggett, *Proc. Natl. Acad. Sci. USA* **91**, 10430 (1994).

35. A. J. Li and V. Daggett, *J. Molec. Biol.* **257**, 412 (1996).

36. A. G. Ladurner, L.S. Itzhaki, V. Daggett, and A. R. Fersht, *Proc. Natl. Acad. Sci. USA* **95**, 8473 (1998).

37. A. R. Fersht, *Proc. Natl. Acad. Sci. USA* **92**, 10869 (1995).

38. A. R. Fersht, *Curr. Opin. Struct. Biol.* **7**, 3 (1997).

39. T. R. Sosnick, M. D. Shtilerman, L. Mayne, and S. W. Englander, *Proc. Natl. Acad. Sci. USA* **94**, 8545 (1997).

40. T. Veitshans, D. Klimov, and D. Thirumalai, *Fold. Des.* **2**, 1 (1997).

41. A. R. Viguera, F. J. Blanco, and L. Serrano, *J. Molec. Biol.* **247**, 670 (1995).

42. A. R. Viguera, L. Serrano, and M. Wilmanns, *Nature Structural Biology* **3**, 874 (1996).

43. T. E. Creighton, *Current Biology* **5**, 353 (1995).

44. Z. Y. Guo and D. Thirumalai, *Biopolymers* **36**, 83 (1995).

45. V. I. Abkevich, A. M. Gutin, and E. I. Shakhnovich, *Biochemistry* **33**, 10026 (1994).

46. A. M. Gutin, V. I. Abkevich, and E. I. Shakhnovich, *Proc. Natl. Acad. Sci. USA* **92**, 1282 (1995).

47. A. M. Gutin, V. I. Abkevich, and E. I. Shakhnovich, *Biochemistry* **34**, 3066 (1995).

48. G. D. Gay and A. R. Fersht, *Biochemistry* **33**, 7957 (1994).

49. J. L. Neira, B. Davis, A. G. Ladurner, A. M. Buckle, G. D. Prat Gay, and A. R. Fersht, *Fold. Des.* **1**, 189 (1996).

50. Y. Mauguen, R. W. Hartley, E. J. Dodson, G. G. Dodson, G. Bricogne, C. Chothia, and A. Jack, *Nature, Lond.* **29**, 162 (1982).

51. H. Yanagawa, K. Yoshida, C. Torigoe, J. S. Park, K. Sato, T. Shirai, and M. Go, *J. Biol. Chem.* **268**, 5861 (1993).

52. J. Sancho, J. L. Neira, and A. R. Fersht, *J. Molec. Biol.* **224**, 749 (1992).

53. J. L. Neira and A. R. Fersht, *Fold. Des.* **1**, 231 (1996).

54. A. D. Kippen, J. Sancho, and A. R. Fersht, *Biochemistry* **33**, 3778 (1994).

55. V. L. Arcus, S. Vuilleumier, S. M. V. Freund, M. Bycroft, and A. R. Fersht, *J. Molec. Biol.* **254**, 305 (1995).

56. S. M. Freund, K. B. Wong, and A. R. Fersht, *Proc. Natl. Acad. Sci. USA* **93**, 10600 (1996).

57. C. J. Bond, K. B. Wong, J. Clarke, A. R. Fersht, and V. Daggett, *Proc. Natl. Acad. Sci. USA* **94**, 13409 (1997).

58. A. Matouschek, J. T. Kellis Jr., L. Serrano, and A. R. Fersht, *Nature* **340**, 122 (1989).

59. L. Serrano, A. Matouschek, and A. R. Fersht, *J. Molec. Biol.* **224**, 847 (1992).

60. A. R. Fersht and L. Serrano, *Curr. Opin. Struct. Biol.* **3**, 75 (1993).

61. J. M. Matthews and A. R. Fersht, *Biochemistry* **34**, 6805 (1995).

62. V. Daggett, A. Li, and A. R. Fersht, *J. Am. Chem. Soc.* **120**, 12740 (1999).

63. R. W. Hartley, *Barnase and barstar*, in *Ribonucleases*, G. Dalessio and J. F. Riordan, eds., Academic Press, 51 (1997).

64. M. J. Lubienski, M. Bycroft, S. M. V. Freund, and A. R. Fersht, *Biochemistry* **33**, 8866 (1994).

65. B. Nolting, R. Golbik, and A. R. Fersht, *Proc. Natl. Acad. Sci. USA* **92**, 10668 (1995).

66. B. Nolting, R. Golbik, J. L. Neira, A. S. Soler-Gonzalez, G. Schreiber, and A. R. Fersht, *Proc. Natl. Acad. Sci. USA* **94**, 826 (1997).

67. E. Lopez-Hernandez and L. Serrano, *Fold. Des.* **1**, 43 (1996).

68. K. A. Dill, K. M. Fiebig, and H. S. Chan, *Proc. Natl. Acad. Sci. USA* **90**, 1942 (1993).

69. K. Yue, K. M. Fiebig, P. D. Thomas, H. S. Chan, E. I. Shakhnovich, and K. A. Dill, *Proc. Natl. Acad. Sci. USA* **92**, 325 (1995).

70. A. Sali, E. Shakhnovich, and M. Karplus, *Nature, Lond.* **369**, 248 (1994).

71. V. I. Abkevich, A. M. Gutin, and E. I. Shakhnovich, *Protein Science* **4**, 1167 (1995).

72. J. D. Bryngelson and P. G. Wolynes, *Proc. Natl. Acad. Sci. USA* **84**, 7524 (1987).

73. P. G. Wolynes, *Proc. Natl. Acad. Sci. USA* **94**, 6170 (1997).

74. C. M. Dobson, A. Sali, and M. Karplus, *Angewandte Chem.* **37**, 868 (1998).

75. J. N. Onuchic, P. G. Wolynes, Z. Luthey Schulten, and N. D. Socci, *Proc. Natl. Acad. Sci. USA* **92**, 3626 (1995).

76. P. E. Leopold, M. Montal, and J. N. Onuchic, *Proc. Natl. Acad. Sci. USA* **89**, 8721 (1992).

77. W. A. Eaton, P. A. Thompson, C. K. Chan, S. J. Hagen, and J. Hofrichter, *Structure* **4**, 1133 (1996).

78. T. R. Sosnick, L. Mayne, R. Hiller, and S. W. Englander, *Nature Structural Biology* **1**, 149 (1994).

79. J. D. Bryngelson and P. G. Wolynes, *J. Phys. Chem.* **93**, 6902 (1989).

80. H. Roder and W. Colon, *Curr. Opin, Struct. Biol.* **7**, 15 (1997).

81. K. W. Plaxco, K. T. Simons, and D. Baker, *J. Molec. Biol.* **277**, 985 (1998).

82. S. E. Jackson, *Fold. Des.* **3**, R81 (1998).

83. R. J. Ellis, *Cell Stress Chaperones* **1**, 155 (1996).

84. J. Martin and F. U. Hartl, *Curr. Opin. Struct. Biol.* **7**, 41 (1997).

85. G. H. Lorimer and M. J. Todd, *Nature Structural Biology* **3**, 116 (1996).

86. K. Braig, Z. Otwinowski, R. Hegde, D. C. Boisvert, A. Joachimiak, A. L. Horwich, and P. B. Sigler, *Nature, Lond.* **371**, 578 (1994).

87. A. M. Roseman, S. X. Chen, H. White, K. Braig, and H. R. Saibil, *Cell* **87**, 241 (1996).

88. Z. H. Xu, A. L. Horwich, and P. B. Sigler, *Nature, Lond.* **388**, 741 (1997).

89. S. Chen, A. M. Roseman, A. S. Hunter, S. P. Wood, S. G. Burston, N. A. Ranson, A. R. Clarke, and H. R. Saibil, *Nature, Lond.* **371**, 261 (1994).

90. A. Azem, S. Diamant, M. Kessel, C. Weiss, and P. Goloubinoff, *Proc. Natl. Acad. Sci. USA* **92**, 12021 (1995).

91. T. E. Gray and A. R. Fersht, *FEBS Letters* **292**, 254 (1991).

92. O. Yifrach and A. Horovitz, *J. Molec. Biol.* **255**, 356 (1996).

93. R. A. Staniforth, S. G. Burston, T. Atkinson, and A. R. Clarke, *Biochemical J.* **300**, 651 (1994).

94. F. J. Corrales and A. R. Fersht, *Proc. Natl. Acad. Sci. USA* **93**, 4509 (1996).

95. L. S. Itzhaki, D. E. Otzen, and A. R. Fersht, *Biochemistry* **34**, 14581 (1995).

96. R. Zahn, S. Perrett, G. Stenberg, and A. R. Fersht, *Science* **271**, 642 (1996).

97. R. Zahn, A. M. Buckle, S. Perrett, C. M. Johnson, F. J. Corrales, R. Golbik, and A. R. Fersht, *Proc. Natl. Acad. Sci. USA* **93**, 15024 (1996).

98. M. M. Altamirano, R. Golbik, R. Zahn, A. M. Buckle, and A. R. Fersht, *Proc. Natl. Acad. Sci. USA* **94**, 3576 (1997).

99. A. M. Buckle, R. Zahn, and A. R. Fersht, *Proc. Natl. Acad. Sci. USA* **94**, 3571 (1997).

100. M. J. Todd, P. V. Viitanen, and G. H. Lorimer, *Science* **265**, 659 (1994).

101. J. S. Weissman, Y. Kashi, W. A. Fenton, and A. L. Horwich, *Cell* **78**, 693 (1994).

102. R. Zahn, S. E. Axmann, K. P. Rucknagel, E. Jaeger, A. A. Laminet, and A. Pluckthun, *J. Molec. Biol.* **242**, 150 (1994).

103. R. Zahn, S. Perrett, and A. R. Fersht, *J. Molec. Biol.* **261**, 43 (1996).

CHAPTER 19

104. J. S. Weissman, H. S. Rye, W. A. Fenton, J. M. Beechem, and A. L. Horwich, *Cell* **84**, 481 (1996).

105. J. Chatellier, F. Hill, P. A. Lund, and A. R. Fersht, *Proc. Natl. Acad. Sci. USA* **95**, 9861 (1998).

106. K. L. Nielson and N. J. Cowan, *Molecular Cell* **2**, 93 (1998).

107. A. Peres Ben-Zvi, J. Chatellier, A. R. Fersht, and P. Goloubinoff, *Proc. Natl. Acad. Sci. USA* **26**, 15275 (1998).

Index